AutoCAD® 2000

ONE STEP AT A TIME
BASICS

AutoCAD® 2000

ONE STEP AT A TIME
BASICS

TIMOTHY SEAN SYKES

North Harris Community College

Houston Community College

Registered Author/Publisher

PRENTICE HALL, Upper Saddle River, New Jersey, 07458

Library of Congress Cataloging-in-Publication Data

Sykes, Timothy Sean.
    AutoCAD 2000-- one step at a time : basics / Timothy Sean Sykes.
      p. cm.
    ISBN 0-13-083210-3
    1. Computer graphics programmed instruction.   2. AutoCAD.
I. Title.
T385.S878   2000
604.2'0285'5369—dc21   99-22319
                                                           CIP

Acquisition editor: Eric Svendsen
Editorial/production supervision: Ann Marie Kalajian
Interior design: Maureen Eide
Cover art and design: Michael Jung
Design director: Ann France
Manufacturing buyer: Pat Brown
Manufacturing manager: Trudy Pisciotti
Assistant managing editor: Eileen Clark
Executive managing editor: Vince O'Brien
Editor-in-chief: Marcia Horton
Vice president of production and manufacturing: David W. Riccardi
Marketing manager: Danny Hoyt
Marketing assistant: Eric Weisser
Editorial assistant: Griffin Cable
Copyeditor: Peter Zurita
Composition: D&G Limited, LLC
Interactive development: Andrew Child, Laurie Cohn,
    Karen Murphy, Jeremy Perkins
Graphic design: Jeremy Perkins
Animation: Pixel Workshop
Still photography: Andrew Child
Media managing editor: Erik Unhjem

© 2000 by Prentice-Hall, Inc.
Upper Saddle River, New Jersey 07458

The author and publisher of this book have used their best efforts in preparing this book. These efforts include the development, research, and testing of the theories and programs to determine their effectiveness. The author and publisher make no warranty of any kind, expressed or implied, with regard to these programs or the documentation contained in this book. The author and publisher shall not be liable in any event for incidental or consequential damages in connection with, or arising out of, the furnishing, performance, or use of these programs.

Trademark info:
AutoCAD® and the AutoCAD® logo are registered trademarks of Autodesk, Inc.
Quicktime® is a registered trademark of Apple, Inc.
Director® is a registered trademark of Macromedia, Inc.
Windows® is a registered trademark of Microsoft.

10 9 8 7 6 5 4 3 2 1

ISBN 0-13-083210-3

Prentice-Hall International (UK) Limited, *London*
Prentice-Hall of Australia Pty. Limited, *Sydney*
Prentice-Hall Canada Inc., *Toronto*
Prentice-Hall Hispanoamericana, S.A., *Mexico*
Prentice-Hall of India Private, Limited, *New Delhi*
Prentice-Hall of Japan, Inc., *Tokyo*
Prentice-Hall (Singapore) Pte., *Singapore*
Editora Prentice-Hall do Brasil, Ltda., *Rio de Janeiro*

LESSON 1

Setup

Following this lesson, you will:
- Know how to use AutoCAD's startup wizards to set up a new drawing
- Know how to use a template to start a new drawing
- Know how to open an existing drawing
- Know how to partially open an existing drawing
- Know how to save a drawing
- Know how to exit a drawing
- Know how to exit a drawing session
- Recognize the various parts of the AutoCAD User Interface

How do you begin a new drafting project?

Over the years, I have heard a variety of replies to this simple question. Some say you begin with the layout. Others say you begin by positioning the drawing on the page. Still others say you fill your coffee cup and turn on the radio! It seems the more experienced the draftsman, the more vague the answer.

This does not mean that I was questioning bad drafters, but simply that experience has engrained the basics so deeply that good drafters do not think about these simple questions anymore.

How do you begin a new drafting project? Well, first you decide on the scale and units (engineering, architectural, etc.) that you will use. Then you decide the size of the page on which you will draw.

This first lesson will introduce you to the AutoCAD User Interface (the screen on which you will draw) and show you how to set up things like units and sheet size.

1.1 The Groundwork: How AutoCAD Handles Scale, Units, and Paper Size

It may surprise you to learn that we will not use a scale when drawing in AutoCAD. *All drawing in AutoCAD is done* FULL SCALE. This means simply that a 3-inch line will actually be drawn 3 inches long. A line 3 miles long will be drawn 3 miles long!

Okay, so if we draw full scale, how can we put a house plan or refinery unit onto a sheet of paper unless the paper is very, very large? Well, we actually *tell* AutoCAD that we have a sheet of paper that is very, very large—a bit larger than the house or unit we will draw. Then later, when we plot the drawing to an actual sheet of paper, we *plot to scale*. We will discuss plotting in Lesson 3, but now let's see how we determine the size of the paper we tell AutoCAD to use.

We know the standard sizes of paper used in the design world. And we know the standard scales used. To determine what we tell AutoCAD about paper size, we need to determine how many feet or inches (or millimeters) will fit onto each size sheet at each scale. The easiest way to do this is to look at the chart called Drawing Scales on the tear out reference card. (You will find a similar chart in the possession of all CAD operators.)

To use the chart, we select the scale at which we will want the final drawing plotted, then select the paper size for the plot. Where row and column meet, we find the width × height *limits* of your drawing (see Section 1.10, for more on *limits*). There is a place for these numbers in the drawing setup.

1.2 Let's Get Started: The Startup Dialog Box

When you loaded AutoCAD, a shortcut that looks like Figure 1.2a was created on your desktop. Double click this with your left mouse button to launch AutoCAD. You will see the image in Figure 1.2b.

FIG. 1.2a
The AutoCAD Shortcut

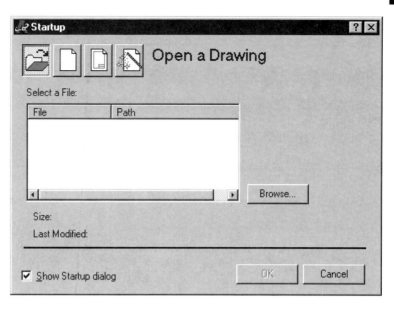

FIG. 1.2b: Startup Dialog Box

AutoCAD provides this Startup *dialog box* as an aid to both new and seasoned CAD operators. It allows the user to select one of four option buttons (across the top of the box) for beginning the drawing session. Let's look at these.

> A *dialog box* resembles a window within a window. It allows for user input in response to AutoCAD prompts or options.

FIG. 1.2c
Open Drawing Button

Open a Drawing. Picking this button (Figure 1.2c) will call up a list of the last few drawings that were loaded into the AutoCAD editor. You will find the list in the list box that appears in the center of the dialog box. You may pick on a drawing name and then click on the **OK** button, or you can simply double-click on the drawing you wish to load. If the desired drawing is not listed, pick the **Browse** button. AutoCAD will present a Select File dialog box similar to the Open File box used by most Windows programs.

FIG. 1.2d
Start from Scratch Button

Start from Scratch. Starting from scratch (Figure 1.2d) requires the user to know some of the older AutoCAD commands used for drawing setup. We will use some of these in later lessons. The user can select to draw using **English** or **Metric** settings. By default, "scratch" drawings use English settings, decimal units and a sheet size of 12 × 9. (A metric setup uses a sheet size of 429mm × 297mm.)

FIG. 1.2e
Use a Template Button

Use a Template. This button (Figure 1.2e) provides a list of drawings that have already been set up. You can use one of these or create one of your own. We will look at templates in more detail in Section 1.8, "Creating Templates," and Section 1.9, "Creating a New Drawing Using a Template."

FIG. 1.2f
Use a Wizard Button

Use a Wizard. This button (Figure 1.2f) provides two options for setting up your drawing. Find them in the list box that appears in the center of the Startup dialog box. The **Quick Setup** option uses some system defaults to get you started faster. The **Advanced Setup** option allows a more detailed setup. We will explore this in Section 1.3.1, "Do This: The Setup Wizard."

> A *Wizard* behaves just like a nanny. It takes you by the hand and leads you *One Step at a Time* through a procedure.

At the bottom of the Startup dialog box, you will find a check box with the words: "Show Startup dialog." Removing the check will cause AutoCAD to "Start from Scratch" using the default settings mentioned earlier. I do not recommend this for beginners. If you start AutoCAD and the dialog box does not show, you can check the box by going to the **General Options** frame of the **System** tab of the Options dialog box (under the Tools pull-down menu).

> *A word of caution:* Experimenting in the Options dialog box can cause AutoCAD to work improperly or stop working altogether. Most of these options should be left to the CAD coordinator on your project.

1.3 Using the Setup Wizard

Now that we have looked at the Startup dialog box, let's set up a drawing using the drawing wizard. We will set up a drawing for a $1/4" = 1'-0"$ scale on a B-size (11 × 17) sheet of paper.

> When using a mouse in the AutoCAD environment, you will *pick* or *select* by clicking once with the left mouse button. You will *confirm* your selection or call a cursor menu by clicking the right mouse button (more on the right button in Section 1.4.1, "Right Click Menus.").

1.3.1 Do This: The Setup Wizard

FIG. 1.3.1a
Autocad Shortcut

I. If you are not already in AutoCAD, double-click on the AutoCAD shortcut on the desktop (Figure 1.3.1a). Your screen should show the Startup dialog box seen in Figure 1.2b.

II. Follow these steps.

TOOLS	COMMAND SEQUENCE	STEPS

1. Click on the **Use a Wizard** button. The center of the box changes—a *list* box appears that looks like Figure 1.3.1.1.

List Box →

FIG. 1.3.1.1: Use a Wizard

2. In the list box, double-click **Advanced Setup**. The Advanced Setup dialog box (Fig. 1.3.1.2) replaces the Startup dialog box.

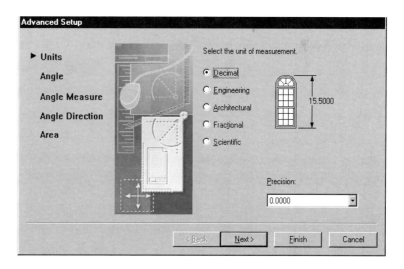

FIG. 1.3.1.2: Advanced Setup Dialog Box

continued

TOOLS	COMMAND SEQUENCE	STEPS

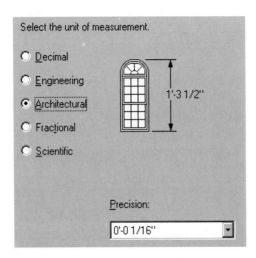

FIG. 1.3.1.3: Unit Setup Options

3. Notice the list down the left side of the dialog box. (Figure 1.3.1.2) This list includes the settings available in the **Advanced Setup Wizard**. The initial screen presents the **Units** options as indicated by the arrow.

Select **Architectural**. A "bullet" appears in the option button to the left (see Figure 1.3.1.3).

You will notice the demo to the right will change showing you a sample of what architectural units look like. You may select the other buttons to see samples of them, then reselect the architectural option.

The **Precision** of the units is already set to $1/16$", but you can change it on this screen as well by picking the down arrow on the control box below the word **Precision**.

An **Option Button** is a round hole. Selecting an option button places a black dot (or bullet) inside the round hole. Option buttons usually come in small groups, but only one button in a group can hold the bullet.

[Next >]

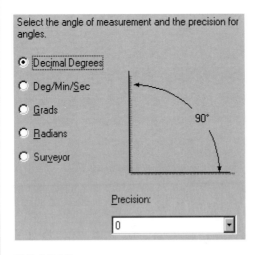

FIG. 1.3.1.5: Angle Options

4. Pick the **Next >** button on the bottom of the dialog box. The arrow on the settings list drops to **Angle** and the options on the right side of the dialog box change accordingly.

5. Select the different option buttons to see how each appears in the sample image to the right, then set the **Decimal Degrees** option (Figure 1.3.1.5). Leave the precision at which you want AutoCAD to measure angles at **0** (for *no decimal places*).

continued

TOOLS	COMMAND SEQUENCE	STEPS

6. Pick the **Next >** button. Your dialog box now looks like Figure 1.3.1.6.

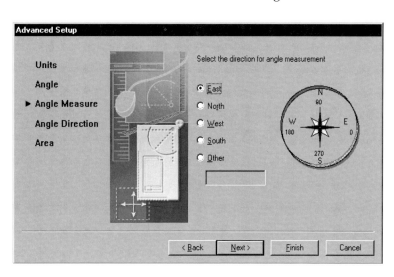

FIG. 1.3.1.6: Angle Measurement Options

Here we control the direction of the angles.

Most of us are familiar with the nautical way of determining degrees (a product of good scout training). Unfortunately, the computer masters that created AutoCAD apparently missed that training. The standard 0° North in the real world has changed to 0° East in the computer world.

You can change it here, if you want. But be warned, the 0° East has become a CAD *convention* (unwritten standard). If any other CAD operator discovers that you have changed the angles, you may be in for some grief. Many third-party programs and AutoLISP routines are based on 0° East.

7. Pick the **Next >** button to go to the **Angle Direction** options. This one also suffers the wrath of the programmers. We must now measure our angles *counterclockwise*! Thus 90° North, 180° West, and 270° South follow 0° East.

continued

TOOLS	COMMAND SEQUENCE	STEPS
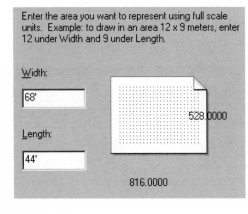 FIG. 1.3.1.8 Finish	Enter the area you want to represent using full scale units. Example: to draw in an area 12 x 9 meters, enter 12 under Width and 9 under Length. Width: 68' Length: 44' 528.0000 816.0000	**8.** In the **Area** options (Figure 1.3.1.8), you will decide the size of the sheet on which you will plot your final drawing. The default is 1' wide by 9" long, shown in the architectural units we set up in Step 4. Look at the Drawing Scales chart on the tear out reference card. Determine what the drawing limits should be for the $1/4$" scale/11" × 17" sheet we want. Did you come up with 44' × 68'? Excellent! Put the 68' in the width text box and the 44' in the height text box. Why 68' and 44' instead of 44' and 68'? This is because we want the drawing to be landscaped (wider than high). (*Note*: Be sure to use the foot mark (') when entering the limits.) **9.** Pick the **Finish** button at the bottom of the Advanced Setup dialog box. The Startup dialog box disappears, replaced by AutoCAD's graphic screen.

Congratulations, you have set up your first AutoCAD drawing session using the **Advanced Setup** option of the Startup Wizard.

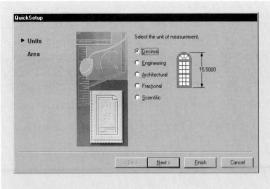

FIG. 1.3g: Quick Setup Wizard

The **Quick Setup** option of the Startup Wizard only has two options, as shown in Figure 1.3g.

These are similar to the same tabs on the Advanced Setup. In the Quick Setup Wizard, AutoCAD uses default settings for all the other options provided by the Advanced Setup Wizard.

1.4 The User Interface

1.4.1 The Screen

AutoDESK made a great effort to make AutoCAD 2000 comply with the standard interface conventions used by most software in the Windows environment. So, if you are at all familiar with IBM clone computers, or even with the Macintosh systems, the interface will be familiar. Let us take a quick look at it. (Refer to Figure 1.4.1a.)

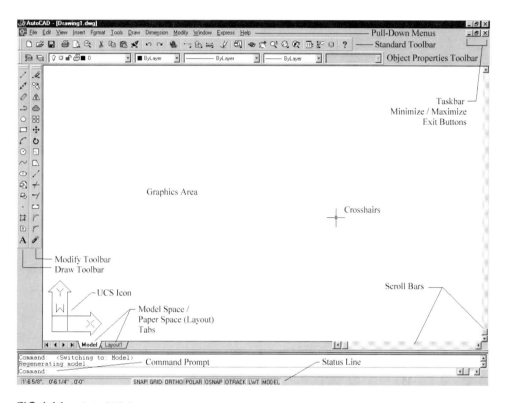

FIG. 1.4.1a: AutoCAD Screen

Except for some software-specific menu items, a user could almost mistake the top of the AutoCAD screen for Word, Excel, Paintbrush, or any other MS product. We find the standard blue Title Bar at the top, with the three buttons on the right. These buttons will, from left to right, send the software to the taskbar (the gray bar at the bottom of the screen), minimize/maximize the AutoCAD window (make it larger or smaller), and exit the software.

Below the Title Bar, we see the Pull-Down Menu Bar. This begins with the conventional Windows menus: **File**, **Edit**, and **View**. Take a minute to explore these and the other pull-down (or drop-down) menus. You will notice that the AutoCAD crosshairs become a cursor (an arrow) when moved out of the graphics area. Place the cursor on one of the pull-down menus and click with the left mouse button. A menu will drop from that location showing various options related to the selection. Picking any of these options will activate an AutoCAD *macro*.

> A *macro* is a series of commands or events responding to a single input. Generally, the user cannot tell this from a single command.

Below the Menu Bar, we see the first of AutoCAD's toolbars. This Standard toolbar carries many of the same buttons, or tools, found in other Windows' software. Reading from the left, you see the following buttons: **New File**, **Open File**, **Save File**, **Print**, **Print Preview**, **Find and Replace**, **Cut**, **Copy**, **Paste**, and so forth. To determine the function of these or other buttons on a toolbar, place your cursor over the button and wait a second. You will see a written description appear in a cheater box.

The next toolbar down is AutoCAD's Object Properties toolbar. This provides useful information about layers and other object properties in the drawing. We will spend some time with these in Lesson 6.

Docked on the left side of the screen, you find the Draw and Modify 1 toolbars. Take a moment to look at each tool and its description in the cheater boxes. Like the menus, selection of a tool button activates a macro command. Do you notice any similarity between the commands found on these toolbars and those found in the pull-down menus? Many of them are the same. You have discovered two different ways you will communicate with AutoCAD—menus and toolbars. But wait, there are others!

> A *toolbar* is a group of buttons. To remove a toolbar from the screen, undock (see what follows) the toolbar, then pick the "**X**" found in the upper right corner of the undocked bar. To add a toolbar, pick the **View** pull-down menu and select **Toolbars** . . . . You will get a dialog box that lists all the toolbars available through the *acad* menu. (Another, quicker method would be to simply right-click on any existing toolbar, then select the name of the desired toolbar from the menu that appears.)
>
> A *docked* toolbar appears "attached" to a side, top, or bottom of the graphics window. An *undocked* toolbar appears to float over the screen. Drawn objects may exist behind an undocked toolbar, but a docked toolbar forces the drawing to move over.
>
> To undock a toolbar, place your cursor on any gray area of the toolbar and drag it free from its dock.
>
> To dock a toolbar, place your cursor in the blue title bar area and drag the toolbar to a docking site (one of the borders of the graphics area).

At the bottom of the screen is the Status Line. On the left of the status line you will see a small box with three numbers in it. This Coordinate Display box shows the X, Y, and Z coordinates of the cursor in your drawing. We will look again at AutoCAD's coordinate system in Lesson 2. Beside this box is a series of toggles that will prove themselves quite handy in time. We will look at these in Lesson 4.

In the lower left corner of the graphics area you will see the User Coordinate System (UCS) icon. The icon serves a very useful purpose in 3-dimensional drafting. For AutoCAD beginners in the 2-dimensional world, however, it serves as a reminder of the X- and Y-axes. We will discuss these in more detail in Lesson 2.

Just above the bottom of the screen, you can see a window with three **Command:** prompts. Like the toolbars, this window can undock, but I do not recommend it. You can also resize it by placing your cursor on the heavy line above it, picking with the left mouse button and "dragging" the window up or down. (The cursor will become a double arrow when properly located, and you must hold the button down for this procedure.) I usually set my command window to be large enough for two command prompts. This allows more room for drawing.

> I cannot overemphasize the importance of becoming familiar with the **Command:** prompt. Look closely. You see the prompt looks like this: **Command:**. AutoCAD speaks to you on the left side of the colon. You respond on the right. Right now, AutoCAD is "prompting" you, or asking you what you want to do. When you respond, either by keyboard entry or mouse selection, your response will appear to the right of the colon.
>
> Once you enter a command, AutoCAD's prompt may change, asking you for more information or input. Just follow the prompts to complete your task.

> AutoCAD actually uses two coordinate systems, the UCS and the World Coordinate System (WCS), but that discussion belongs in the 3-dimensional world. The UCS and WCS are the same for all the lessons in this text.

Above the command line, you will find the **Model Space**/Paper Space (**Layout**) tabs. Paper Space is a more advanced tool covered in our next text—*AutoCAD 2000: One Step at a Time—Advanced Edition*. In this text, we will remain on the **Model Space** tab.

> An advanced drawing environment called *Paper Space* enables the CAD operator to work in multiple scales on the same drawing. It is also a useful plotting tool.

The last part of the screen is the Graphics Area. You will do your drawing here. Normally, you will draw in white (or in a color you select) against a black background. For purposes of this text, I will draw in black or some other color against a white background. This saves printer's ink and the reader's eyes.

The user can toggle between this graphics screen and a text screen using the **F2** function key. (Find the function keys along the top of your keyboard.) The text screen shows the last several lines of command/response exchanges between AutoCAD and the user. (See Appendix B for the use of the other function keys.)

■ 1.4.2 Right-Click Cursor Menus

AutoCAD has continued to develop new and simpler tools to aid its users. With AutoCAD 2000 came expanded use of the right mouse button to access various options (via *cursor menus*) in the program. This allows the user to watch the screen more and spend less time searching for the right thing to press on the keyboard!

> A *cursor menu* is a pop-up screen menu that appears when the user clicks the right mouse button.

We have to give the programmers credit for accomplishing a minor miracle with the right button. They have provided one of the most dynamic of AutoCAD tools. Which cursor menu appears when the right button is picked depends on whether or not there is a command in progress, whether or not there are items selected on the screen, and where the cursor/crosshairs are located when the button is picked.

AutoCAD provides five basic *modes* (or types) of cursor menus: Default, Edit-mode, Dialog-mode, Command-mode, Other menus.

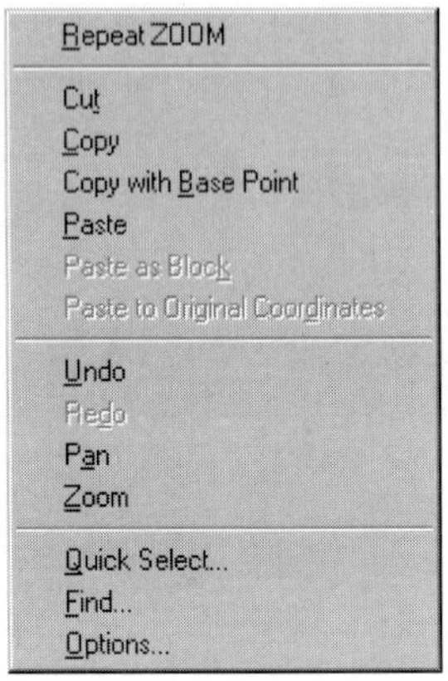

- The Default menu (see Figure 1.4.2a) appears when the user picks anywhere in the graphics area if no command is active and no selection set is available (i.e., nothing has been selected on the screen with which to work). Tools provided are general, frequently used commands.

- The Edit-mode menu appears when objects have been selected on the screen, but no command has been given. The best example of this is the Grips menu discussed in Lesson 17. Generally, you can expect grip-type modifying tools. But object-specific tools will appear for some objects (like dimensions).

- When the user right-clicks in a dialog box, AutoCAD may present a Dialog-mode menu. The tools presented will depend on which dialog box you are using and the cursor's location when you click.

- When the user right-clicks while there is a command in progress, AutoCAD presents a Command-mode menu with tools specific to that command.

- The "Other menus" category includes all the other menus that do not fit easily into the first four categories. Right-clicking in different locations and at different times might present some surprising results. Feel free to experiment while you are in a drawing. Right click over the different buttons on the status bar or over the toolbars.

FIG. 1.4.2a
Default Cursor Menu

We will spend more time exploring the cursor menus as they apply to specific areas throughout this book. But now, let's get back to our drawing.

1.5 Saving and Leaving a Drawing Session

Now that we have created our drawing, we need to save it. We accomplish this using one of three commands: *Save*, *Saveas*, or *Qsave*. Each behaves in a similar manner depending on the status of the drawing. That is, each will present the Save Drawing As dialog box if the drawing has not been previously saved, but the *Qsave* (or QuickSave) command will automatically save the drawing if it has been previously saved (and given a name).

Each of these commands is available at the command prompt or under the File pull-down menu.

How you finish the drawing depends on whether you wish to finish only the current drawing or the entire drawing session. In other words, do you want to close AutoCAD or just close this drawing? To leave the drawing but keep the AutoCAD software active, use the *Close* command. Otherwise, use the *Quit* command or pick the **"X"** button in the upper right corner of the AutoCAD window.

These commands are also available both at the command prompt and under the File pull-down menu.

Let's save and exit our drawing.

1.5.1 Do This: Saving Your Drawing Changes

TOOLS	COMMAND SEQUENCE	STEPS
No Button Available (The **Save** Button on the Standard toolbar will call the Save Drawing As dialog box only if the drawing has not previously been saved.)	**Command:** *saveas*	**1.** Open the Save Drawing As dialog box. To do this, you must choose one of two approaches: type *saveas* at the command prompt, or go to the **File** pull-down menu and select **Save As**. AutoCAD presents the Save Drawing As dialog box shown in Figure 1.5.1.1.

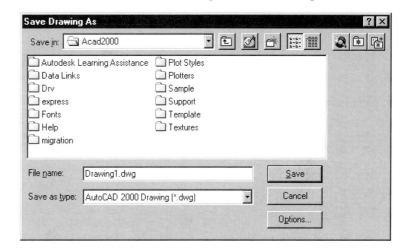

FIG. 1.5.1.1: Save Drawing As Dialog Box

FIG. 1.5.1.2t

2. Next to the **Save in:** box (where you see AutoCAD 2000) you see a downward-pointing arrow. Pick on that arrow to see a path showing where you are on your hard drive (Figure 1.5.1.2t). We are going to save the file to the C:\Steps\Lesson01 folder.

Pick on the **(C:)**. The list box below changes to show all the folders on the C drive.

📁 Steps

Steps Folder

3. Double-click on the folder identified as **Steps**. This folder opens and the list box now shows the contents of the Steps folder.

📁 Lesson01

Lesson01 Folder

4. Double-click on the folder identified as **Lesson01**. This folder now opens. Your dialog box should now look like Figure 1.5.1.4.

continued

TOOLS	COMMAND SEQUENCE	STEPS

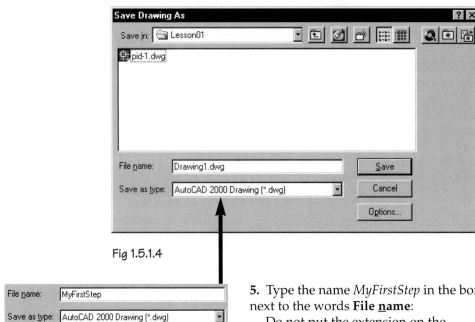

Fig 1.5.1.4

5. Type the name *MyFirstStep* in the box next to the words **File name**:

Do not put the extension on the name. AutoCAD will do that for you.

6. Notice that the type of file you will save has been identified in the **Save as type**: text box. Pick the down arrow to view the different formats available.

The three *.dwg* file types make it possible to save an AutoCAD 2000 drawing so that it may be edited by earlier releases of AutoCAD. The *.dwt* file is an AutoCAD template (we will look more at this in Section 1.8 of this lesson). *.DXF* files are binary (computer programming stuff) files used to exchange AutoCAD drawings with other programs.

7. Pick the **Save** button in the lower right corner of the dialog box.

Command: *quit*

8. Now that your drawing has been saved, you can exit the program. The easiest way to do this is to pick the **X** button in the upper right corner of the program. Alternatively, you can also pick the **File** menu and select **Exit** from the options presented. The keyboard command for leaving the drawing is *Quit*.

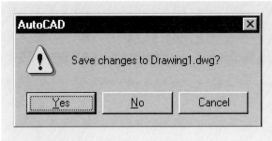

FIG. 1.5a: AutoCAD Warning Box

Note that leaving a drawing that has been changed, without first saving it, will cause AutoCAD to present a warning box (Figure 1.5a). AutoCAD will let you know that you have not saved the drawing and ask if you would like to save it now.

1.6 Opening an Existing Drawing

To open an existing drawing at the beginning of a new drawing session, pick the **Open a Drawing** button on the Startup dialog box (Figure 1.6a). The dialog box will change, revealing the **Select a File:** list box that shows the drawings edited in the last several sessions. (You may have to scroll to the right to see the entire path.)

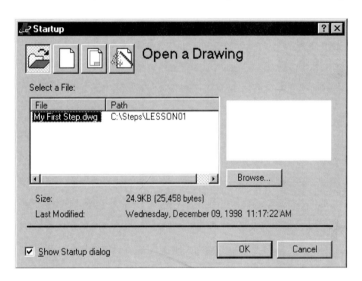

FIG. 1.6a: Startup Dialog Box

If you want to open a listed file, double-click on the drawing name. If your drawing is not listed, click on the **Browse** button. The Select File dialog box appears (Figure 1.6b).

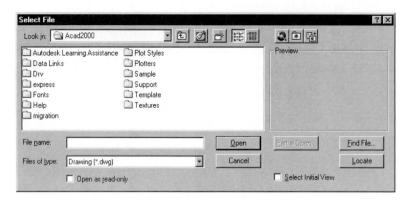

FIG. 1.6b: Select File Dialog Box

Use this box in much the same way you used the Save Drawing As dialog box. Locate the appropriate folder as you did in the Save Drawing As dialog box, then select the desired file in the list box.

As you select a file to open, a preview of the file will appear in the Preview frame to the right. Note that you may also search the web, search the Favorites folder, or add a drawing to the Favorites folder using the three buttons located above the Preview frame (see Figure 1.6c).

Let's open the *MyFirstStep* drawing we created earlier.

FIG. 1.6c:
Web Search, Favorites, Add a Drawing Buttons

1.6.1 Do This: Opening an Existing Drawing

TOOLS	COMMAND SEQUENCE	STEPS
 Open a File Button Open Button		**1.** If you are already in a drawing session, go to step 1a. If you are not in a drawing session, open AutoCAD using the desktop shortcut. At the Startup dialog box, select the **Open a File** button. In the **Select a File** list box that appears, click on the *My First Step* drawing. That is all there is to it! (If the *My First Step* drawing is not listed in the Select a File list box, go to Step 1b.) **1a.** If you are already in AutoCAD, but not at the Startup dialog box, select the **Open** button on the Standard toolbar. In the **Select a File** list box that appears, click on the *MyFirstStep* drawing. That is all there is to it! (If the *MyFirstStep* drawing is not listed in the Select a File list box, go to Step 1b.)

continued

TOOLS	COMMAND SEQUENCE	STEPS

Other methods of displaying the Select File (Open) dialog box from within a drawing session include picking on the **File** menu and selecting **Open**, typing *Open* at the command prompt or holding down the control key while typing *O*.

1b. If the *MyFirstStep* drawing is not listed in the Select a File list box, click on the **Browse** button and go on to Step 2.

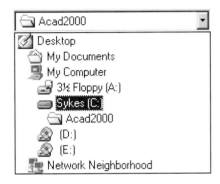

FIG. 1.6.1.2t: **Look in:** Box

2. Pick the down arrow in the **Look in:** box (Figure 1.6.1.2t), and pick on the **(C:)**. (The name in front of the **(C:)** may be different on your computer, or there may be no name at all.)

3. Double-click on the folder identified as *Steps*. This folder opens and the list box now shows the contents of the Steps folder (Figure 1.6.1.3).

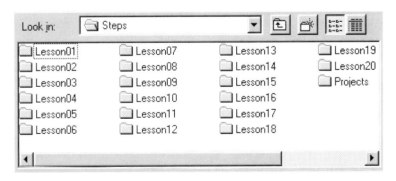

FIG. 1.6.1.3: Steps Folder

4. Double-click on the folder identified as Lesson01. This folder now opens and its contents (including the *MyFirstStep* drawing) are shown in the list box.

continued

TOOLS	COMMAND SEQUENCE	STEPS
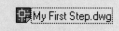		5. Double-click on *MyFirstStep*. AutoCAD will now open your drawing and begin a drawing session.

1.7 Partial Open and Partial Load

A real timesaver has been included with the 2000 package. The *Partialopen* command allows the user to open only that part of a drawing necessary for completion of a simple task. The user can specify what part of the drawing to open based on saved views or layers (more on these in Lessons 5 and 7). This ability minimizes the time spent loading or regenerating a full drawing and increases system performance by reducing the amount of memory required for the editing task.

The command sequence for *Partialopen* looks like this:

Command: *partialopen*
Enter name of drawing to open: *[enter the name of the drawing to open]*
Enter view to load or [?]<*Extents*>: *[enter the name of the desired view]*
Enter layers to load or [?]<none>: *[tell AutoCAD which layers to load or enter a * for all layers]*
Unload all Xrefs on open? [Yes/No] <N>: *[unloading xrefs will save time, if they are not pertinent to your task, type "Y" before hitting enter]*

But the easiest way to partially open a drawing is simply to select the **Partial Open** button on the Select File dialog box presented by the *Open* command. AutoCAD will present the Partial Open dialog box. Let's take a look at how this is done.

WWW 1.7.1 Do This: Partially Opening a File

I. Be sure you are in a drawing session.

II. Follow these steps.

TOOLS	COMMAND SEQUENCE	STEPS
	Command: *open*	1. Enter the *Open* command. AutoCAD presents the Select File dialog box.

continued

TOOLS	COMMAND SEQUENCE	STEPS

2. Follow Steps 2 through 4 in Section 1.6.1, "Do This: Opening an Existing Drawing" to locate the C:\Steps\Lesson01 folder. Select the *PID-1* file.

3. Pick the **Partial Open . . .** button. AutoCAD presents the Partial Open dialog box seen in Figure 1.7.1.3.

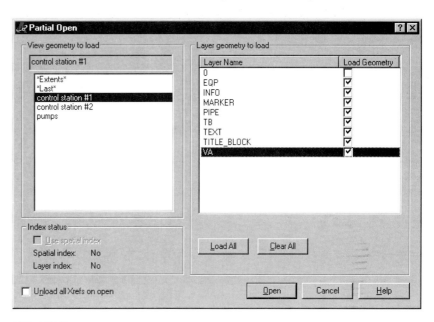

FIG. 1.7.1.3: Partial Open Dialog Box

4. Select the *Control Station #1* view and check the layers indicated.

5. Pick the **Open** button. AutoCAD opens just the part of the drawing shown in Figure 1.7.1.5.

continued

TOOLS	COMMAND SEQUENCE	STEPS

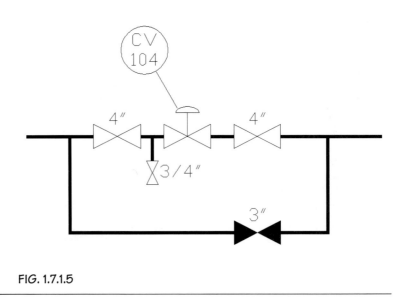

FIG. 1.7.1.5

> Note that when opening a drawing that was saved in the partially opened state, AutoCAD will present a warning box giving the user the opportunity to partially open it using the same settings previously used or to fully open it.

Once the drawing is partially opened, the user can work on the opened section only. But there will be times when the user will discover that he needs more of the drawing than originally anticipated. For this reason, AutoCAD also provided the *Partiaload* command. Using this command, the user has the flexibility to open (or load) additional parts of the partially opened drawing. The *Partiaload* command works with the Partial Load dialog box, which is very similar to the Partial Open dialog box.

Let's give it a try.

WWW 1.7.2 Do This: Partially Loading a Drawing

I. Be sure you are still in the *PID.dwg* file you partially opened in Section 1.7.1 "Do This." If not, repeat that exercise now.

II. Follow these steps.

TOOLS	COMMAND SEQUENCE	STEPS
No Button Available	**Command:** *partiaload*	1. Enter the *Partiaload* command. AutoCAD presents the Partial Load dialog box indicated in Figure 1.7.2.1.

continued

TOOLS	COMMAND SEQUENCE	STEPS

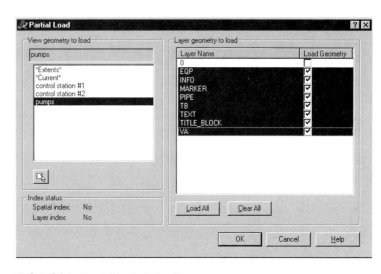

FIG. 1.7.2.1: Partial Load Dialog Box

2. Select the *Pumps* view and the layers indicated.

3. Pick the **OK** button.

Command: *zoom*

Specify corner of window, enter a scale factor (nX or nXP), or [All/Center/Dynamic/Extents/ Previous/Scale/Window] <real time>: *e*

4. Type in the commands shown. (We will look more closely at the *Zoom* command in Lesson 5.)

AutoCAD displays the two views of the *PID* drawing, as seen in Figure 1.7.2.4.

continued

TOOLS	COMMAND SEQUENCE	STEPS

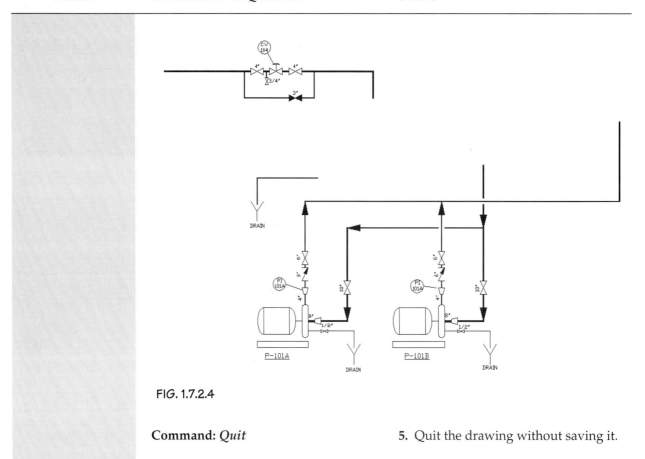

FIG. 1.7.2.4

Command: *Quit*   5. Quit the drawing without saving it.

The *Partiaload* command complements and serves as a fix for oversights in the *PartialOpen* command. Remember these timesaving tools when you are working on very large drawings. Now let's take a look at another timesaver.

1.8 Creating Templates

We have learned how to set up a new drawing from the Startup dialog box. But let's face it, this is a tedious procedure at best. About midway through the exercise, many students ask if "we have to do this every time?" The answer is, "NO!"

You should have to set up a drawing once for each scale and sheet size you will use. So you may have a $1/4" = 1'-0"$ drawing setup for an 11 × 17 sheet of paper, a $3/8" = 1'-0"$ drawing for an 11 × 17 sheet, a $1/4" = 1'-0"$ for a 24 × 36 sheet, and so forth. But once set up, you should not have to set it up again regardless of how many times you may need it.

Here is how this works. Set up your drawing, but save it as a template. Then when you need to create a drawing at that scale and with those *limits* (on that size sheet of paper), use the template to create the drawing. Any drawing created with that template will carry the same setup as that template. But the template will never change unless you overwrite it, so you can use it repeatedly.

Now let's see if we can create a template from the *MyFirstStep* file we just opened. (We will see how to use the template in Section 1.9.1, "Do This: Using a Drawing Template.")

1.8.1 Do This: Creating a Drawing Template

I. Open the *MyFirstStep* file in the C:\Steps\Lesson01 folder. Refer to Section 1.6.1, "Do This: Opening an Existing Drawing" if you need help.

II. Follow these steps.

TOOLS	COMMAND SEQUENCE	STEPS
No Button Available	**Command:** *saveas*	**1.** Enter the *Saveas* command. AutoCAD presents the Save Drawing As dialog box we examined in Section 1.5.1, "Do This: Saving Your Drawing Changes."
		2. Pick the down arrow in the **Save as Type** drop-down box and select the **Drawing Template File** option.
		3. AutoCAD has automatically changed the path to place the template in AutoCAD's Template folder. Change it back to the C:\Steps\Lesson01 folder. (Follow Steps 2, 3, and 4 in Section 1.5.1, "Do This: Saving Your Drawing Changes" if you need help.)
	File name: MyFirstTemplate Save as type: Drawing Template File (*.dwt) Fig 1.8.1.4t	**4.** Change the name to *MyFirstTemplate*, as shown in Figure 1.8.1.4t.
Save		**5.** Pick the **Save** button.
OK		**6.** AutoCAD presents the Template Description dialog box. No description is required, so just pick the **OK** button to complete the command.

1.9 Creating a New Drawing Using a Template

Now let's create a new drawing using the *MyFirstTemplate* you just created.

1.9.1 Do This: Using a Drawing Template

TOOLS	COMMAND SEQUENCE	STEPS
New Button Use a Template Button Browse... 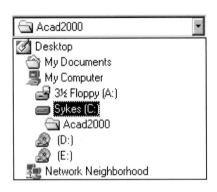	**Command:** *new*	**1.** If you are already in AutoCAD but not at the Startup dialog box, type *new* at the command prompt, hold down the control key and type *n*, or select the **New** button on the Standard toolbar. Any of these will call the Create New Drawing dialog box where you will find the **Use a Template** button. Proceed then to Step 1a. **1a.** At the Startup dialog box, select the **Use a Template** button. The Select a Template list box that appears now shows several standard templates used by AutoCAD (or the CAD guru on your job may have already replaced them with job-specific templates). Normally, you would select one of these. But to find our template, we must click on the **Browse** button. **2.** A Select Template dialog box replaces the Startup box. This box is similar to other dialog boxes we have already encountered. Let's follow the path to *MyFirstTemplate*. Pick the down arrow next to the **Look in:** box and select **(C:)**.

continued

TOOLS	COMMAND SEQUENCE	STEPS
📁 Steps		**3.** In the list box, double-click the *Steps* folder.
📁 Lesson01		**4.** Double-click the *Lesson01* folder. The *MyFirstTemplate* file should now appear in the list box.
My first template.dwt		**5.** Double-click on *MyFirstTemplate*. AutoCAD launches a drawing session using the setup you created for *MyFirstTemplate* as the new drawing's setup.

1.10 Changing the Setup or Starting a New Drawing from Scratch

There is always the possibility that you may need to change your basic setup after you have already begun a drawing session. Maybe you entered the wrong sheet size or used engineering units instead of architectural. You can still adjust, even if you have already begun to draw. (Or you may simply prefer to start all of your drawings *from Scratch*—that is, using AutoCAD's default setup—then change the settings manually.)

To change units, enter **DDUnits** at the command prompt. AutoCAD will provide the Drawing Units dialog box (Figure 1.10a) where you can identify the new units and unit precision, and the angle type, precision, and direction.

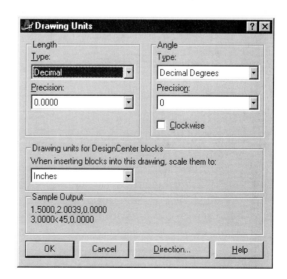

FIG. 1.10a: Drawing Units Dialog Box

To change the sheet size, enter *Limits* at the command prompt. AutoCAD will prompt you for the lower left coordinates of your drawing, using 0,0 as the default (see the following insert). Accept the default by hitting *enter* on your keyboard. AutoCAD will then prompt you for the upper right coordinates. Get these from the Drawing Scales chart and enter them as *width,height* (no spaces). The sequence looks like this:

Command: *limits*
Reset Model space limits:
Specify lower left corner or [ON/OFF] <0.0000,0.0000>: *[enter]*
Specify upper right corner <12.0000,9.0000>: *[enter the new width and height]*
Command:

While you can make these adjustments at any time during your drawing session, I do not recommend waiting too long. It is always best to groom good habits from the very beginning.

We will set up several drawings from scratch over the course of our text, so you will become familiar with the *DDUnits* and *Limits* commands.

Note that most companies will use *0,0* as the lower left corner of their drawings. Occasionally, however, a company may choose to use *true coordinates*. This is best explained by example.

In the petrochem world, plants are divided into units. Each unit may have one or more drawings specific to that unit. Vessels and other items in the unit are located in the plant by overall *east/west* and *north/south* (*X* and *Y*) coordinates. It is often useful to identify the east/west and north/south coordinates with the X and Y coordinates in AutoCAD. this way every item in the plant can be quickly located in the drawings simply by doing an *ID* command. (We will look at the *ID* command in Lesson 9.)

When using true coordinates, the absolute location of the drawing will dictate the lower left as well as the upper right corners of the drawing limits.

■ 1.11 Extra Steps

Something you may have noticed in your first experiments with the AutoCAD User Interface (often called the AUI) is that AutoCAD can open more than one drawing at a time. This ability arrived with the release of AutoCAD 2000 and may be a bit of a shock to experienced AutoCAD users. Its benefits promise to be almost immeasurable as a timesaving device. We will learn more about AutoCAD's MDE (Multiple Document Environment) and its uses as we proceed.

For now, however, the user should be aware that opening multiple drawings is much like opening multiple documents in most other software. The user can split the screen to view more than one document at a time and (most importantly) *use material from one file in another*. The timesaving here is inestimable.

To view more than one drawing at a time, simply go to the Window pull down menu and select an option. Cascading files are placed one atop another with the title bar showing at the top. The user can pick on any title bar (or any visible area of a drawing) to move it to the top of the stack. Tiling Horizontally or Vertically simply places the drawings next to each other (top and bottom or side to side) so the user can simultaneously view more of the area of each drawing.

■ 1.12 What Have We Learned

Items covered in this lesson include:

- *The Startup Dialog Box*
- *The Quick and Advanced Setup Wizards*
- *Templates*
- *The AutoCAD User Interface*
- *Commands:*
  - *QSave*
  - *Save*
  - *SaveAs*
  - *Close*
  - *Quit*
  - *Open*
  - *PartialOpen*
  - *Partiaload*
  - *DDUnits*
  - *Limits*

Well, now you have gotten your feet wet. How was it?

Any road begins with a first step, and any design begins with some basic decisions. What scale should I use? What size sheet of paper do I need? Should I listen to Country or Rock-n-Roll while I draw?

We have looked at how to accomplish most of these tasks on a computer, using AutoCAD as our tool. In the next few lessons, you will see a whole new world of possibilities opening before you as we explore this wonderful drafting tool.

But first, let's get some practice with what we have learned so far.

1.13 EXERCISES

1. Set up a nautical drawing.
    1.1. Create a new drawing with the following parameters:
        1.1.1. You will use engineering units on your drawing.
        1.1.2. Assume a 1" = 25' scale.
        1.1.3. You will plot on a D-size (24 × 36) sheet of paper.
        1.1.4. You will want angles measured clockwise with 0° North.
    1.2. Save the project as *Nautical.dwg* in the C:\Steps\Lesson01 folder.

    Using the Quick Setup, create a few templates to use later in this course.

2. Create a template with these parameters:
        2.1.1. Use architectural units.
        2.1.2. Use a 1 = 1 scale (this requires no setup; it is the default).

2.1.3. Set up for a sheet size of 8 1/2 × 11.
2.2. Save this project as *MyBase1.dwt* in the C:\Steps\Lesson01 folder.
3. Create a new template with the following parameters:
   3.1.1. Use architectural units.
   3.1.2. Use a 1 = 1 scale.
   3.1.3. Set up for a sheet size of 11 × 17.
3.2. Save the project as *MyBase2.dwt* in the C:\Steps\Lesson01 folder.
4. Create a third template with the following parameters:
   4.1.1. Use architectural units.
   4.1.2. Use a 1 = 1 scale.
   4.1.3. Set up for a sheet size of 17 × 22.
4.2. Save the project as *MyBase3.dwt* in the C:\Steps\Lesson01 folder.
5. Create templates for the followings scales and sheet sizes. Save the templates in the C:\Steps\Lesson01 folder.

SETTINGS	UNITS	SCALE	SHEET SIZE	SAVEAS
English	Architectural	1/4" = 1'	34 × 22	D 1_4 Arch
English	Architectural	3/8" = 1'	17 × 11	B 3_8 Arch
English	Decimal	1/4" = 1'	22 × 17	C 1_4 Dec
English	Decimal	3/8" = 1'	17 × 11	B 3_8 Dec
Metric	Decimal	1 = 1'	841 × 594	Met a1
Metric	Decimal	1 = 1'	594 × 420	Met a2

You will use some of these templates and expand on them in later lessons.

1.14 REVIEW QUESTIONS

Please write your answers on a separate sheet of paper.

1. to 3. Name the first three items the draftsman must decide when beginning a new project.

4. A 3" line that will be plotted at 3/8" = 1'−0" will be drawn how long in AutoCAD?

5. I am setting up a drawing that will be plotted at 1/4" = 1'−0" on a D-size (36" × 24") sheet of paper. How large do I tell AutoCAD my paper is?

6. A _____ is a window that allows for user input in response to prompts options.

7. A _____ behaves like a nanny—leading the user through a procedure.

8. and 9. What are the two options provided by the Use a Wizard button on the Startup dialog box?

10. To use a drawing that has already been set up, go to the _____ button in the Startup dialog box.

11. To open an existing drawing, I would pick the _____ button on the Startup dialog box.

12. In the AutoCAD environment, the _____ mouse button serves to confirm a selection, call up a cursor menu, or acts like the enter key on the keyboard.

For Questions 13 to 25, identify the parts of the user interface in Figure 1.14a.

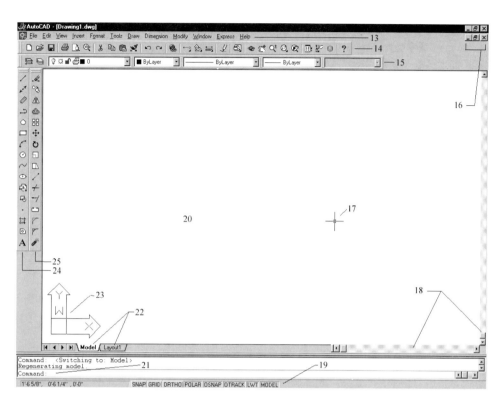

FIG. 1.14a

26. By default, angles in AutoCAD are measured (clockwise, counterclockwise).

27. On which compass point would you find AutoCAD's default 0°?

28. and 29. Which two options are available on the Quick Setup dialog box?

30. A group of buttons together is called a _____.

31. AutoCAD prompts the user on the _____ line.

32. and 33. Name the two coordinate systems used in AutoCAD.

34. and 35. Give two keyboard commands used to save a drawing.

36. and 37. Name two ways to leave a drawing session.

Identify these buttons:

38. 39. 40.

41. To set or reset units after creating a drawing, use the _____ command.

42. Another name for the width and height of a drawing is the _____ of the drawing.

43. 0,0 refers to the lower left corner of a drawing, unless the drawing is set up to use _____ coordinates.

44. To complete a keyboard command, the user must hit the _____ key.

45. An advanced drawing environment called _____ enables the CAD operator to work in multiple scales on the same drawing.

46. The environment that allows the user to view more than one drawing at a time is referred to as the _____.

47. The command used to partially open a drawing is _____.

48. The command used to open more of a partially opened drawing is _____.

LESSON 2

Drawing Basics: Lines and Coordinates

Following this lesson, you will:
- Know how to use the basic draw commands: **Line** and **Rectangle**
- Know how to use the basic modify commands **Erase, Undo,** and **Redo**
- Know how to select objects using object selection, window, and crossing window methods
- Know how to use the display controls: **Redraw** and **Regen**
- Know how to use AutoCAD's Cartesian Coordinate System

We can define drafting as the placing of geometric shapes on paper to represent existing or proposed objects. Of course, we are using a computer rather than the traditional paper or vellum medium, but the results are the same.

All geometric shapes can be formed by the constructive use of two simple objects—lines and circles. Did you know that, in a CAD environment, a draftsman spends only about 30% of his time placing lines and circles? He spends the vast majority of his time modifying what he has drawn and placing text on the drawing. Still, without a basic knowledge of placing lines and circles, the draftsman will never leave the starting gate.

We will begin this lesson learning to use the most basic geometric shape with which we can work—the line. (We will save circles for Lesson 5.) Then we will explore the coordinate system AutoCAD uses to identify locations within the drawing.

2.1 Lines and Rectangles

Being one of the first CAD packages, AutoCAD had the benefit of creating commands that are really quite simple to remember. For example, to draw a line, you type *line* at the command prompt. To draw a circle, you type *circle*. To erase an object, what do you think you would type? Did you say *erase*? That's right!

But wait, it gets simpler yet! Can't type? Most commands have simple abbreviations, or *hotkeys* (also called *aliases* or *shortcuts*). To draw a line, you can type *l*. To erase, type *e*. I will identify the hotkeys for each command as we progress through the book.

Of course, if you just have an aversion to the keyboard, there is also a mouse approach to each command. We will look at those as well.

The line command provides a nice uncluttered approach to seeing the command prompt in action. The sequence involved in drawing a two-point line is as follows:

Command: *line* (or *l*)
Specify first point: *1,1* [*enter the coordinate for a point or you can pick a point on the screen*]
Specify next point or [Undo]: *2,2* [*again, enter the coordinate for a point or you can pick the point*]
Specify next point or [Undo]: [*hit enter or you can click the right mouse button and select Enter from the cursor menu*]

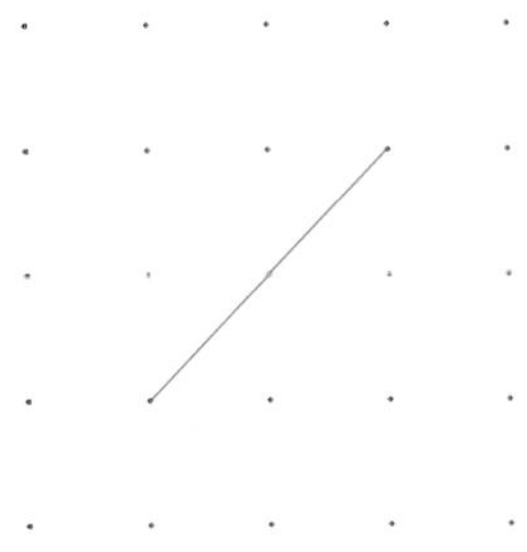

Not too difficult, is it? Try it. Start a new drawing using *Sample Template 02* in the Lesson02 folder. (Notice that I have provided a background grid to help guide you through the drawing.) Enter the sequence. Remember to hit the *enter* key on the keyboard after entering each line of information. Otherwise, AutoCAD will not know you have finished your command. The final *enter* tells AutoCAD you have finished the command. The numbers represent coordinates on the drawing. We will discuss them in Section 2.4, "The Cartesian Coordinate System."

Does the lower left quadrant of your screen look like Figure 2.1a? Excellent! Okay, now erase the line. Use this sequence.

FIG. 2.1a
Line

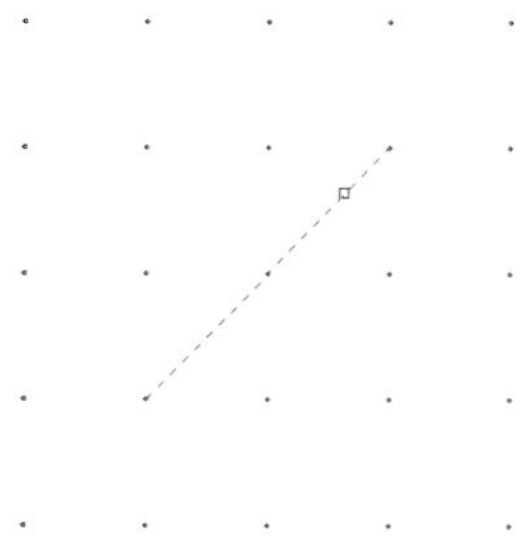

FIG. 2.1b
Selected (Highlighted) Line

Command: *erase* (or *e*)

Select objects: *[The crosshairs become a tiny box. Place the box over the line and pick once with the left mouse button. The line becomes dashed, or highlighted. (see Figure 2.1b)]*

Select objects: *[enter or click the right mouse button—The line disappears]*

Command:

Your first *"Hail Mary"* option: You will frequently find yourself entering a command too fast or out of order. Do not worry. If you make a mistake, use the escape [*ESC*] key in the upper left corner of most keyboards. You will return to the command prompt where you can start over.

Now try an exercise.

2.1.1 Do This: Drawing Lines

I. Try to draw Figure 2.1.1a. Do not worry if you cannot get the lines perfect just yet. This is your first time. Besides, we will soon find much easier and more accurate ways of drawing.

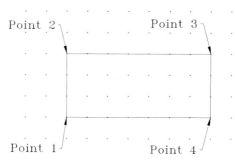

Fig. 2.1.1a: Draw a Rectangle Using the **Line** command

II. Start a new drawing using *Sample Template 02* in the C:\Steps\Lesson02 folder.

III. Follow these steps.

TOOLS	COMMAND SEQUENCE	STEPS
Line Button	Command: *l*	1. Enter the *line* command at the command prompt by typing *line* or *l*. Or you can pick the **Line** button on the draw toolbar or select **Line** from the Draw pull-down menu.

continued

TOOLS	COMMAND SEQUENCE	STEPS
	LINE Specify first point: *[pick a point near 2,1]*	2. Select a point around grid reference 2,1. To help you find this reference point, look at the coordinate display on the left end of the status line. Move your crosshairs around while watching this box. See the numbers change? These are your drawing coordinates. Try to get as close as possible to 2,1 (ignore the third number—the Z coordinate). Do not be surprised if you cannot get it exactly.
Polar: 0'-3" < 90°	**Specify next point or [Undo]:** *[pick a point near 2,4]*	3. Now select a point around 2,4. Notice that AutoCAD presents a cheater box that says "Polar: 0'-3"<90°". This box will help guide you when drawing. We will discuss it more in Lesson 4.
Polar: 0'-7" < 0°	**Specify next point or [Undo]:** *[pick a point near 9,4]*	4. Now select a point around 9,4.
Polar: 0'-3" < 270°	**Specify next point or [Close/Undo]:** *[pick a point near 9,1]*	5. Now select a point around 9,1.
	Specify next point or [Close/Undo]: *c*	6. Now type **c** to *close* the line.

Did you notice that last command? When used in response to a line or polyline prompt command, *c* will *close* the line (or draw a line from the last point selected to the first point selected).

> Some points to remember when drawing lines:
>
> ▌ Hitting *enter* at the *command* prompt will always repeat the last command regardless of what that command was. (You may also right-click in the graphics area and pick the repeat line at the top of the cursor menu.)
>
> ▌ Hitting *enter* at the *Specify first point:* prompt will cause the last point selected to be the first point of the line. In other words, if you leave the line prompt before you actually finish drawing the lines you want, you can hit *enter* to repeat the line command, then hit *enter* again to continue from where you stopped.
>
> ▌ *C* will always draw a line from the last point selected to the first point selected during this command sequence but only during *this* sequence. In other words, if you return to the command prompt, enter the *Line* command again, then continue the same line, "C" will draw a line from the last point selected back to the first point of the current sequence.
>
> All of this will become clearer as we go.

Try some random lines and experiment with these last few statements. When you are comfortable with them, erase all the lines as explained on page 35. Then continue.

Let me show you an easier way to draw the same rectangle, but using only two picks. We will use the *Rectang* command this time.

The *Rectang* command sequence is (refer to Figure 2.1c)

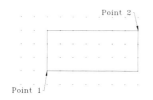

FIG. 2.1c
Rectangle

Command: *rectang* (or *rec*)
Specify first corner point or [Chamfer/Elevation/Fillet/Thickness/Width]: *2,1*
Specify other corner point: *9,4*
Command:

Go ahead and try the preceding sequence. Does it look better than the one you drew using the line command? Was it easier to draw?

Now erase it. Did you notice something different here? You only had to select one line and the entire rectangle highlighted. The rectangle is a *polyline*.

> A *polyline* is a multisegment line, or a line consisting of one or more than one line segment. We will cover this more advanced line technique in Lesson 8.

Did you notice that AutoCAD gave you several options when you drew the rectangle? In the first prompt of the *Rectang* command, AutoCAD tells you what options you have. You may draw a *Chamfered* or *Filleted* rectangle. Or you may use the *Width* option for a rectangle with heavier lines. The other options, *Elevation* and *Thickness*, involve 3-dimensional space. We will cover those in the advanced text. The default choice will always precede the bracketed options. Our response in the preceding sequence accepted the default choice. AutoCAD read *2,1* as the **first corner point** of the rectangle.

To draw a rectangle with ¹/₂" chamfered corners, use this sequence (refer to Figure 2.1d)

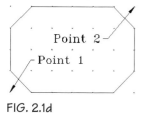

FIG. 2.1d
Chamfered Rectangle

Command: *rectang* (or *rec*)
Specify first corner point or [Chamfer/Elevation/Fillet/Thickness/Width]: *c*
Specify first chamfer distance for rectangles <0'-0">: *.5*
Specify second chamfer distance for rectangles <0'-0 1/2">: *[enter to accept the default]*
Specify first corner point or [Chamfer/Elevation/Fillet/Thickness/Width]: *[select point 1]*
Specify other corner point: *[select point 2]*
Command:

The sequence for drawing a filleted rectangle is essentially the same as that of the chamfered.

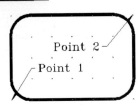

FIG. 2.1e
Filleted Rectangle with Wide Lines

The sequence for drawing a filleted rectangle with 1/16" wide lines is (refer to Figure 2.1e)

Command: *rectang* (or *rec*)
Specify first corner point or [Chamfer/Elevation/Fillet/Thickness/Width]: *w*
Specify line width for rectangles <0'-0">: *1/16*
Specify first corner point or [Chamfer/Elevation/Fillet/Thickness/Width]: *f*
Specify fillet radius for rectangles <0'-0">: *1/2*
Specify first corner point or [Chamfer/Elevation/Fillet/Thickness/Width]: *[select point 1]*
Specify other corner point: *[select point 2]*

> Have you noticed that you do not need to put inch marks on your numbers? We have set up our drawing using architectural units. The inch is the basic architectural unit. AutoCAD knows this, so we do not have to indicate it. We would have to indicate feet, however, with a prime (').
>
> AutoCAD uses a rather different approach to entering feet and inches in response to a prompt. To enter one-foot-seven-and-one-half-inches, for example, type *1'7-1/2*. Notice the lack of a dash between the feet and inch numbers. Separate these only with the foot mark (the prime—'). Use a dash to separate inches from fractions.
>
> The reason for this is simple: AutoCAD reads a space the same way it reads *enter*. So you cannot use a space as a separator. (But you can use it instead of the *Enter* button!)

Now try some rectangles.

 2.1.2 Do This: Drawing Rectangles

I. If you are already in AutoCAD, close the program and then restart it. Start a new drawing using *Sample Template 02* in the C:\Steps\Lesson 02 folder.

II. Follow these steps to draw a simple rectangle.

TOOLS	COMMAND SEQUENCE	STEPS
 Rectang Button	**Command:** *rec*	1. Enter the **Rectang** command by typing **rectang** or **rec** at the command prompt. You can also pick the **Rectang** button from the Draw Toolbar or select **Rectangle** from the Draw pull-down menu.

continued

TOOLS	COMMAND SEQUENCE	STEPS
	Specify first corner point or [Chamfer/Elevation/Fillet/Thickness/Width]: *1,1*	**2.** Type in the lower left coordinate as shown.
	Specify other corner point: *8,4*	**3.** Type in the upper right coordinate as shown. Your rectangle will look like Figure 2.1.2.3 (without the numbers and arrows):

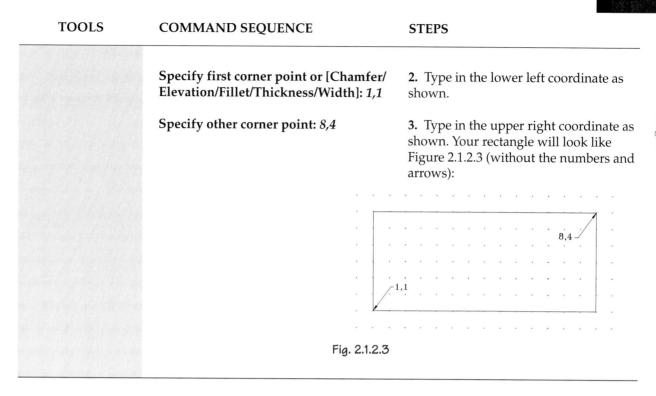

Fig. 2.1.2.3

III. Now draw a ¹/₂" chamfered rectangle as follows.

TOOLS	COMMAND SEQUENCE	STEPS
▢	**Command:** *[enter]*	**4.** Repeat the *Rectang* command.
	Specify first corner point or [Chamfer/Elevation/Fillet/Thickness/Width]: *c*	**5.** Type **c** for the **Chamfer** option.
	Specify first chamfer distance for rectangles <0'-0">: *1/2*	**6.** Tell AutoCAD what the chamfer distances are (how far back on each line of the corner to begin the angle).
	Specify second chamfer distance for rectangles <0'-0 1/2">: *[enter]*	**7.** AutoCAD automatically sets the second chamfer distance the same distance as the first. You can accept by hitting *enter*, or give a different number. Let us accept.
	Specify first corner point or [Chamfer/Elevation/Fillet/Thickness/Width]: *1,5*	**8.** Now tell AutoCAD where to draw. Type in the lower left coordinates.

continued

TOOLS	COMMAND SEQUENCE	STEPS
	Specify other corner point: *8,8*	**9.** Type in the upper right coordinates. Your rectangle will look like Figure 2.1.2.9.

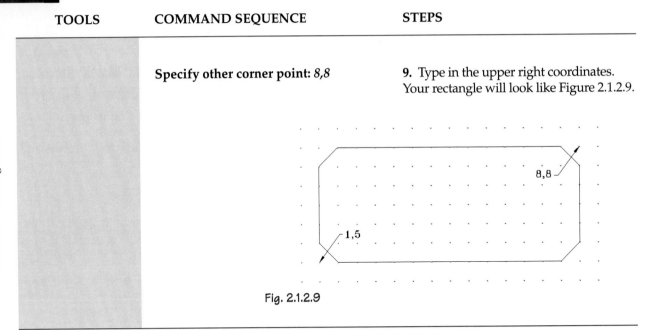

Fig. 2.1.2.9

IV. Now draw a ³/₄" filleted rectangle with 1/16" wide lines.

TOOLS	COMMAND SEQUENCE	STEPS
	Command: *[enter]*	**10.** Repeat the *rectang* command.
	Current rectangle modes: Chamfer=0'-0 1/2" x 0'-0 1/2"	**11.** AutoCAD tells you that you are currently set up to draw a chamfered rectangle. It also tells you the chamfer distances. We will change that now.
	Specify first corner point or [Chamfer/Elevation/Fillet/Thickness/Width]: *w*	Type *w* for the **Width** option.
	Specify line width for rectangles <0'-0">: *1/16*	**12.** Tell AutoCAD how wide to make the lines.
	Specify first corner point or [Chamfer/Elevation/Fillet/Thickness/Width]: *f*	**13.** Type *f* for the **Fillet** option.
	Specify fillet radius for rectangles <0'-0 1/2">: *3/4*	**14.** Tell AutoCAD the radius you want for your fillets (rounded corners).
	Specify first corner point or [Chamfer/Elevation/Fillet/Thickness/Width]: *4,2.5*	**15.** Now tell AutoCAD where to draw. Type in the lower left coordinates.

continued

TOOLS	COMMAND SEQUENCE	STEPS
	Specify other corner point: *10.5,6.5*	**16.** Type in the upper right coordinates. Your rectangle will look like Figure 2.1.2.16.

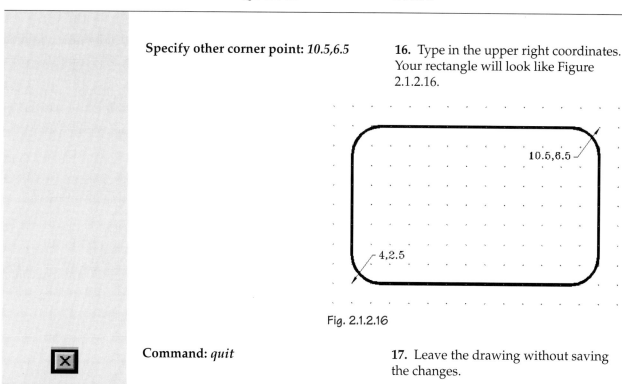

Fig. 2.1.2.16

| | **Command:** *quit* | **17.** Leave the drawing without saving the changes. |

2.2 Fixing the Uh Ohs: Erase, Undo, and Redo

When I first studied drafting—back in the dinosaur days—I thought the electric eraser was the height of laziness. I mean, how spoiled could a professional be?

Then I got my first drafting job. I was assigned to removing revision clouds from old cloth drawings. I had an electric eraser by the second day.

Today, the CAD system makes erasure even easier than that old Bruning did.

The command sequence for erasing a single object, or a group of objects one at a time, is

> **Command:** *erase* (or *e*)
> **Select objects:** *[select an object]*
> **Select objects: 1 found** *[AutoCAD tells you how many objects you have selected, and allows you to select more. Hit enter to confirm that you are done selecting.]*

We should look at another command while we are here. If your erasure was a mistake, you can use the ***Undo*** command. The command sequence is

> **Command:** *undo*
> **Enter the number of operations to undo or [Auto/Control/BEgin/End/Mark/Back]:** *[enter]*
> **REGEN** *[AutoCAD regenerates the drawing]*

Here you find another case of AutoCAD letting you know what options you have. Let us look at each of these.

- **Enter the number of operations to undo** (the default). Undoes the specified number of preceding commands. The default number is one, so an *enter* will undo a single command.
- **Auto.** Undoes a menu selection as a single command. Remember that menu commands and toolbar commands cause a macro to run. If auto is set to *Off*, undo will only undo one step of the macro at a time. If it is *On* (the default setting), *Undo* will undo the entire macro at once.
- **Control.** Controls how *Undo* performs. It has three options: **All / None / One**.

  All. Allows *Undo* to function fully, providing virtually unlimited undos.
  None. Turns *Undo* **Off**.
  One. Limits *Undo* to the last command only.

- **Begin** and **End**. **Begin** begins a group. All commands entered after the **Begin** option will be treated as a single command and are undone by a single *Undo*. **End** ends the **Begin** option.
- **Mark** and **Back**. **Mark** places a mark in the command sequence. **Back** undoes back to the mark.

> About the *Undo* **Command:** Most of these options sound good, but you will never use them. I have never, in the years I have worked with CAD, known in advance that I would be undoing a command. How would I know to **Mark** or **Begin** a sequence for later undoing?
>
> I advise people to learn the *U* command discussed next. It will cover your needs and will not require you to memorize the *Undo* options.

FIG. 2.2a
Undo

FIG. 2.2b
Redo

Undo has a hotkey (sort of)—*U*. Actually, *U* is a command in its own right. Most hotkeys serve as a shortcut to entering a command. *U* acts as a macro for the default *Undo* option, and assumes a number of one.

The **Undo** button on the Standard Toolbar (Figure 2.2a) behaves just like the *U* command.

The opposite of the *Undo* command is, of course, the *Redo* command. Unlike the *Undo* command, however, the *Redo* command is limited to redoing only a single *Undo* command. The button to the right of the **Undo** button on the Standard Toolbar is the **Redo** button (Figure 2.2b).

You will also find the **Undo** (**U**) and **Redo** commands on the Default Cursor menu. Just right-click anywhere in the graphics area of the drawing with nothing selected and the command prompt empty.

Now you have four of AutoCAD's "Hail Mary" procedures—*ESCAPE, ERASE, UNDO,* and *REDO*. Time for an exercise!

WWW 2.2.1 Do This: Practice Erasing Objects One at a Time

I. Practice erasing some lines, then **undo** / **redo**.

TOOLS	COMMAND SEQUENCE	STEPS
Open	**Command:** *open*	**1.** Open drawing *erase-samp*, in the C:\Steps\Lesson02 folder. Refer to Section 1.6 if you need help.
Erase	**Command:** *e*	**2.** Enter the *Erase* command by typing *erase* or *e*. You can also pick the **Erase** button on the Modify I toolbar or select **Erase** from the Modify pull-down menu.
	Select objects:	**3.** Pick one of the lines by placing the selection box on it and clicking the left mouse button as indicated in Figure 2.2.1.3.

continued

TOOLS	COMMAND SEQUENCE	STEPS

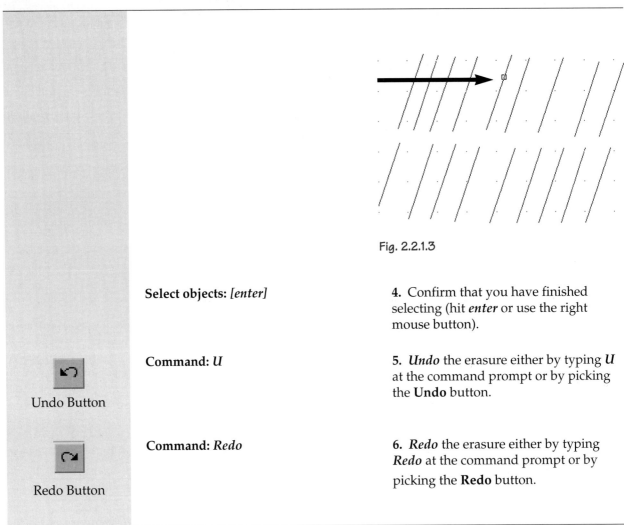

Fig. 2.2.1.3

Select objects: *[enter]*

4. Confirm that you have finished selecting (hit *enter* or use the right mouse button).

Command: *U*

Undo Button

5. *Undo* the erasure either by typing *U* at the command prompt or by picking the **Undo** button.

Command: *Redo*

Redo Button

6. *Redo* the erasure either by typing *Redo* at the command prompt or by picking the **Redo** button.

You can use the Default cursor menu to erase (or *Cut*) an object. But this method moves the erased item to the Windows clipboard. The clipboard requires a bit of system memory, so this is not the preferred method for removing objects from a drawing.

2.3 Multiple-Object Selection Made Easy

We have seen how easily we can erase a single object. But what if we want to erase 20 or 200 objects? Must we select 20 or 200 times?

The answer, of course, is no! Let us look at a couple of options that will make multiple-object selection easier.

The first option places a *window* around the objects to be selected (refer to Figure 2.3a). What do you suppose it is called? What do you think the shortcut will be? If you said "window" and "w," you were quite right. A window selection includes all the objects that are *completely within* a window.

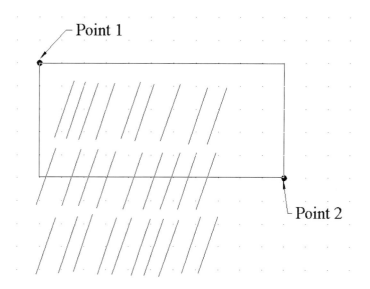

Fig. 2.3a: Window Selection

Watch while we erase the top row of lines. (If you are not already in drawing *erase-samp*, in the C:\Steps\Lesson02 folder, please open it now.)

Command: *e*
Select objects: *w*
Specify first corner: *[pick point 1]*
Specify opposite corner: *[pick point 2]*
Select objects: *[enter]*

Notice that only the lines that were completely encircled by the window were erased.

The second option places a window around *and across* the objects to be selected. This one is called *crossing* and uses *c* as a shortcut. The difference between this and a standard window is simple. A crossing window will select everything within or *touched by* the window.

Watch this one in action (refer to Figure 2.3b). First, undo the last erasure. Then do the following.

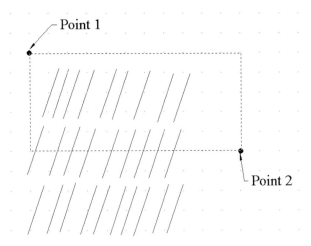

Fig. 2.3b: Crossing Window Selection

Command: *e*
Select objects: *c*
Specify first corner: *[pick point 1]*
Specify opposite corner: *[pick point 2]*
Select objects: *[enter]*

Notice that this time the first two rows of lines were erased, the one within the window and the one crossed by the window.

> To make windowing even easier, AutoCAD includes *Implied Windowing*. This means that you do not actually have to type *w* or *c* to create a window or crossing.
> When you pick an empty place on your drawing at the **Select objects** prompt, AutoCAD assumes that you want to use Implied Windowing.
> To use a window at any **Select objects** prompt, simply pick an empty point to the left of what you want to select, then pick a second point to the right. You will get a window.
> If you pick the first point to the right and the second point to the left, you will get a crossing window.
> Undo the last erasure, then try this by selecting a place near Point 1 then near Point 2 in the current drawing. Then try it by selecting a place near Point 2 then near Point 1.

2.4 The Cartesian Coordinate System

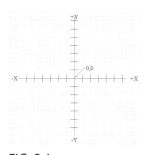

FIG. 2.4a
Cartesian Coordinate System

Remember suffering through plane geometry back in high school math class? That was the one with the crossing number lines, four quadrants, and probably the one that you insisted to your parents, teachers, and anyone else who would listen that you would never use. Look at Figure 2.4a—it is your math teacher's revenge!

On the bright side, it is not nearly as difficult as a math teacher might wish. There are three axes—X, Y, and Z. But we are only concerned with two of them in beginning AutoCAD, so the difficulty is already cut by 1/3!

The *X-axis* runs horizontally (left to right). Everything to the right of a point we call 0 is positive (plus). Everything to the left of the 0 is negative (minus). Thus, we now have +X and -X directions identified. Each number on the X-axis (the axis is the line that runs through the 0) is considered an X-plane. An X-plane runs infinitely up and down, crossing the X-axis.

The *Y-axis* runs vertically (up and down). Everything above the point we call 0 is positive. Everything below the 0 is negative. Each number on the Y-axis is considered a Y-plane. A Y-plane runs infinitely right and left, crossing the Y-axis.

> The X and Y planes also run parallel to the 3-dimensinal Z-axis and might be called the XZ and the YZ planes. But let's stick to two dimensions for now.

Are you still with me?

Okay. Where an X-plane meets a Y-plane, we have a *point*. We identify the point by its *coordinate*—or by the number of the X-plane, followed by a comma, then the number of the Y-plane. Remember this as X,Y (no spaces). For example, point 4,3 is found 4 spaces to the right of zero and 3 spaces above, as shown in Figure 2.4b.

AutoCAD provides three ways to use these coordinates: *absolute*, *relative*, and *polar*. The use of coordinates is your first step toward a drawing precision that was never possible using conventional drafting tools.

➥ Absolute Coordinates

The Absolute System is easiest. You simply enter absolute coordinates as **X,Y** whenever AutoCAD asks for a point.

Open drawing *CCS* in the C:\Steps\Lesson02 folder, then try this (your drawing will look like Figure 2.4b):

Command: *l*
Specify first point: *4,3*
Specify next point or [Undo]: *2,5*
Specify next point or [Undo]: *[enter]*

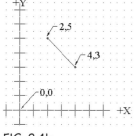

FIG. 2.4b
Absolute Coordinates

Using the absolute coordinate system means that you must know the X and Y value of each point you wish to use. This is not always possible. Let us look, then, at the *Relative* System.

➡ Relative Coordinates

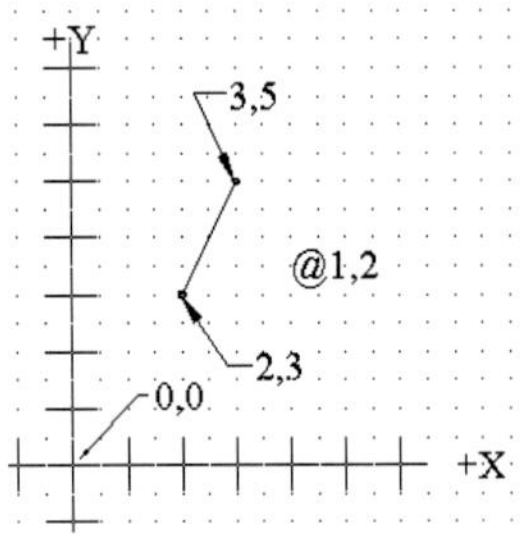

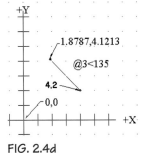

FIG. 2.4c
Relative Coordinates

Use a relative coordinate anytime after identifying the first point. (Either select the first point with the mouse, or use an absolute coordinate.) The syntax for relative coordinates is **@X,Y**.

Erase the line you just drew. Now try this method. (Your drawing will look like Figure 2.4c.)

Command: *l*
Specify first point: *2,3*
Specify next point or [Undo]: *@1,2*
Specify next point or [Undo]: *[enter]*

Notice that the second point is located 1 unit to the right (a positive 1 on the X-axis) and 2 units up (a positive 2 on the Y-axis), or *at 1X and 2Y* from the last point identified. The point is *relative to* the last identified point.

Using the Relative System means you must know how far (in plus/minus terms) along each axis you want to go from where you are. You will find this much easier than having to locate each point in absolute terms, especially on larger drawings.

➡ Polar Coordinates

I find polar coordinates the most useful. But I must admit that I will use the Relative System when needed.

Like relative coordinates, you will use a polar coordinate entry after identifying the first point required. The syntax for polar coordinates is **@*dist<angle*** (at-distance—at an angle of—angle). You must know the distance and direction you wish to go from the last selected point. Remember, measure angles counterclockwise beginning at 0° East.

Erase the last line you drew. Here is how the polar method works (refer to Figure 2.4d).

FIG. 2.4d
Polar Coordinates

Command: *l*
Specify first point: *4,2*
Specify next point or [Undo]: *@3<135*
Specify next point or [Undo]: *[enter]*

Notice that the coordinate of the second point is not an integer. This often happens with polar entries. Remember that you are drawing the *hypotenuse* of an angle. This is why I mix use of polar coordinates with use of relative coordinates. Sometimes I may know how long a line I need (polar coordinates); other times I know the X and Y distances (relative coordinates).

We have seen three approaches to drawing a simple line with a great deal of precision. Time now to practice.

WWW 2.4.1 Do This: Practice Using Cartesian Coordinates

I. Begin a new drawing using *Sample Template 03* found in the C:\Steps\Lesson02 folder.

II. Using the following chart, draw Figure 2.4.1a. (*Note*: Grid marks are 1 unit apart.) Draw the figure first using absolute coordinates. Erase it. Draw it using relative coordinates. Erase it. Draw it using polar coordinates.

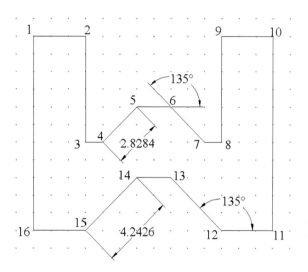

Fig. 2.4.1a

POINT	ABSOLUTE(X,Y)	RELATIVE(@X,Y)	POLAR(@DIST<ANGLE)
1	2,15	2,15	2,15
2	5,15	@3,0	@3<0
3	5,9	@0,-6	@6<270
4	6,9	@1,0	@1<0
5	8,11	@2,2	@2.8284<45
6	10,11	@2,0	@2<0
7	12,9	@2,-2	@2.8284<315
8	13,9	@1,0	@1<0
9	13,15	@0,6	@6<90
10	16,15	@3,0	@3<0
11	16,4	@0,-11	@11<270
12	13,4	@-3,0	@3<180
13	10,7	@-3,3	@4.2426<135
14	8,7	@-2,0	@2<180
15	5,4	@-3,-3	@4.2426<225
16	2,4	@-3,0	@3<180
Back to 1	C	C	C

2.5 Extra Steps

With all the drawing and erasing you have done in this chapter, you may have noticed that you will occasionally "lose" a line (or other object) that you thought was there. Likewise, you may notice that an object may remain highlighted even after you have canceled the command. This problem lies mostly with the video card or monitor you are using. But most programs allow you to *refresh* your screen after doing a great deal of work, to make sure all the screen pixels that should be lit, are lit.

AutoCAD calls its refresh command **Redraw**, but that requires too much typing. When your screen needs refreshing (redrawing), type **R** at the command prompt and hit *enter*. This will redraw the screen.

> By the way, I indicate commands and hotkeys using capital letters as a matter of convention. AutoCAD is *not* case sensitive.

In the event that what you expect to see does not materialize, you have one other option. Every object within an AutoCAD drawing carries with it quite a bit of information. This information includes details about the type of object, its size, location, layer, and so forth. With AutoCAD's **Regen** command (**RE**), AutoCAD reads and redraws every object in the drawing. You might think this would take a lot of time in a larger drawing, and in fact it once did. Now, however, with the advent of faster computers and better programming, regens often go unnoticed.

If the object does not show after a regen, it does not exist as part of the drawing and you will have to redraw it.

So add these two commands to your "Hail Mary" list, but do not cut it off yet. More will follow!

2.6 What Have We Learned?

Items covered in this lesson include:

- *Object Selection*
- *Cartesian Coordinate System*
- *Commands*
  - *Line*
  - *Rectang*
  - *Erase*
  - *Undo / U*
  - *Redo*
  - *Redraw*
  - *Regen*

We began this lesson by describing drafting as "the placing of geometric shapes on paper to represent existing or proposed structures." We have seen how to use the "backbone" of more than half of all geometric structures—the straight line. We have seen how to draw the line with precision and accuracy using the "backbone" of AutoCAD precision—the Cartesian Coordinate System.

But if the Cartesian Coordinate System was solely responsible for precision in AutoCAD, the software might never have survived through 15 releases. Let's face it, at its best, it can be time-consuming and a bit cumbersome. In Lesson 4, we will look at some faster ways to draw that, though not replacements for Cartesian coordinates, accent them nicely. Then, in Lesson 5, we will look at the curved line—arcs and circles—to complete the basic structures of drafting.

First, however, we must look at transferring our computer drawing to paper. Let's face it, no matter how nice it looks on the screen, we need the paper to build it!

Our next lesson looks at plotting.

2.7 EXERCISES

1. Using what you have learned, fill in the blanks with the command entries needed to draw the structure in Figure 2.7.1. Use each of the Cartesian coordinate methods—absolute, relative, and polar. Then draw the object using each of the methods. (Grid marks are 1" apart.) Save the drawing as *MyM.dwg* in the C:\Steps\Lesson02 folder.

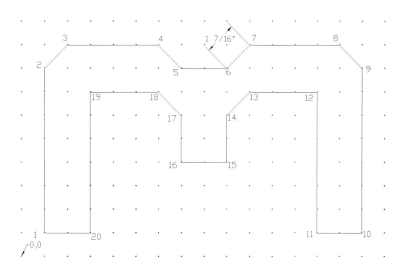

Fig. 2.7.1

POINT	ABSOLUTE(X,Y)	RELATIVE(@X,Y)	POLAR(@DIST<ANGLE)
1	1,1	1,1	1,1
2			
3			
4			
5			
6			
7			
8			
9			
10			
11			
12			
13			
14			
15			
16			
17			
18			
19			
20			
Back to 1			

2. Create a drawing(s) using either templates or wizards. Draw each of the figures shown in Figure 2.7.2. (Grid marks are 1" apart.) Save the drawing as *MyLines.dwg* in the C:\Steps\Lesson02 folder.

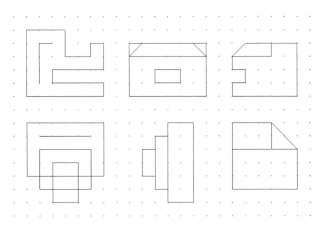

Fig. 2.7.2

3. Create the snowflake in Figure 2.7.3 using what you have learned. Do not try to draw the centerlines or the dimensions.

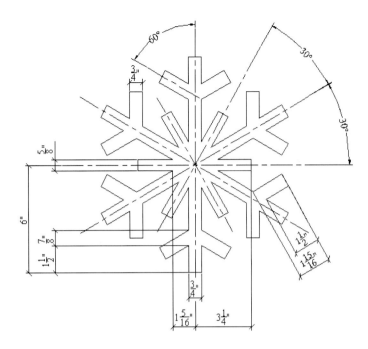

Fig. 2.7.3

4. Now create the drawings in Figure 2.7.4. The grid is $1/4$".

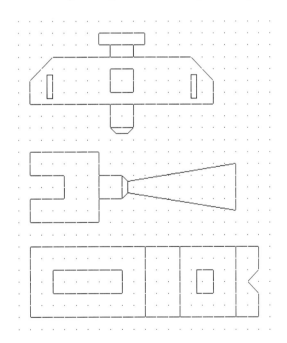

Fig. 2.7.4

2.8 REVIEW QUESTIONS

Please write your answers on a separate sheet of paper.

1. Drafting is the placing of _____ on paper to represent objects.

2. and 3. All geometric shapes can be formed by the constructive use of _____ and _____.

4. and 5. I can draw a line by typing _____ or _____ at the command prompt.

6. and 7. To erase my line, I type _____ or _____ at the command prompt.

8. To exit a command before it is finished, hit the _____ key.

9. To draw a line from the last selected point back to the first selected point, type _____.

10. A _____ is a multisegmented line.

11. Hit _____ on the keyboard to repeat the last command.

12. The _____ option of the Rectang command will cause the rectangle to have rounded corners.

13. How would you enter one-foot-six-and-one-half-inches in AutoCAD?

14. AutoCAD reads a keyboard spacebar entry as an _____.

Identify these buttons:

15. 16. 17. 18. 19.

20. An implied window drawn from right to left is a (window, crossing window).

21. How many planes are in the Cartesian Coordinate System?

22. Name them.

23. The _____ axis runs vertically.

24. The _____ axis runs horizontally.

25. A point located at 3,2 is _____ spaces right of zero.

26. through 28. Name the three ways AutoCAD uses coordinates.

Identify these methods of entering coordinates:

29. X,Y

30. @X,Y

31. @dist<Angle

32. I would type _____ at the command prompt to redraw or refresh AutoCAD's screen.

33. I would type _____ to regenerate an AutoCAD drawing.

34. (T or F) If an entity does not appear after typing regen, it is not part of the drawing.

List three ways to select a group of objects:

35.

36.

37.

LESSON 3

Putting It on Paper

Following this lesson, you will:

➡ Know how to Plot a drawing
  - Setting up the plotter or printer
  - Creating the plot style
  - Setting up the drawing (page)
  - Plotting
➡ Know how to Plot multiple drawings at one time (Batch Plotting)

The ironic thing about CAD is that, after all the wonders of the computer world have created this marvelous drawing, in the end you still need that paper for distribution. But let's face it, it might be a bit clumsy to fold a PC and stick it in your tool belt down at the job site.

Not to worry—AutoCAD has made it fairly painless to make the transfer from computer to paper. We will use dialog boxes that will guide us through the process—*One Step at a Time*.

To begin, power up the printer . . .

. . . now let's get started!

3.1 First Things First: Setting Up Your Printer (or Plotter)

Printing and Plotting, although two different processes in earlier releases of AutoCAD, have used the same procedure in the last several releases. The terms *print* and *plot* are now used interchangeably in industry—and in this lesson.

Before actually printing a document in any software, you must first tell the software what type of printing device you will use. The reasons are technical and involve those driver gizmos about which you have probably read. Most software comes with several printer drivers, and all printers ship with their own drivers. You must help the program—in this case AutoCAD—to figure out which one to use.

A *driver* is actually a file on your computer's hard drive. It contains code—programming language—that helps your computer communicate with your printer. Or more simply, consider the driver as a tiny man living inside your computer. His job is to translate the language your computer speaks into the language your printer speaks (and vice versa). Since each printer speaks its own language, you must help your software decide which little man (translator) you need to do the translating.

Luckily, this is not as complicated as it might seem, and it only needs to be done once for each printing (or plotting) device. AutoCAD provides wizards to help.

Find the shortcut used to access the **Add-a-Plotter Wizard** by typing *Plottermanager* at the command prompt or selecting **Plotter Manager...** from the File pull-down menu. AutoCAD opens a new window (Figure 3.1a) showing the **Add-a-Plotter Wizard** Windows shortcut and the two setups that ship with AutoCAD (these are used for creating web pages and do not actually need a printer). Double-click on the **Add-a-Plotter Wizard** shortcut to begin the wizard.

FIG. 3.1a: Plotters Window

Let's set up a printer together. (I will be using an Epson Stylus Color printer—your screens may vary a bit depending on your printer.)

3.1.1 Do This: Set Up a Printer

I. Open *pid-03.dwg* in the C:\Steps\Lesson03 folder. The drawing looks like Figure 3.1.1a.

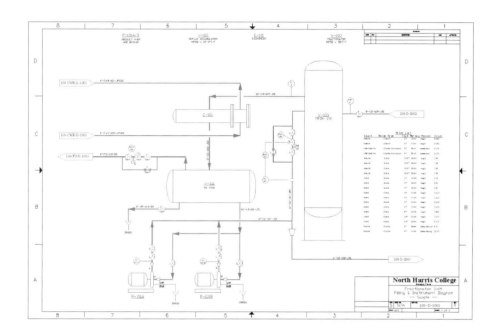

FIG. 3.1.1a: PID-03.dwg

II. Follow these steps.

TOOLS	COMMAND SEQUENCE	STEPS
	**Command:** *plottermanager*	1. Open the **Plotters** window by typing *Plottermanager* at the command prompt or selecting **Plotter Manager** from the File pull-down menu.

continued

TOOLS	COMMAND SEQUENCE	STEPS

2. Double-click on the **Add-a-Plotter Wizard** shortcut.

3. AutoCAD presents the Add Plotter—Introduction Page. Read through this; then pick the **Next >** button.

4. AutoCAD presents the first dialog box (FIG. 3.1.1.4) with the options for selecting the type of printer / plotter you wish to add.

⇨ The **My Computer** option will add a plotter.
⇨ The **Network Plotter Server** option will add a network plotter.
⇨ The **System Printer** will add a local printer (or the printer your computer is already using).

We will add a system printer. Put a bullet in that option, then pick the **Next >** button.

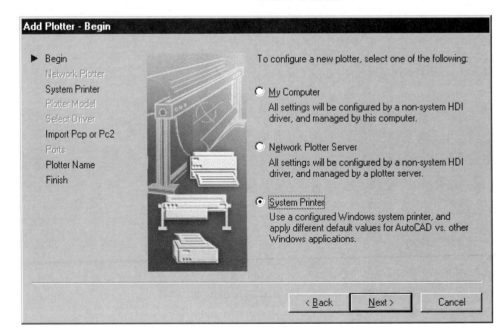

FIG. 3.1.1.4: First **Add-a-Plotter Wizard** Step

continued

TOOLS	COMMAND SEQUENCE	STEPS

[Next >]

5. Now we must select which printer we wish to use from the list box AutoCAD presents in the Add Plotter—System Printer dialog box (Figure 3.1.1.5). (The list comes from the printers your Windows system has already been configured to use.)

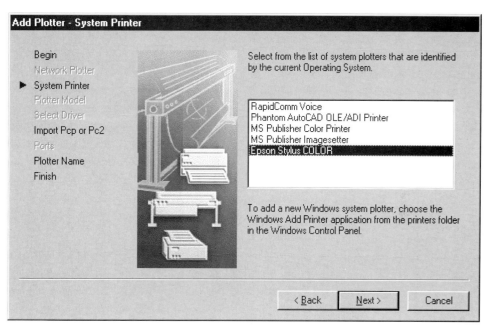

FIG. 3.1.1.5: System Printer

I will select the **Epson**, but you should select the printer your system uses.

Pick the **Next >** button to proceed.

6. Had we saved plotter information from an earlier release of AutoCAD, we could import that information into 2000 (Figure 3.1.1.6). But since we have no files to import, simply pick the next button.

[Next >]

continued

TOOLS	COMMAND SEQUENCE	STEPS

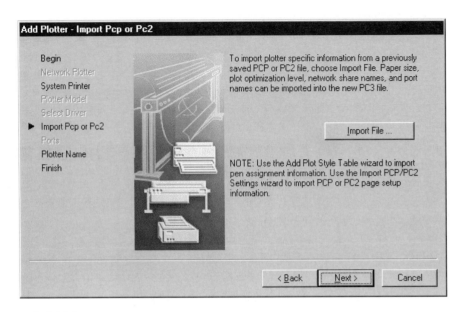

FIG. 3.1.1.6: Import Earlier Release Files

7. Give the plotter a name that you can easily recognize (or accept AutoCAD's suggestion). I have called mine **B&W Epson** (Figure 3.1.1.7) to indicate that I will set up this plotter for black-and-white plots (that part of the setup is still to come).

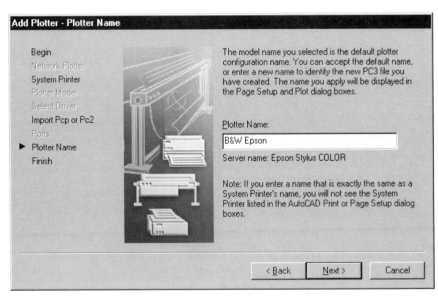

FIG. 3.1.1.7: Name the Plotter

continued

| TOOLS | COMMAND SEQUENCE | STEPS |
|---|---|---|ових
| | | 8. The next dialog box (Figure 3.1.1.8) is a bit deceptive. We are not quite finished until we calibrate our plotter to be sure our "tiny man" is communicating properly. We may also make any changes to our plotter's configuration at this time. |

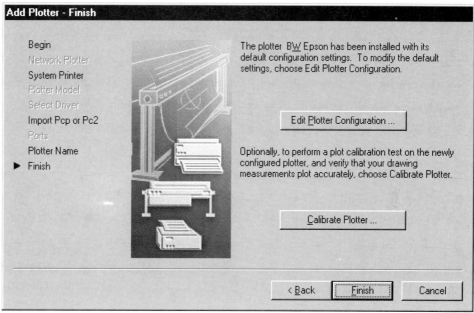

FIG. 3.1.1.8: Not Quite Finished Yet

Let's begin by editing the configuration. Pick the **Edit Plotter Configuration . . .** button.

9. AutoCAD presents the Plotter Configuration Editor (Figure 3.1.1.9) atop the wizard.

continued

TOOLS	COMMAND SEQUENCE	STEPS

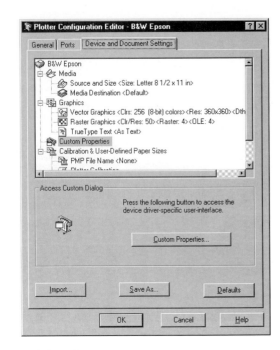

FIG. 3.1.1.9: Plotter Configuration Editor

- We can select different drivers (*not* recommended) using the **General** tab (not shown).
- We can configure a port (best left to the project guru) or tell AutoCAD to plot to a file by default on the **Ports** tab (not shown).
- We can change the printer's settings on the **Device and Document Settings** tab (shown).

We want to tell our printer to print just in black and white, so select the **Custom Properties . . .** button.

(Here Windows presents the printer device's configuration window. This is printer-specific, so you must use your printer's/plotter's manual as a guide when making whatever changes you wish to make. I will configure my printer to create a black-and-white print.)

continued

TOOLS	COMMAND SEQUENCE	STEPS

10. Once you have finished configuring your printer, Windows returns you to AutoCAD's Plotter Configuration dialog box. Pick **OK** to return to the Add Plotter Wizard.

11. Lastly, we must calibrate our printer/plotter. This simply means that we must be sure our "tiny man" knows the units with which we are working.

Pick the **Calibrate Plotter . . .** button.

12. AutoCAD launches the Calibrate Plotter Wizard (Figure 3.1.1.12). The paper size is indicated by default, but you can change the size if necessary. Pick the down arrow to view the options. When done, pick the **Next >** button.

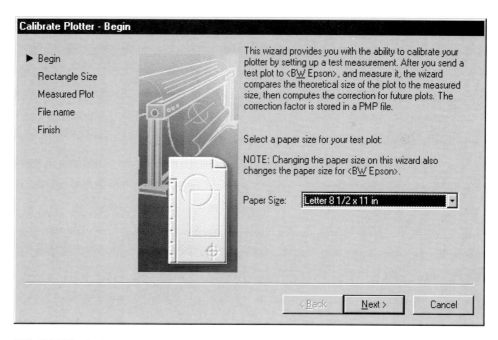

FIG. 3.1.1.12: Calibrate Plotter Wizard

continued

TOOLS	COMMAND SEQUENCE	STEPS

13. On the next screen (Figure 3.1.1.13), tell AutoCAD the largest size rectangle (no fractions) that will fit on the drawing sheet you selected in Step 12.

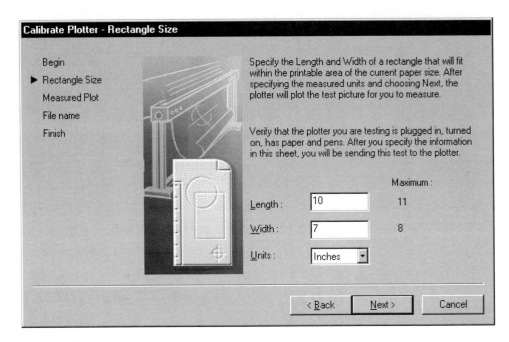

FIG. 3.1.1.13: Rectangle Size

Be sure your printer is connected and turned on so that AutoCAD can print a sample sheet. Then pick the **Next >** button to continue.

14. Measure the rectangle on the printed page. Adjust the numbers on the Measured Plot dialog box (Figure 3.1.1.14) to match those of the printed rectangle.

continued

TOOLS	COMMAND SEQUENCE	STEPS

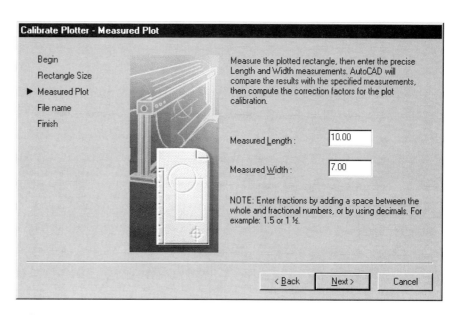

FIG. 3.1.1.14: Indicates the Printed Measurements

Pick **Next >** to continue.

15. Next AutoCAD asks you to name the calibration file (Figure 3.1.1.15) so that the information may be recalled later as needed.

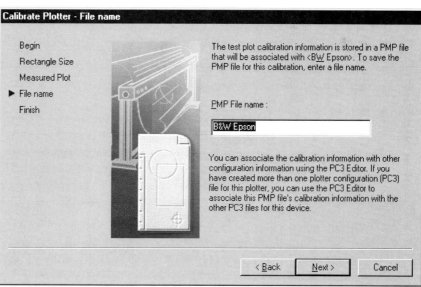

FIG. 3.1.1.15: Name the Calibration File

continued

TOOLS	COMMAND SEQUENCE	STEPS
[Next >] [Finish] [Finish]		Pick **Next >** to continue.

16. On the last screen, you can ask AutoCAD to verify the calibration if you are not comfortable. Pick the **Finish** button when you are done.

17. AutoCAD returns to the Add a Plotter Wizard. Pick the **Finish** button to complete the setup. |

Luckily, the setup procedure only needs to be done once for each plotter! Now your plotter is set up. Next you must set up your Plot Style.

3.2 Plot Styles

Believe it or not, there will be times when you must ask yourself: "Do I want to print it the way I drew it or some other way—different lineweights, linetypes, colors, etc.?" Personally, I have always been a firm advocate of WYSIWYG technology (see the following insert), but there will be times when a change at plot time is necessary. For example, my printer uses color. I like that, but I cannot afford the cost of printer cartridges when I must print working drawings in full color. So I prefer to print in black and white until the final product is ready. With Plot Styles, I can do that without affecting the drawing.

> **WYSIWYG** (pronounced "whiz-ee-wig")—literally "What You See Is What You Get"—simply means that what appears on your screen is what will appear on your paper. This may seem obvious (and generally is for other types of documents). But some project gurus prefer to assign (or change) such things as lineweights, linetypes, or colors at plot time rather than during the drawing setup.

AutoCAD helps create and manage Plot Styles with the **Plot Style Table Wizard**. Let's set up a table to see how it works.

3.2.1 Do This: Set Up a Plot Style

I. If you are not already in the *pid-03.dwg* in the C:\Steps\Lesson03 folder, please open it now.

II. Follow these steps.

TOOLS	COMMAND SEQUENCE	STEPS
	**Command:** *stylesmanager*	**1.** Open the **Plot Style** window by typing *stylesmanager* at the command prompt or by selecting **Plot Style Manager...** from the File pull-down menu.
		2. Start the **Plot Style Table Wizard** by double-clicking on the wizard shortcut.
		3. Read the Add Plot Style Table dialog box that AutoCAD presents. Then pick the **Next >** button to proceed.
		4. Notice the four choices in the Begin dialog box (Figure 3.2.1.4).
		⇒ **Start from scratch** allows the user to start a fresh setup.
		⇒ **Use an existing plot style table** allows the user to create a new setup using the settings in an existing setup as a starting point.
		⇒ The last two (**Use My R14 Plotter Configuration** and **Use a PCP or PC2 file**) allow the user to convert plotter setups from earlier releases of AutoCAD.
		We will start from scratch, as shown in Figure 3.2.1.4. Pick the **Next >** button.

continued

TOOLS	COMMAND SEQUENCE	STEPS

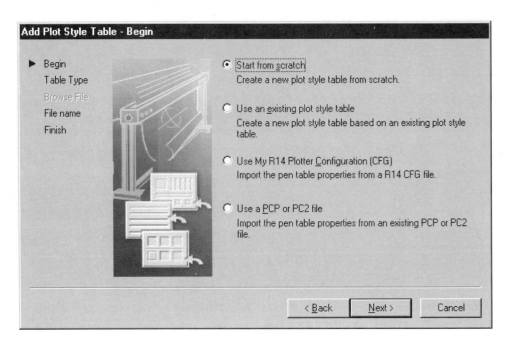

FIG. 3.2.1.4: Table Wizard—Begin Box

5. Now AutoCAD asks for the type of style you will create. Let's opt for the **Named Plot Style Table** (Figure 3.2.1.5) where we can create our own styles.

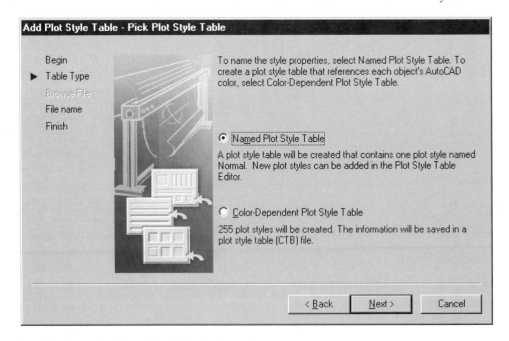

FIG. 3.2.1.5: What Type of Style Will You Create?

continued

TOOLS	COMMAND SEQUENCE	STEPS

Pick the **Next >** button to continue.

6. Name your new table, as indicated in Figure 3.2.1.6. Pick the **Next >** button to continue.

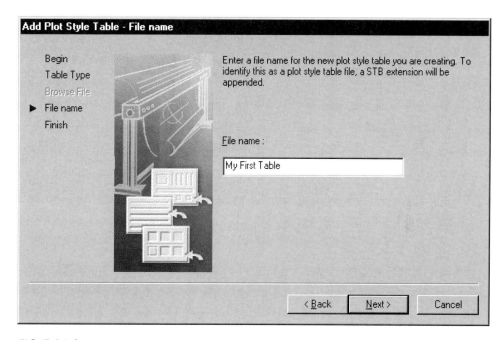

FIG. 3.2.1.6: Name Your New Table

7. AutoCAD tells you that your table has been created (Figure 3.2.1.7). But now, you must set up your table using the Plot Style Table Editor. Pick that button.

continued

TOOLS	COMMAND SEQUENCE	STEPS

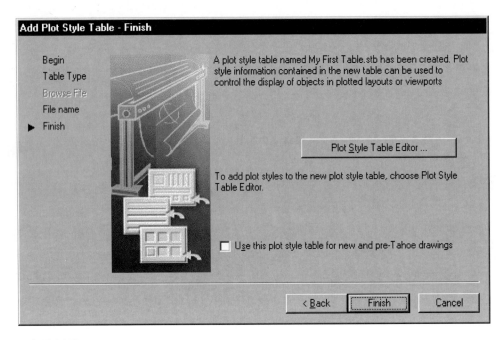

FIG. 3.2.1.7: Your Table Has Been Created

8. AutoCAD presents the Plot Style Table Editor shown in Figure 3.2.1.8. Notice the three tabs available to assist you.

continued

TOOLS	COMMAND SEQUENCE	STEPS

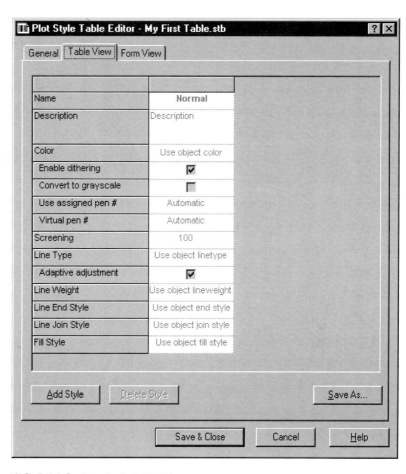

FIG. 3.2.1.8: Plot Style Table Editor

⇒ On the **General** tab (not shown), the user may add a written description of the table being edited or change the linetype global scale factor (Lesson 7).

⇒ The **Table View** and **Form View** tabs offer different formats for doing the same thing—editing the style settings. We will use the columns format on the **Table View** tab shown.

Pick the **Add Style** button.

continued

TOOLS	COMMAND SEQUENCE	STEPS
		9. Notice that AutoCAD adds a new style next to the **Normal** style already shown. We will edit the new style. (AutoCAD will not allow the user to change or delete the **Normal** style.) Let's look at each of our options (refer to Figure 3.2.1.10). *(Note:* We will discuss each variable in more detail over the course of our text; but for now, please leave all settings at their defaults unless instructed otherwise.)

ROW	DEFAULT SETTING	EXPLANATION
Name	**Style 1**	**a)** This is the default name—rename it to *MyStyle*.
Description	**[blank]**	**b)** The user can enter a description of the style if desired.
Color	**Use object color**	**c)** Each object has an assigned color (more on color in Lesson 7). The user can change the color assignment here.
Enable dithering	**[enabled by check box]**	**d)** Dithering is a printer's (plotter's) way of approximating colors by combining the Red / Green / Blue inks in its cartridges. It is best to leave this enabled.
Convert to grayscale	**[disabled by check box]**	**e)** We will enable this one by placing a check in the check box. This way, AutoCAD will send grayscale information (colors interpreted in levels of black and white) to the printer.
Use assigned pen #	**Automatic**	**f)** Here the user can assign colors to specific pens on multipen plotters. When left at its default (Automatic), AutoCAD assigns the drawing color to the pen that most nearly approximates it.

continued

ROW	DEFAULT SETTING	STEPS
Virtual pen #	Automatic	**g)** This is identical to the previous setting, but it is used for nonpen plotters that use virtual settings to mimic pens. See your plotter manual if you are not sure what you are using.
Screening	100	**h)** Screening sets color intensity (how much ink to use when printing). A setting of 100 is most intense; a setting of 0 will reduce the color to white.
Line Type	**Use object linetype**	**i)** The user can set linetypes (Lesson 7) when plotting. This will override linetype settings made in the drawing.
Adaptive adjustment	[enabled by check box]	**j)** The user can insist that linetype patterns (Lesson 7) be completed regardless of drawing settings using this box. It is best to leave this enabled.
Line Weight	**Use object lineweight**	**k)** The user can set lineweights (Lesson 7) when plotting. This will override lineweight settings made in the drawing.
Line End Style	**Use object end style**	**l)** The user can set line end styles when plotting. This will override end style settings made in the drawing.
Line Join Style	**Use object join style**	**m)** The user can set line join styles when plotting. This will override join style settings made in the drawing.
Fill Style	**Use object fill style**	**n)** The user can set the fill style when plotting. This will override fill style settings made in the drawing.

TOOLS	COMMAND SEQUENCE	STEPS
Save & Close		**10.** The new style should now look like Figure 3.2.1.10. Pick the **Save & Close** button.

continued

TOOLS	COMMAND SEQUENCE	STEPS
		[MyStyle properties panel shown]
		FIG. 3.2.1.10: MyStyle
[Finish button]		**11.** You are back in the **Add Plot Style Table Wizard** again. Pick the **Finish** button to complete the procedure.

You have now used two methods of setting up for black-and-white plotting—the **Plot Style Wizard** just shown and the **Add a Plotter Wizard** discussed in Section 3.1. It might fortify you a bit to remember that these two sections need only be done as setup procedures—*not* every time you need to print a drawing!

There is only one more thing to set up before we plot. Let's take a look at setting up our page next.

3.3 Setting Up the Page to Be Plotted

Setting up the page—or the drawing to be plotted—involves telling AutoCAD which printer / plotter to use when you print and how to handle the various colors used in your drawing (i.e., how to translate the colors from AutoCAD definitions to printer definitions). Luckily, we have another wizard to help!

Access the Page Setup Wizard by entering *Pagesetup* at the command prompt or by selecting **Page Setup . . .** from the File pull-down menu. Let's set up our *pid-03* drawing.

3.3.1 Do This: Set Up a Page to Plot

I. Be sure you are still in the *pid-03.dwg* in the C:\Steps\Lesson03 folder. If not, open it now.

II. Follow these steps.

TOOLS	COMMAND SEQUENCE	STEPS
	**Command:** *pagesetup*	1. Open the Page Setup Wizard (Figure 3.3.1.1) by typing *Pagesetup* at the command prompt or selecting **Page Setup . . .** from the File pull-down menu

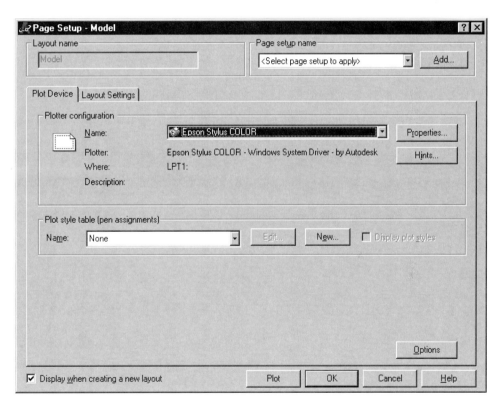

FIG. 3.3.1.1: Page Setup Wizard

Notice that there are two tabs. We will do our page setup on the **Plot Device** tab shown.

continued

TOOLS	COMMAND SEQUENCE	STEPS

There are two buttons available on the **Plotter configuration** frame:

➠ **Properties,** which calls the Plotter Configuration Editor we used in Step 9 of Exercise 3.1.1.
➠ **Hints,** which calls AutoCAD's help files.

Additionally, there is a **Name** control box where we can select the plotter configuration we wish to use.

2. Pick on the down-arrow in the **Name** control box and select the plotter configuration file we created in Exercise 3.1.1 (Figure 3.3.1.2).

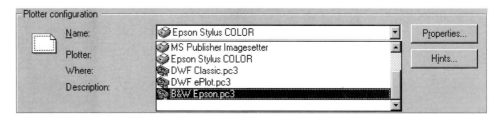

FIG. 3.3.1.2: Select a Printer

3. Notice the **Plot Style Table** frame. No plot style has been assigned. We could assign the plot style we created in Section 3.2. However, in Step 2, we told AutoCAD to use our B&W printer setup so it will not be necessary to assign a plot style.

There is also an **Options** button in the bottom of the dialog box. This calls AutoCAD's configuration menu. The configuration menu is best left to the project guru, so just ignore this button for now.

continued

TOOLS	COMMAND SEQUENCE	STEPS
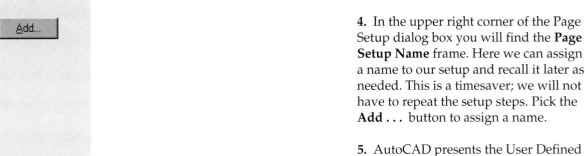		**4.** In the upper right corner of the Page Setup dialog box you will find the **Page Setup Name** frame. Here we can assign a name to our setup and recall it later as needed. This is a timesaver; we will not have to repeat the setup steps. Pick the **Add . . .** button to assign a name.

5. AutoCAD presents the User Defined Page Setups dialog box (Figure 3.3.1.5) that lists any setups already defined. Using the buttons on the right, the user can rename or delete setups, or import them from other drawings.

Call this setup Setup #1 in the **New Page Setup Name** text box (Figure 3.3.1.5). Pick the **OK** button to continue.

FIG. **3.3.1.5:** Name the Setup

6. Pick the **OK** button again to complete the page setup. |

That was certainly easier than setting up the printer, was it not?
Now that your page setup has been created, all that is left is to print the drawing!

3.4 Printing the Drawing

After the last 23 pages, it will hearten you to know that printing is actually the easy part!

Two commands exist to begin our printing/plotting procedure. Can you guess what they are?

If you guessed *Print* and *Plot*, you are absolutely correct. Both commands call the Plot dialog box (Figure 3.4a). In it, you will tell AutoCAD exactly what to plot, on what size sheet of paper and at what scale to plot it, and more. Let's take a look at it.

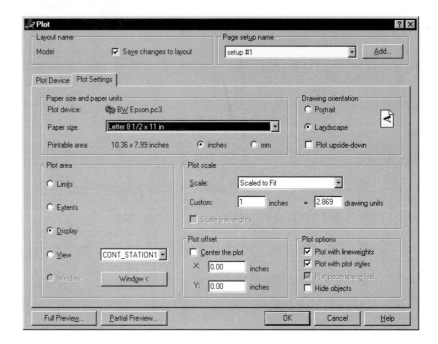

FIG. 3.4a: Plot Dialog Box

- At the top of the dialog box, you will find two frames: **Layout name** and **Page setup name**.
  - The **Layout name** identifies what you will plot. Here we will plot the model (or Model Space). Remember the layout tabs we discussed in Section 1.4.1 (see Figure 1.4.1a on page 13). The user can rename these tabs. The name appears in the **Layout name** frame.
  - The **Page setup name** is what we created in Section 3.3. Once we make our selections on the Plot Settings tab, we should add a new name to save our new settings.

- Below the frames, AutoCAD provides two tabs: **Plot Device** and **Plot Settings**.
  - The **Plot Device** tab resembles the Page Setup dialog box we used in our last exercise (see Figure 3.3.1.1) except that this tab also allows the user to identify which layout tab will be plotted and whether or not to plot to a file.

Some companies will dedicate a computer for plotting—that is, set up a computer specifically for plotting. At these companies, the user will send the plot to a file and then use a program on the dedicated computer to plot the file. As it is considerably faster to plot to a file, this procedure frees the CAD operator's computer from time wasted while a drawing is plotted.

■ The **Plot Settings** tab (the one on top by default) contains the bulk of the information required to actually plot the drawing.

- The **Paper size and paper units** frame indicates:
  - Which device you will be using (you can change this on the **Plot Device** tab, if necessary),
  - The size sheet you are set up to use. This appears in a control box where the user may change it as needed.
  - The amount of area on the sheet (given in inches or millimeters) that the drawing will actually occupy.
- The **Drawing orientation** frame allows the user to orient the drawing on the page. Use the sample page (at the right side of the frame) as a guide.
- In the **Plot Area** frame, the user determines the area of the drawing to plot.
  - A bullet in the **Limits** option tells AutoCAD to plot the drawing limits.
  - A bullet in the **Extents** option tells AutoCAD to plot the drawing extents (plot the drawing as large as possible while being sure to get all drawn objects in the plot).
  - A bullet in the **Display** option tells AutoCAD to plot the current display.
  - A bullet in the **View** option tells AutoCAD to plot the view (Lesson 5) indicated in the control box. Pick the down arrow in the control box to see the defined views available for plotting.
  - Pick the **Window <** button and place a window around the objects in the drawing to be plotted. AutoCAD will automatically place the bullet in the **Window** option.
- In the **Plot scale** frame, the user tells AutoCAD at what scale to plot the drawing. Use the **Scale** control box to select a scale or to select **Custom**. When **Custom** is selected, AutoCAD uses the scale defined in the **Custom** text boxes.
- The **Plot offset** frame offers a check box you can use to center your plot on the sheet of paper. If you do not wish to center the drawing, you can use the X and Y boxes that follow to help locate the drawing as desired. Use the **Partial Preview . . .** button to see how the drawing sits on the page.
- The last frame—**Plot options**—provides the user with four check boxes:
  - **Plot with lineweights:** Check here to use the lineweights defined in the drawing (Lesson 7).
  - **Plot with plot styles:** Check here to use plot styles defined in the drawing.
  - **Plot paper space last:** In earlier releases of AutoCAD, paper space geometry (objects drawn in Paper Space) was always drawn (plotted) first. In 2000, you have a choice what to plot first.
  - **Hide objects:** Check this only when using the third dimension and you wish to avoid plotting objects that are hidden behind other objects.

■ At the bottom of the dialog box, you will see two preview buttons. The **Partial Preview ...** button helps to locate the drawing on the page; the **Full Preview ...** button gives you a view of what your final plot will look like. I strongly recommend using the **Full Preview** prior to actually plotting. This will save time and paper lost to mistakes in the setup.

That is the plot dialog box. Let's print our drawing.

3.4.1 Do This: Printing/Plotting

I. Be sure you are still in the *pid-03.dwg* in the C:\Steps\Lesson03 folder. If not, please open it now.

II. Follow these steps.

TOOLS	COMMAND SEQUENCE	STEPS
	Command: *plot*	1. Begin by entering either *Print* or *Plot* at the command prompt. Or you can select **Plot** from the File pull-down menu. AutoCAD presents the Plot dialog box shown in Figure 3.4a.
	 **FIG. 3.4.1.2:** Plot Scale Frame	2. The paper size and drawing orientation defaults are okay. But as my printer is not set up to plot the D-size sheet of paper (34x22) this drawing was designed to use, I will use the **Scale to fit** setting in the **Plot scale** frame (see Figure 3.4.1.2).
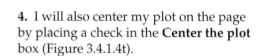 **FIG. 3.4.1.3t** Plot to Extents		3. I want to change the **Plot area** to plot my drawing as large as possible on the selected sheet size. I will put a bullet in the **Extents** option (Figure 3.4.1.3t).
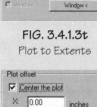 **FIG. 3.4.1.4t** Center the Plot		4. I will also center my plot on the page by placing a check in the **Center the plot** box (Figure 3.4.1.4t).

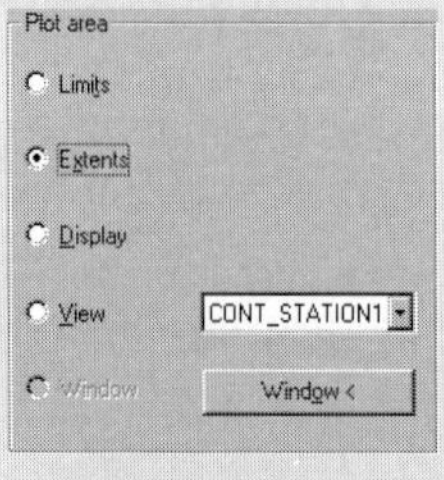

continued

TOOLS	COMMAND SEQUENCE	STEPS

5. Check the plot on the page with the **Partial Preview . . .** button. Adjust the position of the plot, as necessary, using tools in the **Plot offset** frame.

6. When you are satisfied with the position, preview the final plot.

7. Make any adjustments necessary. Then save the layout (refer to Steps 4 and 5 of Exercise 3.3.1 for help).

8. When you are ready to plot, pick the **OK** button. AutoCAD replaces the Plot dialog box with the Plot Progress Indicator shown in Figure 3.4.1.8. When it disappears, the print will be sent to the printer.

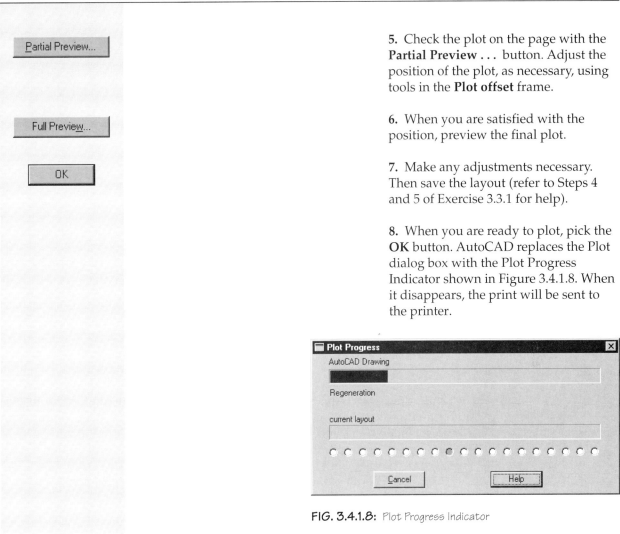

FIG. 3.4.1.8: Plot Progress Indicator

That is it for plotting. You will need to remember the information in Section 3.4 for each print.

3.5 Extra Steps

One thing you will quickly realize on any project is that you seldom need to print just one drawing. For this reason, AutoCAD provided the AutoCAD Batch Plot Utility found in the AutoCAD folder. Access it by following this path from your desktop:

Start—Programs—AutoCAD—Batch Plot Utility

The utility will open AutoCAD and then open the Batch Plot Utility window shown in Figure 3.5a.

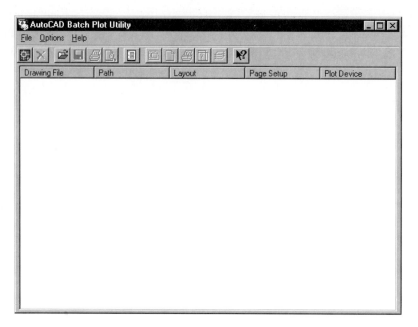

FIG. 3.5a: Batch Plot Utility Window

Add drawings to the batch using the **File** pull-down menu or the AutoCAD button (leftmost button on the toolbar). Change the **Layouts**, **Page setups**, **Plot Devices**, **Plot Settings**, and **Layers** to be plotted using the **Edit** pull-down menu or the appropriate buttons on the toolbar. Use the standard **Print** button to tell AutoCAD to proceed with the batch plotting.

Try plotting both of the drawings in the C:\Steps\Lesson03 folder.

3.6 What Have We Learned?

Items covered in this lesson include:

- *Setting up a printer / plotter*
- *Setting up a Plot Style*
- *Setting up a drawing Page*
- *An Introduction to AutoCAD Batch Plotting*
- *AutoCAD Plot Commands*:
  - *Plotmanager*
  - *Stylesmanager*
  - *Pagesetup*
  - *Plot & Print*

Although printing itself is not difficult, the first three sections of this lesson dealt with setting up printers / plotters, AutoCAD and AutoCAD drawings so that they can all work together to put your work on paper. I strongly suggest printing at least two of the assignments for each of the lessons you will complete throughout this text. Then, when you have finished the course, review this lesson once again. After all, you really cannot roll up the PC and stick it in your tool belt down at the job site. Your drawings must, eventually, be put on paper.

3.7 Exercises

Note: Before trying these exercises, familiarize yourself with your printer and / or plotter by reading the appropriate manuals.

1. If you have a plotter available, plot the *Floor Plan* drawing found in the C:\Steps\Lesson03 folder. Use these parameters:
   - Plot to a $1/4$" = 1'-0" scale.
   - Plot to a C-size (22" × 17") sheet of paper.
2. Using a printer, plot the *Floor Plan* drawing to fit on a 17" × 11" sheet of paper.
3. Using a printer, plot the *Floor Plan* drawing to fit on a 11" × 8.5" sheet of paper.
4. If you have a plotter available, plot the *Pipe Plan* drawing found in the C:\Steps\Lesson03 folder. Use these parameters:
   - Plot to a 3/8"=1'-0" scale.
   - Plot to a D-size (36" × 24") sheet of paper.
5. Using a printer, plot the *Piping Plan* drawing to fit on a 17" × 11" sheet of paper.
6. Using a printer, plot the *Piping Plan* drawing to fit on a 11" × 8.5" sheet of paper.
7. Print each of the drawings you created in Lesson 2 to fit on 11" × 8.5" sheets of paper.

3.8 Review Questions

Please write your answers on a separate sheet of paper.

1. and 2. _____ and _____ are the two commands used to print a drawing.

3. Find the shortcut used to access the Add-a-Plotter Wizard by typing _____ at the command prompt.

4. At the Add-a-Plotter Wizard—Begin dialog box, selecting _____ will add a plotter to AutoCAD's system.

5. To add a local printer at the Add-a-Plotter Wizard—Begin dialog box, select _____.

6. (T or F) You cannot import plotter information used in an earlier release of AutoCAD into AutoCAD 2000.

7. Access the Calibrate Plotter Wizard from the _____ Wizard.

8. WYSIWYG means _____.

9. Use the _____ Wizard to create plot styles.

10. Open Window's Plot Style window by typing _____ at the command prompt.

11. To convert R13 plotter files to 2000 plot styles, use the _____ option on the Add Plot Style Table Wizard—Begin dialog box.

12. **and** 13. The user can change the global linetype scale factor using the _____ tab of the _____.

14. (T or F) Once the user has created a new plot style, he can delete the Normal plot style.

15. To print all drawings in B&W, regardless of color assignments, enable the _____ box of your new plot style.

16. **to** 18. Name the three things that must be set up before plotting.

19. Enter _____ at the command prompt to access the Page Setup Wizard.

20. (T or F) It is not necessary to set up both the printer and a plot style to create B&W prints from a drawing that uses colors. One or the other will suffice.

21. **and** 22. To plot to a file, go to the _____ tab of the _____ dialog box.

23. Look in the _____ frame of the Plot Settings tab (Plot dialog box) to determine the amount of area on the sheet that the drawing will actually occupy.

24. To reorient the drawing on the sheet, go to the _____ frame of the Plot Settings tab (Plot dialog box).

25. Use the _____ option of the Plot Area frame to plot the limits of the drawing.

26. Use the _____ option of the Plot Area frame to plot all of the objects on the drawing as large as possible.

27. Set the scale of the plotted drawing in the _____ frame of the Plot Settings tab (Plot dialog box).

28. Use the _____ button to help locate the drawing on the page.

Identify these:

29. [Add-A-Plotter Wizard]   30. [Add-A-Plot Style Table Wizard]

LESSON 4

Drawing Aids

Following this lesson, you will:

➥ *Know how to use the **Ortho, Grid, Snap,** and **Tracking** tools to your advantage*

➥ *Know how to use **OSNAPS***

➥ *Know how to use the **Direct Distance Options** when you draw*

➥ *Be familiar with the different toggles—keyboard, function keys, and status bar—available to help you*

Lesson 2 showed us how to draw with accuracy. But without some additional help, we will need to know exact coordinates or exact distances every time we draw a line or a circle. What if we want to go to a specific point—say, the end point of an existing line? Must we break out the calculator every time?

No. AutoCAD provides many tools and drawing aids that help you minimize, though not eliminate, the need for calculation. Lesson 2 taught you to draw with *accuracy*. This lesson will show you how to draw with *speed* without sacrificing accuracy.

4.1 The Simple Stuff: Ortho, Grid, Tracking, and Snap

■ 4.1.1 Ortho

The first drawing aid with which you will want to familiarize yourself is the **Ortho** tool. You may recognize the word as an abbreviation of the word *Orthographic*. We learned about orthographic projections in the first days of drafting class. These present views necessary to describe an item from all sides. Orthographic views include the following (Figure 4.1.1a): front, back, right side, left side, top, and bottom. The views are placed above or below, left or right of a primary view (usually the front) depending on which view it is.

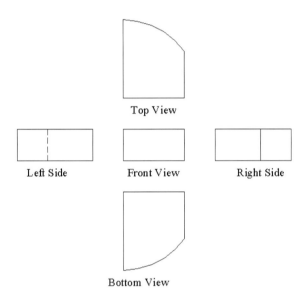

FIG. 4.1.1a: Orthographic Projections

The Ortho tool restricts drawing or editing movement in a drawing to the left/right or up/down directions. When **Ortho** is toggled **On**, your lines will be drawn along the X- or Y-plane. (When drawing in the isometric mode, the X- or Y- planes will be located at 30° and 150°.) This will become clearer during our exercises.

The command sequence for **Ortho** is

Command: *ortho*
Enter mode [ON/OFF] <OFF>: *on*
Command:

■ 4.1.2 Grid

One of the drafting tricks we learned back in the pencil days was to lay a background grid sheet on our drawing board before taping down our drawing sheet. This grid served as a lettering guide and helped in aligning different items.

AutoCAD's grid can be quite useful in aligning things. But as we will see later, it is not necessary for sizing text.

The grid toggles **On** or **Off** quite easily. And controlling the size and shape of the grid, together with some creative use of the *Snap* tool, will often double or triple your drawing speed.

Controlling the grid size is easy. Following is the command sequence:

Command: *grid*
Specify grid spacing(X) or [ON/OFF/Snap/Aspect] <0.5000>: *1*
Command:

AutoCAD provides some options for the *Grid* command. The default option sets the grid spacing. In this example, the spacing defaulted to 0.5000 drawing units. We set the spacing to 1 drawing unit. Other options include:

- **ON/OFF**: Turns the grid on or off
- **Snap**: Sets the grid spacing equal to the snap increments (we will look at *Snap* in a moment)
- **Aspect**: Enables you to set a separate spacing for the X- and Y-planes

4.1.3 Polar Tracking

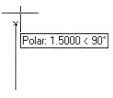

FIG. 4.1.3a
Polar Tracking with Cheater Box

A remarkable and intuitive tool, Polar Tracking was designed to assist the user by placing a temporary construction line from the last point selected. Additionally, Polar Tracking provides a cheater box detailing distance and angle from the last point selected (Figure 4.1.3a). This makes it much easier to locate the next point accurately and quickly with a minimal need for absolute, relative or polar coordinate entry.

Polar Tracking works on the four quadrants (0°, 90°, 180°, and 270°) by default, but the user can set it to track at any angle or to override the settings on the fly. When used with Polar Snap (see Section 4.1.4, "Snap"), the user can snap to any point at any angle with little or no need for the keyboard at all!

> To set your own angles, refer to Section 4.2, "And Now the Easy Way: *DSettings*."

The user can toggle Polar Tracking on or off using the **F10** function key or by clicking the **Polar** toggle on the status bar.

4.1.4 Snap (Polar and Grid)

The Grid and Polar Tracking by themselves are only of minimal use, however. So AutoCAD provides a tool that was not available back in the pencil days. The *Snap* tool actually pulls (or *snaps*) the crosshairs to a grid or polar reference. Controlling the snap while referencing the Grid or Polar Tracking construction lines can provide speed to an otherwise tedious job.

Look at the *Snap* command.

Command: *snap*

Specify snap spacing or [ON/OFF/Aspect/Rotate/Style/Type] <0.5000>: *[enter the desired spacing or select an option]*

What is the default spacing? Did you say 0.5000? Correct. We can see the default inside the < >, as in other commands.

> Conventional wisdom suggests that setting the grid snap to half the grid works best for most drawings. This way, you can snap to each grid, and halfway between each. More than a single snap between grids is difficult to follow.

Snap options include:

- **ON/OFF**: For turning the snap on or off
- **Aspect**: For setting a different snap spacing horizontally and vertically
- **Rotate**: For rotating the snap when drawing at an angle (see **Style**)
- **Style**: For changing from a standard grid/snap style to an isometric style (we will look at this in more detail in Section 4.6)
- **Type:** For setting either **Polar** (to snap to points determined with the Auto-Tracking feature) or to the **Grid** (to snap to grid referenced points)

> *Grid, Snap, Polar Tracking,* and *Ortho* would not be of much use if we had to stop a drawing or editing command every time we needed to turn one **On** or **Off**. Mercifully, AutoCAD gives us toggles that will work even while we are in the sequence of another command. In other words, we can toggle the *Grid, Snap, Polar Tracking,* or *Ortho* **On** or **Off** while drawing a line (or doing some other command).

There are two or three ways to toggle each of these items. They are

	KEYBOARD	FUNCTION KEY	SCREEN
Grid	Ctrl + G	F7	Status bar
Snap	Ctrl + B	F9	Status bar
Ortho	None	F8	Status bar
Polar Tracking	None	F10	Status bar
Object Snap Tracking	None	F11	Status bar

Find the function keys across the top of your keyboard. Identify them by the *F* followed by a number, such as *F8*. Refer to Appendix B for a complete list of AutoCAD's function keys.

Status bar toggles are shown in Figure 4.1.4a.

| SNAP | GRID | ORTHO | POLAR | OSNAP | OTRACK | LWT | MODEL |

FIG. 4.1.4a: Status Bar Toggles

To toggle something **On** or **Off** at the status bar, place your cursor on the item and click with the left mouse button. A "raised" item is currently **Off**; an "indented" item is **On**. To define or redefine the settings of one of these toggles, right-click on it and pick **Settings . . . .**

Let us try an exercise using *Grid*, *Snap*, *Polar Tracking*, and *Ortho*.

WWW 4.1.4.1 Do This: Grid, Snap, Polar Tracking, and Ortho Practice

Try drawing Exercise #1 in Lesson 2 using only the four tools you have just seen.

I. Start a new drawing using *MyBase2.dwt* as your template (in the C:\Steps\Lesson01 folder).

II. Set *Grid* to 1. (You can use the scroll bars to center the grid in the graphics area)

III. Set *Snap* to $^1/_2$ and the snap **Type** to *Grid*.

IV. Draw the "M" toggling the *Ortho* **On** or **Off** as needed during the drawing.

How much faster did you draw? Usually, at this point, my students begin planning my demise for not having shown them this approach first. I am really not that cruel. But once these tools are known, it is exceedingly difficult to convince students to learn the Cartesian Coordinate approaches to drawing. As you will see, both the coordinate system and these drawing tools are quite necessary for speed and accuracy. (Wait till you master OSNAPS!)

Now try drawing the rest of the Lesson 2 exercises using your own settings.

4.2 And Now the Easy Way—DSettings

We have seen the command prompt method of setting **Grid** and **Snap**. Any setting that requires several lines of entry, as these do, is a prime candidate for a dialog box. AutoCAD has not disappointed.

There are several ways to display the Drafting Settings dialog box. The user can enter the command *DSettings* at the command prompt or right-click on any of the toggles on the status bar and select **Settings...** from the cursor menu that appears. AutoCAD will present the Drafting Settings dialog box seen in Figure 4.2a. This box has three tabs available to help you.

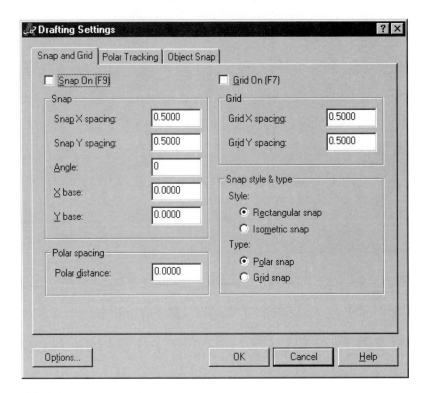

FIG. 4.2a: Drafting Settings Dialog Box—Snap and Grid Tab

- Figure 4.2a shows the **Snap and Grid** tab on top. You will see this when you select **Settings...** from the cursor menu presented when you right-click on the **Snap** or **Grid** toggle on the status bar. This tab presents frames where the user can set increments for the **Snap** spacing (for the grid snap), the **Polar spacing** (for the polar snap), or the **Grid** spacing. This tab also provides check boxes for toggling on / off the **Snap** and **Grid**, and option buttons for setting the type of snap (**Polar** or **Grid**, **Rectangular** or **Isometric**—more on isometric snap in Section 4.6, "Isometric Drafting").
- If you access the Drafting Settings dialog box by right-clicking on the **Polar** toggle, AutoCAD will place the **Polar Tracking** tab on top, as seen in Figure 4.2b.

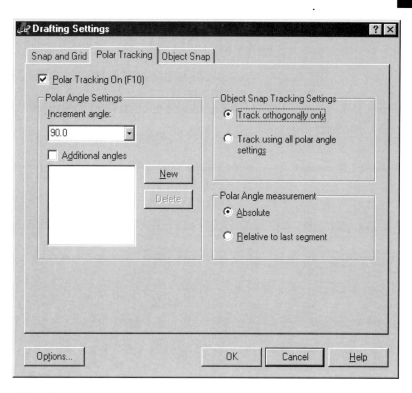

FIG. 4.2b: Drafting Settings Dialog Box—Polar Tracking Tab

Here the user can toggle **Polar Tracking On** (or **Off**) using the check box. But more importantly, the user can adjust the **Incremental Angle Settings** (the angles at which Polar Tracking appears) using a drop-down box or add additional angles (not shown in the drop-down box) by picking the **New** button. **Additional angles** will appear in the list box and be used when the user places a check in the check box.

The user may also determine whether to use Polar Tracking only orthogonally (at the four quadrants—0°, 90°, 180°, and 270°) or using all the angle settings. The **Object Snap Tracking Settings** frame provides option boxes for each.

The **Polar Angle measurement** frame allows the user to show Polar Tracking angles in absolute terms (always showing angles as they relate to AutoCAD's compass points) or relative to the last segment (showing angles as they relate to the last line segment drawn). I recommend using the default Absolute setting to avoid confusion.

- The **Object Snap** tab (Figure 4.2c) will appear on top when the Drafting Settings dialog box is accessed by right-clicking on the **OSNAP** or **OTrack** toggles. Here the user can set Running Object Snaps (OSNAPs). We will look at OSNAPs and Running OSNAPs in Sections 4.3 and 4.4.

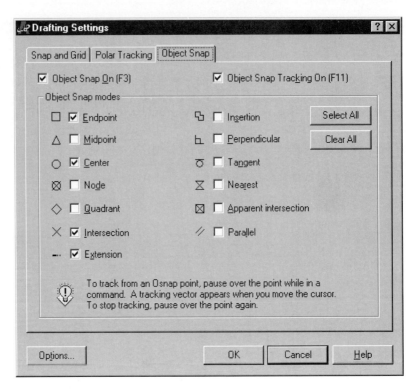

FIG. 4.2c: Drafting Settings Dialog Box—Object Snap Tab

4.3 Never Miss the Point with OSNAPS

We have now seen several ways to draw with precision. What more could we possibly need? Well, not all of these tools lend themselves easily to large drawing environments like those found in disciplines such as architecture or petrochemical. This leads us back to a need for some tools that free us as much as possible from the Cartesian Coordinate System. Enter ObjectSNAPS—*OSNAPS*.

These remarkable tools must and will become second nature to the successful CAD operator. After learning OSNAPS, you must engrain this 11th Commandment into your hearts and minds: *Thou shalt not eyeball!* Because after learning OSNAPS, there will never again be a need to guess about the location of a point.

What are OSNAPS? OSNAPS are a means of responding with precision to any prompt directing you to pick a point in your drawing. They provide the means for precisely locating and selecting a point (endpoint, midpoint, center point, etc.) on (or referenced by) any existing object in the drawing—lines, circles, arcs, and so forth.

Study the following chart. The first column shows the buttons found when selecting the OSNAP Flyout toolbar on the Standard toolbar. (Note: You will find the **Tracking** button on the Standard toolbar. Pick that to access the OSNAP Flyout toolbar.) The middle column shows the cheater symbols AutoCAD shows when trying to use an OSNAP in a drawing. The last column shows the equivalent command found when you call up the cursor menu by clicking the right mouse button (with your cursor in the drawing area), while holding down the shift key on the keyboard.

A *flyout* toolbar button is indicated by a small triangle in the lower right corner of the button. The triangle indicates that there are more buttons below the one shown. To see the other available buttons, pick on the flyout button and hold down the mouse button. A flyout toolbar will appear. To identify what a button will do, rest your cursor on it for a moment. A cheater box will appear telling you the use of the button. To select the button, simply move your cursor to it and release the mouse button.

OSNAP Toolbar	Symbol	Cursor Menu
		Temporary Track Point
		From
	□	Endpoint
	△	Midpoint
	×	Intersection
	⊠	Apparent Intersection
	── --	Extension
	○	Center
	◇	Quadrant
	⊤	Tanget
	⊥	Perpendicular
	//	Parallel
		Insert
	⊗	Node
	⋈	Nearest
		None
		OSNAP Settings

Most of these will become self-explanatory once you have used them. Some will require a bit more explanation. The best way to learn how to use object snaps, though, is through a practice exercise. Follow me!

WWW 4.3.1 Do This: OSNAP Practice

I. Open the *Train.dwg* file found in the C:\Steps\Lesson04 folder. It will look like Figure 4.3.1a (without the numbers and arrows).

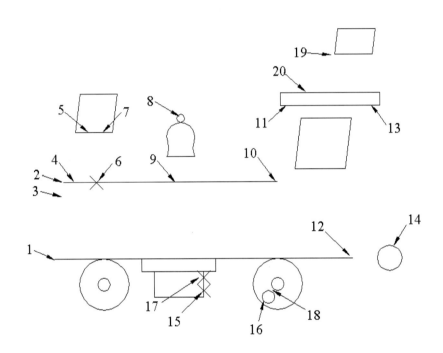

FIG. 4.3.1a: Train.dwg

II. Now follow the command sequence. For this exercise, I will use three methods of selecting OSNAPS, but the methods are completely interchangeable.

TOOLS	COMMAND SEQUENCE	STEPS
		1. Check the status bar and be sure that all the toggles are raised. They should look like Figure 4.3.1.1. If any of the toggles appear depressed, pick on it once to raise it. `SNAP GRID ORTHO POLAR OSNAP OTRACK LWT MODEL` FIG. 4.3.1.1

continued

TOOLS	COMMAND SEQUENCE	STEPS
 Line Button / Endpoint Button	**Command:** *l*	**2.** We will begin at the front of the train. Enter the *Line* command.
	LINE Specify first point: _endp of	**3.** Select the OSNAP flyout toolbar (upper arrow) from the Standard toolbar, hold down the mouse button while sliding down to the endpoint tool (lower arrow); release the mouse button. (Notice that the endpoint button is now on top of the OSNAP flyout.)
	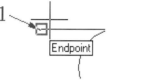 FIG. 4.3.1.4	**4.** Place the cursor at point 1 on Figure 4.3.1a. Notice how a small square (the symbol for endpoint shown on the chart on page 95) appears at the endpoint of the line (see Figure 4.3.1.4)? Pick here. Notice that the line begins at the endpoint of the existing line.
	Specify next point or [Undo]: _endp of	**5.** Select the **Endpoint** button again.
	Specify next point or [Undo]:	**6.** Place the cursor at point 2. Notice the symbol for endpoint appears. Pick here. Your line is drawn to the endpoint.
	Specify next point or [Undo]: *[enter]*	Hit *enter* to complete the command.
	FIG. 4.3.1.7: Snap Toggle Menu	**7.** Next, we will draw a cow catcher using the *Line* command, the **Extension** OSNAP, and Polar Tracking. First, let's set up the Polar Tracking. Right-click on the **Snap** toggle on the status bar. AutoCAD presents the menu shown in Figure 4.3.1.7. Pick on the **Settings . . .** option.
	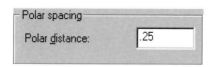 FIG. 4.3.1.8: Set the Polar Distance at .25	**8.** AutoCAD presents the Drafting Settings dialog box (with the **Snap and Grid** tab on top) as seen in Figure 4.2a. Set the **Polar Distance** at **.25** (Figure 4.3.1.8). Check the **Snap On** check box; then pick on the **Polar Tracking** tab.

continued

TOOLS	COMMAND SEQUENCE	STEPS

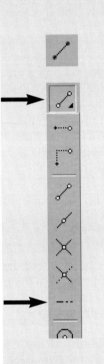

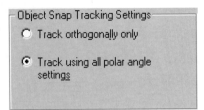

FIG. 4.3.1.9: Object Snap Tracking Settings Frame

9. We will want to use Polar Tracking on all angles, so put a bullet in the **Track using all polar angle settings** option of the Object Snap Tracking Settings frame (Figure 4.3.1.9).

Check the **Polar Tracking On** check box; then pick the **OK** button to complete the setup.

Command: *l*

10. Begin the *Line* command.

11. Pick the OSNAP flyout toolbar (upper arrow), (hold the mouse button down while sliding down) then select the **Extension** button (lower arrow).

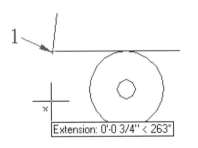

FIG. 4.3.1.12: Polar Tracking—Extension OSNAP

12. Place your crosshairs over point 1, but *do not pick*. AutoCAD will display a small plus symbol indicating that the object has been located. Move your cursor down and slightly to the left. Notice the cheater box as Polar Tracking helps you locate a point off of the extension of the line that you created in Steps 2 through 6.

Pick at the point located $3/4$" at 263° (as seen in Figure 4.3.1.12).

13. Now use the Polar Tracking tool to create a rectangle $1/4$" high and 2" to the left.

continued

TOOLS	COMMAND SEQUENCE	STEPS

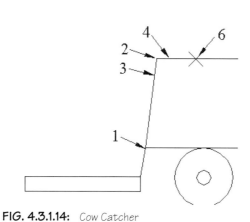

FIG. 4.3.1.14: Cow Catcher

14. Using the **Endpoint** OSNAP you learned in Step 5, draw a line connecting the upper right corner of the cow catcher to point 1, as seen in Figure 4.3.1.14.

Command: *l*

LINE Specify first point: _endp of

15. Use the **Endpoint** OSNAP to begin a line at the upper left corner of the cow catcher.

Specify next point or [Undo]: *nea*

16. At the **Specify next point** prompt, type *nea* to tell AutoCAD you want the point nearest to where you select. Select a point near point 3. Notice the symbol.

Specify next point or [Undo]: *[enter]*

17. Hit *enter* to complete the command.

Command: *[enter]*

18. Repeat the *Line* command. This time we will select a point where two lines would intersect if they were a bit longer. And we will use the cursor menu to enter the OSNAPS.

continued

TOOLS	COMMAND SEQUENCE	STEPS
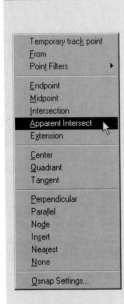 FIG. 4.3.1.19 OSNAP Cursor Menu	LINE Specify first point: _appint of and	19. Hold down the *shift* key on the keyboard and *right-click* in the graphics area of the screen. You will see a menu on the screen that looks like Figure 4.3.1.19. Move your cursor down till the **Apparent Intersection** option highlights. Then pick with the left mouse button. AutoCAD asks you which nonintersecting lines you wish to use. Select the line you drew between the cow catcher and point 3; then select the line at point 4. Notice the symbols. AutoCAD begins the line.
	Specify next point or [Undo]: _par to	20. Now we must identify where to go with the line. Bring up the cursor menu as detailed before; then select **Parallel**. Place your cursor over the line between points 1 and 2 (do not pick). AutoCAD will display a plus sign to let you know it has found the line.
	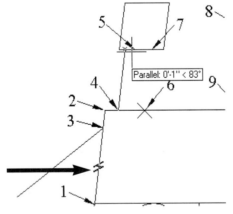 FIG. 4.3.1.21: Parallel Cheater Box	21. Move the cursor to the bottom of the smokestack. Pick when AutoCAD displays the cheater box shown in Figure 4.3.1.21. (Notice that the Parallel symbol appears—see the arrow—when your line is parallel.)
	Specify next point or [Undo]: *[enter]*	22. Hit *enter* to complete the command. Your drawing looks like Figure 4.3.1.22.

continued

TOOLS	COMMAND SEQUENCE	STEPS

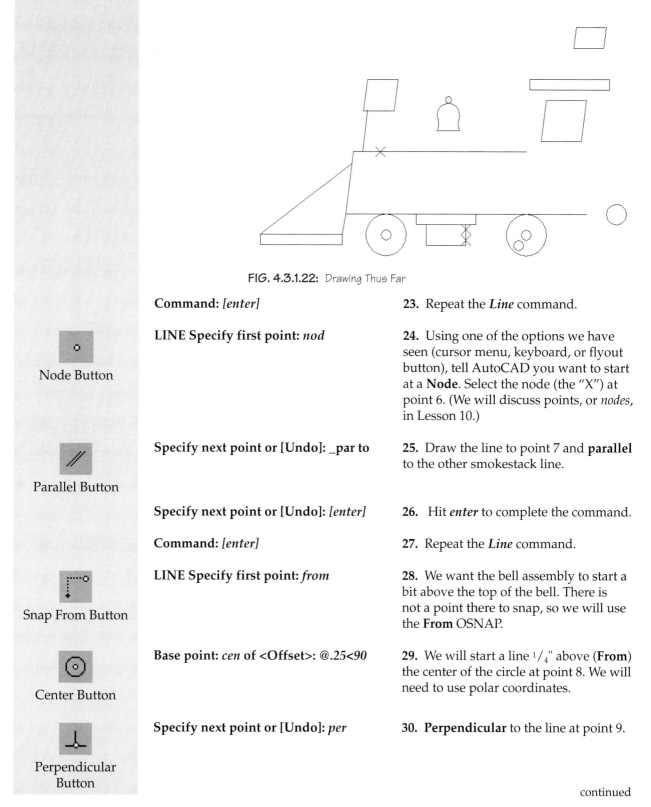

FIG. 4.3.1.22: Drawing Thus Far

Node Button

Command: *[enter]*

23. Repeat the *Line* command.

LINE Specify first point: *nod*

24. Using one of the options we have seen (cursor menu, keyboard, or flyout button), tell AutoCAD you want to start at a **Node**. Select the node (the "X") at point 6. (We will discuss points, or *nodes*, in Lesson 10.)

Parallel Button

Specify next point or [Undo]: *_par to*

25. Draw the line to point 7 and **parallel** to the other smokestack line.

Specify next point or [Undo]: *[enter]*

26. Hit *enter* to complete the command.

Command: *[enter]*

27. Repeat the *Line* command.

Snap From Button

LINE Specify first point: *from*

28. We want the bell assembly to start a bit above the top of the bell. There is not a point there to snap, so we will use the **From** OSNAP.

Center Button

Base point: *cen of* **<Offset>:** *@.25<90*

29. We will start a line $1/4$" above (**From**) the center of the circle at point 8. We will need to use polar coordinates.

Perpendicular Button

Specify next point or [Undo]: *per*

30. **Perpendicular** to the line at point 9.

continued

TOOLS	COMMAND SEQUENCE	STEPS
	Specify next point or [Undo]: *[enter]*	**31.** Complete the command. Your drawing now looks like Figure 4.3.1.31.

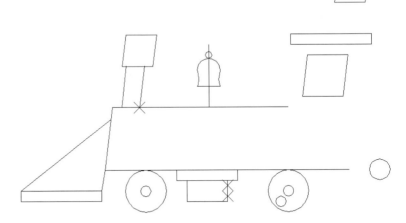

FIG. 4.3.1.31: Drawing Thus Far

	Command: *[enter]*	**32.** Repeat the *Line* command.
	LINE Specify first point: *end*	**33.** From the **endpoint** at point 10 . . .
	Specify next point or [Undo]: *par*	**34.** . . . **parallel** to the vertical cab line to point 11.
	Specify next point or [Undo]: *[enter]*	**35.** Complete the command.
	Command: *[enter]*	**36.** Repeat the *Line* command.
	LINE Specify first point: *end*	**37.** From the **endpoint** at point 12 . . .
	Specify next point or [Undo]: *par*	**38.** . . . **parallel** to the vertical cab line to point 13.
	Specify next point or [Undo]: *[enter]*	**39.** Complete the command.
	Command: *[enter]*	**40.** Repeat the *Line* command.
	LINE Specify first point: *end*	**41.** From the **endpoint** at point 12 . . .
	Specify next point or [Undo]: cen	**42.** . . . to the **center** of the circle at point 14.

continued

TOOLS	COMMAND SEQUENCE	STEPS
	Specify next point or [Undo]: *[enter]*	**43.** Complete the command. Your drawing looks like Figure 4.3.1.43.

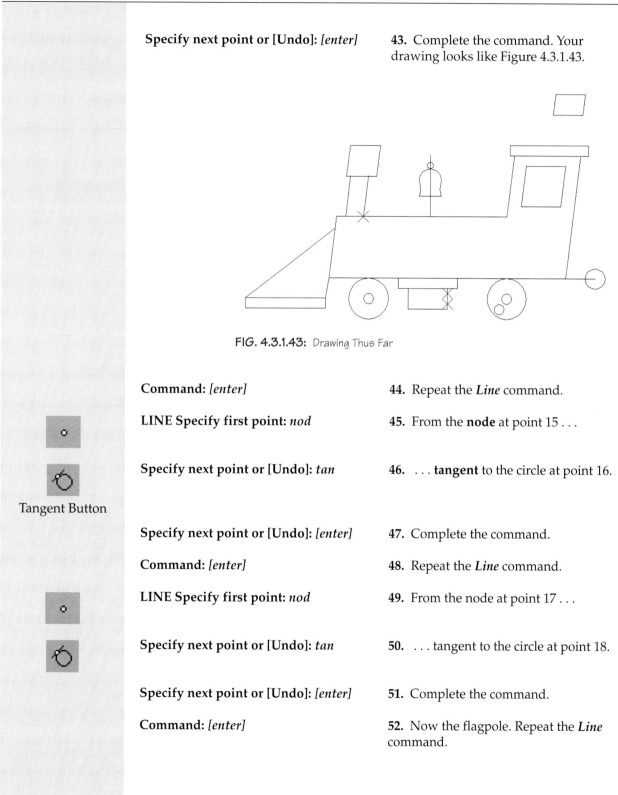

FIG. 4.3.1.43: Drawing Thus Far

	Command: *[enter]*	**44.** Repeat the *Line* command.
	LINE Specify first point: *nod*	**45.** From the **node** at point 15 . . .
Tangent Button	**Specify next point or [Undo]:** *tan*	**46.** . . . **tangent** to the circle at point 16.
	Specify next point or [Undo]: *[enter]*	**47.** Complete the command.
	Command: *[enter]*	**48.** Repeat the *Line* command.
	LINE Specify first point: *nod*	**49.** From the node at point 17 . . .
	Specify next point or [Undo]: *tan*	**50.** . . . tangent to the circle at point 18.
	Specify next point or [Undo]: *[enter]*	**51.** Complete the command.
	Command: *[enter]*	**52.** Now the flagpole. Repeat the *Line* command.

continued

TOOLS	COMMAND SEQUENCE	STEPS
Intersection Button	**From point:** *int*	53. From the **intersection** at point 19 . . .
Midpoint Button	**To point:** *mid*	54. . . . to the **midpoint** of the line at point 20.
	Command: *[enter]*	55. Complete the command. Your drawing now looks like Figure 4.3.1.55.

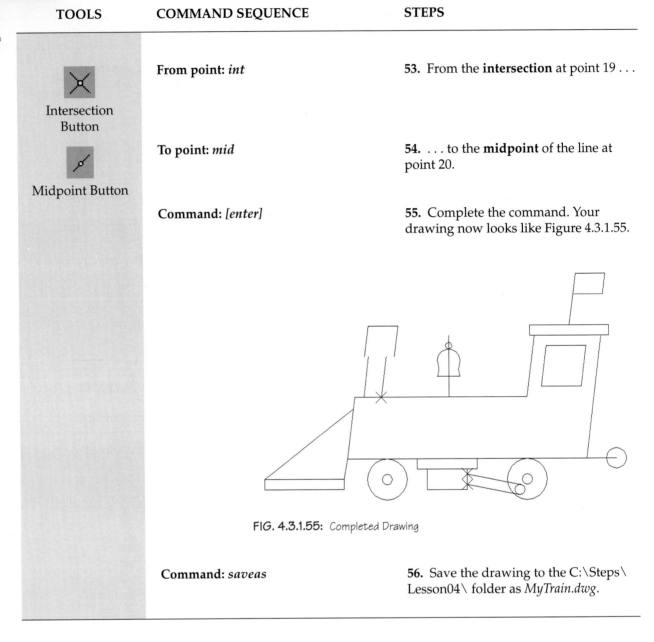

FIG. 4.3.1.55: Completed Drawing

	Command: *saveas*	56. Save the drawing to the C:\Steps\Lesson04\ folder as *MyTrain.dwg*.

You should now be fairly familiar with three ways to call on OSNAPS—the fly-out toolbar, the keyboard, and the cursor menu.

4.4 Running OSNAPS

Let me show you another way to use OSNAPS. We call this one *Running OSNAPS*. Before long, you will discover that having to select an OSNAP every time you want to place a point is a tedious procedure at best (even if it is easier than typing coordinates). Is there not, you might ask, a way to turn OSNAPS on and leave them on? By now, of course, you know that if there were not, I would not ask.

Look at the last button on the OSNAP flyout toolbar (or the last option on the cursor menu). The button (Figure 4.4a) looks like one of those horseshoe magnets with which we played as children. This is the key to setting our running OSNAPS. Pick this button now. The Drafting Settings dialog box appears with the **Object Snap** tab on top (Figure 4.4b).

FIG. 4.4a
Object Snap Settings

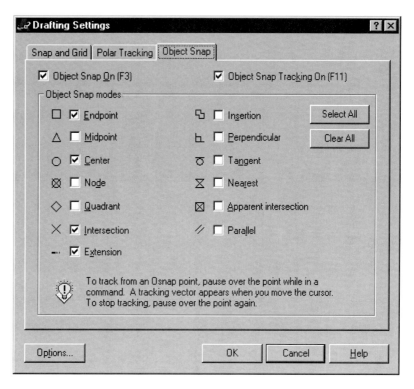

FIG. 4.4b: Object Snap Tab of the Drafting Settings Dialog Box

> Other ways to access the OSNAPS Settings dialog box include: typing **OSNAP** or **OS** at the command prompt or right-click on the OSNAP toggle (on the status bar) and select **Settings**.

Each available OSNAP has a check box beside it. Click in the box to place a check and activate that particular running OSNAP. Look to the left of each box to see a symbol (or marker). We have seen that AutoCAD uses this symbol to indicate that you are selecting a point using this particular OSNAP.

An **Options . . .** button resides at the bottom of the dialog box. This button opens the Options dialog box with the **Drafting** tab on top (Figure 4.4c). Here the user has five frames to help control the behavior of Object Snaps. You can control OSNAP settings from this tab, including:

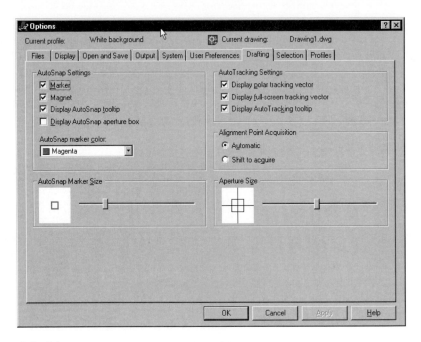

FIG. 4.4c: Options Dialog Box—Drafting Tab

- the size and color of the marker (symbol)
- whether or not you want the marker to show
- whether or not you want a cheater box (tip) when you hesitate over an OSNAP area
- whether or not you want the crosshairs to snap to (magnet) the selected OSNAP

But for optimal performance, I suggest leaving all these options at their default settings.

The *aperture box* was the selection tool used by OSNAPS prior to R14. AutoCAD turns it off by default since it now provides the OSNAP cheater symbols (markers) shown in this dialog box (Figure 4.4b). The aperture box itself is very similar to a normal selection box used with modify commands like *Erase*.

I do not recommend using the aperture box. It is just one more thing to clutter the screen and is not necessary because of the markers. But if you want to use it, check the **Display AutoSnap aperture box** check box.

Try using the Running OSNAPS to redraw the train.

You may notice that it is difficult to select the center or the quadrant of a circle as AutoCAD does not know which you want. If you move your crosshairs over the circle and the center symbol appears instead of the quadrant symbol (or vice versa) hold the mouse steady and pick the tab key on the keyboard. AutoCAD will toggle through the various possibilities till it finds the one you like. This procedure is quite useful for busy drawings in which AutoCAD must choose between several OSNAP possibilities.

On the other hand, selecting the *Quick* OSNAP will cause AutoCAD to grab the first available OSNAP every time, without looking at any other possibilities.

4.4.1 Do This: Use Running OSNAPs

I. Open the *Train.dwg* drawing found in the C:\Steps\Lesson04 folder. If you saved your last work as *MyTrain*, the *Train* drawing should appear unchanged.

II. Follow these steps.

TOOLS	COMMAND SEQUENCE	STEPS
 OSNAP Button	Command: *os*	1. Click on the **OSNAP Settings** button on the OSNAP flyout toolbar, or type *OSNAP* or *os* at the command prompt. 2. Set the Running OSNAPS indicated in Figure 4.4.1.2.

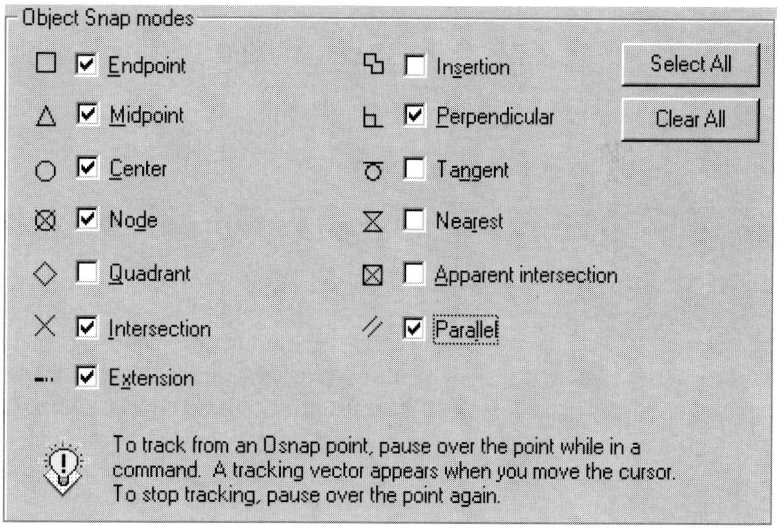

FIG. 4.4.1.2: OSNAP Settings

continued

TOOLS	COMMAND SEQUENCE	STEPS
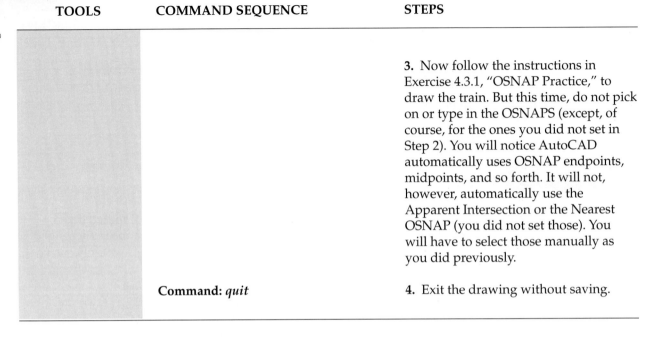		3. Now follow the instructions in Exercise 4.3.1, "OSNAP Practice," to draw the train. But this time, do not pick on or type in the OSNAPS (except, of course, for the ones you did not set in Step 2). You will notice AutoCAD automatically uses OSNAP endpoints, midpoints, and so forth. It will not, however, automatically use the Apparent Intersection or the Nearest OSNAP (you did not set those). You will have to select those manually as you did previously.
	Command: *quit*	4. Exit the drawing without saving.

Was that faster? Easier?

The biggest problem my students have with Running OSNAPS is that they forget to deactivate them. They cannot understand why their lines or circles keep jumping to an endpoint or intersection. If this happens to you, click the OSNAPS toggle on the status bar to deactivate the Running OSNAPS.

> Other ways to activate/deactivate running OSNAPs include clicking on the OSNAP toggle on the status bar, using the F3 function key on the keyboard, or holding down the CTRL key on the keyboard while typing *F*.

4.5 Point Filters and Object Snap Tracking

Often in drafting, we find it necessary to align objects according to the location of other objects. Parallel bars, triangles, and 4H-lead guidelines made this easy on the board. But what does AutoCAD use to substitute for these proven tools?

Actually, there are two things we can use: Point Filters and Object Snap Tracking.

■ 4.5.1 Point Filters

Use Point Filters to tell AutoCAD to use the X, Y, and/or Z coordinate of an existing object and to locate another object. Let's see how they work. We will draw a stick figure in our train's cabin. We will also get a sneak preview of the *Circle* command covered in Lesson 6.

4.5.1.1 Do This: Introducing Point Filters

I. Open the *MyTrain* drawing in the C:\Steps\Lesson04 folder.

II. Follow these steps.

TOOLS	COMMAND SEQUENCE	STEPS
Circle Button	**Command:** *circle*	1. Begin with the *Circle* command.
	CIRCLE Specify center point for circle or [3P/2P/Ttr (tan tan radius)]: *.x*	2. At the circle prompt, type **.X** to tell AutoCAD to use the X coordinate . . .
	mid	. . . of the **midpoint** of the lower horizontal line of the cab (see Figure 4.5.1.1.2).
		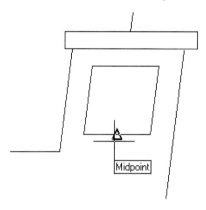 **FIG. 4.5.1.1.2:** Selecting the X Coordinate of the Midpoint of the Horizontal Line
	(need YZ): *mid*	3. Now tell AutoCAD to use the Y and Z coordinates of the midpoint of the vertical line (see Figure 4.5.1.1.3).
		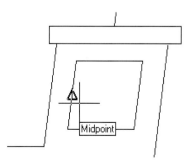 **FIG. 4.5.1.1.3:** Selecting the Y–Z Coordinate of the Midpoint of the Vertical Line

continued

TOOLS	COMMAND SEQUENCE	STEPS
	Specify radius of circle or [Diameter]: .25	4. Tell AutoCAD to use a radius of $1/4"$. The cab now looks like Figure 4.5.1.1.4.

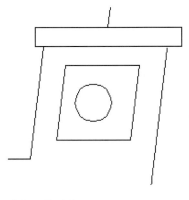

FIG. 4.5.1.1.4: Completed Circle

This is one way of acquiring coordinates from existing geometry in a drawing. But let's look at an easier way.

■ 4.5.2 Object Snap Tracking

AutoCAD produced the tracking feature in response to complaints that most OSNAPS require a direct contact with an existing object and Point Filters were too tedious. However, like Point Filters, Tracking allows you to draw in *relation to* an existing object without actually touching it. Let us see how it works. We will redraw the stick figure in our train's cabin.

 4.5.2.1 Do This: Point Filters the Easy Way—Object Snap Tracking

I. Be sure you are still in the *MyTrain* drawing in the C:\Steps\Lesson04 folder. If not, open it now.

II. Erase the circle you drew in Exercise 4.5.1.1, "Introducing Point Filters."

III. Follow these steps.

TOOLS	COMMAND SEQUENCE	STEPS
	Command: *circle*	1. Begin with the *Circle* command.
 Temporary Tracking Button	**CIRCLE Specify center point for circle or [3P/2P/Ttr (tan tan radius)]: _tt**	2. Select the **Tracking** button from the OSNAPS flyout toolbar.
	Specify temporary OTRACK point: _mid of **Specify center point for circle or [3P/2P/Ttr (tan tan radius)]: _tt Specify temporary OTRACK point: _mid of**	3. Now that the tracking option has been activated, select the midpoint OSNAP. Then select the lower horizontal line of the cab window.
	Specify center point for circle or [3P/2P/Ttr (tan tan radius)]:	4. Repeat Steps 2 and 3, this time selecting the left vertical line of the cab window. 5. Move the crosshairs toward the center of the window. Pick with the left mouse button when the Tracking tool tells you that you are located at 0° from the last point and 90° from the first point (Figure 4.5.2.1.5).

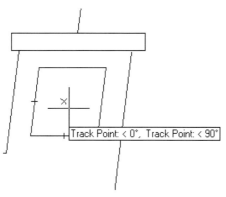

FIG. 4.5.2.1.5: Tracking Location

continued

TOOLS	COMMAND SEQUENCE	STEPS
	Specify radius of circle or [Diameter]: 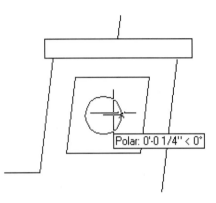	**6.** You will notice that the center of the circle has been placed at a point in the center of the window. Use Polar Tracking to draw a circle with a $1/4$" radius (Figure 4.5.2.1.6).

FIG. 4.5.2.1.6: Polar Tracking to Give a $1/4$" Radius

Command: *l*

7. Draw a line . . .

LINE Specify first point: _qua of

8. . . . from the lower **quadrant** of the circle you just drew . . .

Specify next point or [Undo]: _per to
Specify next point or [Undo]:
[enter]

9. . . . **perpendicular** to the bottom of the window.

Complete the command. Your drawing looks like Figure 4.5.2.1.9.

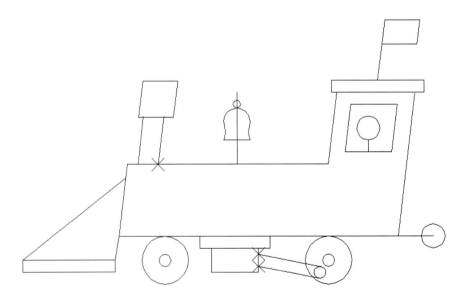

FIG. 4.5.2.1.9: Final Train Drawing

> AutoCAD now combines Running OSNAPs and Object Tracking into a Running Object Tracking tool. Pick the **OTRACK** toggle on the status bar to use both of these timesavers simultaneously!

4.6 Isometric Drafting

After having thoroughly confused us with orthographic projections, my old drafting instructor threw *isometric* drawings at us. (At this point, I started thinking about other ways I could make a living.) I will try to make it a bit easier than he did.

Orthographic drawings (projections) show an object from one aspect—left / right / top / bottom / front / back. An isometric drawing shows three aspects at once—left or right side / top or bottom / front or back. Standard isometrics are drawn so that the faces are seen on a plane running 30° above or below the X-axis, as shown in Figure 4.6a. Of course, there are variations, but the 30° format is standard.

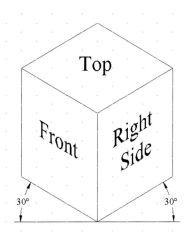

FIG. 4.6a: Standard Isometric

To draw isometrics in AutoCAD requires adjusting the grid and snap to an isometric format. This is easier than it sounds. Remember the options AutoCAD provided for the snap tool? This is where we make the switch from orthographic layout to isometric. Here is how:

Command: *snap*
Specify snap spacing or [ON/OFF/Aspect/Rotate/Style/Type] <0.5000>: *s*
Enter snap grid style [Standard/Isometric] <S>: *i*
Specify vertical spacing <0.5000>: *[enter]*
Command:

That is all there is to it! AutoCAD changes the grid and snap simultaneously. Ortho will now work along the 30°/90° planes. You will notice that even your crosshairs have changed. You will need a new toggle in this mode—the *isometric plane toggle*. This will adjust your crosshairs and help you draw along the 30°/90° plane, the 150°/90° plane, or the 30°/150° plane. The keyboard toggle is **Ctrl + E**, and the function key is **F5**. As with the other toggles we have learned, these will work regardless of the command sequence we are running.

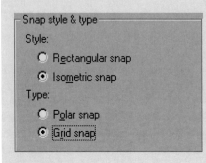

You can also set the isometric snap style on the **Snap and Grid** tab of the Drafting Settings dialog box (see Figure 4.6b). Simply right-click on the **Snap** or **Grid** toggle (on the status bar) and select **Settings** . . .

FIG. 4.6b: Isometric Snap

Let's try a simple isometric drawing.

Try to draw the standard isometric shown in Figure 4.6a. Do not try the text or the dimensions yet. We will look at these in Lessons 5 and 15.

4.6.1 Do This: Isometric Drafting

I. Begin a new drawing using the *MyBase1* template.

II. Follow these steps.

TOOLS	COMMAND SEQUENCE	STEPS
	Command: *sn*	1. Enter the *Snap* command by typing *snap* or *sn* at the command prompt.
	Specify snap spacing or [ON/OFF/Aspect/Rotate/Style/Type] <0'-0 1/2">: *s*	2. Tell AutoCAD you want to change the snap **Style**.
	Enter snap grid style [Standard/Isometric] <S>: *i*	3. Set the **Style** to **Isometric** . . .
	Specify vertical spacing <0'-0 1/2">: *1/4*	4. . . . with a spacing of $1/4$.

continued

TOOLS	COMMAND SEQUENCE	STEPS
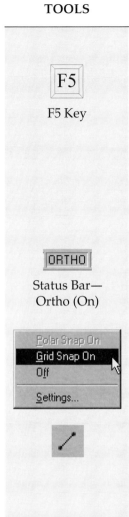 F5 Key	Command:	5. Using the *F5* function key, set your crosshairs to a 90°/30° configuration. The crosshairs will look like Figure 4.6.1.5.

FIG. 4.6.1.5: 90°/30° Isometric Crosshairs

6. Pick on the **Ortho** toggle on the status bar to turn on the ortho tool.

7. Right-click on the **Snap** toggle and set it to **Grid Snap On**.

Command: *l*

8. Draw the right side of the cube. Your drawing looks like Figure 4.6.1.8.

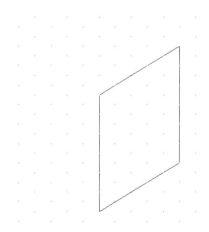

FIG. 4.6.1.8: Beginning of the Isometric Block

continued

TOOLS	COMMAND SEQUENCE	STEPS
F5	FIG. 4.6.1.9: 90°/150° Crosshairs	**9.** Toggle the crosshairs to 90°/150°. The crosshairs look like Figure 4.6.1.9.
	**Command:** *l*	**10.** Draw the left side of the cube. Your drawing now looks like Figure 4.6.1.10.

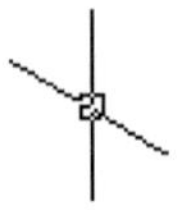

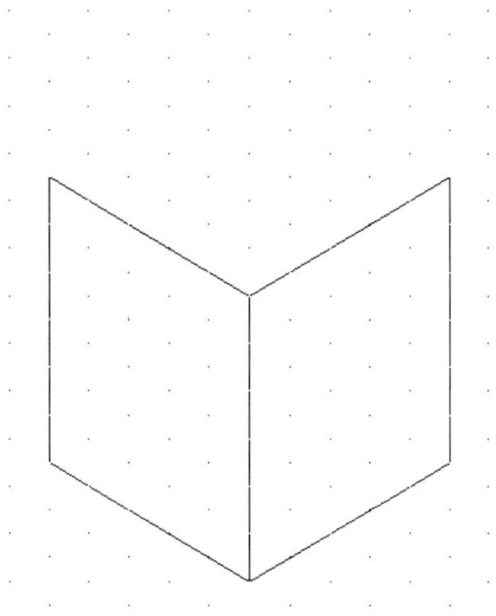

FIG. 4.6.1.10: Isometric Block with Two Sides

F5	FIG. 4.6.1.11: 30°/150° Crosshairs	**11.** Toggle the crosshairs to 30°/150°. The crosshairs look like Figure 4.6.1.11
	**Command:** *l*	**12.** Draw the top of the cube. The drawing now looks like Figure 4.6.1.12.

continued

TOOLS	COMMAND SEQUENCE	STEPS
		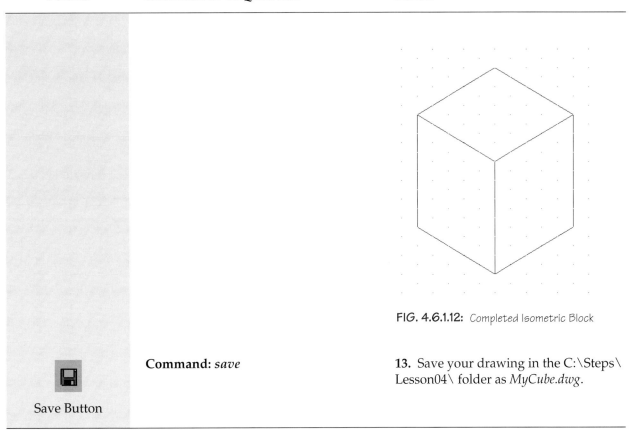 **FIG. 4.6.1.12:** Completed Isometric Block
![Save Button] Save Button	**Command:** *save*	13. Save your drawing in the C:\Steps\Lesson04\ folder as *MyCube.dwg*.

4.7 Extra Steps

One of the lesser-known drawing tools works similarly to the polar coordinate system, but requires less input from the user. This is called the *Direct Distance Option*. Learn this one to amaze and confound older CAD operators.

Simply put, this is how it works. At the **Specify next point or [Undo]:** prompt, you enter a distance at the keyboard, move the crosshairs in the direction you want the line to go, then hit *enter* (or pick the right mouse button then select **Enter**). Let me demonstrate.

4.7.1 Do This: Direct Distance Entry

1. Open a new drawing from scratch.

2. Begin to draw a line from point 2,2.

3. At the **Specify next point or [Undo]:** prompt, type 3. Do not hit *enter* yet.

4. Activate the **Ortho** tool and move your crosshairs to the right.

5. Hit *enter*. Notice that AutoCAD has drawn a line 3 units in the direction you moved the crosshairs?

Pretty neat, huh?

4.8 What Have We Learned?

Items covered in this lesson include:

- *Drafting Settings: Ortho, Snap, Tracking, and Grid tools*
- *OSNAPs and Running OSNAPs*
- *Point Filters*
- *Object Snap Tracking*
- *Direct Distance Entry*
- *Commands*
  - Grid
  - Snap
  - Ortho
  - DSettings
  - OSNAP

In this lesson, we have covered the tools that will make the difference between a computer *doodler* and a CAD *operator*. Anyone can draw lines and circles, but for CAD to be an effective tool in industry, it must have the ability to draw with speed and precision. But do not expect yourself to fly through a drawing yet. First, you must practice the material in this lesson until it becomes second nature—like a draftsman knowing how to begin a drafting project without thinking about it. So take some time to do these exercises. Do them again and again till you are quite comfortable with AutoCAD's Drawing Aids.

4.9 EXERCISES

1. Start a new drawing. Set it up as follows:
   1.1. Units: architectural
   1.2. Lower left limits: 0,0
   1.3. Upper right limits: 17,11
   1.4. Grid: 1
   1.5. Snap: $1/2$
   1.6. Save this as a template file called *MyGrid1.dwt* to the C:\Steps\Lesson04 folder.
2. Start a new drawing. Set it up as follows:
   2.1. Units: decimal
   2.2. Lower left limits: 0,0
   2.3. Upper right limits: 11,8.5
   2.4. Grid: $1/2$
   2.5. Snap: $1/4$
   2.6. Snap style: isometric
   2.7. Save this as a template file called *MyIsoGrid1.dwt* to the C:\Steps\Lesson04 folder.
3. Start a new drawing. Set it up as follows:
   3.1. Units: architectural
   3.2. Lower left limits: 0,0
   3.3. Upper right limits: 17,11
   3.4. Grid: $1/2$
   3.5. Snap: $1/4$
   3.6. Snap style: isometric
   3.7. Save this as a template file called *MyIsoGrid2.dwt* to the C:\Steps\Lesson04 folder.
4. For each of the figures in Figures 4.9.4.1 to 4.9.4.6, start a new drawing using the *MyIsoGrid2* template file created in Exercise 3. (If this file is not available, use the *IsoGrid2* template in the same folder.) Draw the figures and save them as their figure name in the C:\Steps\Lesson04 folder.

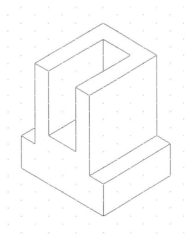

FIG. 4.9.4.1

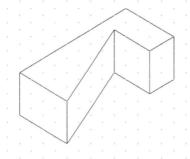

FIG. 4.9.4.2

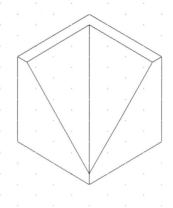

FIG. 4.9.4.3

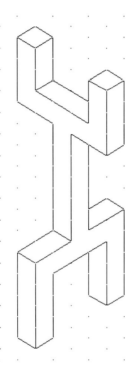

FIG. 4.9.4.4

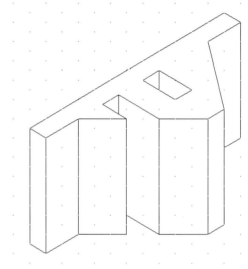

FIG. 4.9.4.5

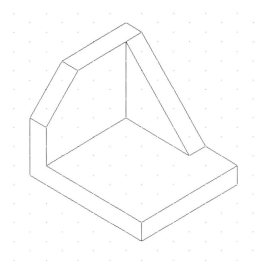

FIG. 4.9.4.6

5. Open the *raygun* drawing in the C:\Steps\Lesson04 folder. Using what you learned about OSNAPS, complete the ray gun in Figure 4.9.5.

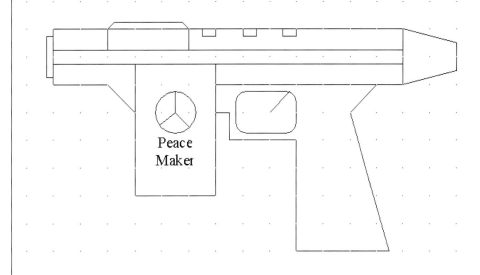

FIG. 4.9.5: Ray Gun

4.10 REVIEW QUESTIONS

Please write your answers on a separate sheet of paper.

1. To see a different X and Y grid spacing, use the _____ option of the grid command.

2. The _____ tool pulls the crosshairs to a grid point or a location between grid points.

3. Conventional wisdom suggests that grid snap increments should be _____ that of the grid increments.

4. and 5. To set up an isometric grid, go to the _____ option of the _____ command.

6. The _____ tool restricts drawing to the left/right or up/down directions.

Identify these function keys:

7. F7   8. F8   9. F9   10. F10   11. F11

12. Verify that these tools are on or off by checking the appropriate site on the _____ bar.

13. to 15. The standard isometric is drawn on which three angles?

16. and 17. Name two ways to toggle between isometric planes.

18. To call the Drafting Settings dialog box, type _____ at the command prompt.

19. Once you have learned grid, snap, ortho, and OSNAPS, what must you *never* do?

20. _____ provide the means for precisely selecting an endpoint, midpoint, or intersection.

Identify:

OSNAP Toolbar	Symbol	Cursor Menu
		35
		36
	22	37
	23	38
	24	39
	25	40
	26	41
	27	42

4.10 Review Questions

OSNAP Toolbar	Symbol	Cursor Menu
◇	28	43
◌	29	44
⊥	30	45
∥	31	46
🗗	32	47
∘	33	48
✎	34	49
✕		50
∩		51

51. To call up the OSNAP screen menu, right-click on the screen while holding down the _____ key.

52. When both the center and quadrant Running OSNAPS are on, AutoCAD may not select the one you want. To toggle between them, place the crosshairs over the circle or arc and hit the _____ key on the keyboard.

LESSON 5

Display Controls and Basic Text

Following this lesson, you will:

➡ Know how to manipulate the screen display of your drawing through:
- The **Zoom** command
- The **Pan** command
- The **View / DDView** commands

➡ Be aware of the **Aerial View** toy

➡ Know how to use Basic Text and Text Editing commands, including:
- **Text**
- **DDEdit**
- **Style**
- **Qtext**
- **Find**

➡ Know how to load a LISP routine

Our last lesson probably made you feel fairly comfortable with your new ability to create some wildly accurate drawings. But have you tried to use some of these tools in larger drawings? If you have, you might have noticed that grid and snap become difficult (if not impossible) to use as the drawing encompasses more space. You might also have noticed that those wonderful OSNAPS do not help if you cannot tell which endpoint you have selected.

We will begin this lesson by addressing these problems. Then we will examine the tools available to help place text in a drawing.

5.1 Getting Closer: The *Zoom* Command

One of the allowances I have had to make as I amassed all this "experience," has been the acquisition of a pair of reading glasses. At first I wore them down on my nose—to look more professorial. Now, however, I wear them closer to my eyes to get a wider field of vision. Ah, age!

For this reason, I really appreciate AutoCAD's **Zoom** command. With this, I can enlarge all or part of my drawing as much as I like, without having to search for my "eyes."

To appreciate the demonstration of the **Zoom** command, we will need to open the *PID* drawing (Piping and Instrument Diagram) in the C:\Steps\Lesson05 folder. The drawing is shown in Figure 5.1a. Note that the *PID* drawing was created for a D-size sheet of paper, so viewing it on the screen (or on our book-size sheet of paper) makes it difficult to read.

The command sequence for the **Zoom** command follows:

Command: *zoom* **(or** *z***)**
Specify corner of window, enter a scale factor (nX or nXP), or
[All/Center/Dynamic/Extents/Previous/Scale/Window] <real time>:
 [select first window corner]
Specify opposite corner: *[select other window corner]*
Command:

FIG. 5.1b
Zoom Window

FIG. 5.1c
Zoom Realtime

Z (the hotkey) will work as well as *zoom*, or you may pick the **Zoom Window** button from the Standard toolbar (Figure 5.1b).

Let's look at the various **Zoom** options. Notice that each option has an equivalent button in the Zoom Window flyout toolbar.

- **Realtime** appears to be the default, but it is accepted only by hitting the enter key. We will look at **Realtime** zooming shortly. If instead you select an empty point on the screen, AutoCAD will assume you are placing a window around the objects you wish to view closer (remember Implied Windowing—Section

5.1 Getting Closer: The Zoom Command

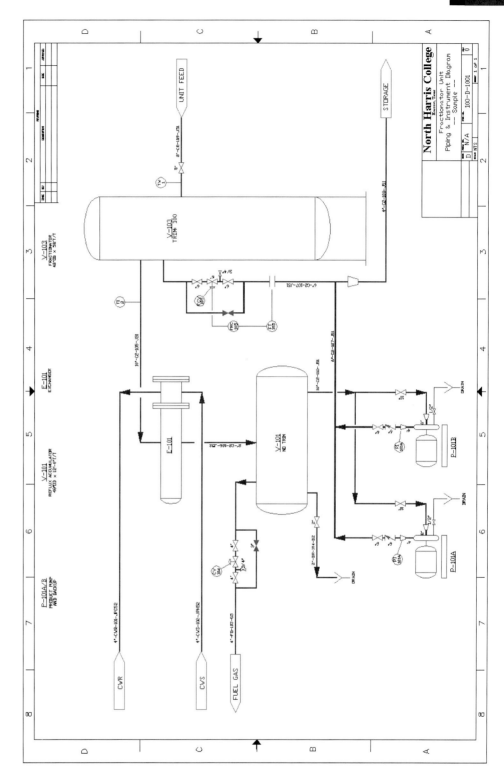

FIG. 5.1a: PID Drawing (Piping and Instrument Diagram)

2.3, "Multiple-Object Selection Made Easy"). It will then prompt you for the other corner.

- **All** displays the limits of the drawing, unless something has been drawn outside the limits. In that case, **All** will display the extents, or *all* the objects on the drawing.
- **Center** prompts the user for the desired center point of the display, then adjusts the display so that the selected point is in the center of the screen.
- One of the most forgotten zoom features, **Dynamic** provides the user the ability to place an adjustable box over that part of the drawing he wants to display. When selected, AutoCAD temporarily replaces the screen with a view of the entire drawing. A view box shows what and where your current view area is. A selection box appears that can be manipulated with the mouse as you would the crosshairs. Move this box over the area to be viewed and hit the right mouse button to confirm your selection. AutoCAD redisplays the drawing with the selected area shown (this will be clearer when we do it in Exercise **5.1.1, "Practice Zooming"**).
- **Extents** brings the user as close as possible to the drawing while showing *all* the objects in the drawing.
- **Previous** displays the last screen viewed.
- **Scale** can be a bit confusing. You do not have to type *s*, but you can. If you do, AutoCAD prompts you for the scale you want. Note that this is *not* the drawing scale, but the size of the drawing in relation to the graphics area. Hence, a scale of **.5** will cause the drawing to occupy half the graphics area of your screen.

    You may want to simply type **.5X** at the **Zoom** prompt (rather than typing *s*). This will cause the drawing to appear half its current size.

    Notice that the **Zoom** prompt suggests the **X** or an **XP** procedure. Ignore the **XP** for now. We will cover that in detail in our discussion of Paper Space (in *AutoCAD 2000: One Step at a Time—Advanced Edition*).

- To simplify the zoom **Scale** option, AutoCAD provides two additional buttons —**Zoom Out** and **Zoom In**. **Zoom In** zooms to a 2X scale and **Zoom Out** zooms to a .5X scale.

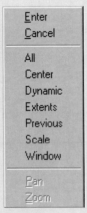

In addition to the command line and toolbars, all the *Zoom* options can be found in the **Zoom** selection of the **View** pull-down menu and on the (right-click) cursor menu once the zoom command has been entered (see Figure 5.1d).

FIG. 5.1d: *Zoom's Cursor Menu*

Let's try some of these now.

5.1.1 Do This: Practice Zooming

I. If *PID.dwg* is not already open, please open it now. It is in the C:\Steps\Lesson05 folder.

II. Follow these steps.

TOOL	COMMAND SEQUENCE	STEPS
 Zoom Window Button	**Command: z**	1. We will begin with the **Window** option of the **Zoom** command. Enter the **Zoom** command or pick the **Zoom Window** button on the Standard toolbar.
	Specify corner of window, enter a scale factor (nX or nXP), or All/Center/Dynamic/Extents/Previous/Scale/Window] <real time>: *[select first window corner]* **Specify opposite corner:** *[select other corner]*	2. Select the corners of the window as shown in Figure 5.1.1.2a. Your display should look like Figure 5.1.1.2b.

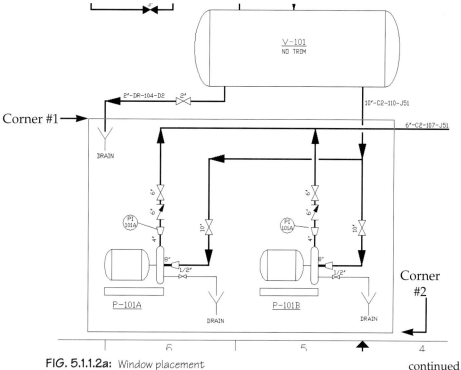

FIG. 5.1.1.2a: Window placement

continued

TOOLS	COMMAND SEQUENCE	STEPS

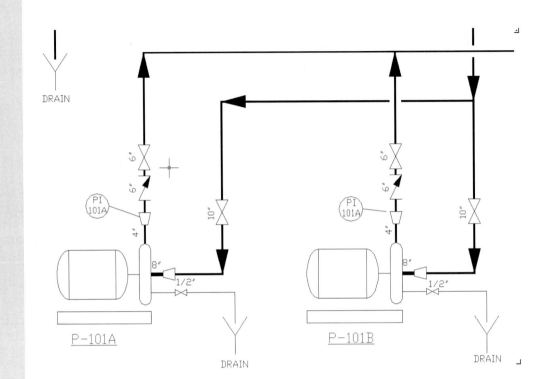

FIG. 5.1.1.2b: Display after *Zoom*

Zoom Previous

Zoom In

Zoom All

Command: *z*
Specify corner of window, enter a scale factor (nX or nXP), or [All/Center/Dynamic/Extents/ Previous/Scale/Window] <real time>: *p*

3. Now let's zoom back to where we started by using the **Previous** option or the **Zoom Previous** button on the Standard toolbar.

4. Zoom to a *.5X* scale using the **Zoom In** button (on the Zoom flyout toolbar).

Command: *z*
Specify corner of window, enter a scale factor (nX or nXP), or [All/ Center/Dynamic/Extents/Previous/ Scale/ Window]<real time>: *a*

5. Now *Zoom All* using either keyboard commands or the *Zoom All* button.

Notice the location of and the spacing around the drawing on the screen. Let's compare that to *Zoom Extents*.

continued

TOOLS	COMMAND SEQUENCE	STEPS
 Zoom Extents	**Command:** *z* **Specify corner of window, enter a scale factor (nX or nXP), or [All/Center/Dynamic/Extents/Previous/Scale/Window] \<real time\>:** *e*	**6.** *Zoom Extents* using keyboard or button. Notice the location of and spacing around the drawing on the screen. How does it compare with the results of the **Zoom All**?
 Zoom Dynamics	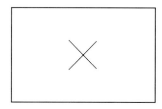 **FIG. 5.1.1.7:** Zoom Dynamic's Location Box	**7.** Now let's play with the **Zoom Dynamics** option. Pick the **Zoom Dynamics** button. Notice the change in the display. There is now a box over the display with an "X" in the middle. This is the *location box* (Figure 5.1.7).
	 **FIG. 5.1.1.8:** Zoom Dynamic's Sizing Box	**8.** Pick with the left mouse button and notice how the box changes. Instead of an "X" in the middle, there is now an arrow pointing toward the right side of the box. This is the *sizing box* (Fig. 5.1.1.8). As you move the mouse back and forth, notice how the size of the selection box changes. When it is just large enough to enclose the exchanger (see Figure 5.1.1.9), pick again with the left mouse button.
	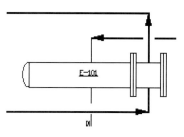 **FIG. 5.1.19:** Exchanger	**9.** Now place the selection box over the exchanger and confirm (pick with the right mouse button). Your display now looks like Figure 5.1.1.9.

continued

TOOLS	COMMAND SEQUENCE	STEPS
 Zoom Center	**Command:** *z* **Specify corner of window, enter a scale factor (nX or nXP), or [All/Center/Dynamic/Extents/Previous/Scale/Window] \<real time\>:** *c* **Enter magnification or height \<X.XXXX\>:** *[enter]*	10. Now let's center our exchanger. Repeat the *Zoom* command and enter the **Center** option. Or you can pick the **Zoom Center** button on the Zoom flyout toolbar. Select a point roughly in the center of the exchanger. 11. Hit *enter* to complete the command.

FIG. 5.1e
Realtime Zoom

FIG. 5.1f
Realtime Zoom Cursor

Before we proceed, there are two other aspects of display manipulation we should discuss—*Realtime Zoom* and *Pan* (or *Realtime Pan*).

Realtime is just a fancy way to say, "Do it while I watch." In other words, you can judge how far you want to zoom by watching the display change as you move your mouse.

The **Realtime Zoom** button looks like Figure 5.1e. You can find it on the Standard toolbar, next to the *Zoom* flyout buttons we have been using. When you pick this button, you will notice the crosshairs change to a cursor that resembles the button image (Figure 5.1f). To use **Realtime Zoom**, pick anywhere in the graphics area of the screen with the left mouse button. While holding the button down (dragging), move the cursor up and down. Notice how the display changes. When you are happy with the display, release the left mouse button, click the right button and pick *Exit* on the cursor menu that appears.

> A trick to remember when using the *Realtime Zoom* command is to do a *Zoom Center* first. You may have noticed that *Realtime Zoom* does not allow you to change *position* during a zoom. In fact, it maintains the same center point on the display, zooming in or out about that point.

You can also access **Realtime Zoom** by entering the *Zoom* command, then hitting *enter* at the first prompt.

A wonderful new trick is included with AutoCAD 2000; AutoCAD will now make use of the wheel located between the two buttons on an *IntelliMouse* pointing device. Rotating the wheel forward or backward works like a **Realtime Zoom**. If you do not have an IntelliMouse, the time savings will more than justify the ~$30 expense!

FIG. 5.1g
Realtime Pan

FIG. 5.1h
Realtime Pan Cursor

Another of AutoCAD's *realtime* features, **Pan** behaves as though you are panning a camera across your display. Located next to the **Realtime Zoom** button on the Standard toolbar, the **Pan** button (Figure 5.1g) will also change the crosshairs to a cursor that resembles the button image (Figure 5.1h). **Realtime Pan** works in much the same was as realtime zoom. Pick and drag with the left mouse button. It is like putting your hand down on your paper and sliding the paper across the drawing table.

Here again AutoCAD makes use of the IntelliMouse. Depressing the wheel between the mouse buttons and dragging the cursor across the screen appears as though you are in the **Realtime Pan** command. But no command line or toolbar entries are needed!

FIG. 5.1i
Realtime Menu

Exit **Realtime Pan** just as you did **Realtime Zoom**.
The hotkey for accessing the *Pan* command is **P**.
One of the really neat aspects of the *realtime* features is the menu that displays with a right-click (Figure 5.1i). I strongly suggest getting comfortable with this menu (if you do not have an IntelliMouse). Expertise with it will enable you to position yourself exactly where you want to be in a drawing, while saving the time you might otherwise spend keyboarding or toolbar clicking.
Before continuing, take a few minutes to play with these last few features in the *PID* drawing. Note the increased speed using the *realtime* features as opposed to the time spent in display manipulation using the methods we learned previously.

The Aerial View Toy

AutoDESK could have saved itself a lot of time and money by not putting so many redundant routines in its software. But if it had, it would not have become the user-friendly benchmark of CAD systems on the market today.

One of the redundancies AutoCAD provides, the **Aerial View** allows the user yet another method of display manipulation.

The **DSViewer** command (or *AV* hotkey) presents the Aerial View window atop the AutoCAD window. The window looks something like Figure 5.1j.

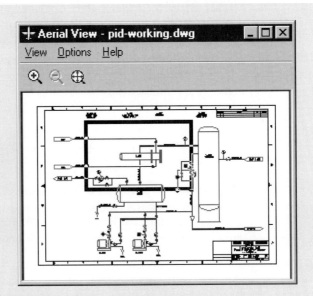

FIG. 5.1j: Aerial View Window

The buttons and methods of display control within the window are just like the ones found in the AutoCAD window. So what is the big difference? Well, the aerial view (AV) permits you to adjust the magnification and position of the view before actually changing the display. There are, however, some unfortunate drawbacks to using the AV. These include:

- The AV window covers part of the display window.
- You must open the AV window each time you want to change the display (or leave it covering a portion of the display), then close it when you are through with it.
- It requires too many steps compared to other, easier methods that produce the same results.

In all honesty, I have worked with CAD for many years and have never seen anyone make constructive use of the AV or its predecessors. It is a nifty toy and might have been a good idea at one time, but we have seen more efficient approaches for manipulating the display screen. Still, this might be just the ticket for you. So take a few minutes to experiment.

5.2 Why Find It Twice: The *View* and *DDView* Commands

One of the things you will discover after drafting on a computer for a while is that you must return frequently to certain areas of your drawing. A beginning CAD operator will use display tools like **Zoom** and **Pan** because these are simple, easily mastered and they work. With so much to learn, who can blame them?

But AutoCAD has provided the *View* command to speed past these display controls. With the *View* command, the user can create and store certain displays (views) then restore these views at any time from any position in the drawing. Thus, using our *PID* drawing as an example, we can go directly from a display of the pumps to a display of the exchanger without the need for panning or zooming in and out.

Let's take a look at the view procedures.

WWW 5.2.1 Do This: The View Command on the Command Line

I. If you are not currently in the *PID.dwg* file, please open it now. It is in the C:\Steps\Lesson05 folder.

II. *Zoom All.*

III. Follow these steps.

COMMAND SEQUENCE	STEPS
Command: *-v*	1. We will begin by entering the *View* command at the command prompt. Type *—view* or *—v* (be sure to use the dash).
?/Delete/Restore/Save/Window: *s*	2. Type *s* to save the current view.
View name to save: *full*	3. Type *full* to give the view that name.
Command: *[enter]*	4. Now let's create a new view around a specific part of the drawing. Repeat the *View* command.
?/Delete/Restore/Save/Window: *w*	5. Type *w* for the **Window** option.
View name to save: *pumps*	6. Give the view the name *pumps*.
First corner: Other corner:	7. Select the same corners as those you selected in Figure 5.1.1.2a (page 129).
Command: *[enter]*	8. Now restore the view. Repeat the *View* command.
?/Delete/Restore/Save/Window: *r*	9. Type *r* for the **Restore** option.
View name to restore: *pumps*	10. Tell AutoCAD which view you want to restore.
Command:	11. Leave the drawing where it is and read on.

Using the *View* approach requires a bit of setup time at first; however, you can see that it will save quite a bit of time later when you might be panning and zooming all over the place.

The *View* command provides some other options—the **?**, **Orthographic**, **Delete**, and **UCS**.

- Using the **?** at the view prompt will tell AutoCAD to list all the views stored in a drawing.
- Can you guess what **Delete** will do? Of course, it will allow you to remove selected views.
- The **Orthographic** and **UCS** views both deal with 3-dimensional space. We will look at those in our next book (*AutoCAD: One Step at a Time—Advanced Edition*).

Did you notice the dash before the command in the last exercise? (Of course you did or the exercise did not work!) Without the dash, the *View* command will call the View dialog box (Figure 5.2a) rather than the *View* prompts at the command line.

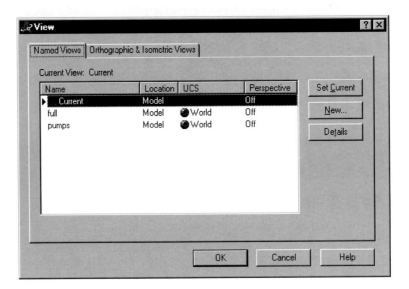

FIG. 5.2.a: View Dialog Box

So what is the difference? Let's see.

The dialog box contains a list box showing all the currently defined views in the drawing, along with information about each (Model Space or Paper Space, UCS, Perspective). Most of the information provided here (as with the **Details** button and the **Orthographic & Isometric Views** tab) relate to 3-dimensional space and are beyond the scope of this text.

The **New . . .** button, however, calls the New View dialog box (Figure 5.2b) that allows the user to create a view just as you did on the command line in our last exercise.

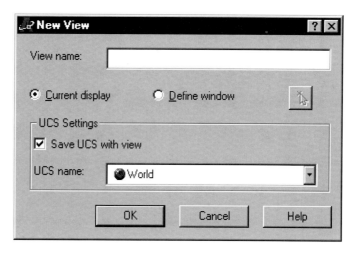

FIG. 5.2.b: New View Dialog Box

Let's try creating and restoring a view using the View dialog box.

5.2.1 Do This: Creating a View Using a Dialog Box

I. If the *PID* drawing is still open, start from where you left off. Otherwise, open it now.

II. *Zoom All*.

III. Follow these steps.

TOOL	COMMAND SEQUENCE	STEPS
Named Views New...	**Command:** *v*	**1.** Enter the *View* command by typing *view* or *v*. You may also pick the **Named Views** button on the Standard toolbar or **Named Views** from the View pull-down menu. **2.** Pick the **New** button on the View dialog box. The New View dialog box will appear (see Figure 5.2b). **3.** Type *exchanger* in the **View name** text box. Place a bullet in the option button beside the words **Define Window**. The box will look like Figure 5.2.1.3.

continued

TOOLS	COMMAND SEQUENCE	STEPS

FIG. 5.2.1.3: Completed Dialog Box

4. Select the **Define View Window** button. The dialog boxes will disappear and AutoCAD will prompt you to locate the window.

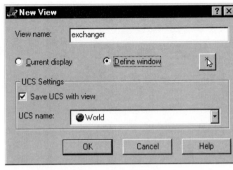

Define View Window Button

Specify first corner:
Specify opposite corner:.

5. Place the window as shown in Figure 5.2.1.5. The dialog box will return

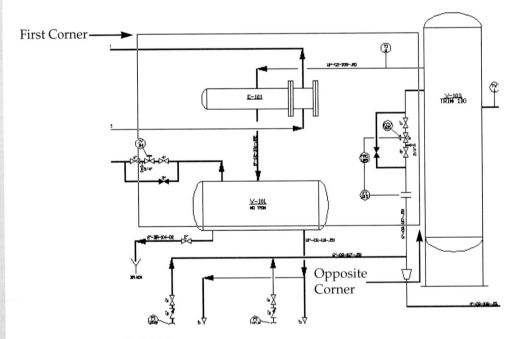

FIG. 5.2.1.5: Window Location

continued

TOOLS	COMMAND SEQUENCE	STEPS

6. Pick the **OK** button.
The View dialog box returns. You see the **Exchanger** view listed in the list box.

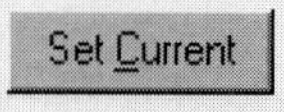

7. Now we will set the exchanger view as current. Pick **Exchanger** in the list box; then pick the **Set Current** button (or you can double-click on **Exchanger**).

8. Pick the **OK** button. We return to the graphics screen where the exchanger view is now displayed (Figure 5.2.1.8).

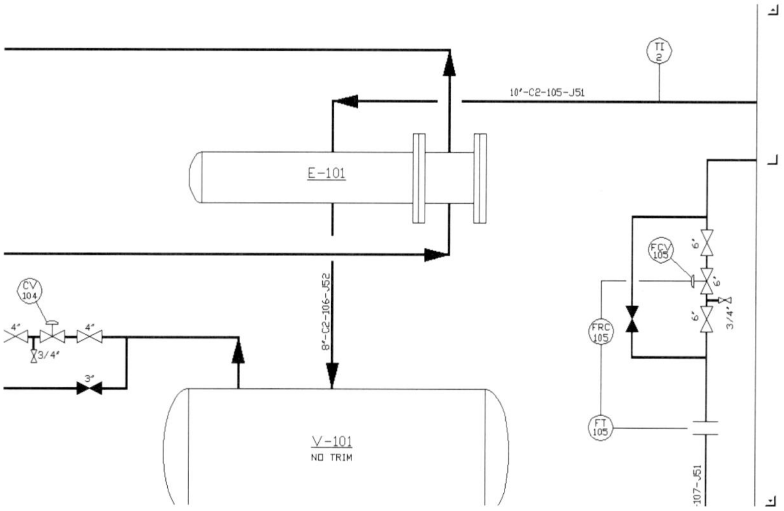

FIG. 5.2.1.8: Exchanger View Restored

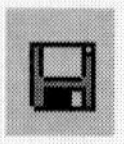

Command: *Qsave*   **9.** Save the drawing and exit.

Of the two approaches—the command line and the dialog box—I must admit to being more accustomed to the former. But that results from years with the software before the inclusion of dialog boxes. I really like the dialog box because it lists the views that are stored in the drawing, and I do not have to be concerned with remembering their names (or their *spellings*!). So, I suppose I will have to make the mental adjustment. Ah, progress! (*Note:* For a nifty trick, see the **Extra Steps** section of this lesson.)

AutoCAD's display commands —including the *View*, *Zoom* and *Pan* commands are *transparent*. That is, they may be used while at another command's prompt (while running another command). The buttons will work as they always do, but to enter a transparent command at a prompt other than the command prompt, precede it with an apostrophe. Thus, entering the *Zoom* command while at the *Line* command's prompt will look like this:

Specify next point or [Undo]: '*z*
>>Specify corner of window, enter a scale factor (nX or nXP), or
[All/Center/Dynamic/Extents/Previous/Scale/Window] <real time>:

The double bracket preceding the zoom prompt indicates that it is operating transparently. When the transparent command is completed, AutoCAD returns to the previous command prompt (in this case, the *Line* command's prompt).

5.3 Simple Text

Entering text into an AutoCAD drawing seems like a simple thing to do. However, it appears quite complicated. Still, once the procedures are mastered, you will see that text is not so difficult after all.

The *Text* command allows multiple lines of text to be entered; and it shows the text on the screen as the user types.

The command sequence seems deceptively simple.

Command: *text* **(or** *dt***)**
Current text style: "Standard" Text height: 0.2000
Specify start point of text or [Justify/Style]: *[pick the starting point]*
Specify height <0.2000>: *[enter the desired text height]*
Specify rotation angle of text <0>: *[enter the desired rotation angle]*
Enter text: *[type in your text]*
Enter text: *[hit enter to complete the command]*
Command:

Let's look at each line.

- The first line shows that *DT* (for Dynamic Text) is the hotkey for *Text*. Note that *T* is a hotkey; but it is a hotkey for the *MText* command, not the *Text* command.

MText is a more advanced text tool that uses a word processor format to enter text. We will discuss it in Lesson 14. (The **Text** button on the Draw toolbar also calls the MText dialog box; so avoid it for now.)

- The second line provides the *Text* options.
  - The default option is **Specify start point**. This simply directs the user to identify the insertion point of the text. The user does this by coordinate input or picking a point on the screen with the mouse.
  - The first option is **Justify**. You do not actually have to select this option (i.e., you do not have to type *J*) to justify your text. Typing *J*, however, will tell AutoCAD to present the various justification options as shown here.

    Enter an option [Align/Fit/Center/Middle/Right/TL/TC/TR/ML/MC/MR/BL/BC/BR]:

    By default, AutoCAD uses the **Bottom Left** option. Refer to Figure 5.3a to see where each of the options will place the text in relation to the insertion point (the "X").

FIG. 5.3a: Text Insertion Points

  - The **Align** and **Fit** justifications behave in much the same manner; however, **Align** will adjust the text height proportionally as it fits the text between the selected points, and **Fit** will maintain the user defined height.
  - The **Style** option enables the user to choose among text styles defined within the drawing. More on this in Section 5.6, "Adding Flavor to Text with Style".
- The user sets the height of the text on the next line. This is one of the places where it is easy to make a mistake. As drafters, we are accustomed to scaling our drawing but not our text. We draw 1/8" text 1/8" high. In CAD, we do just the opposite—or we scale our text but not our drawing!

  You will remember from Lesson 1 that we draw full-size, then adjust for scale when we plot. When we adjust to plot, we downsize our drawing to fit onto a certain size sheet of paper. Well, when we downsize the drawing, the text goes with it. So we must increase the size of the text when we create it to allow for that downsizing.

  Confused? It is really fairly easy to do.

  Look at the Drawing Scales chart on the tear out reference card. The second column contains a Scale Factor. This is the key to creating the proper text size. Simply multiply the size you want the text to be when it is plotted by the Scale Factor for the final plotted scale. Then you use that number as your text height.

Here is an example. I am creating a drawing that I will plot later at 3/8" = 1'-0". The factor for this scale is 32. I want my text size to be 1/8" when I plot. I multiply the 1/8 by 32 to get 4. I use 4 as my text height.

I know that may be a bit difficult to understand at this point, but believe me, it works!

- The next line asks the user for a rotation angle. This means AutoCAD wants to know if your text will be standard read-from-the-bottom-of-the-page (left-to-right) text or something else. Remember how AutoCAD measures angles! Read-from-the-rightside-of-the-page would be entered at 90°.

- Next AutoCAD asks for the actual text. Here the user types the desired text, hitting enter for a return (like the old-fashioned typewriters). When finished, the user hits enter twice and the command prompt returns.

Something to remember: transparent commands—like **Zoom** or **View**—do not work while the **Text** prompt is showing. This way all the keys are available while in text mode.

Let's try some text.

In addition to the command line, you can access the *Text* command by selecting **Text** then **Single Line Text** from the **Draw** pull-down menu.

5.3.1 Do This: Inserting Text

I. Open the *PIDTEXT* drawing in the C:\Steps\Lesson05 folder.

II. Refer to the drawing in Figure 5.1a (page 125) as you follow these steps.

TOOL	COMMAND SEQUENCE	STEPS
 Named Views	**Command:** *v*	**1.** First let's get a bit closer. Restore the **exchanger** view. (See Exercise 5.2.1 or 5.2.2 if you need help.)

continued

TOOLS	COMMAND SEQUENCE	STEPS

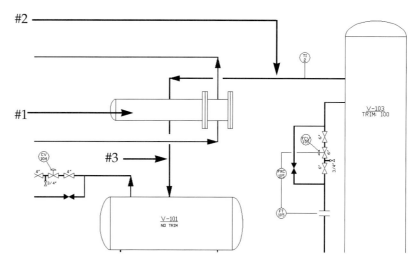

FIG. 5.3.1.1: Exchanger View

	Command: *dt*	2. We will start by identifying the exchanger. Enter the *Text* command by typing *Text* or *dt* at the command prompt.
	Current text style: "STANDARD" Text height: 0.1250 Specify start point of text or [Justify/Style]:	3. Pick a point near the tip of arrow #1 (Figure 5.3.1.1). (Toggle the OSNAP **Off** if necessary.)
	Specify height <0.1250>: *3/16*	4. This drawing was created with no scale on a D-size sheet of paper, so we will not have to scale the text. Assign a text height of *3/16*.
	Specify rotation angle of text <0>: *[enter]*	5. Accept the default rotation angle.
	Enter text: *%%uE-101*	6. Here is your first text trick. Lead the text with *%%u* to underline the text. (The %% trick calls an ASCII code. See the Text Tricks insert for more ASCII symbols.)

continued

TOOLS	COMMAND SEQUENCE	STEPS

> **Text Tricks**
>
> Other text tricks (based on ASCII codes) are
> %%U = underlined
> %%C = diameter symbol
> %%P = plus/minus symbol
> %%D = degrees symbol

	Enter Text: *[enter]*	**7.** Hit *enter* to complete the command. Notice that the text was not underlined until the command was completed.
	Command: *[enter]*	**8.** Now enter a line number by arrow #2 (see Figure 5.3.1.1). Repeat the command.
	Current text style: "STANDARD" **Text height: 0.1250** **Specify start point of text or [Justify/Style]:**	**9.** Pick a point just above the line.
	Specify height <0.1875>: *1/8*	**10.** Set the height to *1/8*.
	Specify rotation angle of text <0>: *[enter]*	**11.** Accept the default rotation angle.
	Enter Text: *10"-C2-105-J51*	**12.** Enter the line number.
	Enter Text: *[enter]*	**13.** Complete the command.
	Command: *[enter]*	**14.** Now let's try some angled text. Repeat the command.
	Current text style: "STANDARD" **Text height: 0.1250**	**15.** Pick a point just to the left of the line at arrow #3 (see Figure 5.3.1.1).
	Specify start point of text or [Justify/Style]:	
	Specify height <0.1250>: *[enter]*	**16.** You see that AutoCAD defaults to the last height setting. Hit *enter* to accept it.

continued

TOOLS	COMMAND SEQUENCE	STEPS
	Specify rotation angle of text <0>: *90*	17. Set the rotation angle to *90°*.
	Enter Text: *8"-C2-106-J52*	18. Type the text.
	Enter Text: *[enter]*	19. Complete the command. Your drawing looks like Figure 5.3.1.19.

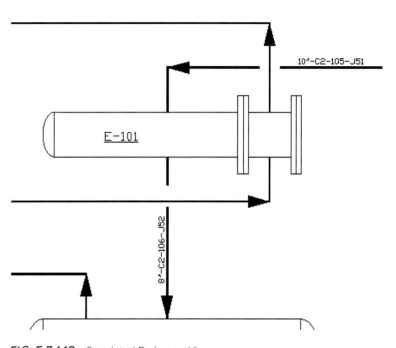

FIG. 5.3.1.19: Completed Exchanger View

	Command: *v*	20. Let's try justifying some text. Change to the **pumps** view (Figure 5.3.1.20).

continued

TOOLS	COMMAND SEQUENCE	STEPS

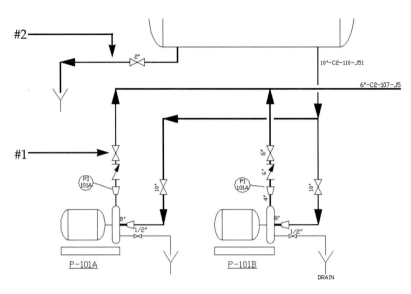

FIG. 5.3.1.20 Pumps View

	Command: *dt*	21. Begin the *Text* command.
	Current text style: "STANDARD" **Text height: 0.1250** **Specify start point of text or** **[Justify/Style]:** *c*	22. We do not have to go to the **Justify** prompt if we know what we want to do. So let's just type *C* for centered text.
	Specify center point of text:	23. AutoCAD asks for the center point of our text. Pick a point two snaps out from the center of the valve—near the tip of arrow #1 in Figure 5.3.1.20. (Toggle *Snap* **On** or **Off** as needed.)
	Specify height <0.1250>: *[enter]* **Specify rotation angle of text <90>:** *[enter]*	24. Accept the defaults for **Height** and **Rotation angle**.
	Enter Text: *6"*	25. Type *6"* for the size of the valve. Notice that the text is *not* centered as you type it.
	Enter Text: *[enter]*	26. Complete the command. The text will now center itself.

continued

TOOLS	COMMAND SEQUENCE	STEPS
		27. Center some text next to the valve just below the last one. This is also a 6" valve (see Figure 5.3.1.29).
	2"-DR-104-D2	**28.** Using the 1/8" text height that should now be the default, place the text indicated above the drain line (indicated by arrow #2 in Figure 5.3.1.20).
		29. Now center the word ***Drain*** below the funnel. This area of the drawing now looks like Figure 5.3.1.29.

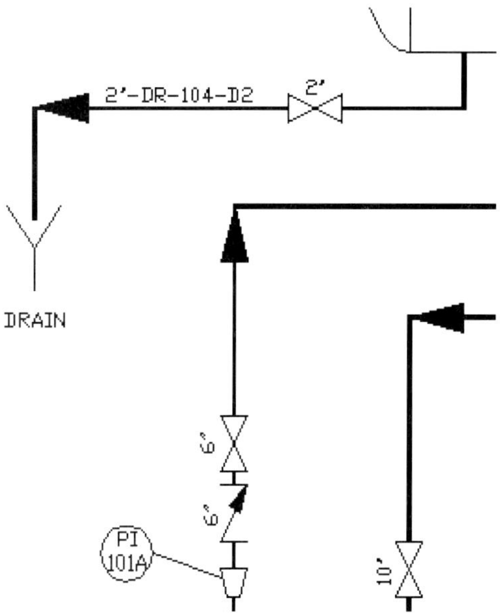

FIG. 5.3.1.29 *Completed Pumps Text*

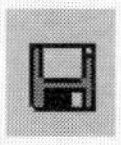	**Command:** *qsave*	**30.** Save the drawing.

One of the ways CAD operators used to save regen time was to tell AutoCAD not to regenerate the text in a drawing. As a great portion of the drawing is text, this saved time with larger drawing files. The command they used was ***Qtext*** and it looked like this:

Command: *qtext*
Enter mode [ON/OFF] <OFF>: *on*

continued

The text in the drawing was replaced with a rectangular locator (to help the operator avoid placing geometry on top of the text).

With the advent of faster computers, the *Qtext* command is not really necessary. But there is always going to be an older operator who likes to *qtext* his drawing before passing it on to a freshman operator as a joke. So if you get a drawing with *Qtext* activated, simply enter the command and turn it **Off**. (Remember to *regen* the drawing afterward.)

5.4 Editing Text—The DDEdit Command

So now you see that creating text is not that difficult. But suppose you make a mistake or just want to change something. Let's look at AutoCAD's text editor— the *DDEdit* command. The command sequence is quite simple:

Command: *ddedit* (**or** *ed*)
Select an annotation object or [Undo]: *[select the text to edit]*

AutoCAD presents the text to edit in the Edit Text dialog box (see Figure 5.4.1.2 on page 149). Make your changes and then simply pick the **OK** button to return to the graphics screen.

FIG 5.4a: Modify II Toolbar

You may notice that the *DDEdit* button is not available on the Modify toolbar where you might expect to find it. AutoCAD has placed this button on the Modify II toolbar (Figure 5.4.a. To access this or additional toolbars, right-click on any existing toolbar. AutoCAD will present a cursor menu listing the toolbars available. Click on the toolbar you wish to use (in this case, the Modify II toolbar). Then dock the toolbar next to the Modify toolbar.

In addition to the command line, you can access the *DDEdit* command by selecting **Object** then **Text** from the **Modify** pull-down menu. Or you can use the preferred method—select the text to edit, right-click, and then select **Text Edit . . .** from the cursor menu.

WWW 5.4.1 Do This: Editing Text

I. Be sure you are still in the *PIDTEXT* drawing in the C:\Steps\Lesson05 folder. If not, open it now.

II. Restore the **TitleB** view shown in Fig. 5.4.1a.

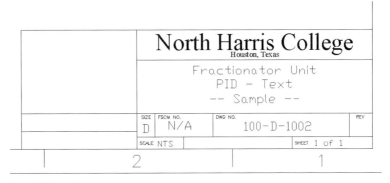

FIG. 5.4.1a: TitleB view

III. Notice that the middle line of the title area reads *PID— Text*. It is inappropriate to abbreviate here, so let's edit this line.

TOOL	COMMAND SEQUENCE	STEPS
 FIG. 5.4.1.2T		1. Select the line of text you will edit (the line that reads *PID-Text*). 2. Right-click in the graphics area and select **Text Edit . . .** from the cursor menu (see Figure 5.4.1.2t). AutoCAD presents the text in the Edit Text dialog box (Figure 5.4.1.2).  FIG. 5.4.1.2: Edit Text Dialog Box

continued

TOOLS	COMMAND SEQUENCE	STEPS
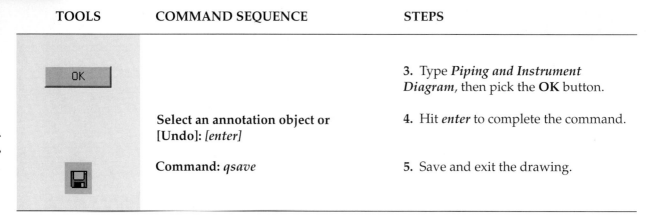		3. Type *Piping and Instrument Diagram*, then pick the **OK** button.
	Select an annotation object or [Undo]: *[enter]*	4. Hit *enter* to complete the command.
	Command: *qsave*	5. Save and exit the drawing.

That is all there is to editing text. Really simple, huh?

5.5 Finding and Replacing Text

One of the things that CAD operators have been demanding is something word processors have used for years — the ability to search for (and replace) text in a document. AutoCAD finally provided this ability with AutoCAD 2000. The command is simply *Find* and it uses an easy-to-follow dialog box (Figure 5.5a).

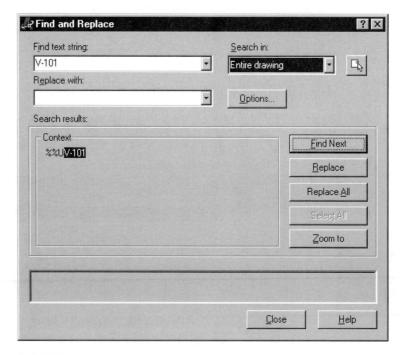

FIG. 5.5a: Text Finder Dialog Box

In our example (Figure 5.5a), we are searching for text *V-101*. We have entered the text in the **Find text string** text box and picked the **Find Next** button to begin our search.

AutoCAD found one instance of the text and displayed it in the **Search Results** list box.

Had we desired to replace the text, we would have placed the replacement text in the **Replace with** text box, done our search (picked the **Find next** button), and then picked either the **Replace** (for a single replacement) or **Replace All** (for a universal replacement).

Note that we can also search part or all of a drawing using the **Search in** drop down box or the **Select objects** button next to the drop-down box.

We can also go to the text's location using the **Zoom in** button once the text has been located.

We will become more familiar with these options in Lesson 14, "Advanced Text: Mtext."

5.6 Adding Flavor to Text with *Style*

Although AutoCAD's default style is called *Standard*, in the last exercise, we edited text that was created using a style called *Times*. I created this style and made it current in the *PIDTEXT* drawing file. I used Windows' Times New Roman True Type Font instead of AutoCAD's TXT font because it shows up better in print.

What exactly is the difference between *style* and *font*? Simply put, *font* refers to the physical shape of a letter or number. *Style* refers to all of the characteristics (including font, size, slant, bold, etc.) of a letter or number.

Since R13, AutoCAD has provided access to the True Type Fonts used by all the other programs on your Windows computer. So when creating a style, your drawing can be consistent with the other documents in your project.

To access the Text Style dialog box (Figure 5.6a), simply type ***Style*** or ***ST*** at the command prompt.

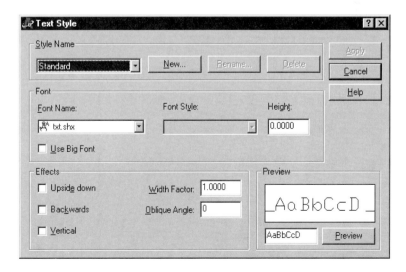

FIG. 5.6a: Text Style Dialog Box

The dialog box looks intimidating, but do not let that throw you. Most of the buttons are self-explanatory— **New** to create a new style, **Rename** to rename a style, and so forth.

> In addition to the command line, you can access the *Style* command by selecting **Text Style** from the **Format** pull-down menu.

Let's create a few styles so we can see how it is done.

5.6.1 Do This: Creating Text Styles

I. Start a new drawing from scratch.

II. Follow these steps.

TOOL	COMMAND SEQUENCE	STEPS
	Command: *ds*	1. Set the *grid* to $1/2$ and the *grid snap* to $1/2$.
	Command: *dt*	2. Accepting the default text size and rotation angle, place the text *Standard* in the upper left corner, as indicated by the arrow in Figure 5.6.1.2. ⟶ Standard Times *Iso30* Iso330 FIG. 5.6.1.2: Text Styles
	Command: *st*	3. Open the Text Style dialog box by typing *style* or *st*.

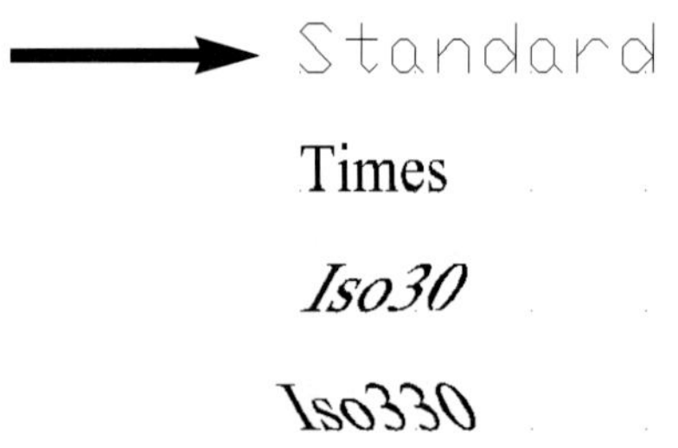

continued

TOOLS	COMMAND SEQUENCE	STEPS

4. Pick the **New** button. The New Text Style dialog box (Figure 5.6.1.4) will appear atop the Text Style dialog box.

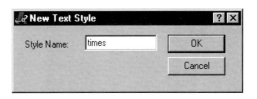

FIG. 5.6.1.4: New Text Style Dialog Box

5. Type in the name: *Times* (as indicated in Figure 5.6.1.4), then pick the *OK* button. The New Text Style dialog box will disappear and the name **Times** appears in the Style Name frame (Figure 5.6.1.5).

FIG. 5.6.1.5: Style Frame

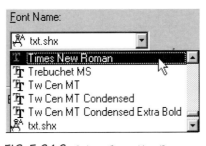

FIG. 5.6.1.6: Select Times New Roman

6. Now let's define the style. Pick the down arrow in the Font Name text box (in the **Font** frame). Scroll as necessary to find **Times New Roman**. Select it. (See Figure 5.6.1.6)

FIG. 5.6.1.7: Font Style Options

7. Notice that the word **Regular** appears in the Font Style text box next to the Font Name text box. Pick the down arrow here to see what your other choices are (Figure 5.6.1.7), but leave it at **Regular** for now.

continued

TOOLS	COMMAND SEQUENCE	STEPS

8. You may set a height for your text in the Height text box. If you do, the text height will be what you have set whenever you use this style and AutoCAD will not prompt you for the height when you enter text. I usually leave the **Height** set to *0* so that I can set the text size when I enter the text.

9. You can set additional physical characteristics for the style in the **Effects** section of the Text Style dialog box (Figure 5.6.1.9). You see the options **Upside down**, **Backwards**, and **Vertical** listed on the left. In all my years using AutoCAD, I have never found a reason for entering text upside down or backward, but the options are available if you find a reason.

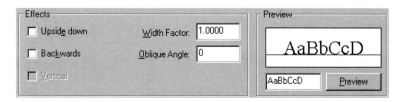

FIG. 5.6.1.9: Text Style Physical Characteristics

FIG. 5.6.1.10: Width Factor

10. The **Width Factor** determines the width of each character in relation to its height (more than one creates a wider character; less than one creates a narrower character). Most people leave this at *1*, but let's set it at *7/8* (Figure 5.6.1.10). I prefer the more narrow characters because it enables me to place more text in a smaller area. The difference is imperceptible when plotted.

11. You can set the slant of the characters in the **Oblique Angle** text box. *0* is straight text. You will use the obliquing angle when you set up isometric text on the next page. We will leave the **Times** style at *0*.

continued

TOOLS	COMMAND SEQUENCE	STEPS
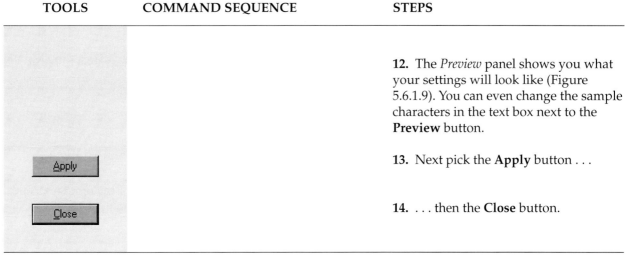		12. The *Preview* panel shows you what your settings will look like (Figure 5.6.1.9). You can even change the sample characters in the text box next to the **Preview** button. 13. Next pick the **Apply** button . . . 14. . . . then the **Close** button.

III. Place the word *Times* just below the word *Standard*. (Zoom in as necessary for a better view.) Can you see the difference?

IV. Now create two more styles with the following settings.

STYLE NAME	FONT	WIDTH	OBLIQUE ANGLE
Iso30	Times New Roman	1	30
Iso330	Times New Roman	1	330

V. Type the names of these styles below the names of the others. (Hint: Type *S* at the first text prompt to set the style.) Your drawing will look like Figure 5.6.1.2 (page 151).

VI. Now let's use the isotext in an isometric setting. Follow these steps.

TOOL	COMMAND SEQUENCE	STEPS
	Command: *sn* **Specify snap spacing or [ON/OFF/ Aspect/Rotate/Style/Type] <0.2500>:** *s* **Enter snap grid style [Standard/Isometric] <S>:** *i* **Specify vertical spacing <0.2500>:** *[enter]*	15. Set your *snap* to iso.

continued

TOOLS	COMMAND SEQUENCE	STEPS
	Command: *dt* **Current text style: "Standard" Text height: 0.2000** **Specify start point of text or [Justify/Style]:** *s* **Enter style name or [?] <Standard>:** *iso30*	16. Set the current text style to *Iso30*.
	Current text style: "iso30" Text height: 0.2000 **Specify start point of text or [Justify/Style]:** **Specify height <0.2000>:** *[enter]* **Specify rotation angle of text <0>:** *30*	17. Pick a point near the other text, and enter 30° as the rotation angle.
	Enter Text: *Iso30* **Enter Text:** *[enter]*	18. Type in the name of the style.
		19. Now repeat the preceding sequence using the *Iso330* style you created and a rotation angle of 330°. Your text looks like Figure 5.6.1.19.

FIG. 5.6.1.19: Isometric Text

5.7 Extra Steps

Perhaps the greatest selling point in AutoCAD's favor is its inclusion of AutoLISP as a customizing agent. Other CAD packages do much of what AutoCAD does, but none is as easily customized.

Certainly, it is too early in our study of AutoCAD to be concerned with customizing it— or at least too early to learn the AutoLISP programming language (although I highly recommend it later). But one thing not included in most textbooks is how to use AutoLISP. You can use it to your advantage quite easily. And as there are zillions of lisp routines available in most CAD environments and the Internet (just ask the guy next to you), you should at least be comfortable with loading the programs.

The command sequence is very simple:

Command: (load "c:/steps/lesson05/views")

That is it! Note that the parentheses are required, as are the quotation marks around the path and file name. Note also that the slashes (normally backslashes) are *front*slashes. (AutoLISP reads backslashes as pauses in routines.)

But in these days of dialog boxes, there is another way. Go to the *Tools* pull-down menu and select *Load Application . . . .* The Load/Unload Applications dialog box will appear (Figure 5.7a). Here is how to use it.

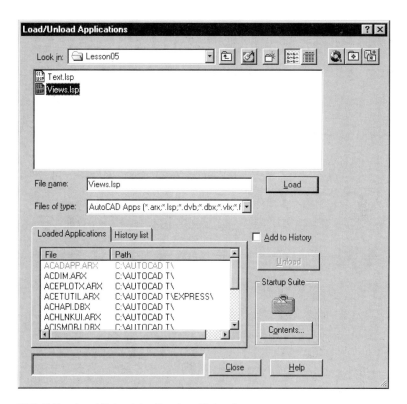

FIG. 5.7a: Load/Unload Applications Dialog Box

1. Use the upper half of the dialog box as you would a typical Windows Open File dialog box. Locate the file you wish to open. In Figure 5.7a, I have located the *Views.lsp* file in the C:\Steps\Lesson05 folder. The name of the selected file will appear in the **File name** text box.
2. Pick the **Load** button. The file appears in the **Loaded Applications** list box in the lower half of the dialog box.
3. Pick the **Close** button to finish the procedure.

You must load each file into the drawing session every time you restart AutoCAD— unless you use the **Startup Suite** to automatically load selected files when AutoCAD begins a new session.

Each file contains one or more *programs* or *routines* intended to shorten or ease your drawing time. Accessing a routine is as easy as typing a command (the command is identified— or programmed— into the routine). Thus, typing **VS** after loading the *Views* routine will enable you to store a view without the dialog box or the many prompts required as part of the normal **View** command. Typing **VR** will restore a view. Try it! You will like it!

You might have noticed another Lisp routine in the C:\Steps\Lesson05 folder — *Text.lsp*. Loading this one will provide two additional commands— **TH** and **CNT**. **TH** enables you to change the height of existing text; **CNT** enables you to continue text below existing text as though you were in the **Text** command.

5.8 What Have We Learned?

Items covered in this lesson include:

- *Display commands*
  - *Zoom*
  - *Pan*
  - *View / DDView*
- *Aerial View*
- *Text Tricks*
- *Text Commands*
  - *Text*
  - *DDEdit*
  - *Qtext*
  - *Style*
- *Loading Lisp Applications*

Well this was quite a lesson! We covered AutoCAD's display commands **Zoom**, **Pan,** and **View**. Then we looked at the basic text command **Text**. After that we covered the **Style** command. We will see that one again when we cover **MText** in Lesson 14. Lastly, we took a quick peek at AutoLISP and how to load and use a Lisp routine. Of all the things covered thus far, mastering these few short paragraphs will go further than any other in proving to an employer that you have mastered AutoCAD.

We covered quite a bit of material; but believe it or not, you will soon be using these tools as second nature (just as you now use a triangle or Ames Lettering Guide).

5.9 EXERCISES

1. Start a new drawing with the following parameters:
   - **1.1.** Grid: 1
   - **1.2.** Snap: 1/2
   - **1.3.** Lower left limits: 0,0
   - **1.4.** Upper right limits: 36,24
   - **1.5.** Text Heights: 3/8", 3/16", 1/4", and 1/8"
   - **1.6.** Create the organizational chart in Figure 5.9.1.

   (Hint: Most of my students spend an hour or so drawing a number of rectangles only to discover that the text will not fit, then must redraw them after entering the text. Enter the text *first*.)
   - **1.7.** Save the drawing as *MyOrg* in the C:\Steps\Lesson05 folder.

2. Create the Isometric Block drawing in Figure 5.9.2, complete with text. Use the *MyIsoGrid1* template file you created in the C:\Steps\Lesson04 folder. (If that file is not available, use the *IsoGrid1* file in the same folder.) Do not do the dimensions. Save the drawing as *MyIsoTxt* in the C:\Steps\Lesson05 folder.

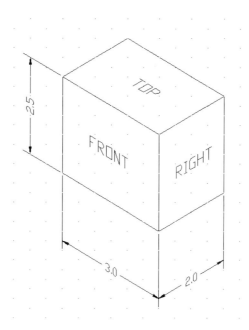

FIG. 5.9.2: Isometric Block

3. Create the Border drawing in Figure 5.9.3b using the following parameters:
   - **3.1.** Grid: start with 1
   - **3.2.** Snap: as needed
   - **3.3.** Lower left limits: 0,0
   - **3.4.** Upper right limits: 36,24
   - **3.5.** Text size: 3/8, 1/2, 1/8
   - **3.6.** Border starts at 1/2, 1/2 and is 1/2" in from the limits on all sides

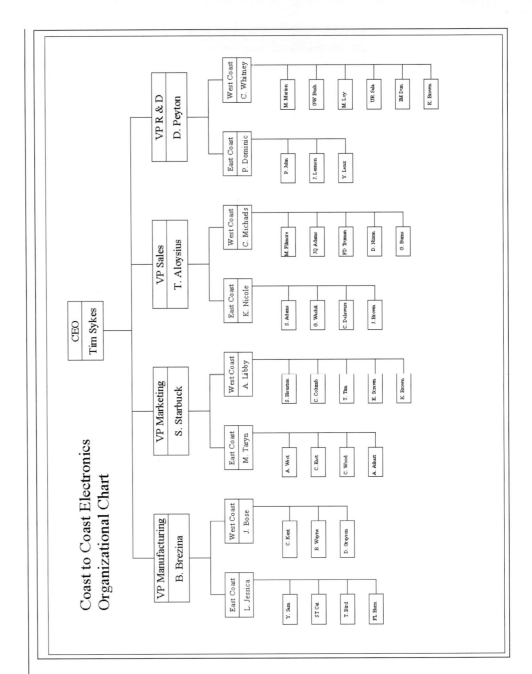

FIG. 5.9.1: Organization Chart

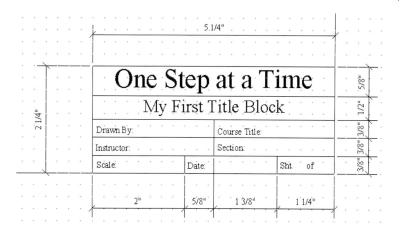

FIG. 5.9.3a: Title Block Details

3.7. Use the Title Block Detail to help (see Figure 5.9.3a)

FIG. 5.9.3b: Border Drawing

4. Using the *MyIsoGrid2* template you created in Lesson 3 (or the *IsoGrid2* template in the Lesson 3 folder), create the drawing in Figure 5.9.4.
   4.1. Text should be 1/4" and use the Times New Roman font
   4.2. Use the grid to guide your dimensions
   4.3. Save the drawing as *MyBlocks.dwg* in the C:\Steps\Lesson05 folder.

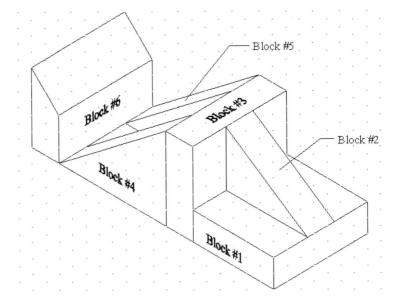

FIG. 5.9.4: Blocks

5. Using what you have learned, create the drawings in Figures 5.9.5b to 5.9.5e. Use a 1:1 scale on an 8 1/2" × 11" sheet of paper for each. I used a 1/2" grid when I drew them. Use the title block (Fig. 5.9.5a) when creating the border.

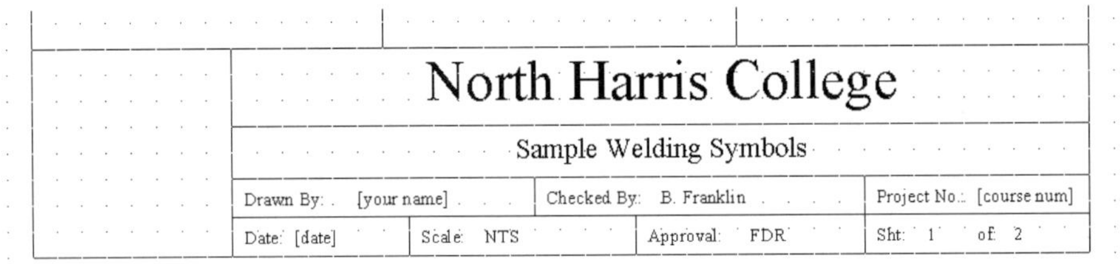

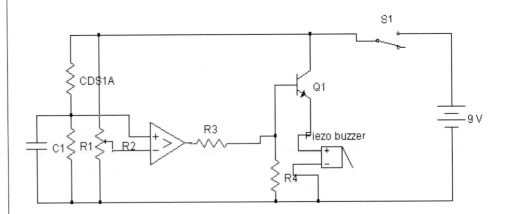

FIG. 5.9.5a: Title Block Detail

FIG. 5.9.5b: Electrical Schematic 4-1

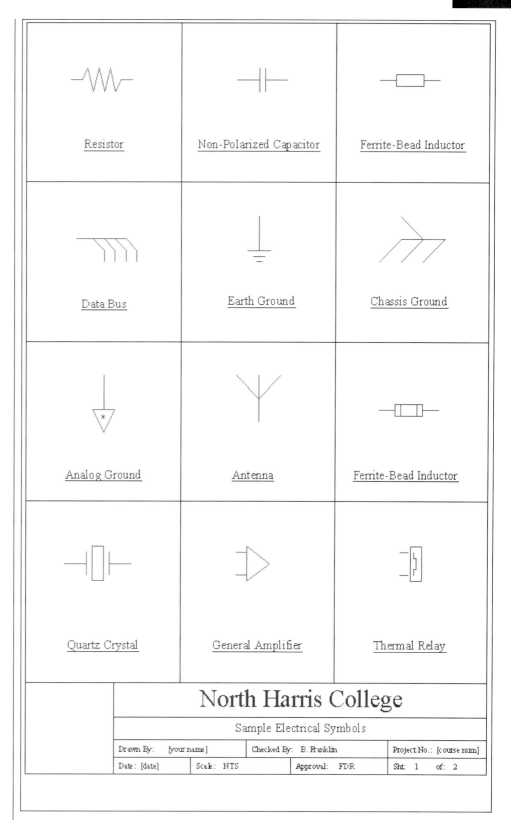

FIG. 5.9.5c: Electrical Symbols

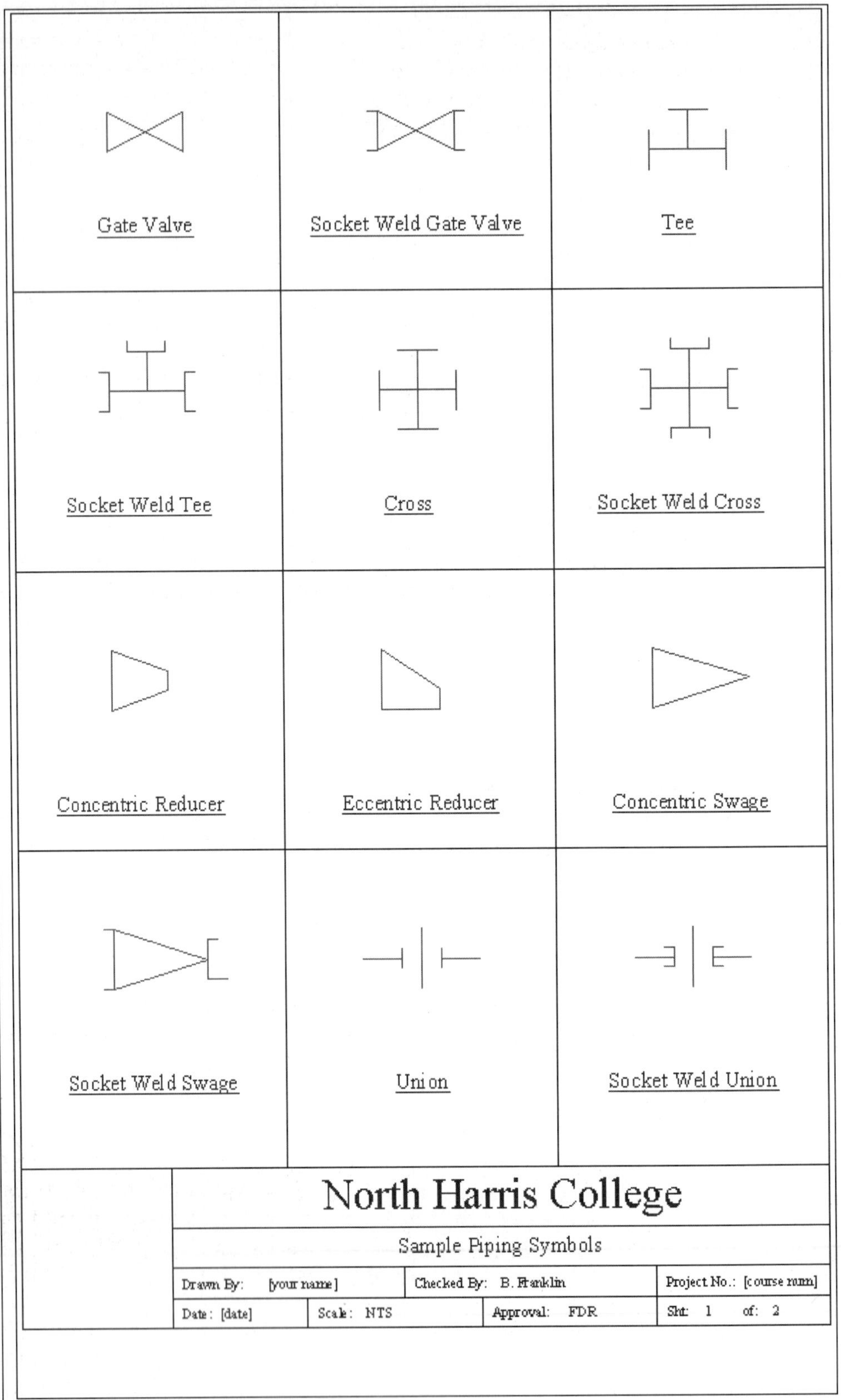

FIG. 5.9.5d: Piping Symbols

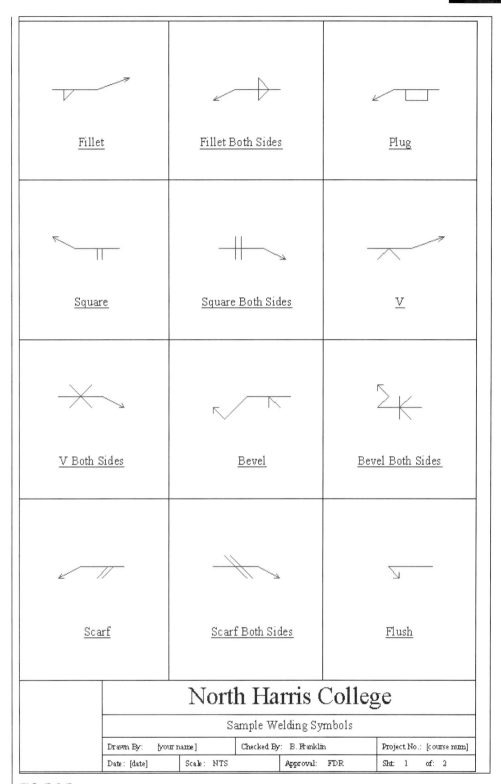

FIG. 5.9.5e: 12 sample Welding Symbols (North Harris College- sht 1 of 2)

5.10 REVIEW QUESTIONS

Please write your answers on a separate sheet of paper.

1. What is the hotkey for the zoom command?

2. What is the default for the zoom command (realtime zoom, implied windowing)?

3. Which zoom option will zoom to the limits of the drawing?

4. Which zoom option enables the user to place a box over the part of the drawing around which he wants to zoom?

5. Which zoom option gets as close as possible to the drawing while showing all the entities in the drawing?

6. Which zoom option will show you the last screen viewed?

7. Zoom in will zoom to a magnification of what scale?

8. Zoom out will zoom to a magnification of what scale?

9. What is AutoCAD's way of saying, "Do it while I watch?"

10. Because realtime zoom will not allow the user to change screen position during the zoom, what zoom option should you use before realtime zooming?

11. What command uses a hand icon and moves the "paper" across your screen?

12. While in the realtime zoom or realtime pan command, what button will you press to get the realtime cursor menu?

13. What is the name of the toy that behaves like a zoom dynamics?

14. and **15.** What are the two commands you may use to change screens without having to use a zoom or pan command?

16. Which of the previous two commands will give the user access to the View Control dialog box?

17. The text button on the draw toolbar calls what command (Text, MText)?

18. (T or F) It is necessary to type *J* at the Start point prompt of the Text command to justify text.

19. (T or F) The difference between the Align and Fit options of the Text command is that Align will adjust the text height proportionally as it fits the text between the selected points.

20. When working on a drawing that will be plotted on a 3/8" = 1'-0" scale, what is the text height you should use for the text to plot at 1/8"?

21. For text to be read from the right side of the page, enter it at _____°.

22. The Text tricks that are used to underline text are based on _____ code.

23. Enter _____ to create a diameter symbol AutoCAD text.

24. Enter _____ at the command prompt to edit text.

25. What is the hotkey for that command?

26. _____ is AutoCAD's default text style.

27. (T of F) AutoCAD cannot use True Type fonts available for other Windows applications.

28. To create a text style that uses a 30° slant on the font, set the oblique angle to _____.

29. The command that replaces text on the screen with rectangles so that the text does not slow regeneration time is _____.

30. _____ is the most common programming language AutoCAD makes available for customization.

31. Write the command sequence to load an AutoLISP routine called *text* located in the C:\Steps\Lesson05 folder.

32. To open the Load/Unload Application dialog box, select Load Application under the _____ pull-down menu.

Identify these buttons:

34.   **35.**   **36.**   **37.**   **38.**

LESSON

6

Geometric Shapes (Other Than Lines and Rectangles!)

Following this lesson, you will:

➥ Know how to draw:
- Circles
- Ellipses
- Arcs
- Polygons

➥ Have mastered most of AutoCAD's basic 2-dimensional drawing commands!

By now, you must have tired of drawing lines and rectangles. After all, back at the beginning of the second lesson, we discussed lines *and circles* as the foundation for drawing geometric shapes. How about those circles?

In this lesson, we will look at drawing not only circles, but arcs and ellipses as well. We will also expand our multisided geometry from simple rectangles to include those *gons* collectively known as *polygons*.

Let's proceed.

6.1 Getting Around to Circles

Next to lines, circles (and parts of circles) are the most frequent factor in geometric drawings. You might think creating such important objects should be complicated, but AutoCAD has made *Circle* one of the easiest of its commands. We draw lines, as you know, using the *Line* command. In keeping with the simple approach, draw circles by using the *Circle* command. Is that not simple enough? Okay, just type *C* at the command prompt!

Here is the command sequence.

Command: *circle* (**or** *c*)
CIRCLE Specify center point for circle or [3P/2P/Ttr (tan tan radius)]: *[pick or identify a point on the screen]*
Specify radius of circle or [Diameter]: *[drag or type the radius]*
Command:

That seems fairly easy. Open the *cir-ell.dwg* file from the C:\Steps\Lesson06 folder and give it a try. Draw a circle in one of the open areas of the screen.
What do you think?
Okay. Let's look at some of the *Circle* command's options.

- The default is the **Specify center point** option. Any point selected on the screen or identified by X and Y coordinates will be the center point of the circle.
- The **3P** option allows the user to draw a circle by selecting three points on the circle.
- The **2P** option allows the user to draw a circle by selecting both ends of an imaginary diameter line.
- **TTR** stands for **Tangent-Tangent-Radius** and allows the user to draw the circle by selecting two objects to which the circle will be tangent and then entering the required radius.

Let's try each option.

All of the options are also available in the **Circle** selection under the **Draw** pull-down menu.

WWW 6.1.1 Do This: Circle Practice

I. If you have not yet opened *cir-ell.dwg*, open it now. It is in the C:\Steps\Lesson06 folder and looks like Figure 6.1.1a.

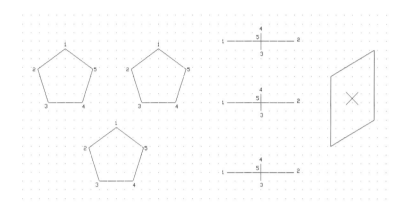

FIG. 6.1.1a Cir-Ell.dwg

II. Restore the **Circles** view.

III. Follow these steps.

TOOLS	COMMAND SEQUENCE	STEPS
Running OSNAP button	**Command:** *os*	1. Set your running OSNAP to **endpoint**. Clear all other settings.
Circle Button	**Command:** *c*	2. Type the *Circle* command or the *c* hotkey at the command prompt. Or you may pick the **Circle** button on the Draw toolbar.
	Specify center point for circle or [3P/2P/Ttr (tan tan radius)]: *3p*	3. Type *3p* for the **three-point** option.

continued

TOOLS	COMMAND SEQUENCE	STEPS
	Specify first point on circle: **Specify second point on circle:** **Specify third point on circle:**	4. Select any three endpoints on the upper left polygon. The results should look like Figure 6.1.1.4.

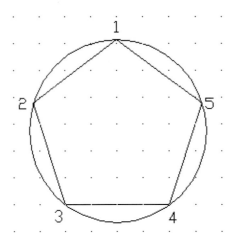

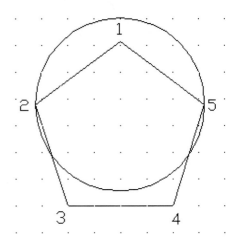

FIG. 6.1.1.4: Three-Point Circle

FIG. 6.1.1.7: Two-Point Circle

Command: *[enter]*

5. Repeat the **Circle** command.

Specify center point for circle or [3P/2P/Ttr (tan tan radius)]: *2p*

6. Type **2p** for the **two-point** option.

Specify first end point of circle's diameter: *[select point 2]*
Specify second end point of circle's diameter: *[select point 5]*

7. Select points **2** and **5** on the upper right polygon. The results will look like Figure 6.1.1.7.

Command: *[enter]*

8. Repeat the *Circle* command.

Specify center point for circle or [3P/2P/Ttr (tan tan radius)]: *ttr*

9. Type *TTR* for the **Tangent-Tangent-Radius** option.

Specify point on object for first tangent of circle:

10. Select any point on line **1–2** of the remaining polygon. Notice how the OSNAP symbol for **tangent** appears when you cross the line.

Specify point on object for second tangent of circle:

11. Now select any point on line **2–3**.

continued

TOOLS	COMMAND SEQUENCE	STEPS

Specify radius of circle <1.6180>: *1*

12. Enter a radius of *1*. Your drawing will look like Figure 6.1.1.12. Notice that AutoCAD draws a 1"R circle tangent to the lines.

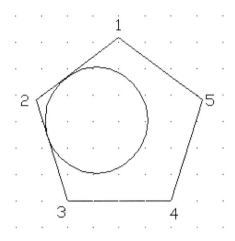

FIG. 6.1.1.12: TTR Circle with a 1" radius

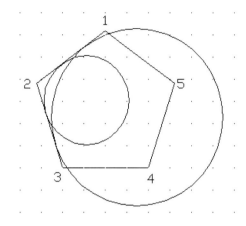

FIG. 6.1.1.17: TTR Circles with 1" and 2" Radii

Command: *[enter]*

13. Let's try a larger circle. Repeat the *Circle* command.

CIRCLE Specify center point for circle or [3P/2P/Ttr (tan tan radius)]: *ttr*

14. Type *TTR* for the **Tangent-Tangent-Radius** option.

Specify point on object for first tangent of circle:

15. Select any point on line **1–2** of the same polygon that you used in Step 10.

Specify point on object for second tangent of circle:

16. Now select any point on line **2–3**.

Specify radius of circle <1.0000>: *2*

17. Enter a radius of *2*. Your drawing will look like Figure 6.1.1.17. Notice that AutoCAD draws a 2"R circle. Both circles are tangent to the same lines, but appear in different places to accommodate the difference in radii.

Command: *qsave*

18. Save your drawing.

Save Button

I know. Had you known it was this easy you would have learned AutoCAD long ago!

6.2 "Squished" Circles and Isometric Circles: *Ellipse* Command

I have often been amazed—and frequently aggravated—by the number of incomplete or oddball circles required in drafting. Back in my board days, arcs were seldom a problem. I just used my circle template and drew as much as I needed. Ellipses, however, required the purchase of specific templates—often several! There were, of course, templates that tried to provide almost every dimensional ellipse the draftsman might need—from 25° to 80°, and from $1/4$" to 6". But the ellipse I needed would inevitably fall into an oddball degree or size.

Ellipses are one of those things that AutoCAD makes quite a bit easier than plastic templates or clumsy compass attempts to create oddball shapes. Let's look at the command sequence. Refer to Figure 6.2a.

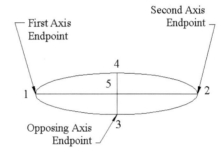

FIG. 6.2a: Sample Ellipse

Command: *ellipse* (or *el*)

Specify axis endpoint of ellipse or [Arc/Center]: *[select the first axis endpoint]*

Specify other endpoint of axis: *[select the opposite axis endpoint]*

Specify distance to other axis or [Rotation]:Arc/Center/<Axis endpoint 1>: *[select the opposing axis endpoint]*

Command:

The basic ellipse is really very easy to draw regardless of rotation or dimension. But as you can see, we have some options to consider.

- **Specify axis endpoint** is the default. It allows the user to draw the ellipse by selecting three axis endpoints, or two axis endpoints and a rotation angle.
- The **Arc** option allows the user to draw partial ellipses.
- The **Center** option allows the user to create an ellipse using a center point and two axis endpoints (rather than three axis endpoints).

All of the options are also available in the **Ellipse** selection under the **Draw** pulldown menu or on the cursor menu once the *Ellipse* command has been entered.

Let's look at each of these.

WWW 6.2.1 Do This: Drawing Ellipses

I. Be sure you are still in the *cir-ell* drawing. If not, open it now. It is in the C:\Steps\Lesson06 folder.

II. Restore the *Ellipses* view.

III. Be sure the running OSNAP is still set to **endpoint**.

IV. Follow these steps.

TOOLS	COMMAND SEQUENCE	STEPS
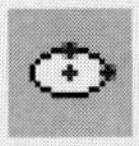 Ellipse Button	Command: *el*	1. Type the *Ellipse* command or *el* hotkey at the command prompt. Or you may pick the **Ellipse** button on the Draw toolbar.
	Specify axis endpoint of ellipse or [Arc/Center]:	2. Select the endpoint at point 1 of the top set of lines. (The lines are not necessary. We use them as guides only.)
	Specify other endpoint of axis:	3. Select the endpoint at point 2.
	Specify distance to other axis or [Rotation]:	4. Select the endpoint at either point 3 or point 4. Your ellipse looks like Figure 6.2.1.4.

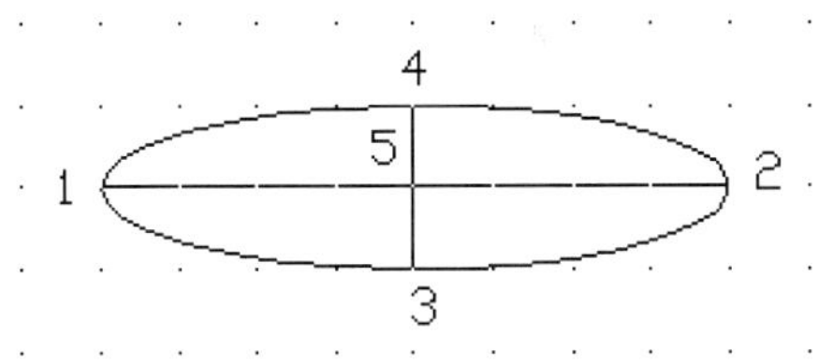

FIG. 6.2.1.4: Completed Ellipse

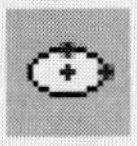 Ellipse Button	Command: *[enter]*	5. Repeat the *Ellipse* command.
	Specify axis endpoint of ellipse or [Arc/Center]: *c*	6. Type *c* to access the **Center** option.

continued

TOOLS	COMMAND SEQUENCE	STEPS
 Intersection Button	Specify center of ellipse: int	7. When AutoCAD prompts you for the center of the ellipse, select the OSNAP button for **intersection**, then select point 5 of the second set of lines.
	Specify endpoint of axis:	8. Select the endpoint at either point 1 or point 2.
	Specify distance to other axis or [Rotation]: *r*	9. Let's try the **Rotation** option. Type *R*. (**Rotation** refers to the angle at which the viewer sees the circle.)
	Specify rotation around major axis: 75	10. Type in a rotation angle of 75°. Your ellipse will look like the one in Step 4 (Figure 6.2.1.4).
	Command: *[enter]*	11. Repeat the *Ellipse* command.
	Specify axis endpoint of ellipse or [Arc/Center]: *a*	12. Type *a* to access the **Arc** option.
	Specify axis endpoint of elliptical arc or [Center]:	13. Notice that you again have the option to select either an axis endpoint or the center of the ellipse. Let's use the default—**axis endpoint**—and select the endpoint at point 1 on the bottom set of lines.
	Specify other endpoint of axis:	14. Select the endpoint at point 2.
	Specify distance to other axis or [Rotation]:	15. Here again your options are repeated. You may select an axis endpoint of the opposing lines or enter *R* for a rotation angle. Select point 3.
	Specify start angle or [Parameter]: *0*	16. Now you have some new options. The **parameter** option will yield the same result as the **start angle** option, but AutoCAD will use a different method of calculation. We will use the default—**start angle**. Type *0*. (*0* refers to the first point you selected on the ellipse *not* the WCS's 0°.)

continued

TOOLS	COMMAND SEQUENCE	STEPS
	Specify end angle or [Parameter/Included angle]: Specify end angle or [Parameter/Included angle]: *180*	**17.** We started our arc at angle 0. An included angle is an angle measured (counterclockwise) from that point. We can type *I* followed by the angle we wish to use to create our arc. Or we can use the **end angle** approach and simply type the angle we want for the other end of the ellipse. Let's use the **end angle** default and type in *180°*. Your ellipse looks like Figure 6.2.1.17. **FIG. 6.2.1.17:** Elliptical Arc
Save Button	**Command:** *qsave*	**18.** Save the drawing, but do not exit.

You may now see that drawing ellipses is not difficult—although mastering the various approaches may take some time.

The ellipses we have drawn thus far have all existed in a true 2-dimensional plane. But the user may also use ellipses to draw *isometric circles*. The procedure is simple but requires that you be in *isometric mode*. Otherwise, the *Ellipse* command will not provide the **Isocircle** option.

Let's look at this.

> By default, AutoCAD draws true ellipses. This enables you to find the center with an OSNAP.
>
> The system variable **PELLIPSE** allows the user to change the way AutoCAD draws ellipses. When set to the default *0*, ellipses work the way we have seen. But set it to *1* and ellipses are drawn as polylines (just as rectangles are drawn). Both ellipses look the same, but the latter can be given width using the *PEdit* command, and the user cannot easily find the center of the ellipse.
>
> We will learn about **Polylines** and the *Pedit* command in Lesson 9.

WWW 6.2.2 Do This: Drawing Isometric Circles

I. Be sure you are still in the *cir-ell.dwg* drawing file in the C:\Steps\Lesson06 folder. If not, open it now.

II. Restore the *Iso-Ellipse* view.

III. Follow these steps.

TOOLS	COMMAND SEQUENCE	STEPS
	Command: *sn*	1. First, set your drawing to isometric mode. Your crosshairs will change (Figure 6.2.2.1.), as will the grid.
		FIG. 6.2.2.1: Isometric Crosshairs
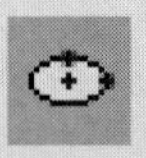 Ellipse Button	**Command:** *el*	2. Repeat the *Ellipse* command.
	Specify axis endpoint of ellipse or [Arc/Center/Isocircle]: *i*	3. Type *i* to select the **Isocircle** option.
 Node Button	**Specify center of isocircle: _nod of**	4. Pick the **node** button from the OSNAP flyout toolbar; then select the node in the center of the isometric rectangle.
 F5 Key		5. See how the ellipse drags with the cursor in the current isometric plane? Toggle the plane using the *F5* key to see how the ellipse changes. Stop toggling when the crosshairs return to the 90°/30° position as shown in Figure 6.2.2.1 (Step 1).

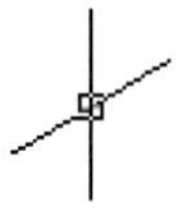

continued

TOOLS	COMMAND SEQUENCE	STEPS
	Specify radius of isocircle or [Diameter]: *1*	**6.** Notice that you may specify a radius or diameter. Let's use the default (radius) and type *1*. Your drawing looks like Figure 6.2.2.6.

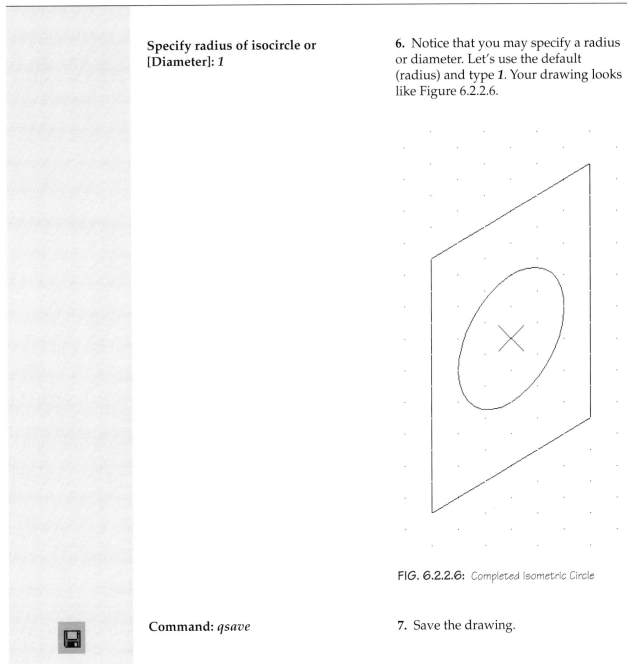

FIG. 6.2.2.6: Completed Isometric Circle

| | **Command:** *qsave* | **7.** Save the drawing. |

We have looked at circles and "squished" circles, partial ellipses (ellipse arcs), and isometric circles. Let's move on then to partial circles—***arcs***.

6.3 Arcs: The Hard Way!

How is that for a section title to make you want to skip past this part of the lesson?! But wait! "There's gold in them thar pages!"

Although it will frequently be quicker and easier to create a circle and trim away the part you do not want, there are times when there is simply no substitute for the *Arc* command. (Besides, we have not learned the *Trim* command yet!)

Drawing an arc is not difficult—providing the user knows which of the eleven available procedures to use.

The default command sequence is

Command: *arc* (or *a*)
Specify start point of arc or [CEnter]: *[select or identify the starting point]*
Specify second point of arc or [CEnter/ENd]: *[select or identify a point on the arc]*
Specify end point of arc: *[select or identify the endpoint]*

> You can find the various options of the *Arc* command under the Draw pull-down menu.
>
> However, you may have noticed by now that I do not always provide details about procedures using the pull-down menus. There is a reason for this.
>
> Most commands available on a pull-down menu are also available on a toolbar (draw pull-down = draw toolbar, etc.). However, one of the easiest (and certainly the first) parts of AutoCAD to be customized is the pull-down menu system. In the years I have worked AutoCAD in various capacities, I have never seen two companies use the same pull-down menu system. Additionally, I have never seen a company use AutoCAD's default pull-down menus.
>
> For this reason, I shy away from the pull-down menus and discourage my students from becoming too reliant on them (or any one method of doing things).
>
> You can also find the various options in the cursor menu once the *Arc* command has been entered.

The best way to see each of the options is through an exercise. So fire up the computer and full speed ahead!

WWW 6.3.1 Do This: Arcs, Arcs, Arcs

I. Open the *arcs* drawing in the C:\Steps\Lesson06 folder.

II. Restore the *top* view.

III. Set the Running OSNAP to **Node** and clear any other settings.

IV. Follow these steps.

TOOLS	COMMAND SEQUENCE	STEPS
Arc Button	**Command:** *a*	**1.** Enter the *Arc* command by typing *arc* or *a* at the command prompt. Or you can select the **Arc** button on the Draw toolbar.
	Specify start point of arc or [CEnter]:	**2.** Select node **a** in square #1.
	Specify second point of arc or [CEnter/ENd]:	**3.** The **second point** is any point on the arc between the **start point** and **end point**. Select node **b**.
	Specify end point of arc:	**4.** Select node **c**. Your arc should look like Figure 6.3.1.4.

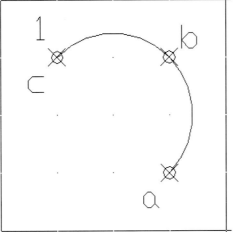

FIG. 6.3.1.4: Three-Point Arc

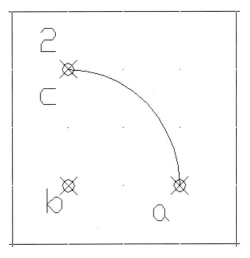

FIG. 6.3.1.9: Start—Center—End

	Command: *[enter]*	**5.** Repeat the *Arc* command.
	Specify start point of arc or [CEnter]:	**6.** Select node **a** in square #2.
	Specify second point of arc or [CEnter/ENd]: *c*	**7.** Type *c* (or *ce*) to select the **CEnter** option.
	Specify center point of arc:	**8.** Select node **b**.
	Specify end point of arc or [Angle/chord Length]:	**9.** What!? More options?! We will discuss these next. Let's just go with the default **end point** and select node **c**. Your arc should look like Figure 6.3.1.9.

continued

TOOLS	COMMAND SEQUENCE	STEPS
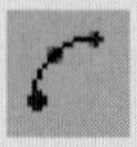	**Command:** *[enter]*	10. Now let's look at the **Angle** option of the **Start Center** method. Repeat the *Arc* command.
	Specify start point of arc or [CEnter]:	11. Select node **a** in square #3.
	Specify second point of arc or [CEnter/ENd]: *c*	12. Type *c* to select the **CEnter** option.
	Specify center point of arc:	13. Select node **b**.
	Specify end point of arc or [Angle/chord Length]: *a*	14. Type *a* for the **Angle** option.
	Specify included angle: *45*	15. Type *45* to draw a 45° arc. Your arc should look like Figure 6.3.1.15.

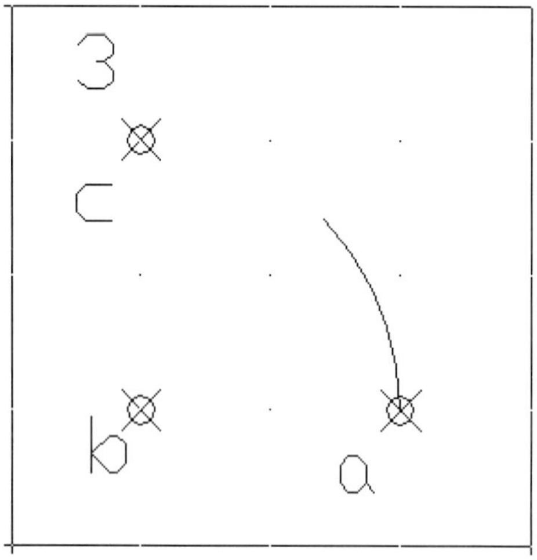

FIG. 6.3.1.15: Start—Center—Angle

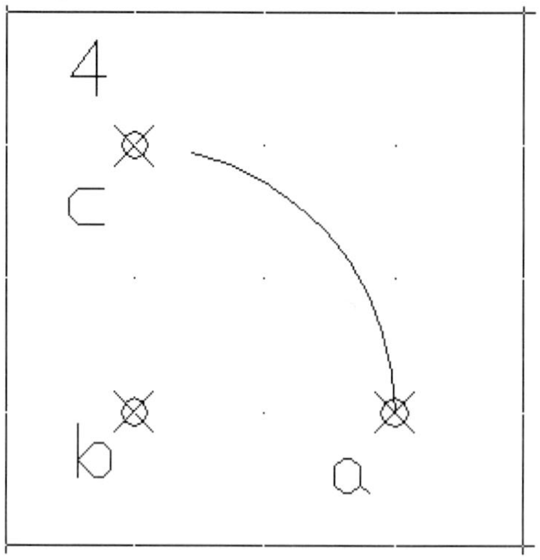

FIG. 6.3.1.21: Start—Center—Length

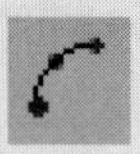	**Command:** *[enter]*	16. Okay. Now let's look at the **chord Length** option of the **Start—Center** method. Repeat the *Arc* command.
	Specify start point of arc or [CEnter]:	17. Select node **a** in square #4.
	Specify second point of arc or [CEnter/ENd]: *c*	18. Type *c* to select the **CEnter** option.
	Specify center point of arc:	19. Select node **b**.

continued

TOOLS	COMMAND SEQUENCE	STEPS
	Specify end point of arc or [Angle/chord Length]: *l*	**20.** Type *l* for the **chord Length** option.
	Specify length of chord: *1.25*	**21.** Type in the desired true length of the arc—*1.25*. Your arc will look like Figure 6.3.1.21.
	Command: *[enter]*	**22.** Let's look at the **Start—End** method. Repeat the *Arc* command.
	Specify start point of arc or [CEnter]:	**23.** Select node **a** in square #5.
	Specify second point of arc or [CEnter/ENd]: *e*	**24.** Type *e* (or *en*) to select the **ENd** option.
	Specify end point of arc:	**25.** Select node **b**.
	Specify center point of arc or [Angle/Direction/Radius]: *a*	**26.** You see that **center point** is the default option, but let's use the **Angle** option. Type *a*.
	Specify included angle: *45*	**27.** AutoCAD asks for the angle. Type *45* to draw an arc to fill a 45° angle. Your arc will look like Figure 6.3.1.27. Notice that AutoCAD has drawn a 1/8-circle arc. Remember 45° is 1/8 of a circle!

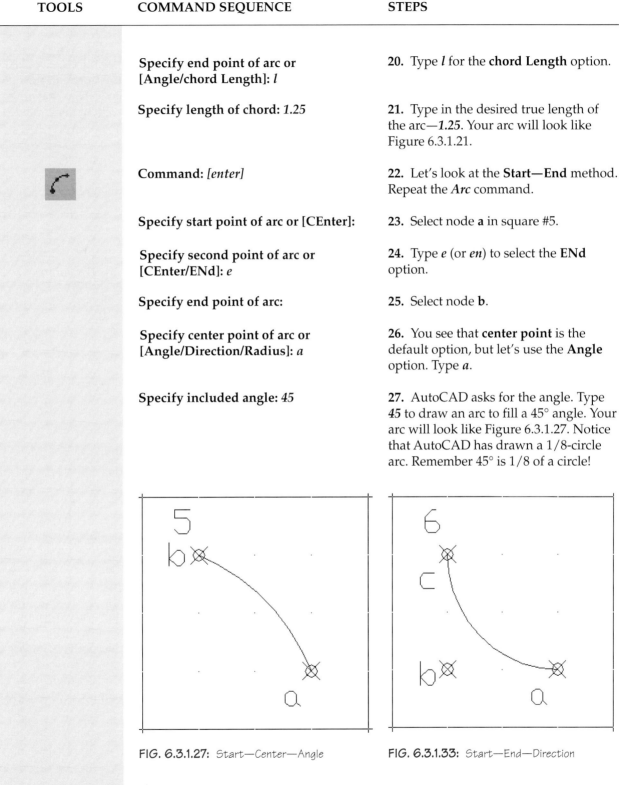

FIG. 6.3.1.27: Start—Center—Angle

FIG. 6.3.1.33: Start—End—Direction

continued

TOOLS	COMMAND SEQUENCE	STEPS
	Command: *[enter]*	**28.** As you may have noticed, AutoCAD draws arcs *counterclockwise*. This next method will tell AutoCAD to draw the arc in a different direction. Repeat the *Arc* command.
	Specify start point of arc or [CEnter]:	**29.** Select node **a** in square #6.
	Specify second point of arc or [CEnter/ENd]: *e*	**30.** Type *e* to select the **ENd** option.
	Specify end point of arc:	**31.** Select point **c**.
	Specify center point of arc or [Angle/Direction/ Radius]: *d*	**32.** Type *d* to select the **Direction** option.
	Specify tangent direction for the start point of arc:	**33.** Select point **b**. Your arc will look like Figure 6.3.1.33.
		34. Save your drawing, then restore the *bottom* view.
Named Views		
	Command: *a*	**35.** Repeat the *Arc* command.
	Specify start point of arc or [CEnter]:	**36.** Select node **a** in square #7.
	Specify second point of arc or [CEnter/ENd]: *e*	**37.** Type *e* to select the **ENd** option.
	Specify end point of arc:	**38.** Select point **b**.
	Specify center point of arc or [Angle/Direction/ Radius]: *r*	**39.** Type *r* to select the **Radius** option.
	Specify radius of arc: *1*	**40.** Assign a radius of 1—type *1*. Your arc should look like Figure 6.3.1.40.

continued

TOOLS	COMMAND SEQUENCE	STEPS

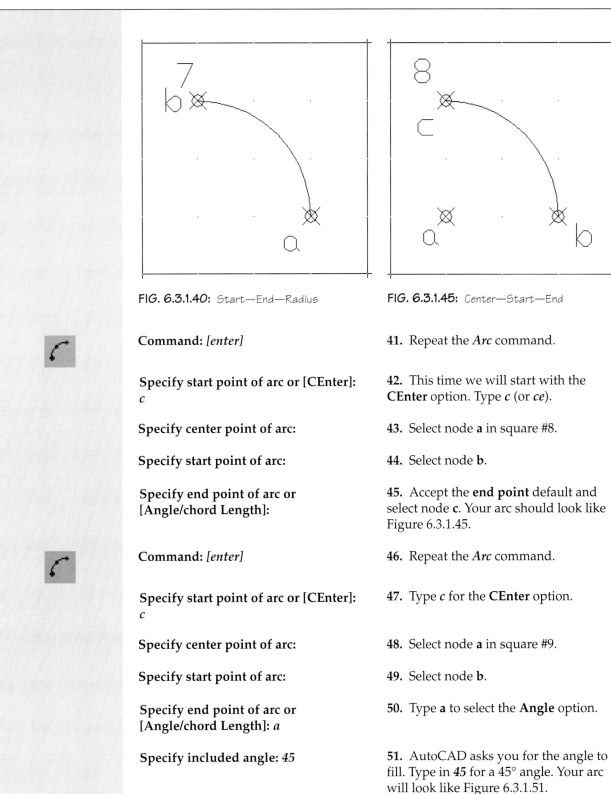

FIG. 6.3.1.40: Start—End—Radius

FIG. 6.3.1.45: Center—Start—End

	Command: *[enter]*	41. Repeat the *Arc* command.
	Specify start point of arc or [CEnter]: *c*	42. This time we will start with the **CEnter** option. Type *c* (or *ce*).
	Specify center point of arc:	43. Select node **a** in square #8.
	Specify start point of arc:	44. Select node **b**.
	Specify end point of arc or [Angle/chord Length]:	45. Accept the **end point** default and select node **c**. Your arc should look like Figure 6.3.1.45.
	Command: *[enter]*	46. Repeat the *Arc* command.
	Specify start point of arc or [CEnter]: *c*	47. Type *c* for the **CEnter** option.
	Specify center point of arc:	48. Select node **a** in square #9.
	Specify start point of arc:	49. Select node **b**.
	Specify end point of arc or [Angle/chord Length]: *a*	50. Type **a** to select the **Angle** option.
	Specify included angle: 45	51. AutoCAD asks you for the angle to fill. Type in **45** for a 45° angle. Your arc will look like Figure 6.3.1.51.

continued

TOOLS	COMMAND SEQUENCE	STEPS
	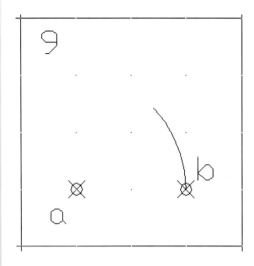 FIG. 6.3.1.51: Center—Start—Angle	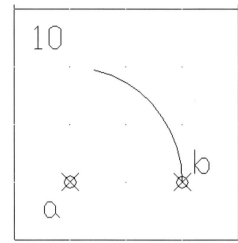 FIG. 6.3.1.57: Center—Start—Length

Command: *[enter]*

52. Repeat the *Arc* command.

Specify start point of arc or [CEnter]: *c*

53. Type *c* for the **CEnter** option.

Specify center point of arc:

54. Select node **a** in square #10.

Specify start point of arc:

55. Select node **b**.

Specify end point of arc or [Angle/chord Length]: *l*

56. Type *l* to select the **chord Length** option.

Specify length of chord: *1.25*

57. Tell AutoCAD the desired true length of the arc. Type *1.25*. Your arc should look Figure 6.3.1.57.

58. In square #11, draw the arc (shown in Figure 6.3.1.58) using the *Start—Center—End* method detailed in Steps 5 to 9.

Command: *[enter]*

59. Repeat the *Arc* command by hitting *enter*.

continued

TOOLS	COMMAND SEQUENCE	STEPS

Specify start point of arc or [CEnter]: *[enter]*
Specify end point of arc:

60. Now hit *enter* again. This is a *continue* procedure for the *Arc* command. Notice how AutoCAD assigns the first point as the endpoint of the last arc. AutoCAD also reverses the direction and prompts you for the endpoint. Your drawing should look like Figure 6.3.1.60.

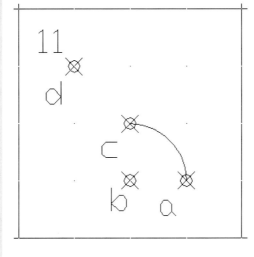

FIG. 6.3.1.58: Start—Center—End

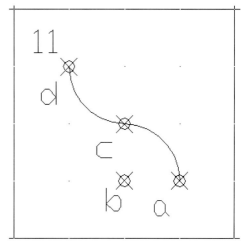

FIG. 6.3.1.60: Continued Arc

Command: *qsave*

61. Save the drawing.

Wow! That was a chore! But as you can see, drawing arcs is not difficult if you just know which method to use.

6.4 Drawing Multisided Figures: The *Polygon* Command

The polygon has long been one of the foundations on which our world is designed. Everything from the simple triangles of the pyramids to the hex head bolts that hold our automobiles together relies on the mathematics associated with multisided objects. Fortunately, I can leave the mathematics to those better qualified to confuse. All I must do is explain the three simple methods for drawing polygons.

Consider the chart in Figure 6.4a. It is really as simple to draw polygons as this chart suggests. Let's look at the command sequence:

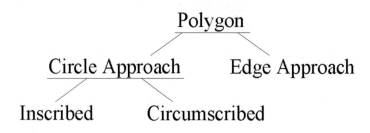

FIG. 6.4a: Polygon approaches

Command: *polygon* (or *pol*)

Enter number of sides <4>: *[enter the desired number of equal sides from 3 to 1024]*

Specify center of polygon or [Edge]: *[either identify the location of the center of the polygon (the Circle Approach), or type E to use the Edge Approach]*

Enter an option [Inscribed in circle/Circumscribed about circle] <I>: *[let AutoCAD know if you will draw your polygon* **inside** *(inscribed) or* **outside** *(circumscribed) an imaginary circle)*

Specify radius of circle: *[tell AutoCAD the radius of the imaginary circle in which or around which you want to draw the polygon]*

Command:

You see the first branch of the chart at the second command prompt. Find the second branch at the third prompt.

You can also access the ***Polygon*** command under the Draw pull-down menu.

 ### 6.4.1 Do This: Drawing Polygons

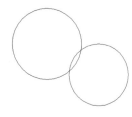

I. Open the *polygons* drawing in the C:\Steps\Lesson06 folder. It looks like Figure 6.4.1a.

II. Follow these steps.

FIG. 6.4.1a
The Polygons drawing

TOOLS	COMMAND SEQUENCE	STEPS
 Polygon Button	**Command:** *pol*	1. Enter the *Polygon* command by typing *polygon* or *pol* at the command prompt. You may also select the **Polygon** button on the Draw toolbar.
	POLYGON Enter number of sides <4>: *6*	2. AutoCAD asks for the number of sides needed to create the polygon. Type *6*.
 Snap to Center	**Specify center of polygon or [Edge]:** _cen of	3. Now AutoCAD needs to know how to draw the polygon. We will accept the **center of polygon** option. Use OSNAPS to select the center of the larger upper circle.
	Enter an option [Inscribed in circle/Circumscribed about circle] <I>: *[enter]*	4. We will draw our polygon *inside* the circle (**Inscribed**). (Note: The circle is not necessary for drawing the polygon. We are just using it for demonstration purposes.) Hit *enter*.
	Specify radius of circle: *1.5*	5. Type *1.5* as the radius of the circle. Your drawing should look like Figure 6.4.1.5.

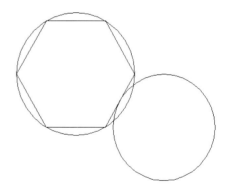

FIG. 6.4.1.5: Inscribed Six-Sided Polygon

	Command: *[enter]*	6. Repeat the *Polygon* command.

continued

TOOLS	COMMAND SEQUENCE	STEPS
	Enter number of sides <6>: *[enter]*	**7.** This time AutoCAD defaults to six sides as that was the last number used. Hit *enter*.
	**Specify center of polygon or [Edge]:** _cen of	**8.** Use OSNAPS to select the center of the smaller circle.
	Enter an option [Inscribed in circle/Circumscribed about circle] <I>: *c*	**9.** Type *c* to draw our polygon *Circumscribed* (*outside* the circle).
 Snap to Quadrant	**Specify radius of circle:** _qua of	**10.** When prompted for the radius, use the OSNAP to select the bottom quadrant of the circle. Your drawing looks like Figure 6.4.1.10.

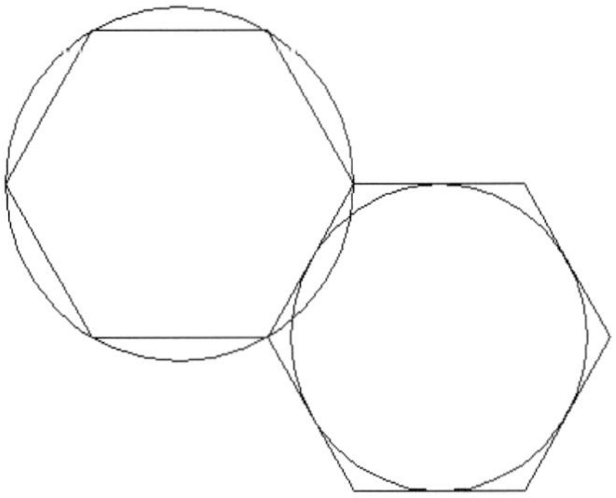

FIG. 6.4.1.10: Inscribed and Circumscribed Six-Sided Polygons

	**Command:** *[enter]*	**11.** Now let's look at the **Edge Approach**. Repeat the *Polygon* command.
	Enter number of sides <6>: *[enter]*	**12.** Hit *enter* to accept six as the number of sides.
	Specify center of polygon or [Edge]: *e*	**13.** Type *e* to select the **Edge** option.
	Specify first endpoint of edge: _endp of	**14.** Select Point **1** as indicated in Figure 6.4.1.14. (Use the endpoint OSNAP.)

continued

TOOLS	COMMAND SEQUENCE	STEPS
		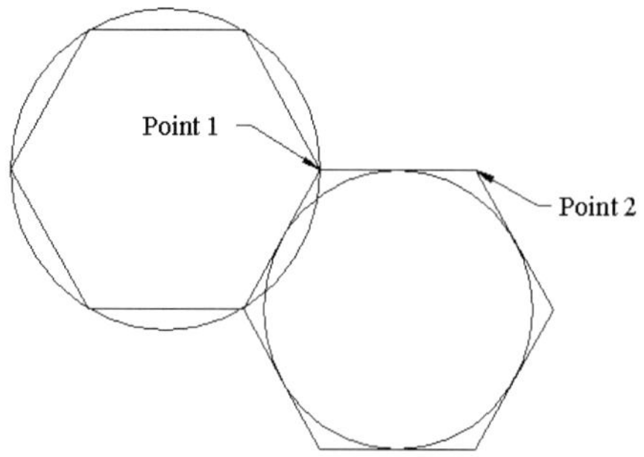 **FIG. 6.4.1.14:** Endpoint Location
	Specify second endpoint of edge: _endp of	**15.** Select point **2**. Your drawing looks like Figure 6.4.1.15.
		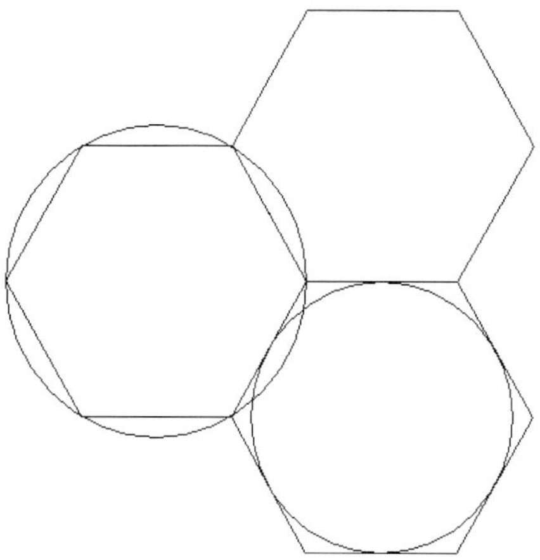 **FIG. 6.4.1.15:** Completed Exercise
	Command: *qsave*	**16.** Save the drawing.

That is it for polygons! Remember, you can draw polygons with anything from 3 to 1024 sides.

6.5 Putting It All Together

Let's try a project using what we have learned. We will draw the Ring Stand Base shown in Figure 6.5a.

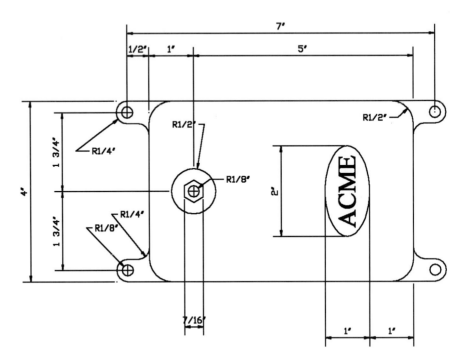

FIG. 6.5a: Ring Stand Base

6.5.1 Do This: Polygons, Arcs & Circles—The Project

I. Create a new drawing with the following setup:

- **Lower left limits:** 0,0
- **Upper right limits:** 17,11
- **Units:** Architectural
- **Grid:** $1/2$
- **Snap:** $1/4$
- **Textsize:** 3/8
- **Font:** Times New Roman

II. Follow these steps.

TOOLS	COMMAND SEQUENCE	STEPS
	Command: *z*	1. **Zoom All** to see the entire drawing.
	Command: *l*	2. Save the drawing as *MyStand* to the C:\Steps\Lesson05 folder.
	Command:	3. Enter the **Line** command.
	Command: *[enter]* From point: *4,2-1/2* To point: *@3<90* To point: *[enter]*	4. Start at point *4,2½* and draw a 3" line upward.
	Command: *[enter]* From point: *4-1/2,2* To point: *@5-1/2<0* To point: *[enter]*	5. Repeat the **Line** command. Then draw the second line.
	Command: *[enter]* From point: *4-1/2,6* To point: *@5-1/2<0* To point: *[enter]*	6. Repeat the **Line** command. Then draw the third line.
	Command: *[enter]* From point: *10-1/2,2-1/2* To point: *@3<90* To point: *[enter]*	7. Repeat the **Line** command. Then draw the fourth line. Your drawing now looks like Figure 6.5.1.7

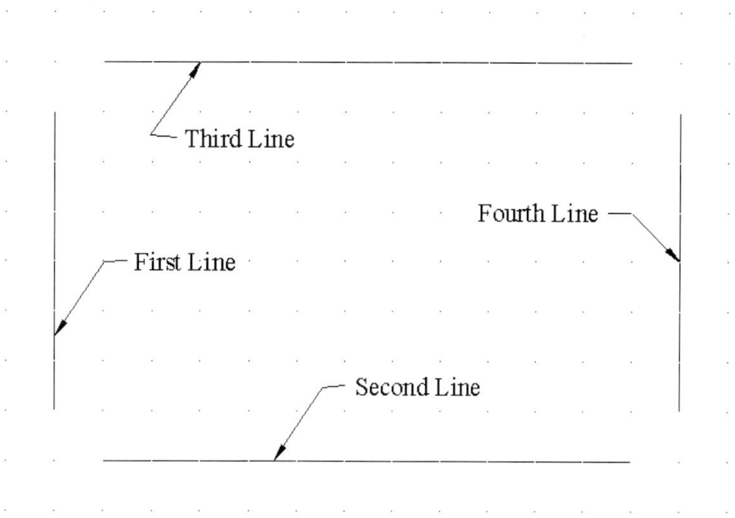

FIG. 6.5.1.7: Four Lines

continued

TOOLS	COMMAND SEQUENCE	STEPS
	Command: *os*	**8.** Set the Running OSNAP to **endpoint**. Clear all other settings. (Remember to toggle the Running OSNAP **On** or **Off** as needed!)
	Command: *a*	**9.** Enter the *Arc* command.
	ARC Specify start point of arc or [CEnter]:	**10.** Select the endpoint of the second line nearest the fourth line.
	Specify second point of arc or [CEnter/ENd]: *c*	**11.** Type *c* for the **CEnter** option.
	Specify center point of arc:	**12.** Snap to the grid mark directly above the point you selected in Step 11.
	Specify end point of arc or [Angle/chord Length]:	**13.** Select the lower endpoint of the fourth line.
		14. Repeat Steps 9 through 13 to draw the other three arcs. Your drawing now looks like Figure 6.5.1.14.

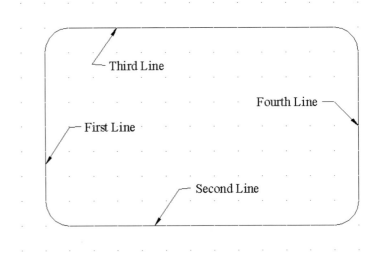

FIG. 6.5.1.14: Lines and Arcs

	Command: *l*	**15.** Repeat the *Line* command.

continued

TOOLS	COMMAND SEQUENCE	STEPS
	Specify first point: **Specify next point or [Undo]:** *@1<180* **Specify next point or [Undo]:** *[enter]*	16. (Refer to Figure 6.5.1.14.) Draw a 1" line from the left endpoint of the second line in the 180° direction.
		17. Repeat Step 16 at all four corners. Your drawing looks like Figure 6.5.1.17.

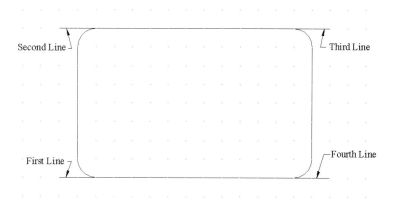

FIG. 6.5.1.17: Extended Lines

	Command: *a* **ARC Specify start point of arc or [CEnter]:** *3-1/2,2-1/2*	18. (Refer to Figure 6.5.1.17.) Draw an arc at the end of the first line.
	Specify second point of arc or [CEnter/ENd]: *c*	19. Type *c* for the **CEnter** option.
	Specify center point of arc:	20. Select a point one snap down from the point selected in Step 18.
	Specify end point of arc or [Angle/chord Length]:	21. Select the **endpoint** of the first line.

continued

TOOLS	COMMAND SEQUENCE	STEPS
		22. Repeat Steps 18 through 21 at the other four lines. Your drawing will look like Figure 6.5.1.22.

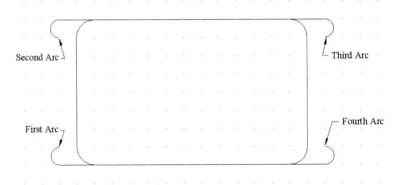

FIG. 6.5.1.22: More Lines and Arcs

	Command: l **LINE Specify first point: <Osnap on> Specify next point or [Undo]: @1/4<0** **Specify next point or [Undo]:** *[enter]*	**23.** Now add a $1/4$" line at the end of the first arc. Repeat for each arc. Your drawing will look like Figure 6.5.1.23.

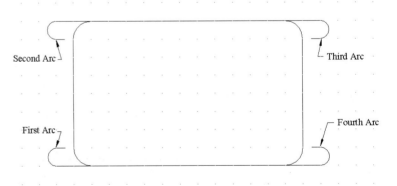

FIG. 6.5.1.23: Horizontal Lines Added

continued

TOOLS	COMMAND SEQUENCE	STEPS
	Command: *a* **ARC Specify start point of arc or [CEnter]:** *[select the endpoint of the line at the first arc]*	**24.** Now draw arcs at the end of the new lines (refer to Figure 6.5.1.29).
	Specify second point of arc or [CEnter/ENd]: *e*	**25.** Select the **ENd** option.
	Specify end point of arc:	**26.** The endpoint is one snap up and one snap to the right.
	Specify center point of arc or [Angle/Direction/Radius]: *d*	**27.** Select the **Direction** option.
	Specify tangent direction for the start point of arc:	**28.** Select a point to the right of the point selected on Step 24.
		29. Repeat Steps 24 through 28 for the other arcs. Your drawing looks like Figure 6.5.1.29.

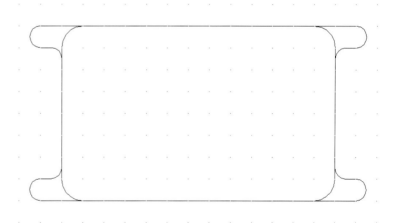

FIG. 6.5.1.29: More Arcs Added

	Command: *qsave*	**30.** It is a good idea to save your drawings occasionally!
	Command: *c*	**31.** Draw a circle in the center of the first anchor leg.

continued

TOOLS	COMMAND SEQUENCE	STEPS
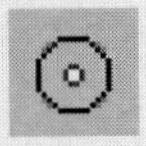	**Specify center point for circle or [3P/2P/Ttr (tan tan radius)]:** *[select a point in the center of the first arc]*	**32.** Use the **Center** OSNAP to locate the center of the first arc.
	Specify radius of circle or [Diameter]: *1/8*	**33.** Use a 1/8" radius.
		34. Repeat Steps 31 through 33 for each of the anchor legs. Your drawing will look like Figure 6.5.1.34.
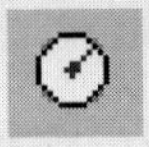	**Command:** *c* **Specify center point for circle or [3P/2P/Ttr (tan tan radius)]:** *5,4* **Specify radius of circle or [Diameter] <0"-0 1/8">:** *1/2*	**35.** Now draw the 1" washer using the *Circle* command.

FIG. 6.5.1.34: Circles Added

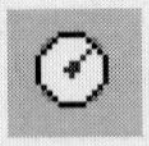	**Command:** *c* **Specify center point for circle or [3P/2P/Ttr (tan tan radius)]:** *5,4* **Specify radius of circle or [Diameter] <0"-0 1/2'>:** *1/8*	**36.** Draw the 1/4" bolt center using the *Circle* command.
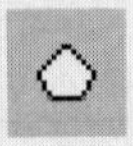	**Command:** *pol*	**37.** Draw the bolt using the *Polygon* command.
	Enter number of sides <4>: *6*	**38.** Give it 6 sides.

continued

TOOLS	COMMAND SEQUENCE	STEPS
	Specify center of polygon or [Edge]: _cen of	**39.** Place it in the center of the last circle you drew.
	Enter an option [Inscribed in circle/Circumscribed about circle] <I>: *c*	**40.** Since we have a dimension on the bolt from flat side to flat side, we will draw the polygon around a circle with that diameter. Type *c* for **Circumscribed**.
	Specify radius of circle: *7/32*	**41.** Type in *7/32* (half the given diameter). Your drawing now looks like Figure 6.5.1.41.

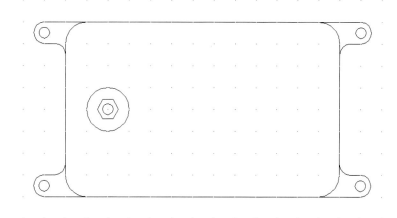

FIG. 6.5.1.41: Bolt Drawing Added

	Command: *el*	**42.** Now let's draw the logo plate using the *Ellipse* command.
	Specify axis endpoint of ellipse or [Arc/Center]: *8-1/2,3*	**43.** Start at point $8\,1/2, 3$.
	Specify other endpoint of axis: *@2<90*	**44.** The ellipse is 2" along the long axis.
	Specify distance to other axis or [Rotation]: *@1/2<0*	**45.** The ellipse is 1" along the short axis.
	Command: *dt*	**46.** Add the *ACME* text to finish the project.

continued

TOOLS	COMMAND SEQUENCE	STEPS
	Current text style: "times" Text height: 0'-0 3/16" **Specify start point of text or [Justify/Style]:** *m* **Specify middle point of text:** *8-1/2,4*	47. We will middle-justify the text in the center of the ellipse.
	Specify height <0'-0 3/16">: *3/8*	48. If you set the **textsize** to *3/8* during setup, it will default to that now. Otherwise, set it to *3/8*.
	Specify rotation angle of text <0>: *90*	49. Set the **rotation angle** so the text may be read from the bottom of the stand.
	Enter text: *ACME* **Enter text:** *[enter]*	50. Enter the text.
	Command: *qsave*	51. Save your drawing. It should now look like the sample in Figure 6.5a.

6.6 Extra Steps

These may be the most important paragraphs in the text!

AutoCAD provides several *system variables* including one called **Savetime**.

> SYSVARS (system variables) are one of the ways AutoCAD provides for the user to configure the software for optimal user performance.

The *Savetime* command sequence is

Command: *savetime*
Enter new value for SAVETIME <120>: *10*

The default time is 2 hours. I prefer 10 minutes, but you may set it to whatever makes you feel comfortable. Remember that whatever number you assign to *Savetime* is the amount of drawing time you may lose in case of a system crash.

As just set, AutoCAD will automatically save the current file every 10 minutes. But you do not have to worry about it overwriting a file if you do not want your changes saved. AutoCAD saves the drawing to a file called *ACAD.SV$* and puts it in the \Windows\Temp folder.

If you lose your drawing (power failure, forget to save, etc.), simply rename the *ACAD.SV$* to your drawing name (be sure to use a .DWG extension). You may then open it as you would any other drawing.

6.7 What Have We Learned?

Items covered in this lesson include:

- *Commands*
  - *Circle*
  - *Ellipse*
  - *Polygon*
  - *Savetime*

Lines and circles (and their various complements) shape our world. With this lesson, we wrap up the basic drawing tools. You are now able to draw quite a few things in the 2-dimensional world. But remember that only about 30% of CAD is drawing. The rest is modifying what you have drawn and entering text. We will consider many of AutoCAD's modifying tools in Lesson 8, but first we will look at adding a bit of flavor to our work in Lesson 7.

Work on the exercises until you are more comfortable with what you have learned so far. Then go on to the next lesson.

6.8 EXERCISES

1. Create the drawing in Figure 6.8.1. Save the drawing as *MyDrillGizmo* to the C:\Steps\Lesson06 folder. (Do not draw the dimensions.) Begin with the following settings:
   1.1. **Lower Left Limits:** 0,0
   1.2. **Upper Right Limits:** 12,9
   1.3. **Units:** architectural
   1.4. **Grid:** $1/8"$
   1.5. **Snap:** $1/16"$
   1.6. **Textsize:** $1/8"$
   1.7. **Font:** Times New Roman

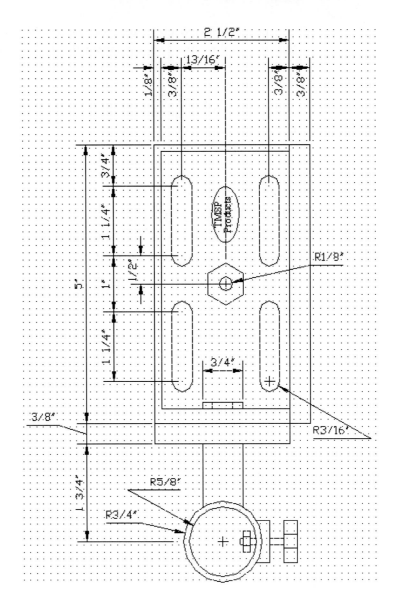

FIG. 6.8.1: Drill Gizmo

2. Use the *MyIsoGrid2* template you created in Lesson 4 (or the *IsoGrid2* template in the Lesson04 folder) to create the drawing in Figure 6.8.2. Save the drawing as *MyIsoBlockwithEllipses* in the C:\Steps\Lesson06 folder.

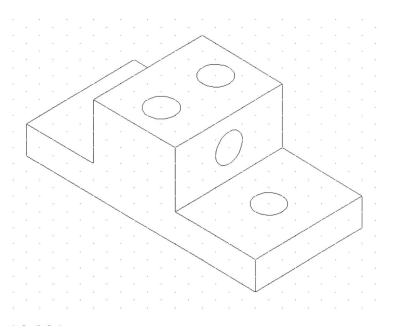

FIG. 6.8.2: Isometric Block with Isometric Circles

3. Using the *MyBase3* template you created in Lesson 1 (or the *Base3* template in the Lesson01 folder), create the drawing in Figure 6.8.3. Save the drawing as *MyHolder* in the C:\Steps\Lesson06 folder.

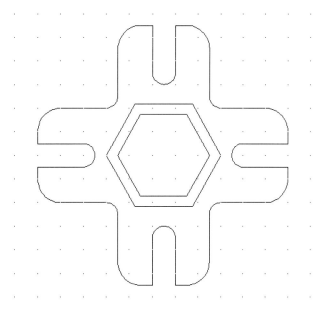

FIG. 6.8.3: Slotted holder

4. Using the *MyGrid* template you created in Lesson 4 (or the *Grid* template in the Lesson04 folder), create the drawing in Figure 6.8.4. Save the drawing as *MySlideGuide* in the C:\Steps\Lesson06 folder. (Note: Change the grid spacing to $1/2$".)

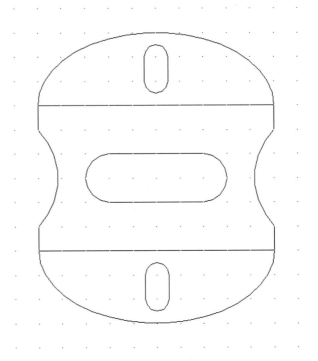

FIG. 6.8.4 Slide Guide

5. Using what you have learned, create the drawing in Figure 6.8.5. The grid is $1/8$".

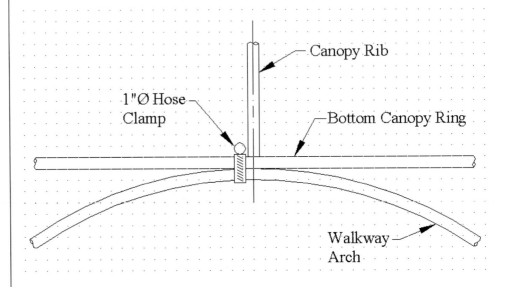

FIG. 6.8.5: Arch / Canopy Assembly Detail

6. Now create the detail sheets in Figures 6.8.6a to 6.8.6c. Use the title block you used in Lesson 5, Exercise 5 (again, I used a $1/4$" grid to create the detail sheets).

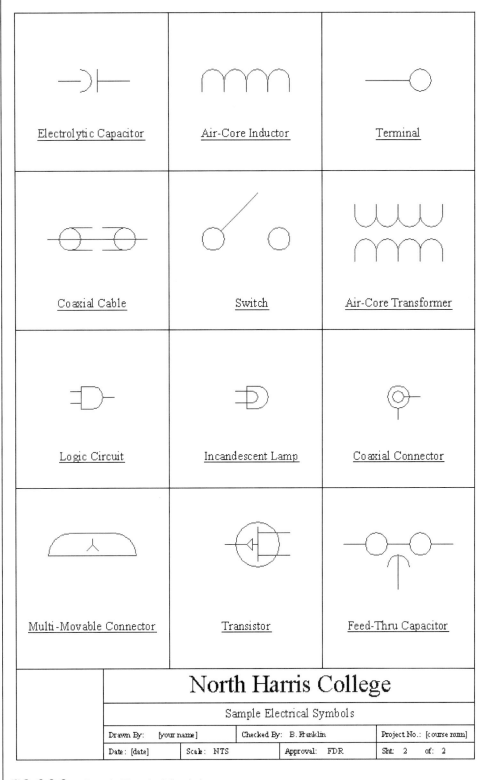

FIG. 6.8.6a: Sample Electrical Symbols

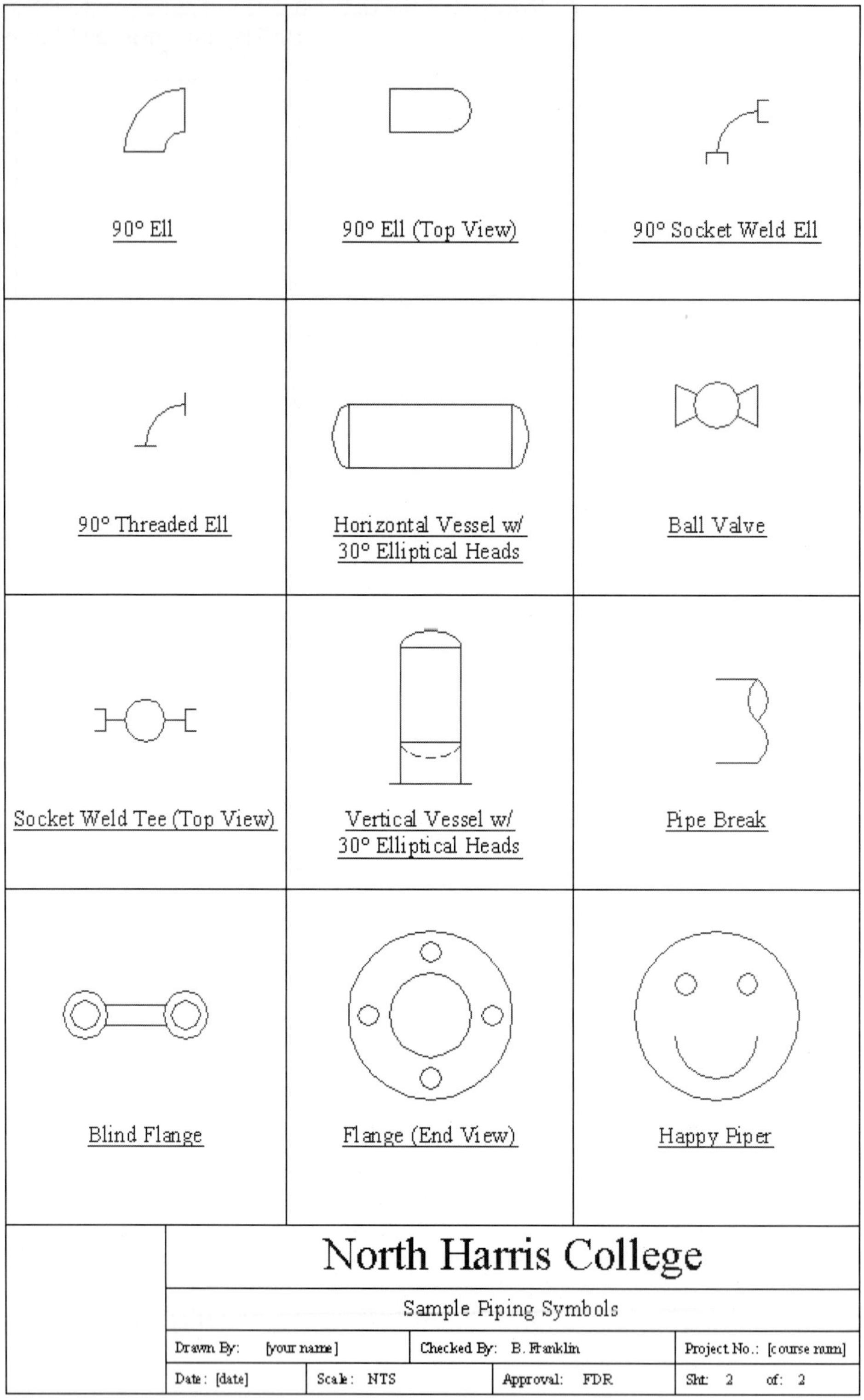

90° Ell	90° Ell (Top View)	90° Socket Weld Ell
90° Threaded Ell	Horizontal Vessel w/ 30° Elliptical Heads	Ball Valve
Socket Weld Tee (Top View)	Vertical Vessel w/ 30° Elliptical Heads	Pipe Break
Blind Flange	Flange (End View)	Happy Piper

North Harris College
Sample Piping Symbols

Drawn By: [your name]		Checked By: B. Franklin	Project No.: [course num]
Date: [date]	Scale: NTS	Approval: FDR	Sht: 2 of: 2

FIG. 6.8.6b: Sample Piping Symbols

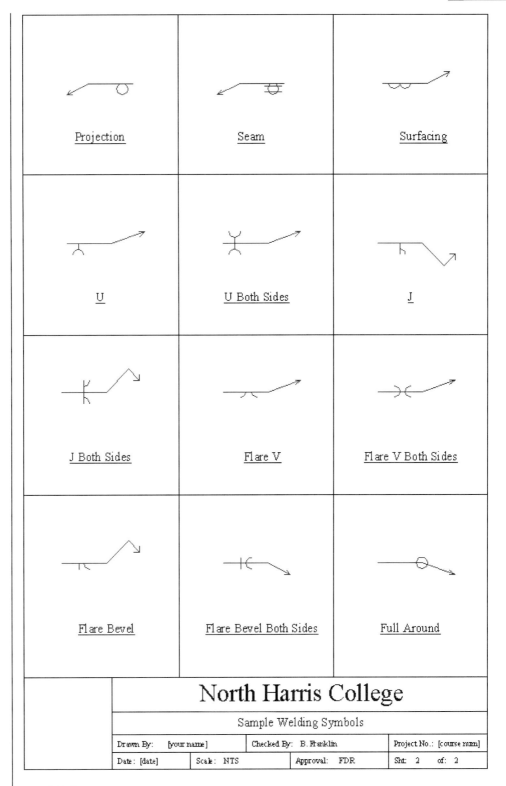

FIG. 6.8.6c: Sample Welding Symbols

7. Create the electrical schematics below using the same border and title block you used in Exercise 6.

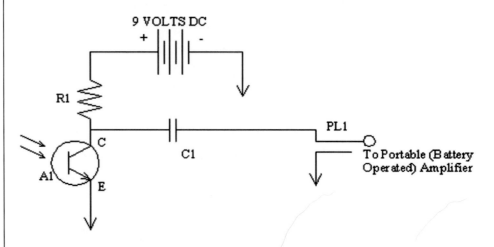

FIG. 6.8.7a: Electrical Schematic

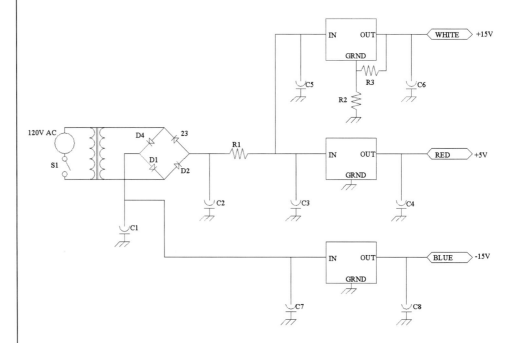

FIG. 6.8.7b: Electrical Schematic

6.9 REVIEW QUESTIONS

Please write your answers on a separate sheet of paper.

1. _____ is the hotkey for the circle command.

2. The 2P option of the circle command allows the user to draw a circle by identifying two points on the circle's _____.

3. _____ identifies the size of the circle on the TTR option of the circle command.

4. To repeat the last command, hit _____ on the keyboard.

5. to 7. Using the ellipse command defaults, what three points must the user identify to draw the ellipse?

8. The _____ option of the ellipse command allows the user to draw partial ellipses.

9. The drawing must be in _____ mode to draw isocircles.

10. By default, AutoCAD draws true ellipses. However, the user may draw polyline ellipses by changing the _____ sysvar to 1.

11. (T or F) It is easy to find the center of a polyline ellipse.

12. What is the first and easiest part of AutoCAD to customize?

Identify these buttons:

13. [icon]   14. [icon]   15. [icon]   16. [icon]   17. [icon]

18. to 20. What are the three ways the user can end a Start—Center arc?

21. By default, AutoCAD draws arcs in the _____ direction.

22. To draw an arc in a different direction, use the _____ option.

23. What must the user do at the first arc prompt to continue the last arc drawn?

24. and 25. What are the two approaches to drawing a polygon?

26. How many sides can an AutoCAD polygon have?

27. and 28. Using the circle approach, the user can place the polygon either _____ in or _____ around an imaginary circle.

29. _____ is the hotkey for the polygon command.

30. The command used to reset the interval between automatic saves in AutoCAD is _____.

31. By default, AutoCAD automatically saves drawings to a file called _____.

32. _____ are AutoCAD's way of allowing the user to manually configure the software for optimal performance.

PART

Beyond the Basics

This part of our text contains these lessons:

7. Adding Flavor to Your Drawings with Layers
8. Editing Your Drawing: Modification Procedures
9. More Complex Lines: Polylines
   (and Light Weight Polylines)
10. More Editing Tools

LESSON 7

Adding Flavor to Your Drawings with Layers

Following this lesson, you will:
➦ Know how to add color to a drawing using the **Color** command
➦ Know how to use linetypes in a drawing using the **Linetype** and **Ltype** commands
➦ Know how to use lineweights in a drawing using the **Lweight** command
➦ Know how to use layers in a drawing
➦ Know how to modify layers, linetypes, lineweights, and colors in a drawing using:
  • **CHProp/DDCHProp**
  • **Properties** (and the Object Property Manager)
  • **Matchprop**
➦ Know how to use the Autodesk Design Center

This is one of my favorite lessons to teach. It may be that I appreciate the respite from new drawing routines. It may be that I like dialog boxes. But most probably, it is just that by this point in the course I am really tired of looking at black-and-white drawings!

In the drafting world, we learn to differentiate between objects by well-established uses of linetypes and lineweights (widths). The number and spacing of dashes in a line, the width of a line, or combinations of dashes and width say a lot about what is being represented on the drawing. (For more information about specific representations, look in any basic drafting text.) We will learn to use these tools in a CAD environment as well. But in the CAD environment, you will have an additional tool at your disposal—color.

This lesson will lead you through the different methods of using linetype, lineweight, and color to differentiate between objects in your drawing. Each method should be considered exclusive. That is, the user should not combine them as the results will no doubt aggravate someone.

7.1 The Simple Command Approach

■ 7.1.1 Adding Color with the Color Command

Changing the color in which you draw can be as simple as typing:

> **Command:** *-color* (or *cecolor*)
> **Enter default object color <BYLAYER>:** *red [or any of several available colors]*
> **Command:**

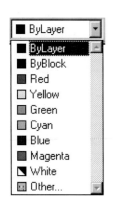

FIG. 7.1.1a
Color Control Box

Of course, if you do not like to type, you can select the down arrow in the Color Control box on the Object Properties toolbar (Figure 7.1.1a), then select your color. Here you will find the nine basic options including the seven basic colors AutoCAD provides. The other two are **ByLayer,** which assigns colors according to the layer setting (more on this in Section 7.2), and **ByBlock. ByBlock** assigns colors that change according to the AutoCAD settings in effect when a block is inserted. We will learn more about blocks in Lesson 19.

You may notice that **White** has a white/black color assignment. The results of using this setting will depend on the background color you have set for the graphics area of your screen.

If you need more colors from which to choose, select the **Other** option. Or you can type *Color* (or *Col*) at the command prompt and bypass the Object Properties toolbar altogether. AutoCAD will prompt you with the Select Color dialog box (Figure 7.1.1b). This box provides the opportunity to select any of AutoCAD's seven primary colors (across the top of the box), or one of the 255 other colors or gray shades available. (Note: AutoCAD will not provide this many choices unless your monitor's resolution will support them.)

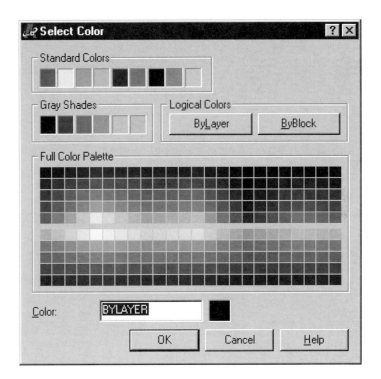

FIG. 7.1.1b: Select Color Dialog Box

To select a color from the Select Color dialog box, simply double-click on the desired color, or click on the desired color and pick the **OK** button.

You will notice that when you pick on a color from the **Full Color Palette**, AutoCAD puts a number in the **Color** text box below. This is the code assigned to that color (thank goodness we do not have to remember names for all 255 colors!). You can, of course, type in the number yourself if you know it.

Okay, let's try to draw some lines and circles in color!

> In addition to the command line and toolbar, you can access the *Color* command by selecting **Color** from the Format pull-down menu.

7.1.1.1 Do This: Drawing in Color

I. Start a new drawing from scratch.
II. Follow these steps.

TOOLS	COMMAND SEQUENCE	STEPS
No Button Available	**Command:** *-color* ***Enter default object color <BYLAYER>:*** *red*	**1.** Set **Red** as the current color using the keyboard method. Notice that **Red** shows as the current color in the Color Control box (Figure 7.1.1.1.1) in the center of the Object Properties toolbar. 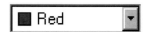 FIG. 7.1.1.1.1: Color Control box
	Command: *c* **Specify center point for circle or [3P/2P/Ttr (tan tan radius)]:** *4,3* **Specify radius of circle or [Diameter]:** *1.5*	**2.** Draw a circle with a *1 1/2"* radius at coordinates *4,3*.
FIG. 7.1.1.1.3 Color Control Box		**3.** Set **Blue** as the current color using the Color Control box (Figure 7.1.1.1.3) on the Object Properties toolbar. To do this, pick the down arrow next to the box; then pick **Blue**.

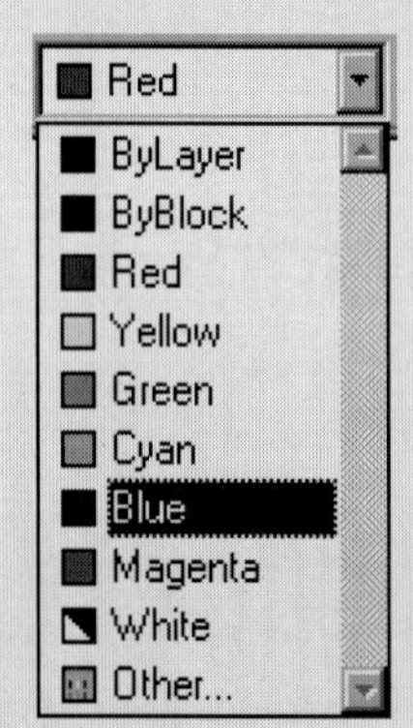

continued

TOOLS	COMMAND SEQUENCE	STEPS

Polygon Button

Center Osnap

Command: *pol*
Enter number of sides <4>: *5*

Specify center of polygon or [Edge]: _ cen of

Enter an option [Inscribed in circle/Circumscribed about circle] <I>: *[enter]*

Specify radius of circle: *1.5*

4. Draw a five-sided polygon inscribed in the circle you just drew.
   Be sure to use the **Center** OSNAP . . .
   Your drawing looks like Figure 7.1.1.1.4.

FIG. 7.1.1.1.4: Circle & Polygon

Command: *col*

5. Call up the Select Color dialog box by typing *Color* or *col* at the command prompt.

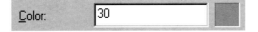

FIG. 7.1.1.1.6: Color Text Box

6. Set the **Color** to number *30* by typing the number into the **Color** text box toward the bottom of the dialog box, as shown in Figure 7.1.1.1.6.

7. Pick the **OK** button.

Line Button

Command: *os*

Command: *l*

8. Set the running OSNAP to **endpoint**, then draw lines connecting the corners of the polygon to form a star. Your drawing looks like Figure 7.1.1.1.8.

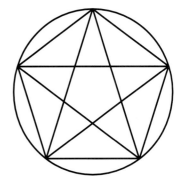

FIG. 7.1.1.1.8: Star

continued

TOOLS	COMMAND SEQUENCE	STEPS
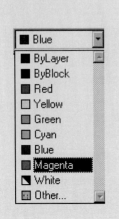	**Command:** *col*	**9.** Now set the **color** to *magenta* using one of the methods mentioned.
	Command: *dt*	**10.** Create the text shown in Figure 7.1.1.1.10.
	Current text style: "Standard" Text height: 0.2000	
	Specify start point of text or [Justify/Style]: *m*	The text should be middle-justified . . .
	Specify middle point of text: cen of	In the center of the circle.
	Specify height <0.2000>: *3/16*	Make it 3/16" high.
	Specify rotation angle of text <0>: *[enter]*	Your drawing now looks like Figure 7.1.1.1.10.
	Enter text: *Star*	
	Enter text: *[enter]*	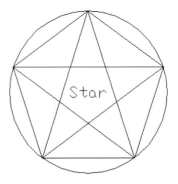 **FIG. 7.1.1.1.10:** Completed Exercise
	Command: *Save*	**11.** Save your drawing as *MyStar* in the C:\Steps\Lesson07 folder.

7.1.2 Drawing with Linetypes

What would drafting be without linetypes? *Boring!* And builders would not be able to tell hidden lines from centerlines. I suppose we would have some pretty interesting buildings out there (not to mention some OSHA nightmares).

Luckily, AutoCAD has continued all the traditional drafting tools—and even added a few. But let's look at linetypes.

The command to bring up the Linetype Manager dialog box (Figure 7.1.2a) is, you guessed it, *Linetype*, although *Ltype* will work as well. The hotkey for *Linetype* is *lt*.

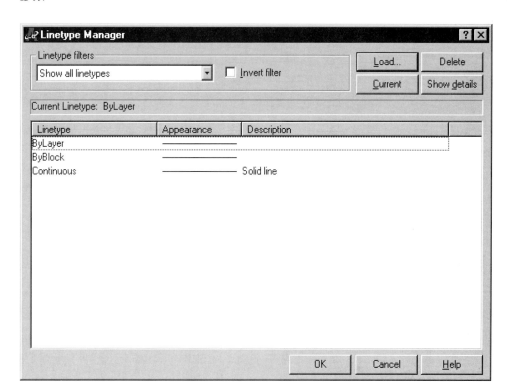

FIG. 7.1.2a: Linetype Manager Dialog Box

As you can see in the list box, there are only three options available by default. **ByLayer** and **ByBlock** work in the same way they did with the *Color* command. The only other option is **Continuous**, which provides a solid line. To select one of the options, simply pick on the desired **Linetype**, pick the **Current** button, then pick the **OK** button. You will notice that the **Linetype** control box on the Object Properties toolbar shows your choice as current.

> Why are there two commands to do the same thing? Well, in previous releases, *Linetype* delivered its options to the command prompt line. This is still quite useful for Lisp programmers.

The command line options are still available by typing a dash in front of the command, as in *—linetype*. The command line sequence to set a linetype looks like this:

Command: *—linetype*
Current line type: "ByLayer"
Enter an option [?/Create/Load/Set]: *s*
Specify linetype name or [?] <ByLayer>: *hidden*
Enter an option [?/Create/Load/Set]: *[enter]*
Command:

You will see that there are easier ways in 2000. The command line method is rarely useful; although by using this approach, the user can set a linetype "current" without first loading it. (AutoCAD will load it automatically.)

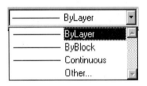

FIG. 7.1.2b
Linetype Control Box

An easier way of setting a *loaded* linetype to current is to simply pick the down arrow next to the Linetype control box (Figure 7.1.2b) on the Object Properties toolbar, and select the linetype of your choice (much as you did with the Color control box).

So what if the linetype I need is not shown?

To avoid using a great deal of memory to hold information that may not be used, AutoCAD does not automatically load all the more than 45 available linetypes. But you can load them all or just the ones you want. We will see how in our next exercise.

In addition to the command line and toolbar, you can access the **Linetype** command by selecting **Linetype** from the Format pull-down menu.

7.1.2.1 Do This: Drawing with Linetypes

I. Open *Va.dwg* from the C:\Steps\Lesson07 folder. This is a valve attached to a vessel wall (Figure 7.1.2.1a). We will add a centerline to the valve, and some hidden lines to indicate pipe inside the vessel.

II. Follow these steps.

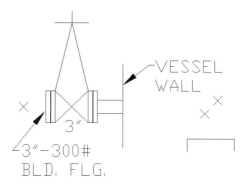

FIG. 7.1.2.1a: Valve & Vessel Wall

TOOLS	COMMAND SEQUENCE	STEPS
No Button Available	**Command:** *lt*	**1.** Open the Linetype Manager (Figure 7.1.2.1.1) by typing *Linetype*, *Ltype*, or *lt*.

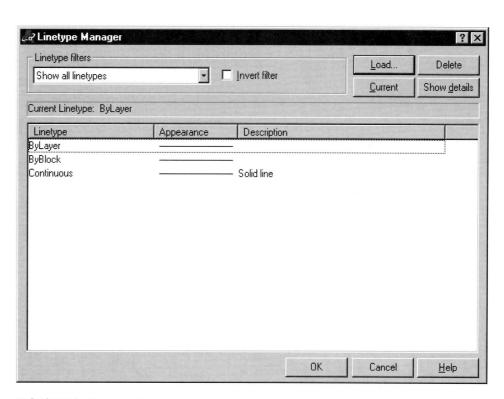

FIG. 7.1.2.1.1: Linetype Manager

continued

TOOLS	COMMAND SEQUENCE	STEPS
		2. Pick the **Load** button. The Load or Reload Linetypes dialog appears as shown in Figure 7.1.2.1.2.

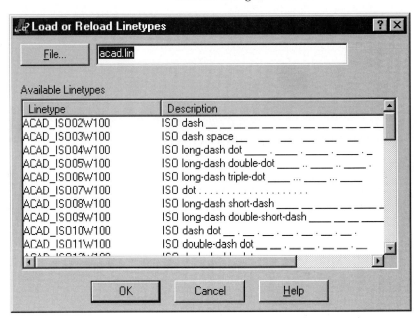

FIG. 7.1.2.1.2: Load or Reload Linetypes Dialog Box

3. Notice the **File** button at top. AutoCAD stores its linetype definitions in a file called *Acad.lin* (although there is another file—the *Acadiso.lin* file used for metrics—with additional definitions). Selecting the **File** button will open a standard Windows Select . . . File dialog box where you may select a different file. We will stick with the *Acad.lin* file.

Notice also the list of linetype names under the **Linetype** heading in the list box. A description and sample appears to the right of each. Scroll down the list as necessary to find the linetype(s) you want to load.

For now, scroll down till you see **Center**. Select it by picking on the word **Center**.

Scroll down a bit more till you see **Hidden**. Holding down the **Ctrl** key on the keyboard (to enable you to select more than one linetype at a time) select **Hidden**.

continued

TOOLS	COMMAND SEQUENCE	STEPS

4. Pick the **OK** button. AutoCAD closes the dialog box and lists **Center** and **Hidden** in the Linetype Manager, as indicated in Figure 7.1.2.1.4.

FIG. 7.1.2.1.4: Linetype Manager List Box Showing the Center and Hidden Linetypes

5. Pick the **OK** button to complete the command.

Command: *col*

6. Set the current color to **Magenta**.

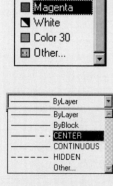

FIG. 7.1.2.1.7
Linetype Drop Down Box

7. Now let's draw some lines. First, set **Center** as the current linetype. Pick the down arrow in the Linetype control box on the Object Properties toolbar; then select the **Center** linetype, as shown in Figure 7.1.2.1.7.

continued

TOOLS	COMMAND SEQUENCE	STEPS
	Command: *l*	**8.** Draw a line from the node at Point 1 to Point 2 (refer to the drawing in Figure 7.1.2.1.9). Notice that you have drawn a centerline.

9. Now draw a line from the handwheel —Point 3—(be sure to use the appropriate OSNAP and **Ortho** settings) through the center of the valve to Point 4. Your drawing looks like Figure 7.1.2.1.9. |

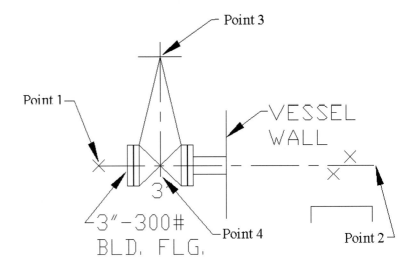

FIG. 7.1.2.1.9: Drawing Centerlines

	**Command:** *col*	**10.** Set the current color to **Blue**.

continued

TOOLS	COMMAND SEQUENCE	STEPS
		11. Reset the linetype to **Hidden**.
	Command: *l*	12. Now draw the **Hidden** lines to show the pipe inside the vessel (refer to Figure 7.1.2.1.12). Start at the intersection (use OSNAPS) at Point 1 and draw to the node at Point 2. Continue the line perpendicular to the basin below. Repeat the *line* command and draw a line from Point 3 to Point 4 and down to the basin.

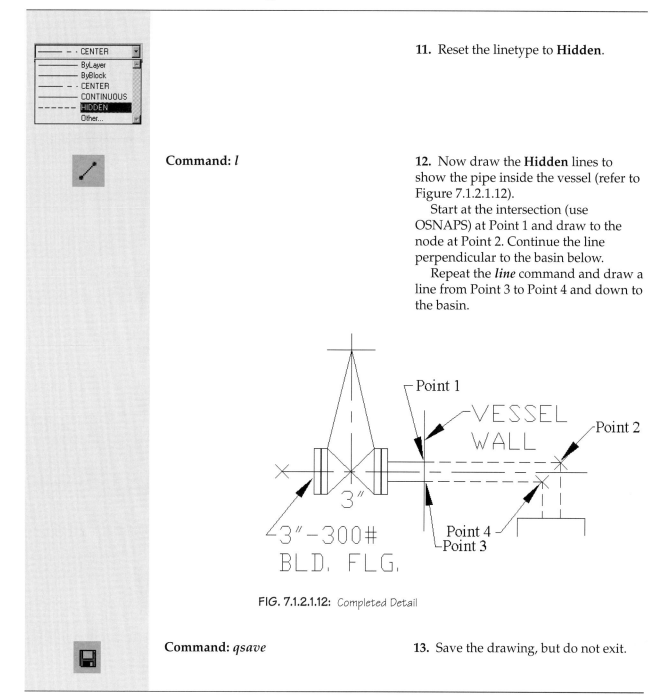

FIG. 7.1.2.1.12: Completed Detail

| | Command: *qsave* | 13. Save the drawing, but do not exit. |

You have seen that drawing with **linetypes** is fairly simple. But there are some other considerations of which the user must be aware. The first of these is a system variable called **LTScale**.

AutoCAD defines dashed and dotted lines by a code that details how long to make each dash and each space. A user can define his own lines by learning that code—but that is a topic for a customization guide. At this level, the user must know how to adjust the line definitions (much as you learned to adjust the text size) to appear as dashes or dots on a larger drawing. Otherwise, a dashed line defined as having 1/4" dashes separated with 1/8" spaces may appear as a solid line on a scaled drawing.

Try the *LTScale* command now. You will see that it has already been set to **8** (the normal default is **1**). I set it here so you could see the hidden and centerlines as you drew them. Set it back to *1* and note the change to your drawing. Here is the command sequence:

Command: *LTSCALE*
Enter new linetype scale factor <8.0000>: *1*

Your drawing will look like Figure 7.1.2c—with dashes and spaces too tiny to see.

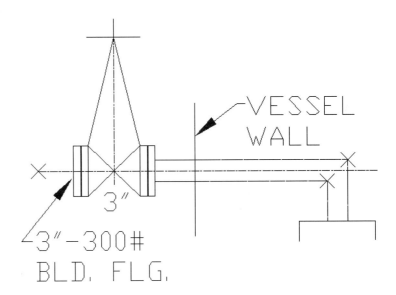

FIG. 7.1.2c: LTScale Reset to 1

The best setting for the **LTScale** is the scale factor for the drawing (use the Drawing Scales chart on the tear out reference card).

The *LTScale* command sets the universal linetype scale factor for the drawing. All lines defined with dashes and spaces are affected. You can, however, adjust the **LTScale** specifically to a single line or linetype if you find it necessary.

We can set the **LTScale** universally or specifically from the **Details** section of the Linetype Manager dialog box. Let's look at this now.

7.1.2.2 Do This: Setting the LTScale

I. If you are not still in the *va.dwg* file, open it now. It is in the C:\Steps\Lesson07 folder.

II. Open the Linetype Manager dialog box.

III. Follow these steps.

TOOLS	COMMAND SEQUENCE	STEPS
		1. Pick on the **Show Details** button to open the **Details** section at the bottom of the box. The **Details** section (Figure 7.1.2.2.1) appears at the bottom of the dialog box.

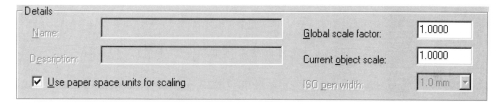

FIG. 7.1.2.2.1: Details Section of the Linetype Manager Dialog Box

		2. Notice that the **Global scale factor** is currently set to *1.0000*. Set it back to *8* (double-click in the text box next to the words: **Global scale factor:** then type *8*).
	Command: *re*	3. Pick the **OK** button to return to the graphics screen, then regenerate your drawing. Notice that the dashes reappear on the drawing.
	Command: *lt*	4. Reopen the Linetype Manager dialog box.
		5. Repeat steps 2 and 3 to reset the **Global scale factor** to *1*. Notice that the dashes again disappear from the drawing.

continued

TOOLS	COMMAND SEQUENCE	STEPS
	Command: *e*	**6.** Erase lines 1 and 2, as indicated in Figure 7.1.2.2.6.

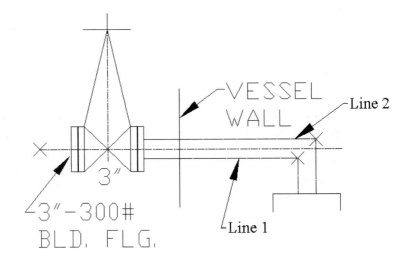

FIG. 7.1.2.2.6: Erase These Lines

	Command: *lt*	**7.** Reopen the Linetype Manager dialog box.
		8. Select the **Hidden** linetype from the list box. The word **Hidden** appears in the **Name** text box in the **Details** area. Now set the **Current object scale** to **8** (see Figure 7.1.2.2.8). Pick the **OK** button to return to the screen.

continued

TOOLS	COMMAND SEQUENCE	STEPS

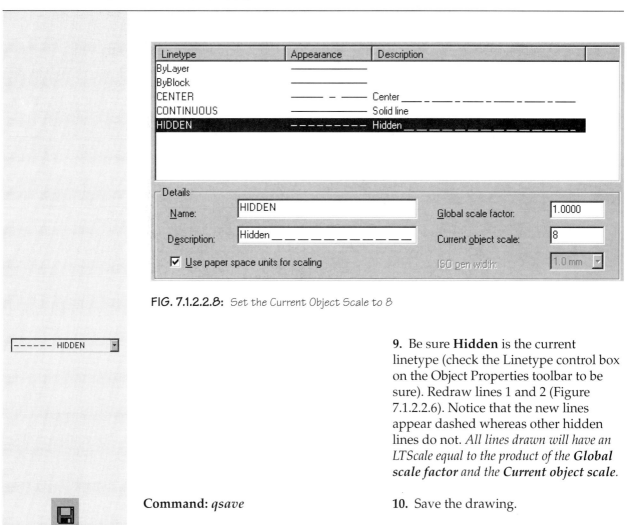

FIG. 7.1.2.2.8: Set the Current Object Scale to 8

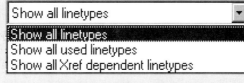

9. Be sure **Hidden** is the current linetype (check the Linetype control box on the Object Properties toolbar to be sure). Redraw lines 1 and 2 (Figure 7.1.2.2.6). Notice that the new lines appear dashed whereas other hidden lines do not. *All lines drawn will have an LTScale equal to the product of the **Global scale factor** and the **Current object scale***.

Command: *qsave*

10. Save the drawing.

At the top of the Linetype Manager, there is a control box with the words **Linetype filters** above it (Figure 7.1.2d). This is a *filter*—a box used to help the user determine the linetype status of the current drawing.

FIG. 7.1.2d: Linetype Filters Box

> **Show all linetypes** refers to all the linetypes loaded into the drawing; **Show all used linetypes** refers to only those linetypes currently in use; and **Show all Xref dependent linetypes** refers to linetypes dependent on their Xref status (more on Xrefs in our advanced book).
>
> The user may, at the end of a project, choose to see all the unused linetypes so he can delete them to free some memory. Do this by selecting the **Show all used linetypes** option, then placing a check in the **Invert filters** check box next to the control box.

■ 7.1.3 Using Lineweights

A new addition to AutoCAD 2000 is the ability to add lineweight to *all* lines. Earlier releases allowed lineweight only on polylines or by plotter manipulation at print time, so be aware that the material in this section will not be available if you work on releases other than 2000. (That is a good point to bring up when you argue for the upgrade!)

Caution should be exercised when using lineweights, as they are not true WYSIWYG objects—that is: What You See Is (not necessarily) What You Get. In Model Space, AutoCAD displays lineweight by using a relationship between *pixels* and lineweight. This enables the user to tell that an object has weight even if he cannot tell how much weight until he plots the drawing (or views the drawing in Paper Space).

> Pixels are those little lights that make your monitor work. Think of them as tiny flashlights shining through colored lenses in the back of your monitor screen. The number of tiny flashlights depends on the *resolution* of your monitor (see your Windows manual). A standard resolution for AutoCAD would be 800 × 600—or 800 columns of tiny flashlight and 600 rows of tiny flashlight in every square inch of your screen.

The user may set lineweight as easily as he set color and linetypes. The command sequence is

   Command: —*lweight*
   Current lineweight: BYLAYER
   Enter default lineweight for new objects or [?]: *[enter the desired lineweight —AutoCAD will round your number to the nearest preset number available]*
   Command:

AutoCAD also provides the Lineweight Settings dialog box (Figure 7.1.3a) to assist the user in setting up lineweights. Let's take a look at it.

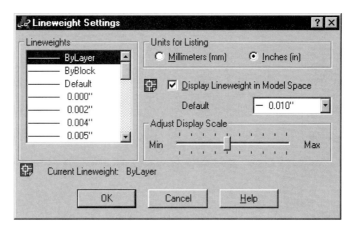

FIG. 7.1.3a: Lineweight Settings Dialog Box

- The **Lineweights** frame on the left provides a choice of predetermined lineweights from which to select—from 0.000 to 0.083.
- The **Units for Listing** frame allows the user to list the lineweight options in **Inches** or **Millimeters**. The user may also set units at the command line by using the *LWUnits* command.
- The check in the box next to the line: **Display Lineweight in Model Space** means that lineweights will be shown in Model Space. The weight will be determined in pixels as explained previously. The user may remove the check (or type **Off** at the prompt after the *LWDisplay* command) to prevent seeing lineweights.
- The slider bar in the **Adjust Display Scale** frame allows the user to control the pixels to lineweight ratio. In other words, slide the bar to the left to minimize (or to the right to maximize) the number of pixels used to show lineweight.

Of course the easiest way the user can set the lineweight is just as he set the color or linetype—using the Lineweight control box on the Object Properties toolbar.

An interesting thing to note about the control drop down boxes—the user can also use them to *change* (as well as set) lineweight, color, and linetype! Let's take a look at this.

7.1.3.1 Do This: Changing the Lineweight

I. Be sure you are still in the *Va.dwg* file in the C:\Steps\Lesson07 folder. If not, open it now.

II. Follow these steps.

TOOLS	COMMAND SEQUENCE	STEPS
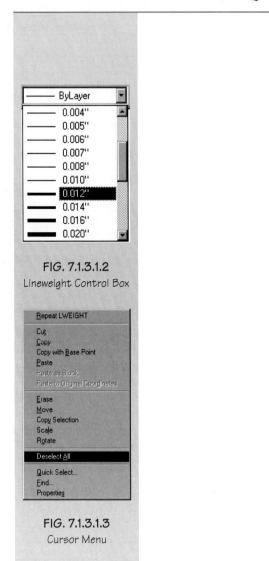 FIG. 7.1.3.1.2 Lineweight Control Box FIG. 7.1.3.1.3 Cursor Menu		**1.** Without entering a command, select the Vessel Wall. **2.** Pick the down arrow on the Lineweight control box; then scroll down and select 0.012, as shown in Figure 7.1.3.1.2. **3.** Right-click in the graphics area and select **Deselect All** from the cursor menu (Figure 7.1.3.1.3). This will clear the selection set on the screen. Notice the change in the line you selected. (Remember: You can use this technique to change linetypes, colors, and layers as well!)

Now that we have spent so much time discovering the **Color**, **Linetype**, and **Lineweight** approaches to adding flavor to a drawing, I must break the bad news. Only rarely will the user find it appropriate to use any of them. But before you get your knickers in a twist, let me tell you that AutoCAD incorporates each into a considerably more powerful tool that the user will find inescapable! Let's look at **Layers**.

7.2 Color, Linetype, and So Much More: Layers

When I was a child (back when drafting was done on shovel blades with charcoal from the fire), one of the most coveted possessions of our household was a set of *Encyclopedia Britannica*. I spent hours exploring the world through those books, but my favorite site was the picture of the human body. I was not all that interested in anatomy—but I was fascinated by the way the body was shown. There was one page with an outline of the body. Then there were successive pages made of clear plastic overlays with the skeletal, reproductive, digestive, and circulatory systems. As these folded down atop each other, the body took shape. If one system was in the way, all I did was fold that sheet back. This is the idea behind **Layers** in AutoCAD.

The user assigns each **Layer** a specific color, lineweight, and linetype. The user also assigns a specific name to the layer—like *dim* for dimensions, *txt* for text, or *obj* for objects. All objects referenced by that name are drawn on that layer much as everything related to the skeletal system was found on a single plastic sheet. If something gets in the viewer's way during subsequent drawing sessions or discussions, the appropriate layer can be toggled **Off** or **Frozen** much as I could fold the unwanted sheet back when viewing the body.

Use the *Layer* command to call the Layer Properties Manager dialog box. But, of course, the two-letter *la* hotkey is quicker and easier.

Let's take a look at the Layer Properties Manager.

First, open the *flrpln.dwg* file (Figure 7.2a) from the C:\Steps\Lesson07 folder.

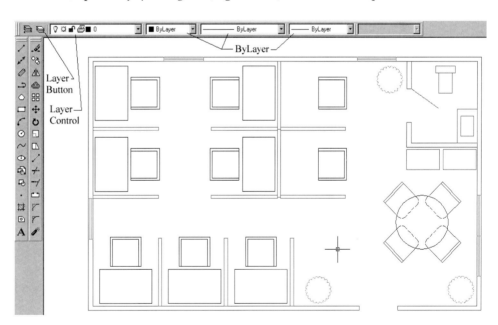

FIG. 7.2a: *flrpln.dwg*

Notice that the Color, Linetype, and Lineweight control boxes each indicate the **ByLayer** setting. This will be important if layers are to work properly, as the Color, Linetype, and Lineweight settings will override the Layer settings. Note also where the Layer control box and **Layer** button are located.

Now open the Layer Properties Manager (Figure 7.2b) by clicking the **Layer** button (see Figure 7.2a for its location on the Object Properties toolbar).

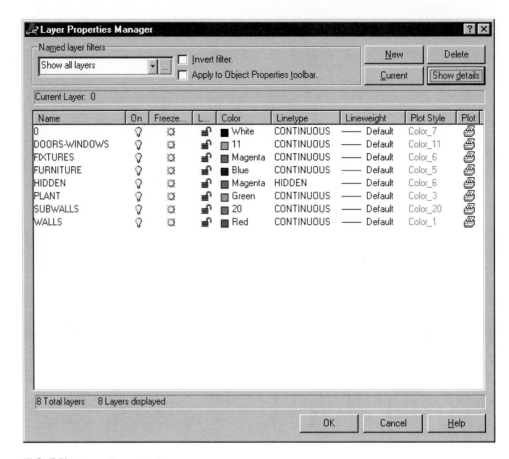

FIG. 7.2b: Layer Properties Manager

For such a simple looking box, there is a lot of information here. Each symbol in the list box represents a specific setting for the layer.

Let's look at each:

The **Name** column is just what it implies—the name of the layer.

FIG. 7.2c
On/Off

The **On** column (Figure 7.2c) shows a light bulb that is either lit (yellow) or not lit (gray). If lit, the layer is **On**; if not lit, the layer is **Off**. When **Off**, all objects on the layer will disappear from the screen but remain part of the drawing and be regenerated.

FIG. 7.2d
Freeze /Thaw

The **Freeze** column (Figure 7.2d) shows an image of the sun when **Thawed** or a snowflake when **Frozen**. When **Frozen**, all objects on the layer will disappear from the screen but remain part of the drawing and *not* be regenerated. Most people prefer **Freeze** to **Off** as it speeds regeneration time.

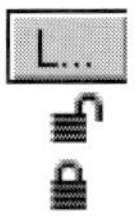

FIG. 7.2e
Lock/Unlock

The **Lock** column (Figure 7.2e) shows an image of a lock. When a layer is **Locked**, the user cannot edit or erase any objects on that layer. This is a useful tool when working on a crowded drawing and selecting objects may be difficult, or when you just want to be sure that you do not accidentally move or delete something.

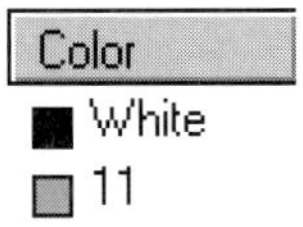

FIG. 7.2f
Color

The **Color** column (Figure 7.2f) shows a box that indicates the current color setting for that layer.

FIG. 7.2g
Linetype

The **Linetype** column (Figure 7.2g) lists the specific linetype assigned to that layer.

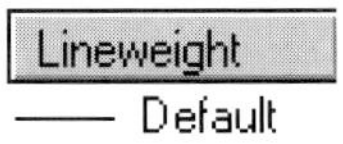

FIG. 7.2h
Lineweight

The **Lineweight** column (Figure 7.2h) shows the specific lineweight assigned to the layer.

FIG. 7.2i
Plot Style

The **Plot Style** column (Figure 7.2i) shows how this layer will be plotted in the current plot style.

FIG. 7.2j
Plot

The **Plot** column (Figure 7.2j) allows the user to plot the layer or to remove all objects on the layer from the plot. (*Note:* This does *not* affect the layer's visibility.)

To create a new layer, simply pick the **New** button. A new layer will appear at the bottom of the list box with the **Name** highlighted so you can immediately name it. The default color and linetype will be **White** and **Continuous**; the default lineweight will be **0.010**. To reset these, simply select the column's symbol or word (i.e., the box for color or the word *continuous* for linetype). AutoCAD will show the appropriate dialog box and you can follow the same procedures you followed for setting Color and Linetype.

Remember the dash in front of a typed command? This is one of the places where it can be useful. The old *Layer* command provided command line prompts. Using them, you can freeze (or thaw) all but the current layer at one time. Here is how:

Command: —*la*

Current layer: "0" Enter an option [?/Make/Set/New/ON/OFF/Color/Ltype/LWeight/Plot/Freeze/Thaw/LOck/Unlock]: *f*

Enter name list of layer(s) to freeze: *

Cannot freeze layer "0". It is the CURRENT layer. Enter an option [?/Make/Set/New/ON/OFF/Color/Ltype/LWeight/Plot/Freeze/Thaw/LOck/Unlock]: *[enter to confirm that you are done]*

Command:

The asterisk (*) you enter at the **Layer name(s) to Freeze** prompt is a *wildcard*. It means *all*, or *freeze all the layers*. This is a handy tool if you just want to work on objects on a single (or a few) layer(s).

You will probably not want to **Freeze** or turn **Off** any layers at this point, but you can do it later within the drawing.

Prior to R14, it was necessary to *purge* a drawing to get rid of unused layers. This was a tedious and often unreliable procedure. To remove a layer now, simply select the layer (pick on the layer's name). It will highlight. Then pick the **Delete** button. If there are no objects on that layer, it will be removed. Don't you just love simplicity?!

You can select multiple layers for formatting at one time by holding down the **Ctrl** or **Shift** key while making your selections.

Setting a layer current (making it the active layer) is just as easy. Just select the one you want, then pick the **Current** button.

To the left of the **Current** button, you will find a control box called **Named layer filters**. This is a filter much like the **Linetype filters** box on the Linetype Manager. This one provides the choices you see in Figure 7.2k. (These choices are also available on the cursor menu when you right-click within the Layer Manager.)

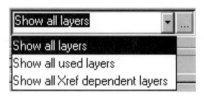

FIG. 7.2k: Named Layer Filters

You can set the filters to **Show all layers** in the drawing, **All used layers** in the drawing, or **All Xref dependent layers**. Of course, the user can also invert the filters, as you saw in the Linetype Manager. But here, by placing a check in the appropriate check box, the user can also control which layers appear in the Layer control box on the Object Properties toolbar.

The button next to the control box will call the Named Layer Filters dialog box (Figure 7.2l). Here you may set the layer filters individually. For example, if you wanted to see only the layers that are **Locked** and **Thawed** but use the **Color** red, you would:

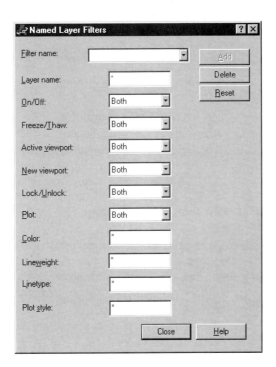

FIG. 7.2l: Named Layer Filters Dialog Box

1. Pick the button next to the **Named layer filters** control box . . . .
2. Give your filter a name in the **Filter name** text box . . . .
3. Pick the down arrow in the control box beside the **Freeze/Thaw** option and select **Thawed** . . . .
4. Pick the down arrow in the control box beside the **Lock/Unlock** option and select **Locked** . . . .
5. Type *Red* in the **Color** control box.
6. Pick the **Add** button to add the filter to the options in the **Named layer filters** control box.
7. Pick the **OK** button.
8. Back on the **Named layer filters** control box, pick the down arrow. The list box will show the name of the filter you just created. Select that one.
9. To reset the list box, repeat Step 1, and then pick the **Reset** Button. Pick **OK** to return to the Layer Properties Manager.

The last item to consider on the Layer Properties Manager will open the **Details** section (Figure 7.2m) just as it did on the Linetype Manager. Pick on the **Walls** layer, then select the **Show details** button. AutoCAD will present the Details frame.

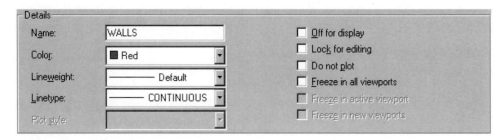

FIG. 7.2m: Details Frame of Layer Properties Manager

Here you find another, more descriptive way to set **Color**, **Linetype**, or **Lineweight** (via control boxes), or to toggle the other options **On** or **Off** (via check boxes). Remember AutoCAD's love for redundancy.

Now close the Layer Properties Manager (pick the **OK** button).

We must explore one other item before trying a layer exercise.

Notice the Layer control box (Figure 7.2n) located to the left on the Object Properties toolbar. Once the layers are set up, you will do most of your layer controlling from this box. It is quicker than having to open the Layer Properties Manager dialog box every time. Simply pick the down arrow; then select the layer you want to be current. The drop-down list box will disappear and the current layer will appear in the box. You can also **Freeze/Thaw**, **Lock/Unlock** or turn a layer **On/Off** from this box simply by picking on the appropriate icon.

FIG. 7.2n
Layer Control Box

Okay. That was a lot of detail to cover. Let's try putting it to use.

7.2.1 Do This: Using Layers

I. Be sure you are in the *flrpln.dwg* file in the C:\Steps\Lesson07 folder.

II. Follow these steps.

TOOLS	COMMAND SEQUENCE	STEPS
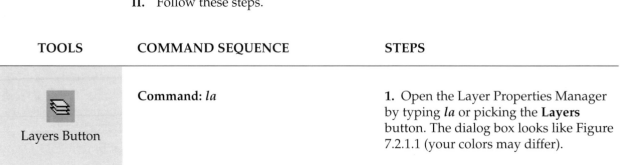 Layers Button	Command: *la*	1. Open the Layer Properties Manager by typing *la* or picking the **Layers** button. The dialog box looks like Figure 7.2.1.1 (your colors may differ).

continued

TOOLS	COMMAND SEQUENCE	STEPS

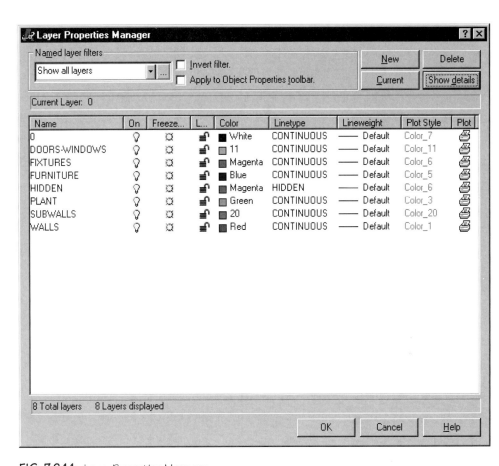

FIG. 7.2.1.1: Layer Properties Manager

[New button]

2. Notice that several layers already exist. They were created at the beginning of the drawing. Let's create a new one. Pick the **New** button. A new entry appears at the bottom of the list box.

3. Type in the name of our new layer—*Text*. The line now looks like Figure 7.2.1.3.

FIG. 7.2.1.3: The New Text Layer

continued

TOOLS	COMMAND SEQUENCE	STEPS
[OK]		**4.** To set the color for the text layer, click the square in the **Color** column. The Select Color dialog box appears. Select the color **Yellow**. Pick the **OK** button.
[OK]		**5.** AutoCAD returns you to the Layer Properties Manager. Notice that the color that is shown for the new **Text** layer is now **Yellow**. Pick the **OK** button here as well.
[save icon]	**Command:** *qsave*	**6.** Save your drawing.

III. You must set a layer *current* before you can use it. You will notice on the Layer Properties Manager shown in Figure 7.2.1.1 that there is a **Current** button. We could have used this, but you will more often want to set a layer current without having to access the Layer Properties Manager. Let's use the control box.

TOOLS	COMMAND SEQUENCE	STEPS
[Layer list showing: 0, 0, DOOR...DOWS, FIXTURES, FURNITURE (selected), HIDDEN, PLANT, SUBWALLS, Text, WALLS]		**7.** Pick the down arrow next to the Layer Control box. Select the **Furniture** layer.
[rectangle icon]	**Command:** *rec*	**8.** Now draw the two 2'-6" × 4' desks missing from the center cubicles. Use the grid and snap as needed. Notice that the desks assume the blue color and continuous linetype associated with the **Furniture** layer. Your drawing looks like Figure 7.2.1.8.

continued

TOOLS	COMMAND SEQUENCE	STEPS

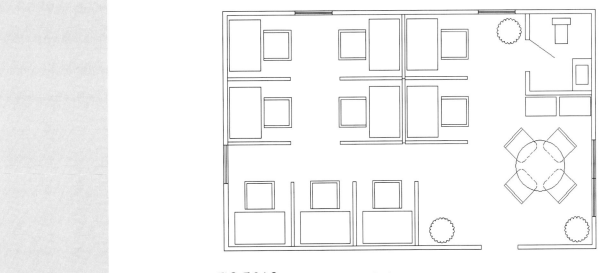

FIG. 7.2.1.8: Drawing with New Desks

9. Set your new **Text** layer current. Using the default text size, add the text as shown in Figure 7.2.1.9.

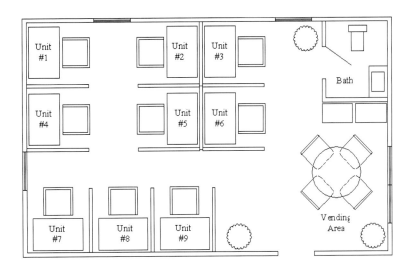

FIG. 7.2.1.9: Completed Drawing

Command: *qsave*

10. Save and close your drawing.

And now I will show you an easier way.

You have seen how to set a layer as current (the preceding exercise). But this procedure requires that you know which layer you want to be current. In some drawings, there may be dozens of layers from which to choose, and remembering the name of the one you want may be impossible. AutoCAD comes to the rescue again!

Pick the **Make Object's Layer Current** button (Figure 7.2o) at the left end of the Object Properties toolbar. AutoCAD prompts:

Command: *_ai-_molc*
Select object whose layer will become current: *[select a chair]*
FURNITURE is now the current layer.
Command:

FIG. 7.2o: Make Object's Layer Current Button

Using this method, you can set a layer current by selecting something already on the desired layer—you do not even have to know which layer it is!

7.3 Swapping Setups: The AutoCAD Design Center

Prior to 2000, the only way to use a drawing setup in more than one drawing was to use templates. This was, and is, a valid approach when creating many drawing. However, it may require too much effort when the user simply wants to use part of the setup—say, the layers—from another drawing, or when the other drawing has already been created.

Rather than force the user to recreate several layers, AutoCAD has provided a new tool to expedite matters. This tool—the AutoCAD Design Center (ADC)—closely resembles the Windows Explorer (see Figure 7.3a). But the user uses it to *mine* another drawing. That is, the user can dig into another drawing and find and retrieve the pieces he wants. The mined drawing remains unaffected by the task.

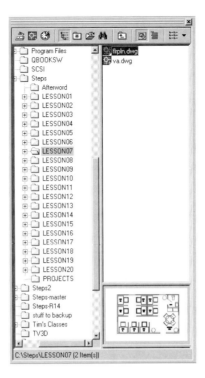

FIG. 7.3a: AutoCAD Design Center

Access the ADC from the command line by entering *ADCenter* (or *ADC*). AutoCAD presents the ADC docked to the left side of the screen. You can undock it or dock it somewhere else, but this location is as good as any other.

Let's look at each section.

- We see the ADC in *tree* view. That is, like Windows Explorer, we see the path to our folder on the left (here we are showing the C:\Steps\Lesson07 path). The right shows the contents of our folder.
- The lower right corner of the ADC presents a preview of the highlighted drawing.
- A toolbar resides along the top of the ADC. The function of each tool follows:

TOOL			FUNCTION
 FIG. 7.3b Desktop	 **FIG. 7.3c** Open Drawings	 **FIG. 7.3d** History	These are toggles. **Open Drawings** changes the display to view only the currently opened drawings. **History** changes the display to show the drawing (and path) of the last several drawings edited in AutoCAD. **Desktop** returns the display to view the full path (as seen in Figure 7.3a).

TOOL	FUNCTION

FIG. 7.3e
Tree View

This is also a toggle. It changes the display between tree view (the view seen in Figure 7.3a) and a single window display. While in the single window display, you must double-click a folder to open it and use the **Up** button (Figure 7.3i) to follow the path backward.

FIG. 7.3f
Favorites

Favorites opens the Windows\Favorites folder.

FIG. 7.3g
Load

Load provides a standard File . . . Open window. With that, you can navigate to the desired folder and open the *palette* for the desired drawing. (The palette is simply a list of things available to you—layers, blocks, text styles, etc.)

FIG. 7.3h
Find

Find presents a dialog box similar to the Windows **Find** program. Using this, the user can search for drawings by date modified or included text.

FIG. 7.3i
Up

Up is simply a navigation button. It changes the display to a step back along the path.

FIG. 7.3j
Preview

The **Preview** button toggles on/off the drawing display in the lower right corner of the ADC.

FIG. 7.3k
Description

The **Description** button toggles on/off a description display in the lower right corner of the ADC.

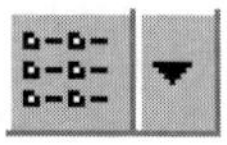

FIG. 7.3l
Views

The **Views** button works just like the views button in Windows—it allows the user to determine how he will see items in a folder (large icons, small icons, listed, or with details).

> In addition to the command line and toolbar, you can access the *ADCenter* command by selecting **AutoCAD DesignCenter** from the Tools pull-down menu.

Let's use the ADC to copy the layers from our *flrpln.dwg* drawing file to a new file.

7.3.1 Do This: Using the AutoCAD Design Center

I. Start a new drawing from scratch.

II. Follow these steps.

TOOLS	COMMAND SEQUENCE	STEPS
 ADC Button	**Command:** *adc*	**1.** Open the AutoCAD Design Center by typing **ADCenter** or **ADC** at the command prompt. Or you may pick the **AutoCAD DesignCenter** button on the Standard toolbar.
	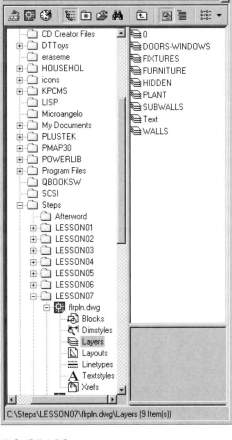 FIG. 7.3.1.2T: The ADC Path to the Flrpln Drawing	**2.** Navigate to the C:\Steps\Lesson07 folder (just as you would using Windows Explorer), and pick on the plus sign next to the *flrpln* drawing. Next, pick on **Layers**. The ADC will look like Figure 7.3.1.2t. Notice the palette shown in the right window. Since **Layers** has been selected in the left window, the palette shows the layers in the *flrpln* drawing.

continued

TOOLS	COMMAND SEQUENCE	STEPS
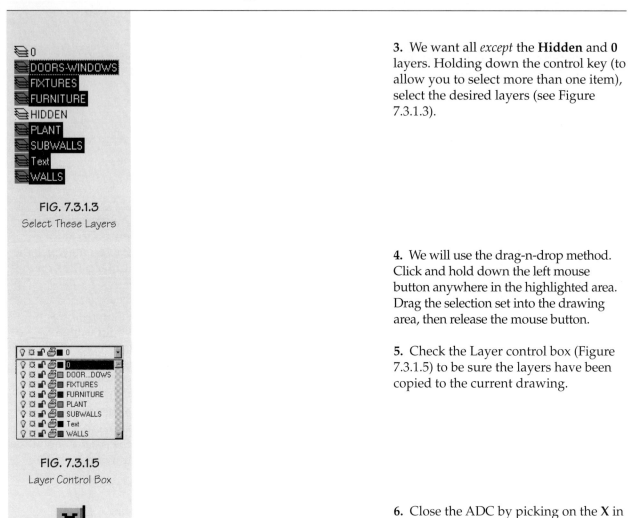 FIG. 7.3.1.3 Select These Layers FIG. 7.3.1.5 Layer Control Box		**3.** We want all *except* the **Hidden** and **0** layers. Holding down the control key (to allow you to select more than one item), select the desired layers (see Figure 7.3.1.3). **4.** We will use the drag-n-drop method. Click and hold down the left mouse button anywhere in the highlighted area. Drag the selection set into the drawing area, then release the mouse button. **5.** Check the Layer control box (Figure 7.3.1.5) to be sure the layers have been copied to the current drawing. **6.** Close the ADC by picking on the **X** in the upper right corner *of the ADC*.

The ADC can be used as easily to copy blocks, dimstyles, text styles, or linetypes.

7.4 The Scenario

You have created the *flrpln* drawing—a furniture location plan for your office. Your boss wants to review your work. You want to give him a step-by-step rundown of what you have done.

7.4.1 Do This: Layers Practice

I. Be sure you are still in the *flrpln.dwg* file in the C:\Steps\Lesson07 folder. If not, open it now.

II. Follow these steps.

TOOLS	COMMAND SEQUENCE	STEPS
		1. Pick the down arrow next to the Layer control box. Set layer *0* as current. (Remember that you cannot freeze the current layer.)
		2. Pick the down arrow next to the Layer control box again. Freeze the following layers: *furniture*, *hidden*, *plant*, *subwalls*, and *text*. To do this, pick on the image of the sun next to each. Notice that the images become snowflakes indicating that the layers are frozen. Pick anywhere on the screen to close the control box. Your drawing looks like Figure 7.4.1.2.

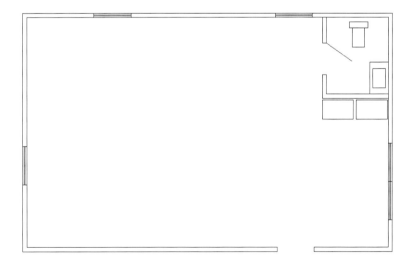

FIG. 7.4.1.2: Frozen Layers

continued

TOOLS	COMMAND SEQUENCE	STEPS

3. Tell your boss that this is the floor plan of the office area with which you have been working.

Explain that you began by laying out the partitions for as many cubicles as possible. ***Thaw*** the *subwalls* layer—pick the down arrow next to the Layer control box again, then pick on the snowflake next to the *subwalls* layer.

Your drawing will look like Figure 7.4.1.3.

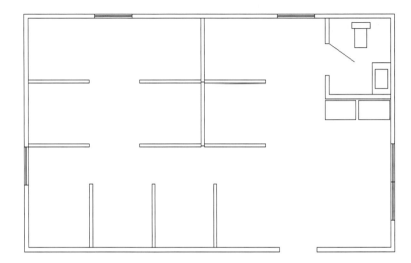

FIG. 7.4.1.3: Subwalls Thawed

4. Next, you placed the furniture within the cubicles and the break area. ***Thaw*** the *furniture* and *hidden* layers—pick the down arrow next to the Layer control box again, then pick on the snowflakes next to the *furniture* and *hidden* layers.

Your drawing will look like Figure 7.4.1.4.

continued

| TOOLS | COMMAND SEQUENCE | STEPS |

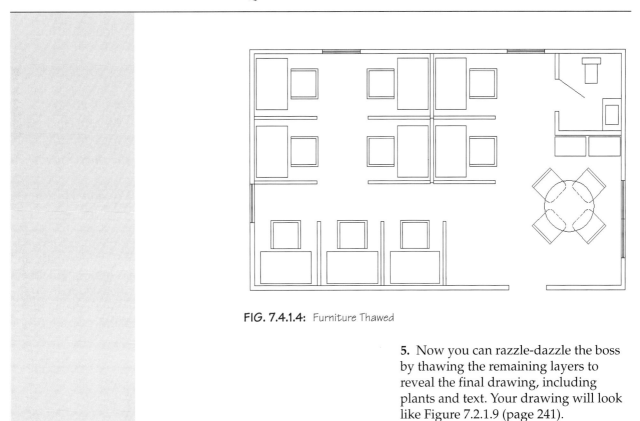

FIG. 7.4.1.4: Furniture Thawed

5. Now you can razzle-dazzle the boss by thawing the remaining layers to reveal the final drawing, including plants and text. Your drawing will look like Figure 7.2.1.9 (page 241).

Review this section several times, if you wish. You will find that getting comfortable with layers is fundamental in the world of AutoCAD. Then go on to see how to adjust for the inevitable outcry—"Oh, God! I drew all that on the wrong layer!!!"

7.5 Uh Ohs, Boo Boos, Ah $%&#$s: The Miracles of CHProp and The Object Property Manager

■ 7.5.1 The Old Way: Changing Properties with CHProp and DDCHProp

You will soon discover that perhaps the easiest mistake to make in AutoCAD is drawing an object on the wrong layer. You get into a flow with your drawing and simply forget or overlook the necessity of changing layers. The same will hold true if you are using the **Color/Linetype** method in your drawing. A good drawing flow simply does not allow time for changing layers / colors / linetypes.

Fortunately, AutoCAD has simple and wonderful tools to fix oversights. These include *CHPprop*, *DDCHProp*, and *Properties*.

Let's begin with *CHProp*. The command prompt approach looks like this:

Command: *chprop*

Select objects: *[select an object—AutoCAD reports:]* **1 found**

Select objects: *[hit enter to confirm completion of the selection process]*

Enter property to change [Color/LAyer/LType/ltScale/LWeight/Thickness]: *[tell AutoCAD what you want to change]* **la**

Enter new layer name <FURNITURE>: *[tell AutoCAD the change you want to make]* **0**

Enter property to change [Color/LAyer/LType/ltScale/LWeight/Thickness]: *[hit enter to confirm that you are finished]*

As you can see, this approach allows the user to change the *properties* of an object—**Color**, **Layer**, **Linetype(LType)**, **Linetype Scale (ltScale)**, **Lineweight**, and **Thickness**. You are familiar with all these except the **Thickness** option. **Thickness** refers to 3-dimensional depth and should not be reset in a 2-dimensional drawing.

But to answer your next question: yes, AutoCAD does provide a dialog box to make this process easier. Access it with the *DDCHProp* command. See the dialog box in Figure 7.5.1a.

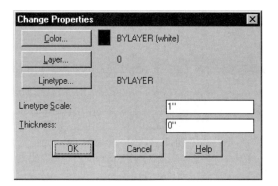

FIG. 7.5.1a: Change Properties Dialog Box

Notice that AutoCAD provides most of the same options available at the command line. The difference is that each button in this dialog box will call a Selection dialog box. (The **Color** button calls the Select Color dialog box, etc. See Figures 7.5.1b and 7.5.1c.)

FIG. 7.5.1b: Layer Control Dialog Box

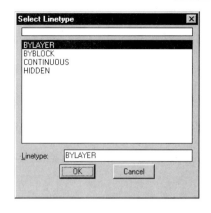

FIG. 7.5.1c: Linetype Control Dialog Box

The user can change the **Linetype Scale** (or the 3-dimensional **Thickness**) for a specific object in the appropriate text box provided.

Notice that Lineweight is not changeable using the Change Properties dialog box. This dialog box is a holdover from previous releases and was not updated for 2000. It will probably be eliminated completely in the next release. But its replacement—the Object Property Manager—is a remarkable tool so the Change Properties dialog box will not be missed.

■ 7.5.2 And Now the Modern Approach: The Object Property Manager (Properties)

AutoCAD has created one of the most useful—and user-friendly—tools to come along in the last several releases with the introduction of the Object Property Manager (OPM). It takes advantage of the new Visual Basic interface to provide an on-screen, editable listing of properties associated with drawing objects. For those of you who are newly computer literate, this simply means that it's really easy to change things (like layers, color, linetypes, and lineweights).

The OPM appears docked on the left side of the screen, just as the ADC did. It begins looking like Figure 7.5.2a, but the properties listed change according to the selected object. Let's take a look at the OPM.

■ Notice the selection drop-down box at the top of the window. The user can use this box to select the type of objects in the drawing to edit. Or the user can select an object in the drawing and that object will show in the selection box.

■ The **Quick Select** button next to the drop-down box allows the user to filter the drawing for objects to edit.

■ In the large list box, you see all the editable properties of the selected object. The two tabs allow the user to view the properties by category (the default) or alphabetically.

Let's take a look at the OPM in action.

FIG. 7.5.2a
The Object Property Manager

7.5.2.1 Do This: Using the Object Property Manager

I. Open the *flrpln.dwg* file in the C:\Steps\Lesson07 folder. The dashed lines under the table have been drawn on the **Furniture** layer with a hidden linetype. Let's change them to the **Hidden** layer.

II. Follow these steps.

TOOLS	COMMAND SEQUENCE	STEPS
 Properties Button	**Command:** *props*	1. Open the OPM by typing **Properties**, or **Props**, at the command line or picking the **Properties** button on the Object Properties toolbar. Or you can pick **Properties** from the Tools pull-down menu.
 Quick Select Button		2. Pick the **Quick Select** button. AutoCAD presents the Quick Select dialog box.
 Select Objects Button	**Select objects: Specify opposite corner:** **Select objects:** *[enter]*	3. Pick the **Select objects** button. AutoCAD returns to the graphics screen. Select the area around the table, as shown in Figure 7.5.2.3. Hit *enter* to complete the selection.

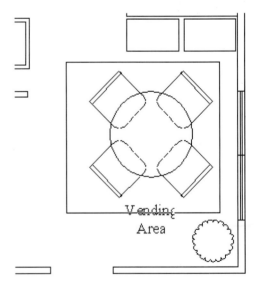

FIG. 7.5.2.3: Selection Window (indicated by the arrow)

continued

TOOLS	COMMAND SEQUENCE	STEPS

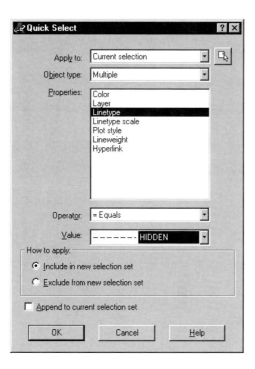

FIG. 7.5.1.4: Quick Select Dialog Box

4. Pick **Linetype** in the **Properties** list box, then pick the down arrow in the **Value** control box and select **Hidden** (Figure 7.5.1.4).

5. Pick the **OK** button to exit the Quick Select dialog box.

Notice that the objects—the parts of the chairs that are under the table and drawn with hidden linetypes—have been selected. Notice also that the OPM has changed.

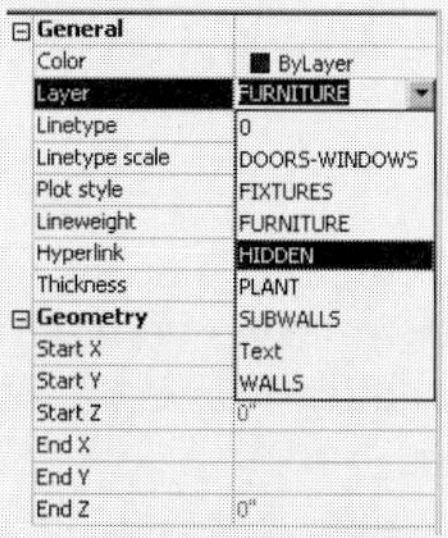

6. (Refer to Figure 7.5.2.1.6t.) Change the layer of the objects by:

- picking in the right column of the **Layer** row
- picking the down arrow and selecting the **Hidden** layer

FIG. 7.5.2.1.6t
Change the Layer

continued

TOOLS	COMMAND SEQUENCE	STEPS
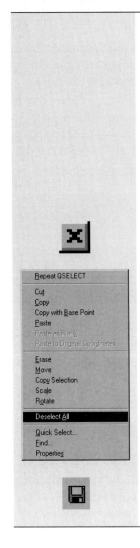	**FIG. 7.5.2.1.7t:** Change the Linetype	7. Follow the procedure in Step 6 to change the linetype to **ByLayer** (Figure 7.5.2.1.7t).
		8. Close the OPM.
		9. Deselect the objects using the cursor menu.
	Command: *qsave*	10. Save and close the drawing.

You can leave the OPM open all the time if you wish; but unless you have a fairly large monitor, it may take up too much of your drawing area.

In this exercise, we used the **Quick select** button to select the objects we were going to modify. Frankly, I think the **Quick select** approach takes longer than simply selecting the objects on the screen using a window or a pick (selection) box. **Quick select** works well when you must filter a large number of objects to get what you want. But otherwise, just do this:

- Select what you want to modify.
- Right-click in the graphics area.
- Select **Properties** from the cursor menu.
- Make your changes in the OPM.

7.6 And Again, the Easy Way to Modify: *Matchprop*

Let's say you have drawn a chair but used the wrong layer by mistake. The chair consists of a rectangle and three lines. All you have to know is on what layer the chair should have been drawn. But you cannot remember. You can do a *DDCHProp* or a *Properties* on one of the other chairs just to read the layer setting, then do the *DDCHProp* or *Properties* on the chair you just drew.

This is getting too complicated.

Enter the *Matchprop* command. Pick the **Match Properties** button (Figure 7.6a) on the Standard toolbar, type it at the command prompt, or select **Match Properties** in the Modify pull-down menu. AutoCAD prompts:

FIG. 7.6a
Match Properties Button

Command: _matchprop (or *ma*)
Select source object: *[select an object that was drawn on the correct layer]*
Current active settings: Color Layer Ltype Ltscale Lineweight Thickness PlotStyle Text Dim Hatch
Select destination object(s) or [Settings]: *[select the object you want to change]*
Select destination object(s) or [Settings]: *[enter to complete the command]*

Who could ask for anything more? But wait! There is more!

Notice that the *Matchprop* command gives you a **Settings** option. The line above the **Settings** option tells you what the current settings are. These are the properties AutoCAD will match. Selecting the **Settings** option will produce the Property Settings dialog box (Figure 7.6b) in which you can tell AutoCAD the properties you do or do not want to match: **Color**, **Layer**, **Linetype**, **Linetype Scale**, and more!

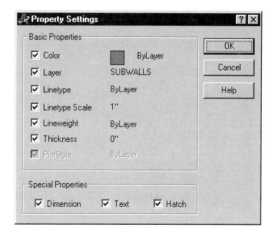

FIG. 7.6b: Properties Settings Dialog Box for the Match Property Command

Personally, I like to leave all the boxes checked. I rarely run into a situation where I might need some properties matched but not others. Still, it is nice to know that I have this option.

7.7 Extra Steps

A side benefit to using AutoCAD has to be the abundance of third-party software and support magazines, books, and bulletin boards.

I would like to recommend one of the best of these—a monthly magazine called *CADENCE: The World's Largest Independent CAD Magazine*. *CADENCE* sports some of the finest technical writers in the field today. Lynn Allen does a column called "Circles and Lines," which is a must read for the CAD novice. She writes well and has the enviable talent of being able to take a complicated subject and explain it in simple terms. I use a couple of her pieces in my classes to help explain things like multilines! Bill Kramer writes a column for the more advanced user and programmer. Other columns include "Tips and Tricks," which supplies just that, and "The Support Column," which answers reader's questions.

I am not affiliated with the magazine, but I try to recommend it to my more serious students. Pick up a copy of *CADENCE* at your local newsstand or magazine rack. Look through it. See what the issues are in the CAD world today. It sells for $3.95 at the newsstands, or you can call them at (800) 289-0484 for their subscription rates.

7.8 What Have We Learned

Items covered in this lesson include:

- *Commands*
  - *Color*
  - *CeColor*
  - *Linetype & Ltype*
  - *LTScale*
  - *LWeight*
  - *LWUnits*
  - *LWDisplay*
  - *Layer*
  - *CHProp & DDCHProp*
  - *Properties*
  - *Matchprop*
  - *ADCenter*

This lesson has challenged you more than any other thus far. We have taken AutoCAD from a simple drawing toy to a real drafting tool. We have seen how to use colors in our drawing (yea!—no more b&w), use linetypes and lineweights, and use layers to put more into a drawing than was ever possible in the world of graphite and paper. (Do you feel like Dorothy walking through the door into Oz?)

Next we will look at that 70% of CAD work that is *not* drawing—modifying what we have drawn. But first, let's try some exercises.

7.9 EXERCISES

1. Create the drawing in Figure 7.9.1 using the following setup. Save the drawing as *MyRemote* to the C:\Steps\Lesson07 folder.

    1.1. Layers

		Color	Linetype
1.1.1.	Button	magenta	continuous
1.1.2.	Dim	cyan	continuous
1.1.3.	Obj	red	continuous
1.1.4.	Text	yellow	continuous
1.1.5.	Toggle	blue	continuous
1.1.6.	Up-1/2down	red	continuous

    1.2. Lower left limits: 0,0
    1.3. Upper right limits: 6,10
    1.4. Architectural units
    1.5. Grid: 1/4"
    1.6. Snap: as needed
    1.7. Font: Times New Roman
    1.8. Text Height: 1/8"

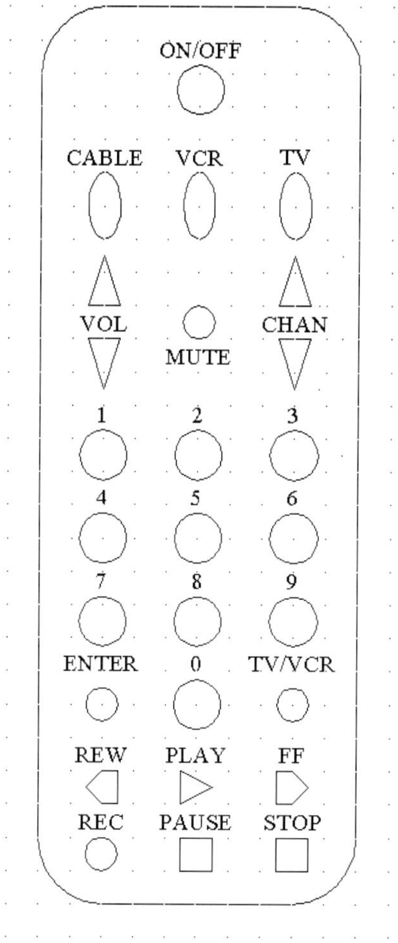

FIG. 7.9.1: MyRemote

2. Set up a drawing to be used in Lesson 10. Use the following parameters:
   2.1. Lower left limits: 0,0
   2.2. Upper right limits: 17,11
   2.3. Architectural units
   2.4. Grid: 1/2"
   2.5. Snap: 1/2"
   2.6. <u>Layer Name</u>   <u>color</u>   <u>linetype</u>
       2.6.1.   Border   magenta   continuous
       2.6.2.   Gradient   cyan   continuous
       2.6.3.   Ruler   red   continuous
       2.6.4.   Text   yellow   continuous
       2.6.5.   Cl   blue   center
   2.7. Draw a border 1/4" in from the limits on all sides (16 $\frac{1}{2}$" × 10 $\frac{1}{2}$")
   2.8. Save the drawing as *MyRuler* to the C:\Steps\Lesson10 folder.
3. Using the *MyBase3* template created in Lesson 1 (or the *Base3* template in the Lesson01 folder), create the next three drawings (Figures 7.9.3a to 7.9.3c). Save the drawings to the C:\Steps\Lesson07 folder using the name indicated in the title block of each. All drawings use the layers indicated in the Layer 1 drawing. (Hint: Change the snap setting as needed.)

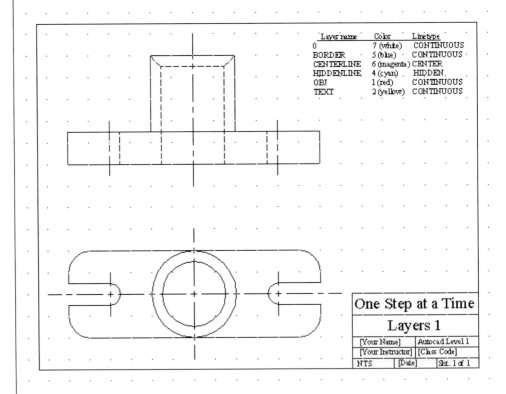

FIG. 7.9.3a: Layers 1

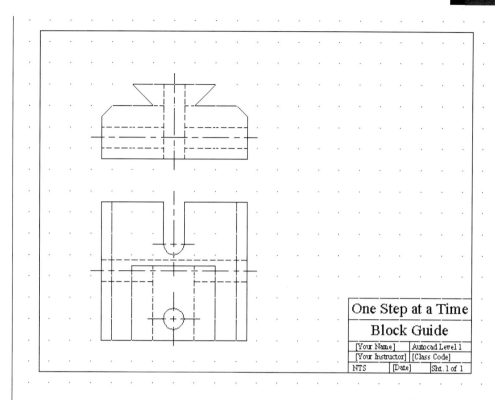

FIG. 7.9.3b: Block Guide

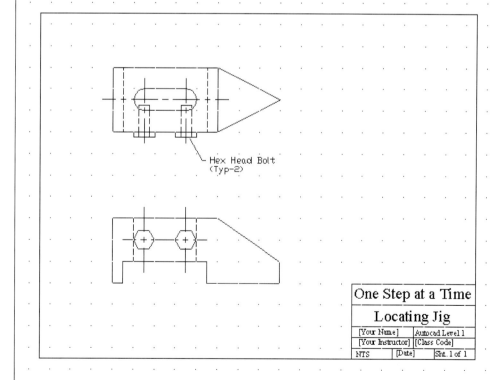

FIG. 7.9.3c: Locating Jig

4. Using what you know, create the drawings in Figures 7.9.4a to 7.9.4e. The grid on each, when shown, is 1/4" (except the hook and catch assembly, which is 1/8").

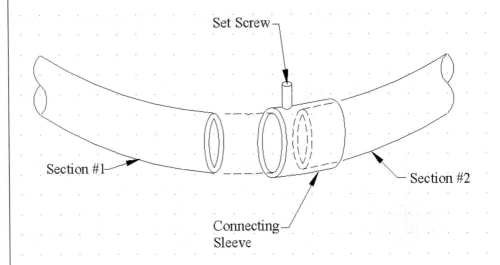

FIG. 7.9.4a: Assembly Detail

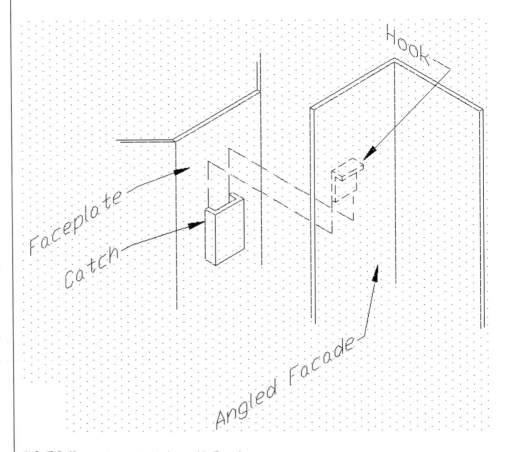

FIG. 7.9.4b: Hook and Catch Assembly Detail

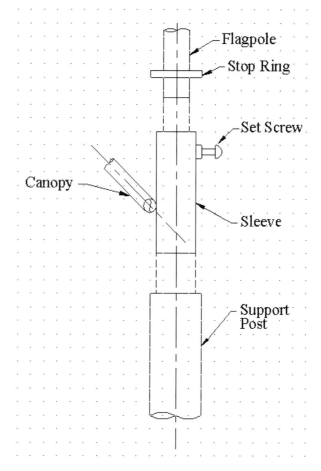

FIG. 7.9.4c: Flagpole Assembly Detail

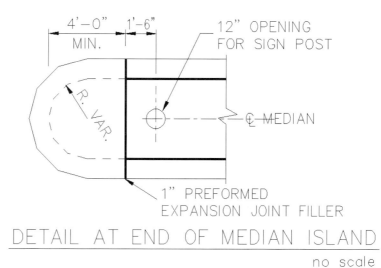

FIG. 7.9.4d: Median Island Detail (Special thanks to Randy Behounek at the Sarpy County Surveyors Office in Papillion, Nebraska, for permission to use this detail.)

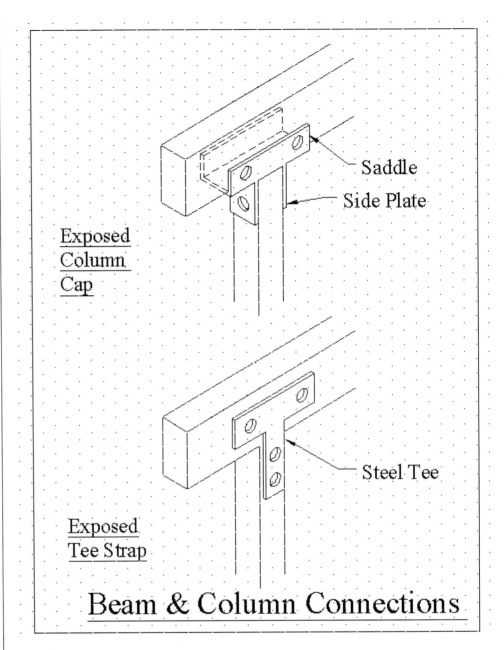

FIG. 7.9.4e: Beam and Column Connections

7.10 REVIEW QUESTIONS

Please write your answers on a separate sheet of paper.

1. (T or F) As a rule, the user should not combine the simple color and linetype command approach with the layer approach.

2. _____ is the system variable that stores the drawings current color.

3. The color selection drop-down box is located on the _____ toolbar.

4. The color selection _____ assigns colors according to the current layer setting.

5. and 6. To reach the Select Color dialog box, pick the _____ option from the color selection box, or type _____.

7. to 13. List AutoCAD's seven primary colors.

14. and 15. To assign a color that is not one of AutoCAD's primary colors, select it from the _____ or type its number code in the _____ box of the Select Color dialog box.

16. To get command line options instead of a dialog box for the linetype or layer command, put a _____ before the command.

17. To get to the Load or Reload Linetypes dialog box, pick the _____ button on the Linetype Manager.

18. What is the name of the default file in which AutoCAD's linetype definitions are stored?

19. To select more than one linetype at a time to load, hold down the _____ key while selecting the linetypes.

20. and 21. To adjust the size of dashes and spaces in dashed lines to the scale of the drawing, set the _____ sysvar to the scale _____ of the drawing.

22. (T or F) Layer settings override color and linetype settings.

Identify these layer symbols:

23. On **24.** Freeze... **25.** L... **26.** Color ■ White □ 11 **27.** Plot

28. and 29. An asterisk entered at a command prompt is a _____, which means _____.

30. You must set a layer _____ before you can use it.

31. To set a layer current, select it in the _____ box.

Identify these buttons:

32. **33.** **34.** **35.**

36. to 38. Name three commands you can use to change an object's properties.

39. Use the _____ to copy layer definitions from one drawing to another.

LESSON 8

Editing Your Drawing: Modification Procedures

Following this lesson, you will:
➡ *Be familiar with several modification procedures, including:*
  - **Trim, Extend,** & **Change**
  - **Break**
  - **Fillet** & **Chamfer**
  - **Move, Align,** & **Copy**

➡ *Be familiar with the Multiple Document Environment*
➡ *Be familiar with Windows' procedures for moving and copying objects between drawings*
➡ *Have mastered the Basic Section of our Text!*

Although we will cover some more advanced drawing commands and techniques in later chapters, the first seven lessons have made you comfortable with AutoCAD's basic approach to 2-dimensional drawing. But you probably still feel a bit clumsy with some of your work. This derives partly from the newness of CAD to you, partly from the need to erase and redo an effort because of small mistakes, and partly from the need to draw the same thing over and over as you did the remote's buttons in our last lesson's exercise. Additionally, you cannot create some more complex drawings that might be simple on the drawing board—given templates and an electric eraser.

In this lesson, we will tackle several commands meant to save drawing time and effort. There are many commands to cover, but you can relax. They are not difficult.

We will divide the basic modification routines into two groups: the *Change Group*, which will include commands designed to change an object's appearance or basic properties, and the *Location & Number Group*, which will include commands designed to move or duplicate existing objects.

8.1 The Change Group

■ 8.1.1 Cutting It Out with the *Trim* Command

One of the first questions I usually hear as students begin to get a handle on what they are doing is, "Is there a way to erase just part of the line?" Unfortunately, they usually start asking this back in Lesson 3 or 4, and they have to wait for the answer.

The answer is, "Yes, you can *remove* part of a line or other object." Let's look at how.

You will often find it easier to draw one long line across an area and then cut away the extra pieces than to draw several shorter lines. AutoCAD designed the *Trim* command to remove the "extra" bits of lines and circles. Here is the command sequence:

Command: *trim* (or *tr*)
Current settings: Projection=UCS Edge=None
Select cutting edges ...
Select objects: *[select the cutting edge]* **1 found**
Select objects: *[enter to confirm completion of this selection set]*

Select object to trim or [Project/Edge/Undo]: *[select the part of which you want to be rid]*

Select object to trim or [Project/Edge/Undo]: *[enter to confirm completion of this selection set]*

Command:

- The **cutting edges** are usually lines or circles to which you want to trim. In other words, as you select a **cutting edge**, say to yourself, "I want to cut back *to here.*"
- The **object to trim** is the piece of a line or circle you want to remove from the drawing. When selecting the **object to trim**, say to yourself, "I want to get *rid of* this."
- The **Project** refers to a UCS projection—something we will cover in our discussion of 3-dimensional drafting in our next book. Ignore it for now.
- **Edge** refers to one of the more useful innovations AutoCAD has provided—the ability to trim an object even if the object does not touch our **cutting edge**.
- The **Undo** *option*, of course, will undo the last modification within the command. Remember, the *Undo command* will undo the entire *Trim* command modification (all the changes made by the command).

Let's experiment.

> In addition to the command line and toolbar, you will find all of the commands in this lesson in the Modify pull-down menu.

WWW 8.1.1.1 Do This: Using the *Trim* Command

I. Open *trim-extend.dwg* in the C:\Steps\Lesson08 folder. The drawing looks like Figure 8.1.1.1a.

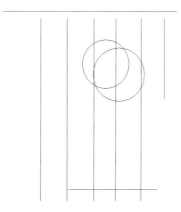

FIG. **8.1.1.1a:** Trim-Extend.dwg

II. Follow these steps.

TOOLS	COMMAND SEQUENCE	STEPS
Trim Button	**Command:** *tr*	1. Enter the *Trim* command—by typing *trim,* or *tr,* at the command prompt. Or you can select the **Trim** button from the Modify toolbar.
	Current settings: Projection=UCS Edge=None Select cutting edges ... **Select objects:**	2. Select the circles at points 1 and 2, and the line at point 3 as shown in Figure 8.1.1.1.2.

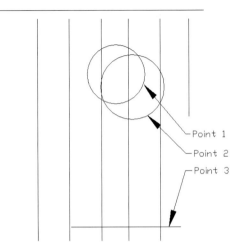

FIG. 8.1.1.1.2: Select Cutting Edges

	Select objects: *[enter]*	3. Confirm that you have completed selecting the **cutting edges** by hitting *enter*.
	Select object to trim or [Project/Edge/ Undo]:	4. Select what you want to trim—the circles at points 1 and 2, and everything that extends below the line you selected in Step 2, as shown in Figure 8.1.1.1.4. Notice that the leftmost line you selected above does *not* trim. We will fix that.

continued

TOOLS	COMMAND SEQUENCE	STEPS

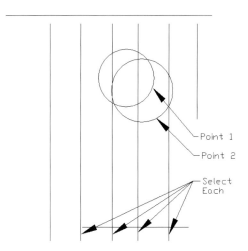

FIG. 8.1.1.1.4: Select Object to Trim

	Select object to trim or [Project/Edge/ Undo]: *e*	**5.** Without leaving the *Trim* command, type *e* to select the **Edge** option.
	Enter an implied edge extension mode [Extend/No extend] <No extend>: *e*	**6.** Type *e* again to select the **Extend** option.
	Select object to trim or [Project/Edge/Undo]:	**7.** Now again select the piece that did *not* trim in Step 4 and the bottom of the leftmost vertical line. Notice the difference? In Steps 5 and 6, you told AutoCAD to *extend* the cutting edge as an invisible plane in both directions. AutoCAD then used that invisible *edge* to trim the objects.
	Select object to trim or [Project/Edge/Undo]: *[enter]*	**8.** Hit *enter* to complete the command. Your drawing looks like Figure 8.1.1.1.8.

continued

TOOLS	COMMAND SEQUENCE	STEPS

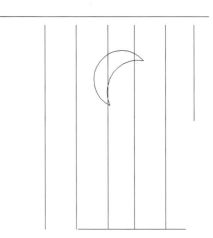

FIG. 8.1.1.1.8: Trimmed Objects

Command: *tr*

9. Now follow the preceding procedures to trim the lines inside the moon. Your drawing will look like Figure 8.1.1.1.9.

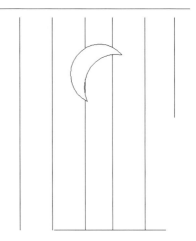

FIG. 8.1.1.1.9: Completed Exercise

Command: *qsave*

10. Save your drawing, but do not exit.

The *Trim* command, as useful as it is, is only one side of a coin. The opposite side is the *Extend* command, which is just as useful even if performing the opposite task.

■ 8.1.2 Adding to It with the Extend Command

Perhaps more frustrating than realizing that your line has gone too far is realizing that you have come up short. The *Trim* command handles the former; the *Extend* command handles the latter.

Here is the command sequence:

Command: *extend* (or *ex*)
Current settings: Projection=UCS Edge=Extend
Select boundary edges . . .
Select objects: *[select the boundary edge]* **1 found**
Select objects: *[enter to confirm completion of this selection set]*
Select object to extend or [Project/Edge/Undo]: *[select the object to extend]*
Select object to extend or [Project/Edge/Undo]: *[enter to confirm completion of this selection set]*
Command:

Each step corresponds to the same step in the *Trim* command sequence. The only difference is that, with the *Extend* command, you will select a **boundary edge** —the place to which you want to extend a line or arc. (Say to yourself, "I want to extend *to here*.")

Notice that the **Edge**, while defaulting to **None** in the *Trim* command, now defaults to **Extend**. This is because *Trim* and *Extend* share the **Edgemode** system variable. When you set the **Edge extension mode** to **Extend** in Steps 5 and 6 of the last exercise, you set it to **Extend** for the *Extend* command as well. This will become clearer as we do the next exercise.

WWW 8.1.2.1 Do This: Using the Extend Command

I. Be sure you are still in the *Trim-extend.dwg* file. If not, please open it now.

II. Follow these steps.

TOOLS	COMMAND SEQUENCE	STEPS
Extend Button	**Command:** *ex*	1. Enter the *Extend* command by typing *extend*, or *ex*, at the command prompt. Or you can select the **Extend** button from the Modify toolbar.
	Select boundary edges: (Projmode = UCS, Edgemode = Extend)	2. Select the **boundary edges** (Figure 8.1.2.1.2).

continued

TOOLS	COMMAND SEQUENCE	STEPS
		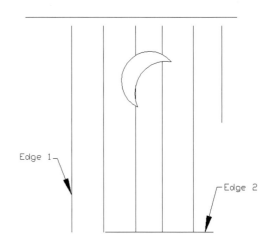
FIG. 8.1.2.1.2: Select the Boundary Edges
	Select objects: *[enter]*	3. Hit *enter* to complete the selection set.
	<Select object to extend>/Project/Edge/Undo:	4. Select the two objects to extend (Figure 8.1.2.1.4).
		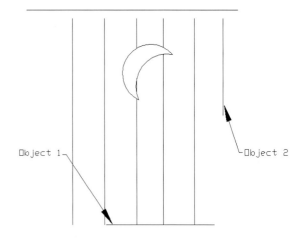
FIG. 8.1.2.1.4: Select the Objects to Extend
	<Select object to extend>/Project/Edge/Undo: *[enter]*	5. Hit *enter* to confirm that you are finished. Your drawing looks like Figure 8.1.2.1.5.

continued

TOOLS	COMMAND SEQUENCE	STEPS

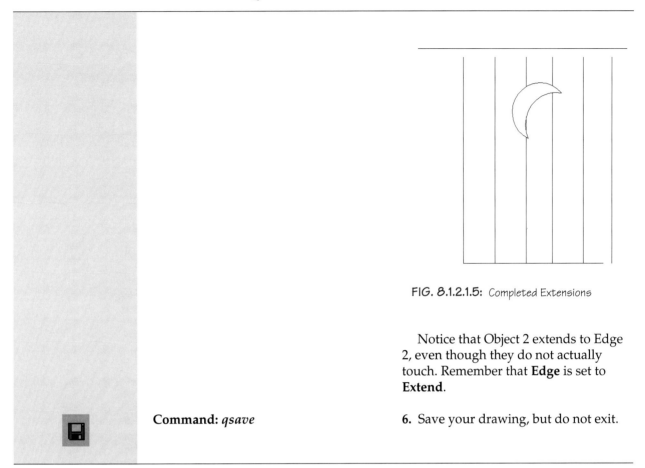

FIG. 8.1.2.1.5: Completed Extensions

Notice that Object 2 extends to Edge 2, even though they do not actually touch. Remember that **Edge** is set to **Extend**.

Command: *qsave* 6. Save your drawing, but do not exit.

You have probably already recognized the value of these two simple commands. The ability to trim away unneeded material or extend a line or arc will save a lot of redraw time.

■ 8.1.3 The Forgotten Change Command

I would like to include a command that has taken a back seat to most of the newer innovations in recent AutoCAD releases—the *Change* command. Remember the *CHProp* command from our last lesson? The *Change* command is its father. That is, the *Change* command has an option called **Properties**. Over time, AutoCAD shortened the sequence:

Command: *change* (or —*ch*)
Select objects: *[select an object]*

Select objects: *[enter]*
Specify change point or [Properties]: *p*

to

Command: *chprop*
Select objects:

But AutoCAD never eliminated the *Change* command. And if you accept the **Change point** default rather than going with **Properties**, you can lengthen or shorten many items at once. Here is how.

WWW 8.1.3.1 Do This: Using the Change Command

I. Be sure you are still in the *Trim-extend.dwg* file. If not, please open it now.

II. Follow these steps.

TOOLS	COMMAND SEQUENCE	STEPS
F8 Ortho Toggle	Command: *ortho* Enter mode [ON/OFF] <ON>: *Off*	1. Be sure **Ortho** is toggled **Off**. (Refer to the status bar to be sure **Ortho** is raised.)
	Command: *-ch*	2. Enter the *change* command by typing *change* or —*ch*—there are no buttons for this one (ah, the fate of the forgotten command).
	Select objects:	3. Select the six vertical lines with a crossing window, as shown in Figure 8.1.3.1.3.

continued

TOOLS	COMMAND SEQUENCE	STEPS
		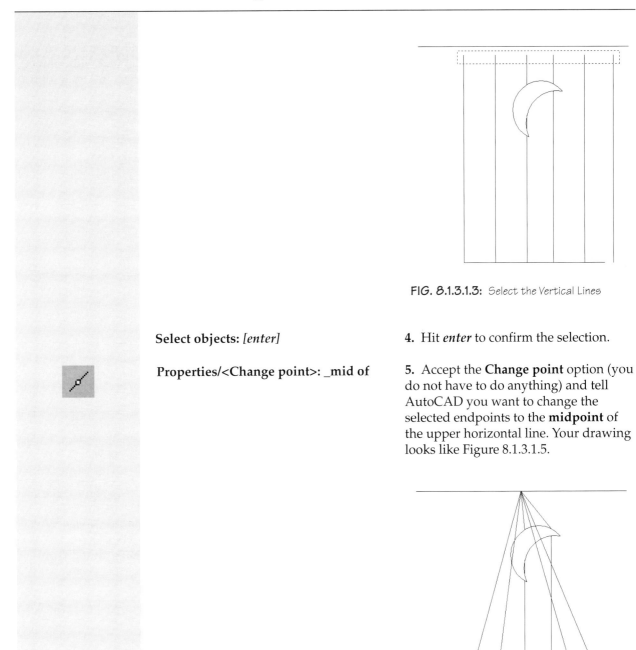 FIG. 8.1.3.1.3: Select the Vertical Lines
	Select objects: *[enter]*	4. Hit *enter* to confirm the selection.
	Properties/<Change point>: **_mid of**	5. Accept the **Change point** option (you do not have to do anything) and tell AutoCAD you want to change the selected endpoints to the **midpoint** of the upper horizontal line. Your drawing looks like Figure 8.1.3.1.5.

FIG. 8.1.3.1.5: Results of the *Change* Command with Ortho Toggled **Off**

continued

TOOLS	COMMAND SEQUENCE	STEPS
↶	**Command:** *u* **Command:** *ortho*	**6.** That shows what happens when you use the *Change* command without the **Ortho**. Let's *undo* that, and try it with the **Ortho** toggled **On**.
F8	**Enter mode [ON/OFF] <OFF>:** *On* **Command:** *-ch*	**7.** Repeat Steps 2 through 5. Your drawing now looks like Figure 8.1.3.1.7.
		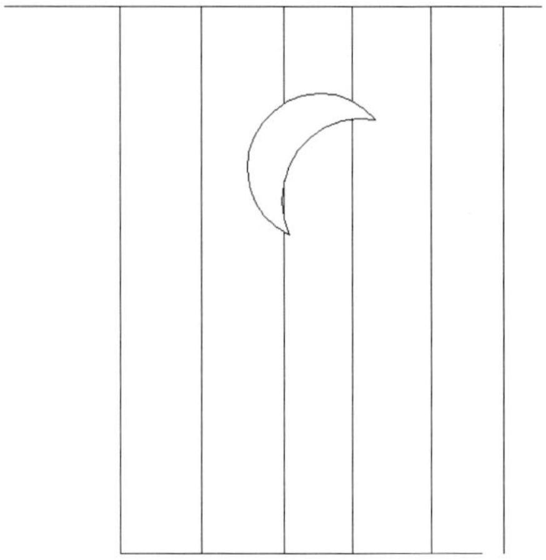 **FIG. 8.1.3.1.7:** Results of the *Change* Command with Ortho Toggled **On**
💾	**Command:** *qsave*	**8.** See the difference? See how different aspects of AutoCAD—in this case, Ortho, OSNAPS, and the *Change* command—work together? Save your drawing, but do not exit.

■ 8.1.4 Redundancy—Thy Name Is AutoCAD: The Break Command

AutoCAD provides yet another way to remove parts of lines or circles. But this one is a bit more tedious than the *Trim* command. The *Break* command sequence is

Command: *break* **(or** *br***)**
Select object: *[select the object to be broken]*
Specify second break point or [First point]: *f*

Specify first break point: *[select the first end of the break]*
Specify second break point: *[select the other end of the break]*
Command:

The section between the first and second selected points is removed.

There is only one option in this sequence—that of entering *F* for the **First point**. If the user enters *F*, AutoCAD prompts for the first end of the break. The user can then use OSNAPS for precise selection of both first and second points. If the user does not enter *F*, AutoCAD assumes the point at which the user selected the line or circle is the first point. As no OSNAP is used to select the line, there is no precision in selecting the first point. So we discover one of those unwritten rules: *always type* **F** *for* **First point** *when using the* **Break** *command*. (Of course, you know what they say about rules.)

One of AutoCAD's forgotten tricks concerns the *Break* command. The user can enter an @ symbol at the **Enter second point:** Prompt to break the object into two pieces without removing anything. The object is broken at the point selected at the **Specify first break point:** prompt.

Let's use the *Break* command to add a handle to our moon door.

WWW | 8.1.4.1 Do This: | Using the Break Command

I. Be sure you are still in the *Trim-extend.dwg* file. If not, please open it now.

II. Follow these steps.

TOOLS	COMMAND SEQUENCE	STEPS
🔲	Command: z	1. *Zoom* in to the area indicated by the arrow in Figure 8.1.4.1.1.

continued

TOOLS	COMMAND SEQUENCE	STEPS

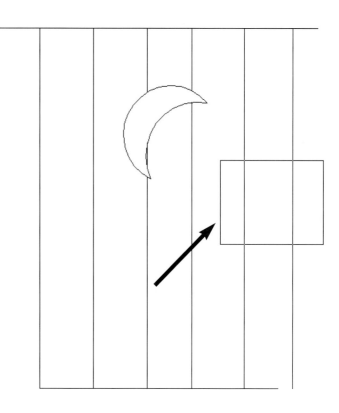

FIG. 8.1.4.1.1: Zoom In

Command: *l*

Specify first point: *7.75,5*

Specify next point or [Undo]: *@2<0*

Specify next point or [Undo]: *[enter]*

Command: *[enter]*

LINE Specify first point: *7.75,4.75*

Specify next point or [Undo]: *@2<0*

Specify next point or [Undo]: *[enter]*

2. On the **obj** layer, draw two lines as indicated. Your screen will look like Figure 8.1.4.1.2.

continued

TOOLS	COMMAND SEQUENCE	STEPS

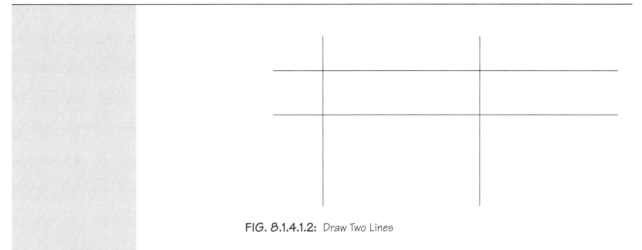

FIG. 8.1.4.1.2: Draw Two Lines

Break Button

Command: *br*

3. Enter the *Break* command by typing **break** or **br**. Or you can select the **Break** button from the Modify toolbar.

Select object:

4. Select the left vertical line.

Specify second break point or [First point]: *f*

5. Type *f* for the first point.

Specify first break point: _int of

Specify second break point: _int of

6. Place the first end of the break on the vertical line at the Point 1 **intersection** (Figure 8.1.4.1.6); select the other end of the break at the Point 2 **intersection**. (Be sure to use OSNAPS.)

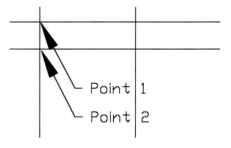

FIG. 8.1.4.1.6: Break between These Lines

Command: *[enter]*

7. Repeat Step 7 for the other line. Your drawing looks like Figure 8.1.4.1.7.

continued

TOOLS	COMMAND SEQUENCE	STEPS

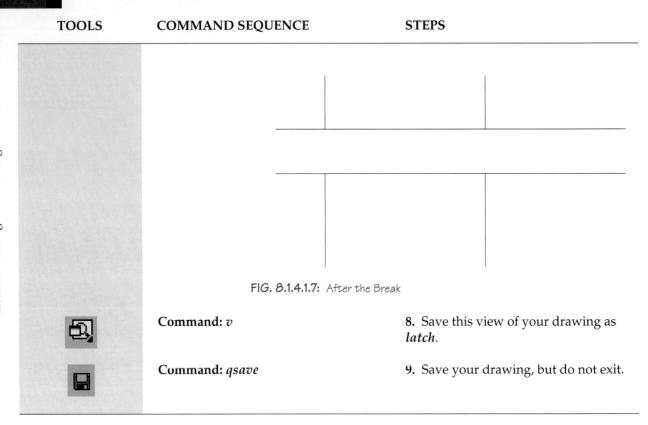

FIG. 8.1.4.1.7: After the Break

	Command: *v*	**8.** Save this view of your drawing as *latch*.
	Command: *qsave*	**9.** Save your drawing, but do not exit.

■ 8.1.5 Now We Can Round That Corner: The Fillet Command

The *Fillet* command provides an easy way to round corners without the need for any of the *Arc* routines; or you can use it to *square* corners! The command sequence is

> **Command:** *fillet* (or *f*)
> **Current settings: Mode = TRIM, Radius = 0.5000**
> **Select first object or [Polyline/Radius/Trim]:** *[select the first object]*
> **Select second object:** *[select the second object]*
> **Command:**

- The default radius for the *Fillet* command is **0.5**, but you can change that by typing *R* for the **Radius** option. AutoCAD will prompt

  Enter fillet radius <0.5000>:

  You can type a different radius, or you can type *0* for a square corner.
- The **Polyline** option allows the user to fillet all the corners of a polyline at once.
- The **Trim** option allows the user to decide whether or not to automatically trim away the excess line, circle, arc, etc., when the *Fillet* operation is done.
- Another option (one that AutoCAD does not show here) allows the user to fillet parallel lines. This procedure draws an arc connecting the endpoints of both lines. We will see how it works in the next exercise.

8.1.5.1 Do This: Using the Fillet Command

I. Be sure you are still in the *Trim-extend.dwg* file. If not, please open it now.
II. *Zoom all*.
III. Follow these steps.

TOOLS	COMMAND SEQUENCE	STEPS
 Fillet Button	**Command:** *f*	1. Enter the *Fillet* command by typing *fillet* or *f* at the command prompt. Or you can pick the **Fillet** button from the Modify toolbar.
	Current settings: Mode = TRIM, Radius = 0.5000 **Select first object or [Polyline/Radius/Trim]:** **Select second object:**	2. Accept the default radius for now and select the upper and right lines at points 1 and 2, as shown in Figure 8.1.5.1.2.

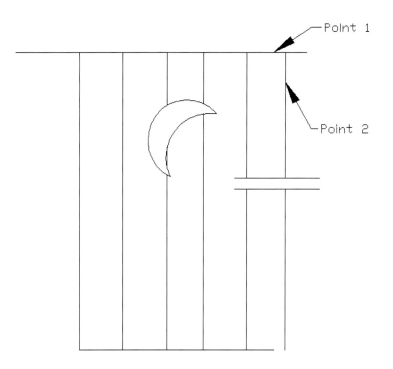

FIG. 8.1.5.1.2: Select These Points to Fillet the Lines

continued

TOOLS	COMMAND SEQUENCE	STEPS
		Notice that the edge is rounded (with a .5 radius arc) and the excess lines are automatically trimmed away.
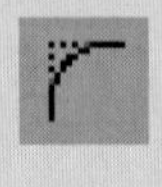	**Command:** *[enter]*	**3.** Repeat Steps 1 and 2 at the lower right corner of the door. Your drawing will look like Figure 8.1.5.1.3.

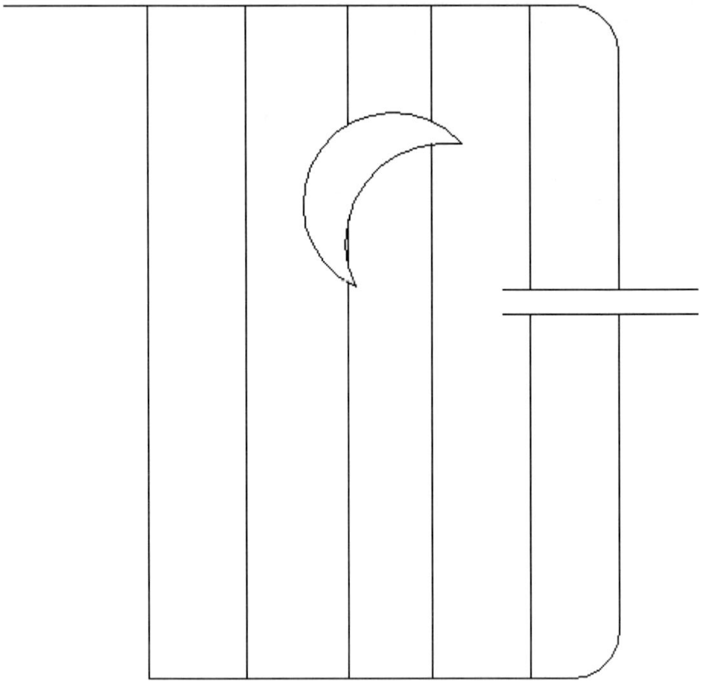

FIG. 8.1.5.1.3: Filleted Corners

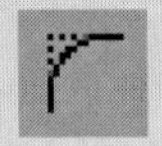	**Command:** *[enter]*	**4.** Now let's square the other corners. Repeat the *Fillet* command.
	Current settings: Mode = TRIM, Radius = 0.5000	**5.** Type *R* for the **Radius** option.
	Select first object or [Polyline/Radius/Trim]: *r*	
	Specify fillet radius <0.5000>: *0*	**6.** Enter *0* for a zero-degree radius. Notice that AutoCAD returns you to the **Command** prompt.

continued

TOOLS	COMMAND SEQUENCE	STEPS
	Command: *[enter]*	**7.** Hit *enter* to repeat the command.
	Command: *[enter]*	**8.** Notice that the **Current fillet radius** is now *0*. Fillet the other two corners. Your drawing looks like Figure 8.1.5.1.8.
	Current settings: Mode = TRIM, Radius = 0.0000	
	Select first object or [Polyline/Radius/Trim]:	
	Select second object:	

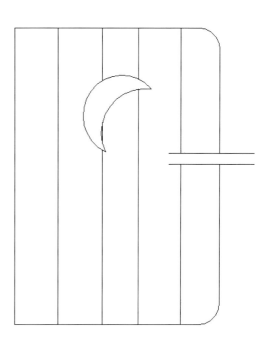

FIG. 8.1.5.1.8: *Four Filleted Corners*

	Command: *v*	**9.** Restore the *latch* view you created at the end of the last exercise.
	Command: *f*	**10.** Fillet the two lines on both ends. Do not worry about the radius; AutoCAD will calculate it from the distance between the parallel lines.
	Current settings: Mode = TRIM, Radius = 0.0000	(Add a couple of 0.05"R circles for an anchor and a lever, if you want —see Figure 8.1.5.1.11.)
	Select first object or [Polyline/Radius/Trim]:	
	Select second object:	*continued*

TOOLS	COMMAND SEQUENCE	STEPS
 Zoom Limits Button	**Command:** *zoom* All/Center/Dynamic/Extents/Previous/ Scale(X/XP)/Window/<Realtime>: *a*	**11.** Zoom back out again. Your drawing will look like Figure 8.1.5.1.11.
		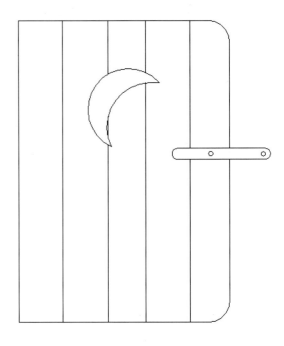 **FIG. 8.1.5.1.11:** Completed Door
	Command: *qsave*	**12.** Save the drawing.

The **Trim** option, available for both the *Fillet* and *Chamfer* commands was added to AutoCAD after users complained that they do not always want to trim away the excess part of the lines during *Fillet* or *Chamfer* procedures. When selecting the **Trim** option, AutoCAD prompts:

Enter Trim mode option [Trim/No trim] <Trim>:

Notice that **Trim** is the default. AutoCAD will automatically trim away the excess. If the user wants to keep the excess, he enters *N*. AutoCAD sets the **Trimmode** (the system variable that controls the **Trim/No trim** option) accordingly, and the objects are filleted but the excess remains.

8.1.6 Fillet's Cousin: The Chamfer Command

Fillet and *Chamfer* are so similar, you might get confused as to which one you want. Look at the pictures on the buttons on the Modify toolbar if you get confused.

Where *Fillet* rounds corners, *Chamfer* provides a mitre—a flat edge at a corner —much like a carpenter achieves with a hand plane. The user controls the size and angle of the edge through responses to the prompts, which look like this:

> **Command:** *chamfer* (or *cha*)
> **(TRIM mode) Current chamfer Dist1 = 0.5000, Dist2 = 0.5000**
> **Select first line or [Polyline/Distance/Angle/Trim/Method]:** *[select the first chamfered line]*
> **Select second line:** *[select the second chamfered line]*
> **Command:**

- The default **Method** is the **Distance** method. That is the one shown in the sample sequence. Using the **Distance** option means that AutoCAD measures a user-defined distance from the corner (real or apparent) on both lines, puts a line between these two points, and either **Trims** or does not **Trim** the lines (depending on the **Trimmode** setting).
- Another **Method** is the **Angle** method. Here, the user defines one distance (as in the **Distance** method) and an angle of cut.
- The user can switch between methods (**Distance** or **Angle**) using the **Method** option.
- The **Polyline** method can be used to chamfer entire polylines, but the results can be surprising. I tend to shy away from chamfering polylines. (We will look at **Polylines** in Lesson 9.)
- These methods will become clearer with an exercise.

WWW 8.1.6.1 Do This: Using the Chamfer Command

I. Be sure you are still in the *Trim-extend.dwg* file. If not, please open it now.

II. *Zoom all*.

III. Erase the arcs we created using the *Fillet* command.

IV. Follow these steps.

TOOLS	COMMAND SEQUENCE	STEPS
 Chamfer Button	**Command:** *cha*	1. Enter the *Chamfer* command by typing *chamfer* or *cha* at the command prompt. Or you can select the **Chamfer** button from the Modify toolbar.

continued

TOOLS	COMMAND SEQUENCE	STEPS
	(TRIM mode) Current chamfer Dist1 = 0.5000, Dist2 = 0.5000 **Select first line or [Polyline/ Distance/Angle/Trim/ Method]:** *d*	**2.** Let's change the **Dist1** and **Dist2** settings so that we can better tell what is happening. Type *D* to select the **Distance** option.
	Specify first chamfer distance <0.5000>: .25 **Specify second chamfer distance <0.2500>:** .75	**3.** AutoCAD prompts for each distance. Enter **.25** for the **first chamfer distance**, and **.75** for the **second chamfer distance**. AutoCAD returns to the command prompt.
	Command: *[enter]*	**4.** Repeat the command.
	(TRIM mode) Current chamfer Dist1 = 0.2500, Dist2 = 0.7500 **Select first line or [Polyline/ Distance/Angle/Trim/Method]:** *[select the first chamfered line at point 1]* **Select second line:** *[select the second chamfered line at point 2]*	**5.** Now AutoCAD shows the new distances. Select the top line of our door then the right vertical line where indicated in Figure 8.1.6.1.5.

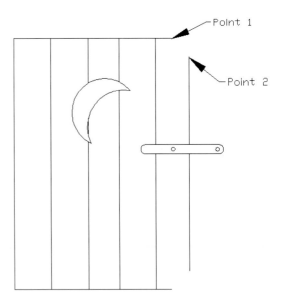

FIG. 8.1.6.1.5: Chamfer this Corner

continued

TOOLS	COMMAND SEQUENCE	STEPS
	Command: *[enter]*	6. Repeat Step 5 at the bottom corner. Your drawing will look like Figure 8.1.6.1.6.

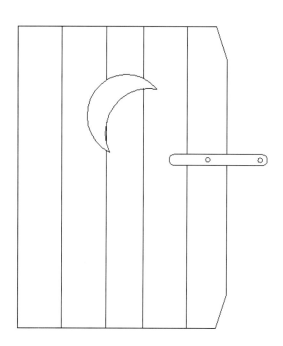

FIG. **8.1.6.1.6:** Chamfered Corners

	Command: *[enter]*	7. Let's try the **Angle** method. Repeat the command.
	(TRIM mode) Current chamfer Dist1 = 0.2500, Dist2 = 0.7500	8. Type *a* to select the **Angle** option.
	Select first line or [Polyline/ Distance/Angle/Trim/Method]: *a*	
	Specify chamfer length on the first line <1.0000>: *.5*	9. Enter *.5* as the **chamfer length on the first line**; then enter *60* as the **chamfer angle from the first line**. AutoCAD returns to the command prompt.
	Specify chamfer angle from the first line <0>: *60*	
	Command: *[enter]*	10. Repeat the command.

continued

TOOLS	COMMAND SEQUENCE	STEPS
	(TRIM mode) Current chamfer Length = 0.5000, Angle = 60 **Select first line or [Polyline/ Distance/Angle/Trim/Method]:** *[select the upper line at Point 1]* **Select second line:** *[select the vertical line at Point 2]*	**11.** AutoCAD now reports the **Length** and **Angle** settings, and prompts you for the lines to chamfer. Select the points indicated in Figure 8.1.6.1.11.

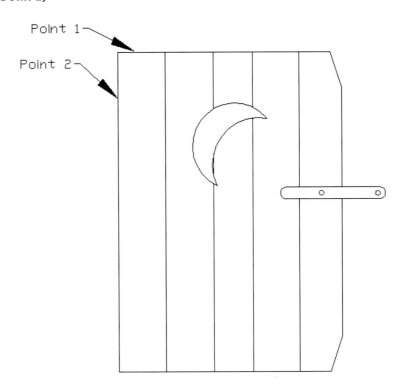

FIG. 8.1.6.1.11: Chamfer This Corner

| | **Command:** *[enter]* | **12.** Repeat Step 11 at the bottom-left corner of the door. Your drawing looks like Figure 8.1.6.1.12. |

continued

TOOLS	COMMAND SEQUENCE	STEPS

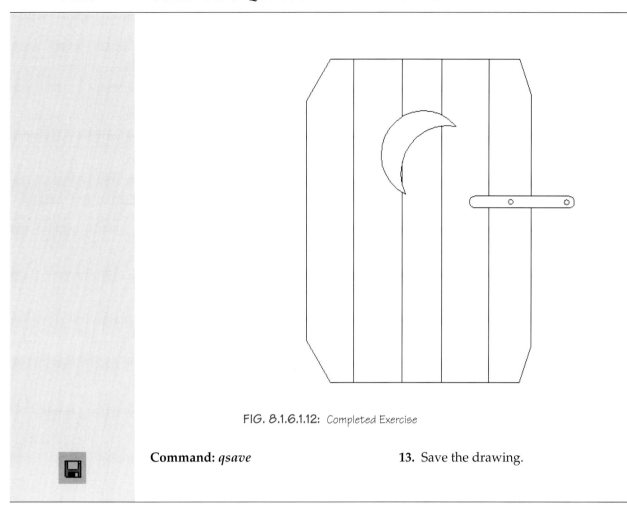

FIG. 8.1.6.1.12: Completed Exercise

	Command: *qsave*	**13.** Save the drawing.

We have completed the six commands of the first part of this lesson. This is a lot of material to absorb with more to come. If you feel you need to practice what you have learned so far, go to the **Exercises** section at the end of the lesson and do number 1 (the **Checkers** exercise). Then return to Section 8.2.

8.2 The Location and Number Group

■ 8.2.1 Here to There: The Move Command

The *Move* command allows the user to *move* one or more objects from one place to another. It has one of the easiest sequences to remember:

Command: *move* **(or *m*)**
Select objects: *[select one or more objects]* **1 found**
Select objects: *[enter to confirm completion of this selection set]*

Specify base point or displacement: *[pick a starting point]*
Specify second point of displacement or <use first point as displacement>: *[pick a target point]*
Command:

The only options here involve the **Base point** and the **displacement**.

- The easiest way to explain **Base point** is this: Imagine the object(s) you are moving as a solid object sitting on a table. To *move* the object, you must first pick it up. The **Base point** is the place you grab (a corner, an edge, the middle, etc.). The **Second point of displacement** is where you put it down—or the point at which you will place the corner or edge (or whatever) you grabbed.

- Notice the **Base point or displacement** prompt. One mistake of even seasoned CAD operators is to read this as base point **of** displacement, which is wrong. They miss an easier way to move the object(s). Using the **displacement** method, the sequence would look like this:

Command: *move* (or *m*)
Select objects: *[select one or more objects]* **1 found**
Select objects: *[enter to confirm completion of this selection set]*
Specify base point or displacement: **2,0**
Specify second point of displacement or <use first point as displacement>: *[enter]*
Command:

Here we have moved the object(s) two units in the X-direction and zero in the Y-direction. We did this by entering a displacement in terms of X and Y rather than selecting a base point.

WWW 8.2.1.1 Do This: Using the Move Command

I. Open the *star.dwg* in the C:\Steps\Lesson08 folder. The drawing looks like Figure 8.2.1.1a.

II. Follow these steps.

FIG. 8.2.1.1a: *Star.dwg*

TOOLS	COMMAND SEQUENCE	STEPS
Move Button	**Command:** *m*	**1.** Enter the *Move* command by typing *move* or *m* at the command line. Or you can select the **Move** button from the Modify toolbar.

continued

TOOLS	COMMAND SEQUENCE	STEPS
Snap to Quadrant	Select objects: Specify opposite corner: 13 found Select objects: *[enter]*	2. Put a selection window around the entire circle and all its contents. Then hit *enter* to confirm completion of the selection set.
	Specify base point or displacement: _qua of **Specify second point of displacement or <use first point as displacement>:** *@4<0*	3. Select the lower quadrant of the circle. (You may select any point, but this one is convenient.) Use polar coordinates to move the circle 4 units to the right. Notice how the circle changes position.
	Command: *[enter]*	4. We will now use the **displacement** method to move it back. Repeat the command.
	Select objects: *p* Select objects: *[enter]*	5. Type *p* to use the **previous** selection set, then confirm the selection.
	Specify base point or displacement: *-4,0* **Specify second point of displacement or <use first point as displacement>:** *[enter]*	6. Rather than selecting a **Base point**, type in the units you want to move. At the **Second point of displacement** prompt, hit *enter*. Notice that the objects move back.

As you can see, there really is not much to the *Move* command. But you do need to remember things like OSNAPS to ensure precision in your modifications.

■ 8.2.2 Okay, Move It—But Then Line It Up: The Align Command

A variation of the *Move* command, *Align* will move objects and then *align* them with something else. The command sequence is

> **Command:** *align* (or *al*)
> **Select objects:** *[select the objects to be moved]*
> **Select objects:** *[enter to confirm completion of selection set]*
> **Specify first source point:** *[this is the first point where you "grab" the objects]*
> **Specify first destination point:** *[this is where you put the first grabbed point]*
> **Specify second source point:** *[this is a second point where you grab the objects—you need at least two points so you can align the objects]*

Specify second destination point: *[this is where you put the second grabbed point]*

Specify third source point or <continue>: *[if a third point is necessary (usually not unless you are aligning a 3-dimensional object), you may select a third point, otherwise hit enter to continue]*

Scale objects to alignment points? [Yes/No] <No>: *[do you want to adjust the size of the object you are moving so that the grabbed points fit exactly on the destination points?]*

It is not as complicated as it looks.

WWW 8.2.2.1 Do This: Using the Align Command

I. Open the *align.dwg* from the C:\Steps\Lesson08 folder.

II. Follow these steps.

TOOLS	COMMAND SEQUENCE	STEPS
(osnap icon) No Button Available	**Command:** *os*	1. Turn on the **endpoint** Running OSNAP. Clear all other OSNAPs.
	Command: *al*	2. Enter the *Align* command by typing *align* or *al* at the command prompt.
	Select objects: **Select objects:** *[enter]*	3. Select box 1 and the text inside. Confirm completion of the selection set.
	Specify 1st source point: *[select Point 1a]* **Specify 1st destination point:** *[select Point 1b]* **Specify 2nd source point:** *[select Point 2a]* **Specify 2nd destination point:** *[select Point 2b]* **Specify 3rd source point or <continue>:** *[enter to complete]*	4. Using the **endpoint** OSNAP, select the points, as indicated in Figure 8.2.2.1.4.

continued

TOOLS	COMMAND SEQUENCE	STEPS

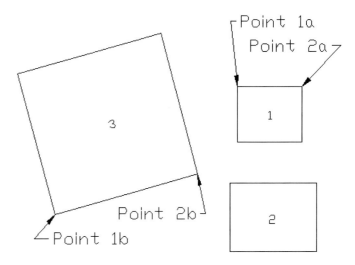

FIG. 8.2.2.1.4: Source and Destination Points

	Scale objects to alignment points? [Yes/No] <No>: *[enter]*	**5.** Do not **Scale objects** this time. Your drawing looks like Figure 8.2.2.1.5.

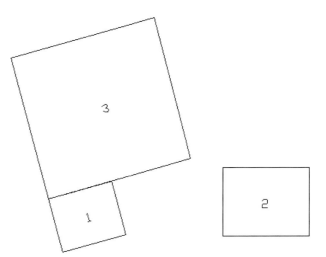

FIG. 8.2.2.1.5: Aligned Boxes

continued

TOOLS	COMMAND SEQUENCE	STEPS
	Command: *[enter]*	6. Repeat the *Align* command. Let's try it again using box 2; this time, we will scale the objects.
	Select objects: *[select box 2 and the text inside]* **2 found**	
	Select objects: *[enter]*	
	Specify 1st source point: *[select Point 1a]*	7. Select the points as indicated in Figure 8.2.2.1.7.
	Specify 1st destination point: *[select Point 1b]*	
	Specify 2nd source point: *[select Point 2a]*	
	Specify 2nd destination point: *[select Point 2b]*	
	Specify 3rd source point or <continue>: *[enter to complete]*	

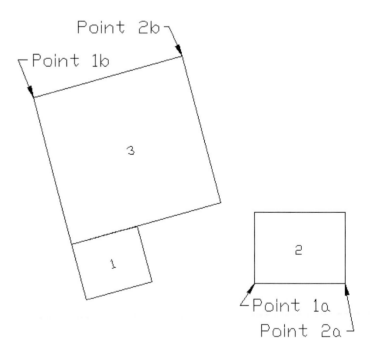

FIG. 8.2.2.1.7: More Source and Destination Points

continued

TOOLS	COMMAND SEQUENCE	STEPS

Scale objects to alignment points? [Yes/No] <No>: *y*

8. Scale the objects. Your drawing looks like Figure 8.2.2.1.8.

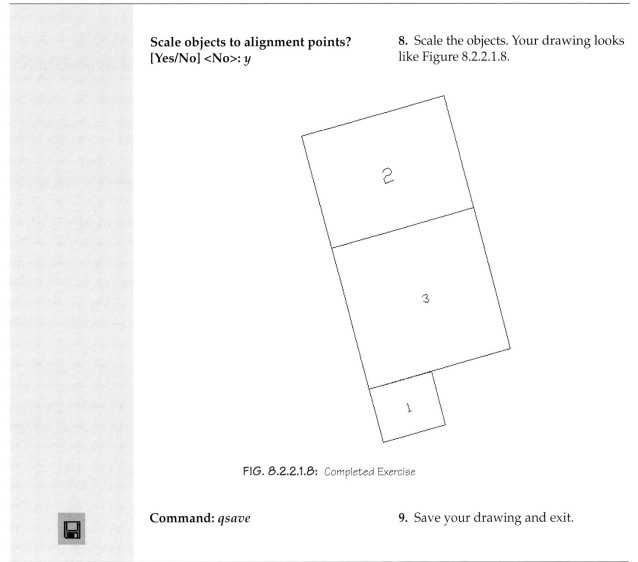

FIG. 8.2.2.1.8: Completed Exercise

Command: *qsave*

9. Save your drawing and exit.

You can see, then, that the *Align* command is related to the *Move* command, although it is a bit more complex. You might say it is also related to two commands we have not yet studied—*Rotate* and *Scale.* Wow! Three commands in one! We will look at the *Rotate* and *Scale* commands in Lesson 10.

■ 8.2.3 The Copy Command: From One to Many

Another command that is related to *Move*, *Copy* has a command sequence that closely resembles the *Move* command sequence. Compare the following sequence with the *Move* sequence in Section 8.2.1.

Command: *copy* (or *co* or *cp*)
Select objects: *[select one or more objects]* 1 found
Select objects: *[enter to confirm completion of this selection set]*
Specify base point or displacement, or [Multiple]: *[pick a starting point]*
Specify second point of displacement or <use first point as displacement>: *[pick a target point]*
Command:

All the options are the same including the **displacement** method. But *Copy* provides one additional option—**Multiple**. The **Multiple** option allows the user to make more than one copy of an object(s).

WWW 8.2.3.1 Do This: Using the Copy Command

I. Reopen the *star.dwg* from the C:\Steps\Lesson08 folder.

II. Follow these steps.

TOOLS	COMMAND SEQUENCE	STEPS
 Copy Button	**Command:** *co*	**1.** Enter the *Copy* command by typing *copy*, *cp*, or *co* at the command prompt. Or you can select the **Copy** button from the Modify toolbar.
	Select objects: **Select objects:** *[enter]*	**2.** Put a selection window around the circle and all its contents. Then hit *enter* to confirm completion of the selection set.
	**Specify base point or displacement, or [Multiple]:** _qua of **Specify second point of displacement or <use first point as displacement>:** @4<0	**3.** Select the lower quadrant of the circle as the **base point**. Use polar coordinates to make a copy of the circle 4 units to the right. Your drawing looks like Figure 8.2.3.1.3.

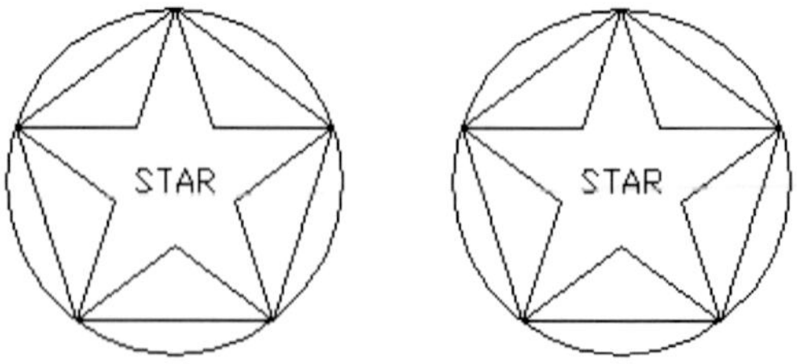

FIG. 8.2.3.1.3: Copied Circle

continued

TOOLS	COMMAND SEQUENCE	STEPS
	Command: *e*	4. Erase the new circle.
	Command: *co*	5. Now let's make **Multiple** copies. Repeat the *Copy* command.
	Select objects:	6. Select the objects as you did in Step 2.
	Select objects: *[enter]*	
	Specify base point or displacement, or [Multiple]: *m*	7. Type *m* for the **Multiple** option.
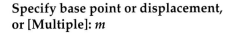	**Specify base point:** _cen of	8. The next prompt changes only slightly. Select the center of the circle (just to be different) as the **Base point**.
	Specify second point of displacement or <use first point as displacement>: *@4<0*	9. Make three copies of the objects, as indicated. Your drawing will look like Figure 8.2.3.1.9.
	Specify second point of displacement or <use first point as displacement>: *@8<0*	
	Specify second point of displacement or <use first point as displacement>: *@3<90*	
	Specify second point of displacement or <use first point as displacement>: *[enter]*	

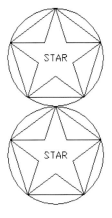

FIG. 8.2.3.1.9: Multiple Copies

continued

TOOLS	COMMAND SEQUENCE	STEPS
💾	Command: *qsave*	10. Save the drawing, but do not exit.

As you can see Copy, like Move, requires little effort to master. But the benefits can be wondrous in terms of time and effort saved.

> Like so many other users, I was thrilled to find the *Align* command when AutoCAD included it with the software. Combining *Move* with *Rotate* was a stroke of genius. Unfortunately, there is no equivalent for the *Copy* command.
>
> Here again is where AutoCAD has provided the ability to adjust its CAD package to the user's need. I have included another Lisp routine in the C:\Steps\Lesson08 folder that will enable the user to copy and rotate with one command.
>
> Loading *Coprot.lsp* will provide you with a command called *CR*. When accessed, you will be prompted for the object(s) to be copied, the base point, the new location, and the desired rotation angle. AutoLISP will do the rest!

8.3 Moving and Copying Objects *between* Drawings

AutoCAD has given us the ability to move or copy objects from one drawing to another as well as within a single drawing. This feature—new with AutoCAD 2000—makes use of the Windows *Cut/Paste* or *Copy/Paste* commands. It also takes advantage of AutoCAD's new Multiple Document Environment (MDE), which enables the user to open more than one drawing at a time. (Oh, the wonders that can happen with a little cooperation!)

You have probably seen that the *Copy* and *Move* commands can be found in the Modify pull down menu. These, or course, are AutoCAD commands. But did you notice that the *Copy* command is also located in the Edit pull down menu? This command (and the other commands in this menu) belongs to Windows. It takes advantage of the Windows clipboard (Windows method of copying and moving objects and files within and between Windows' documents). It also makes it possible to copy and move objects between AutoCAD drawing files.

The Windows method of copying requires two steps: *Copy* and *Paste*. *Copy* places the item(s) on the clipboard (an *imaginary* clipboard—a location in your computer's memory); *Paste* takes it from the clipboard and puts it into your document.

The Windows method of moving also requires two steps: *Cut* and *Paste*. *Cut* (like *Copy*) places the item(s) on the clipboard and removes the item(s) from the source location. *Paste* puts the item(s) into your document.

AutoCAD's command equivalents for Windows commands are

WINDOWS	AUTOCAD
Copy	*copyclip*
Cut	*cutclip*
Paste	*pasteclip*

Of course, the best way to understand all of this is to see it in action. Let's do an exercise.

8.3.1 Do This: Using the MDE to Move and Copy Objects between Drawings

I. Be sure you are still in the *star.dwg*. If not, please open it now (it is in the C:\Steps\Lesson08 folder).

II. Follow these steps.

TOOLS	COMMAND SEQUENCE	STEPS
 New Button	**Command:** *new*	1. Without closing the *star* drawing, open a new drawing from scratch.
 FIG. 8.3.1.2		2. The new drawing opens atop the already open *star* drawing. We must position them so that we can see both at one time. Pick the Window pull-down menu and select **Tile Vertically,** as indicated in Figure 8.3.1.2. Notice that the two open drawings are listed at the bottom of the menu. Had you not wished to tile your drawings, you would have used these to toggle between drawings.
		3. The drawings now appear side by side as seen in Figure 8.3.1.3. Pick anywhere in the right drawing to make it active. Notice that the title bar of the active (current) drawing is blue and the other one is gray.

continued

TOOLS	COMMAND SEQUENCE	STEPS

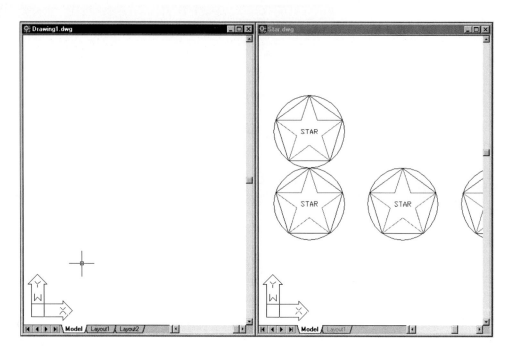

FIG. 8.3.1.3: Two Open Drawings Tiled Vertically

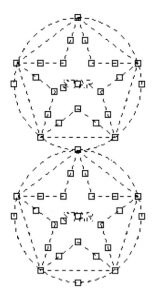

FIG. 8.3.1.4: Selected Objects

4. Without entering a command, place a selection window around the two circles on the left. They now appear highlighted as seen in Figure 8.3.1.4.

continued

TOOLS	COMMAND SEQUENCE	STEPS
	**Command:** *copybase*	5. We could use the *Copy* command in the Edit pull-down menu, but the cursor menu will be more convenient. Right-click anywhere in the active document and select **Copy with Base Point** on the menu (the command line equivalent is *Copybase*). (Had we used the *Copy* command, AutoCAD would have assumed a base point in the lower-left corner of the objects selected.)
	Specify base point: _cen of	6. Select the center of the lower circle as the base point.
	Command:	7. AutoCAD returns to the command line. It does not appear that anything has been done, but the objects are now recorded on the Windows clipboard. Click anywhere in the drawing on the left to activate it. Notice that its title bar turns blue.
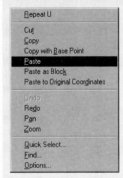	**Command:** *pasteclip*	8. Right-click to access the cursor menu. Notice that you have three **Paste** options. The simple **Paste** (*Pasteclip*) will add the objects to the new drawing as though you had copied them as you did in the last section. **Paste as Block** (*pasteblock*) will add theobjects as a block (more on blocks in Lesson 19). **Paste to Original Coordinates** (*pasteorig*) will add the objects to the new drawing at the same coordinates at which they existed in the source drawing. Select **Paste**.
	Command: _pasteclip Specify insertion point: 7,3.5	9. Paste the objects at coordinates: **7,3.5**. The objects now appear in the new drawing.
	Command: *quit*	10. Exit both drawings without saving.

Some things to remember about this procedure include:

- To move objects, use the **Cut** (*Cutclip* command) option rather than the **Copy** or **Copy with Base Point** option we used in Step 5.
- To clear the highlighting, either hit the **Esc** key twice or pick the **Deselect all** option from the cursor menu.
- Using the clipboard requires a small amount of computer memory. Keep this in mind if you use a particularly slow system (not much RAM) or have problems with your computer's memory.
- You can also use the *Matchprop* command from one drawing to another.
- Hitting *enter* to repeat a command will repeat the last command given in the active drawing (i.e., commands are active document-specific).
- *Remember to* close *a drawing when you are finished with it.* (Type **Close** or select it from the File pull-down menu.)

This wraps up Lesson 8. There are more nifty and wondrous modification tools that we will discuss in Lesson 10. But let's do some exercises to get comfortable with what we have learned so far.

8.4 Extra Steps

After completing this lesson, spend some time experimenting with the new Lisp routines mentioned. All the lisp routines included in this book are yours to keep, play with, pass around, and so forth. Some are quite primitive; others are more advanced. But the value of each has been proved countless times since I wrote them.

Other Lisp routines will be available to you, some with this text, others where you go to work. One thing about the AutoCAD world—there is no shortage of these programs from which to pick and choose. When you get to your job site, ask around. CAD operators love to share!

8.5 What Have We Learned?

Items covered in this lesson include:

- *AutoCAD Modifying Commands*
  - *Trim*
  - *Extend*
  - *Change*
  - *Break*
  - *Fillet*
  - *Chamfer*
  - *Move*
  - *Align*
  - *Copy*

- *AutoCAD's Multiple Document Environment (MDE)*
- *Window's Commands*
  - Cut
  - Copy
  - Paste

This has been a very good lesson. You have learned the basics of manipulating drawing objects to your advantage. Your drawing time will be noticeably lessened as you become more adept with these tools.

You have completed **Part 1: Getting Started with AutoCAD**. Relax for a moment and think about what you have learned. Sure, it is all still quite new to you. But when you began, did you think you would be able to draw the log cabin (Exercise 3) this soon? Draw it, then pat yourself on the back and go on to **Part 2: Beyond the Basics**!

8.6 EXERCISES

1. Open the *Checkers* drawing in the C:\Steps\Lesson08 folder.
   1.1. Using the commands you have learned in this lesson, create the drawing in Figure 8.6.1 from the objects provided. (Be sure to use different layers and colors to mark the different sides.)

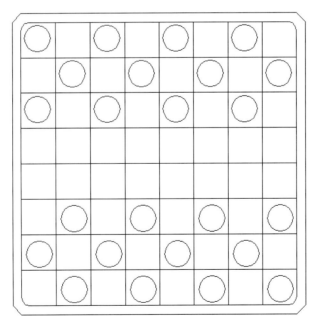
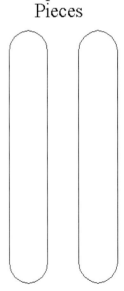

FIG. 8.6.1a: Completed Checkers Project

2. Using the *MyGrid3* template you created in Lesson 1 (or the *Grid3* template in the Lesson01 folder), create the drawing in Figure 8.6.2. Reset the grid to $1/4$" and the snap accordingly. Use the same layers you used in Exercise 3 of Lesson 7. Do not draw the dimensions.

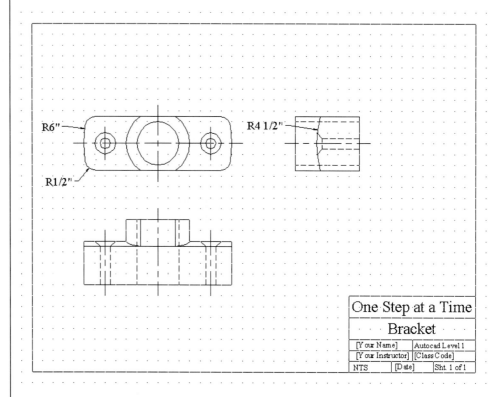

FIG. 8.6.2: Bracket Drawing

3. Using the following setup, create the *log cabin* drawing shown in Figure 8.6.3d
   3.1. Lower left limits: **0,0**
   3.2. Upper right limits: **36,24**
   3.3. Grid and Snap: *[as needed]*
   3.4. Layers:

	Layer name	State	Color	Linetype
3.4.1.	Layer name	State	Color	Linetype
3.4.2.	---------------	-------	---------------	--------------------
3.4.3.	0	On	7 (white)	CONTINUOUS
3.4.4.	CURTAINS	On	6 (magenta)	CONTINUOUS
3.4.5.	DIM	On	5 (blue)	CONTINUOUS
3.4.6.	GABLE	On	30	CONTINUOUS
3.4.7.	LOG	On	23	CONTINUOUS
3.4.8.	ROOF	On	3 (green)	CONTINUOUS
3.4.9.	STEEL	On	62	CONTINUOUS
3.4.10.	TEXT	On	42	CONTINUOUS

The details are shown in Figures 8.6.3a to 8.6.3c.

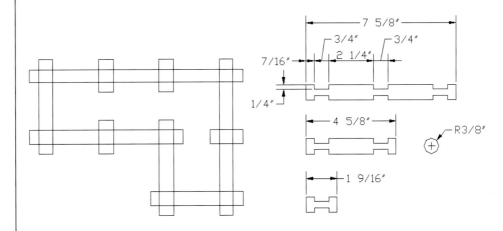

FIG. 8.6.3a: Floor Plan and Pieces

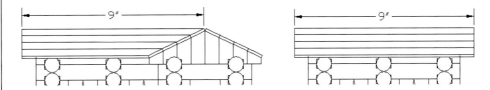

FIG. 8.6.3b: Roof Dimensions

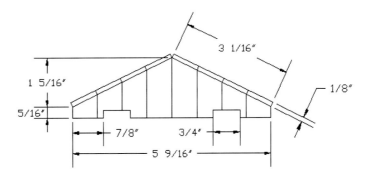

FIG. 8.6.3c: More Roof Dimensions

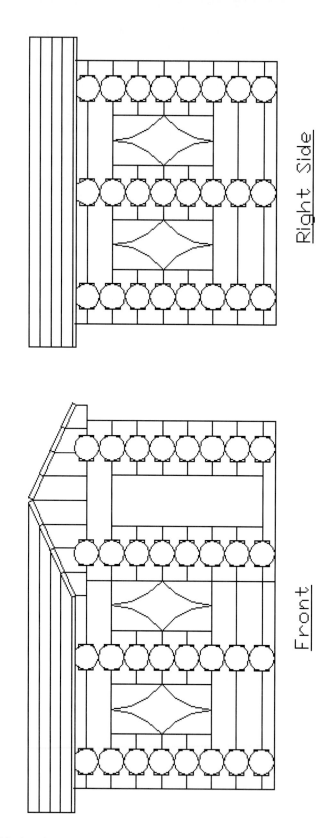

FIG. 8.6.3d: Completed Log Cabin

4. Using what you have learned, create the drawing in Figure 8.6.4. Create separate layers for the different objects and linetypes, and for the text. The grid is 1".

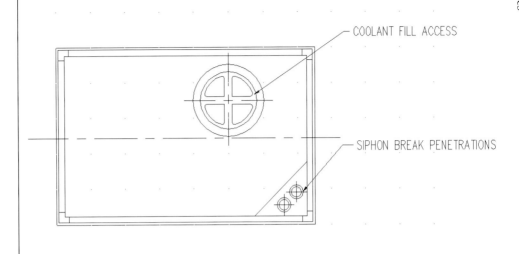

FIG. 8.6.4a: Sound Shield Detail (Special thanks to Alaska Diesel Electric of Seattle, Washington, and Mr. Lee Pangilinan for allowing me to use this sound shield detail.)

5. Now create the drawings in Figures 8.6.5a to 8.6.5d. Be sure to set up the drawing limits, layers, and whatever grid and snap you think will be useful. Each drawing should be plotted onto an $8 \frac{1}{2}$" × 11" sheet of paper, and should be drawn to scale when dimensions are provided. (Do not draw the dimensions.)

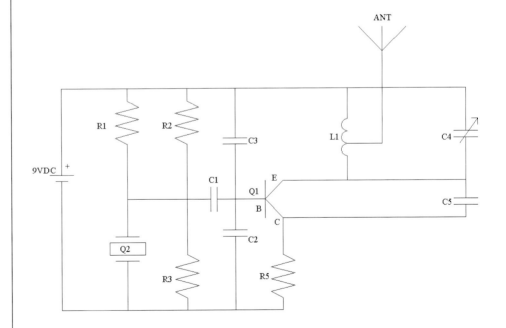

FIG. 8.6.5a: Electrical Schematic 7-1

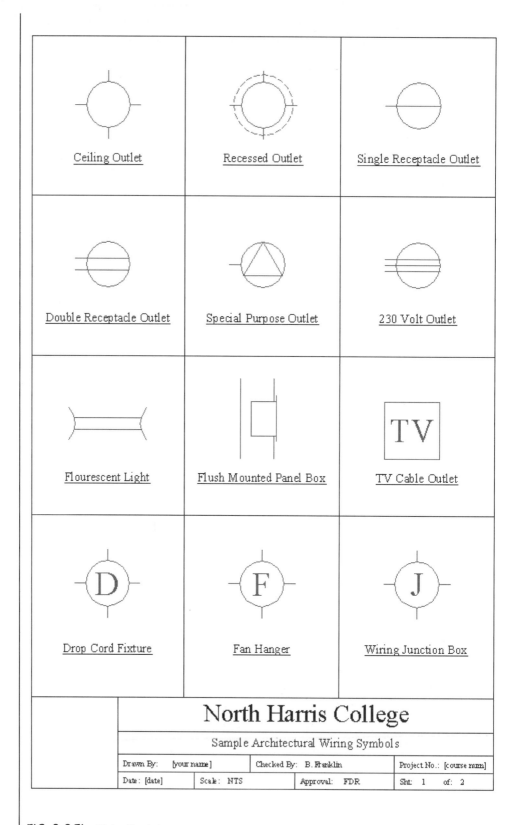

FIG. 8.6.5b: Wiring Symbols

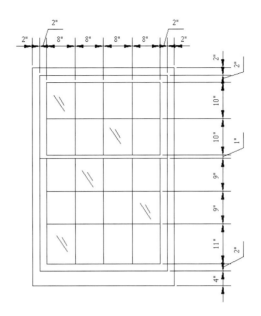

FIG. 8.6.5c: Window Detail

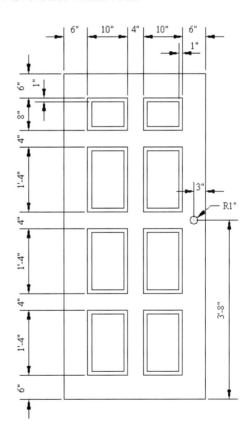

FIG. 8.6.5d: Door Detail

8.7 REVIEW QUESTIONS

Please write your answers on a separate sheet of paper.

1. AutoCAD designed the _____ command to remove parts of lines or circles.

2. _____ are usually lines or circles to which you want to trim.

3. (T or F) In the Trim command, the cutting edge must touch the object to be trimmed.

4. Set the _____ to "extend" the plane of the cutting edge in the Trim command.

5. Use the_____ command to handle a problem where a line is short of where it should reach.

6. and 7. The Trim command prompts the user for a _____ edge while the Extend command prompts the user for a _____ edge.

8. The _____ command can be used to lengthen or shorten several objects at once without a boundary or cutting edge.

9. To remove part of a line, circle, or arc without using a cutting edge, use the _____ command.

10. To break an object into two pieces without removing one of the pieces, type _____ at the Enter second point: prompt of the Break command.

11. Use the _____ command to create rounded or squared corners.

12. To remove the extra portions of line or circles during the Fillet or Chamfer command, set the _____ to No.

13. To angle a corner, use the (chamfer, mitre) command.

14. and 15. Name the two methods of drawing a chamfer.

16. and 17. To use the displacement option to move an object three units along the X-axis and 0 units along the Y-axis, type _____ at the Base point or displacement prompt and _____ at the Second point of displacement prompt.

18. (T or F) You can only move one object at a time, but you can copy several objects at once.

19. To move and rotate an object at one time, use the _____ command.

20. (T or F) You can resize an object with the Align command.

21. To make more than one copy of an object, use the _____ option of the copy command.

Identify these buttons:

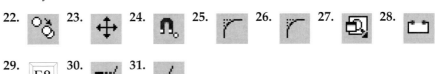

32. To copy an object from one drawing to another, use the (Windows, AutoCAD) copy command.

33. To move an object from one drawing to another, use the Windows (Cut, Paste, Cut and Paste) command(s).

LESSON 9

More Complex Lines: Polylines (and Light Weight Polylines)

Following this lesson, you will:
- Know how to draw Polylines: The **Pline** command
- Know how to edit Polylines: The **Pedit** command
- Know how to convert Lines to Polylines: The **Pedit** command
- Know how to convert Polylines to Lines: The **Explode** command
- Be familiar with AutoCAD's Inquiry Commands
  - *List*
  - *Dist*
  - *Area*
  - *ID*

Back in Lesson 7, we discussed ways of differentiating among the various parts of a drawing using colors, linetypes and layers. We also used lineweights to add width, but we had problems with them because lineweight is not a true WYSIWYG property. In that lesson, I promised to show you how to create lines with true WYSIWYG width. Once you have conquered width—through *polylines*—you can combine wide lines with linetypes and layers for more complex drawings.

The user can draw lines with width using the *Pline* command. We call these lines *polylines*, not for their ability to show width, but for their ability to be drawn as multisegmented lines. That is, polylines can contain many lines and still be treated as a single unit (much as you have already seen with polyline rectangles and polygons).

> Additionally, polylines can carry information. These are known as *smart* polylines but are beyond the scope of this course. For more information on smart polylines, see an AutoCAD customization text.

Prior to R14, polylines often confused the AutoCAD user because of the massive amounts of information they contained. They also dramatically increased the size of a drawing with stored information that was often never used or needed. For this reason, AutoCAD created the *light weight polyline* or *lwpolyline*. Capable of containing most of the information available to the polyline, the lwpolyline is much more condensed, presents its information via the *List* command (see Section 9.2.1) in a much more logical and understandable manner, and takes up much less drawing memory. The *Pline* command will actually draw an *lwpolyline* (pronounced "ell-double-u-polyline") and convert automatically to a *polyline* if the need arises to function as the older polylines did.

9.1 Using the *Pline* Command for Wide Lines and Multisegmented Lines

You have already used two commands that make use of polylines—***Rectang*** and ***Polygon***; so you have some idea of how polylines behave. Each group of lines acts as a single object when modified or manipulated. With ***Rectang***, you even saw that you could control the line width. Now we will see how to create objects with a little more freedom than these commands allowed—by using the ***Pline*** command.

One of AutoCAD's more difficult commands, ***Pline*** contains several options. We will look at each, but the basic sequence is

>Command: *pline* (or *pl*)
>Specify start point: *[select a starting point]*
>Current line-width is 0.0000
>Specify next point or [Arc/Close/Halfwidth/Length/Undo/Width]: *[select the next point]*
>Specify next point or [Arc/Close/Halfwidth/Length/Undo/Width]: *[select the next point]*
>Specify next point or [Arc/Close/Halfwidth/Length/Undo/Width]: *[you can continue to create line segments or hit enter to complete the command]*
>Command:

The options appear more intimidating than they actually are.

- **Close** works just like it does with the ***Line*** command. That is, it closes the polyline. But remember, unlike lines, the entire polyline will be treated as a single unit.
- **Halfwidth** and **Width** do the same thing. These are your options for telling AutoCAD how wide to make the polyline. **Width** is how wide you want the polyline whereas **Halfwidth**, of course, is half of the width. Why two options to achieve the same end? AutoCAD redundancy again! Each of these options provides its own prompts. The Halfwidth prompts are

  >Specify starting half-width <0.0000>:
  >Specify ending half-width <0.0000>:

  The **Width** prompts are the same, but ask for **width** instead of **halfwidth**. Notice that the starting and ending widths can be set separately. This will come in handy when creating arrowheads, as we will see in our first exercise.
- **Length** allows the user to enter the length of a line segment. The user who ignores this option in favor of one of AutoCAD's many other methods of defining line length (direct distance option, cartesian coordinates, polar tracking, etc.) will probably be happier. The **Length** option provides this prompt:

  >Specify length of line:

- **Undo** will do just that. Here at the *Pline* prompt, it undoes the last segment drawn. (Of course, if *undo* is entered at the *command* prompt after the polyline is drawn, the entire command is undone.)
- Notice that AutoCAD repeats the list of options, always with the **Specify next point** option as the default. This makes it simple to continue drawing—selecting points—until finished while making all the other options available for *each* line segment. In other words, the user can change the width, close the polyline or switch to an arc at any time during the creation process.
- **Arc** allows the user to create *polyline arcs*—that is, sequential arcs or arcs with width. The **Arc** option provides a second *tier* of prompts that behave remarkably like the options of the *Arc* command:

    Specify endpoint of arc or [Angle/CEnter/CLose/Direction/Halfwidth/Line/Radius/Second pt/Undo/Width]:

    - The **Angle** option allows the user to specify an included angle to define the arc. To more easily understand this, think of *pieces* of a circle. A circle is 360°. 45° would be $1/8$ of the circle, so specifying an angle of 45° will tell AutoCAD to draw $1/8$ of a circle. AutoCAD will prompt as it did in the *Arc* command:

        Specify included angle: *[specify the angle]*
        Specify endpoint of arc or [Center/Radius]: *[tell AutoCAD how you want to draw the arc]*

    - The **CEnter** option allows the user to specify the center point of the arc followed by an **Angle**, the **Length** of the arc, or the **Endpoint** of the arc. Again, pline arc prompts resemble the *Arc* command:

        Specify center point of arc:
        Specify endpoint of arc or [Angle/CEnter/CLose/Direction/Halfwidth/Line/Radius/Second pt/Undo/Width]:

    - The **CLose** option in this tier will close the polyline using an arc. You can use this method to create circles with line-width.
    - AutoCAD normally draws an arc in a counterclockwise direction. Once you draw an arc using the *Pline* command, AutoCAD reverses direction but stays in the **Arc** mode until told to switch back to **Lines** or to **CLose** (or exit) the *Pline* command. If you wish to draw a series of arcs in the same direction—like tiles on a roof—the **Direction** option allows the user to specify the direction of the next arc. AutoCAD prompts

        Specify the tangent direction for the start point of arc:
        Specify end point of the arc:

    - **Halfwidth** and **Width** provide a way to change the width of the polyline from within the **Arc** option. AutoCAD prompts the same way it did for these options in the first tier of *Pline* options.
    - The **Line** option returns the user to the first tier of *Pline* options. From there, the user can continue the polyline or exit the command.

- One of the easier options is the **Radius** option. With this, AutoCAD allows the user to specify the radius of the arc followed by the **Angle** or **Endpoint**. R**adius** options are

    Specify radius of arc:
    Specify endpoint of arc or [Angle]:

- The user can draw a three-point arc, just as he did with the *Arc* command, by using the **Second point** option. AutoCAD prompts for the second point on the arc, then the endpoint, as seen here:

    Specify second point on arc:
    Specify endpoint of arc:

- **Undo** performs just as it did in the upper tier.
- The default option is simply to **Specify endpoint of arc**. When that is done, AutoCAD repeats the options until the user either selects the **Line** option or exits the command by hitting *enter*.

Do you see why we say this command is **Beyond the Basics**? It took several pages to explain the various parameters of the *Pline* command. It is a powerful tool. But do not let the scope of the *Pline* command deter you from its use. Most students get the hang of it fairly quickly. Let's look at some of the options in an exercise.

> In addition to the command line and toolbar, you can access the *Polyline* command by selecting **Polyline** in the Draw pull-down menu.
> The various options of the *Pline* command are also available in the cursor menu once the *Pline* command has been entered.

WWW 9.1.1 Do This: Drawing with Polylines

I. Start a new drawing from scratch.

II. Set the *grid* to 0.5 and the *grid snap* to 0.25. Turn polar tracking **On**.

III. Follow these steps.

TOOLS	COMMAND SEQUENCE	STEPS
Polyline Button	**Command:** *pl*	1. Enter the *Pline* command by typing *pline* or *pl*. Or you can select the **Polyline** button from the Draw toolbar.
	Specify start point: 6,5	2. Select a starting point at **6,5**.

continued

TOOLS	COMMAND SEQUENCE	STEPS
	Current line-width is 0.0000	3. AutoCAD tells you what the **Current line-width** setting is, then prompts you for the next step. Type *w* to change the line width.
	Specify next point or [Arc/Close/ Halfwidth/Length/Undo/Width]: *w*	
	Specify starting width <0.0000>: $1/16$	4. Set the **starting width** to $1/16$.
	Specify ending width <0.0625>: *[enter]*	5. AutoCAD has set the **ending width** to match the **starting width**, but asks you if want to change it. Hit *enter* to accept the setting.
	Specify next point or [Arc/Close/ Halfwidth/Length/Undo/Width]: *@1.5<90*	6. Draw the first line segment *1.5* units upward.
	Specify next point or [Arc/Close/ Halfwidth/Length/Undo/Width]: *a*	7. Tell AutoCAD you want to draw an **Arc** next.
	Specify endpoint of arc or [Angle/ CEnter/CLose/Direction/Halfwidth/ Line/Radius/Second pt/Undo/ Width]: *@1<180*	8. Accept the **endpoint of arc** option and place the endpoint *1* unit to the left.
	Specify endpoint of arc or [Angle/ CEnter/CLose/Direction/Halfwidth/ Line/Radius/Second pt/Undo/ Width]: *l*	9. Type *L* for the **Line** option. (Note: Remember, AutoCAD is *not* case-sensitive—that is, it does not matter if you type a capital or small letter.)
	Specify next point or [Arc/Close/ Halfwidth/Length/Undo/Width]:	10. Continue the polyline *1.5* units downward using polar tracking. At this time, your polyline will look like Figure 9.1.1.10.

continued

TOOLS	COMMAND SEQUENCE	STEPS
		FIG. 9.1.1.10: Use Polar Tracking
	Specify next point or [Arc/Close/Halfwidth/Length/Undo/Width]: *@1.5<180*	11. Continue *1.5* units to the left.
	Specify next point or [Arc/Close/Halfwidth/Length/Undo/Width]: *a*	12. Use the **Arc** option again.
	Specify endpoint of arc or [Angle/CEnter/CLose/Direction/Halfwidth/Line/Radius/Second pt/Undo/Width]: *r*	13. Draw the arc with a $1/_4$ unit **Radius** downward.
	Specify radius of arc: *.25*	
	Specify endpoint of arc or [Angle]: *[select a point one grid mark due south of the current point]*	
	Specify endpoint of arc or [Angle/CEnter/CLose/Direction/Halfwidth/Line/Radius/Second pt/Undo/Width]: *d*	14. Let's use the **Direction** option to repeat the arc.
	Specify the tangent direction for the start point of arc: *[pick a point as shown in Figure 9.1.1.15]*	15. Pick a point due west of the current point, as indicated in Figure 9.1.1.15.

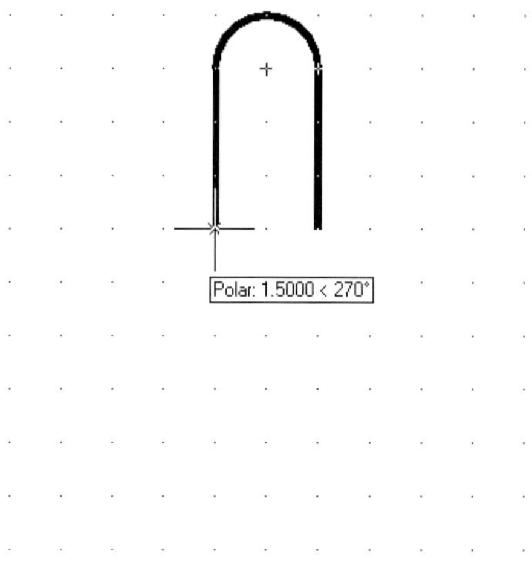

continued

TOOLS	COMMAND SEQUENCE	STEPS

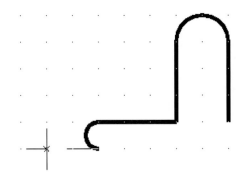

FIG. 9.1.1.15: Pick a Point Due West

Specify endpoint of the arc: @½<270

16. Pick a point ½" to the south.

Specify endpoint of arc or [Angle/CEnter/CLose/Direction/Halfwidth/Line/Radius/Second pt/Undo/Width]: *l*

17. Return to the **Line** option.

Specify next point or [Arc/Close/Halfwidth/Length/Undo/Width]:

18. Draw the line **1.5** units to the right using polar tracking. Your drawing now looks like Figure 9.1.1.18.

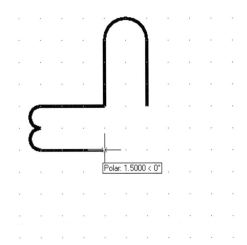

FIG. 9.1.1.18: Use Polar Tracking to Draw This Line

continued

TOOLS	COMMAND SEQUENCE	STEPS
		19. Using the preceding details, see if you can complete the drawing. (Remember to type *c* to complete the polyline.) It will look like Figure 9.1.1.19 when you are finished.

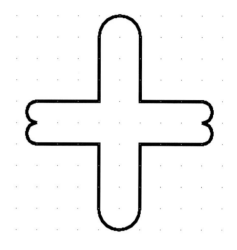

FIG. 9.1.1.19: Completed Figure

 Command: *z*

20. And now let's try something completely different. Zoom in around the center of the object you just drew. Your screen should look like Figure 9.1.1.20.

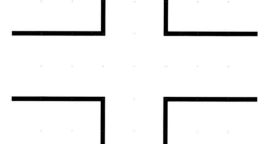

FIG. 9.1.1.20: Zoom In Like This

 Command: *pl*

21. Repeat the *Pline* command.

continued

TOOLS	COMMAND SEQUENCE	STEPS
	Specify start point: *4.75,4.5*	22. Pick the starting point indicated at left.
	Current line-width is 0.0625	23. Tell AutoCAD you want to change the **Width**.
	Specify next point or [Arc/Close/Halfwidth/Length/Undo/Width]: *w*	
	Specify starting width <0.0625>: *0*	24. Set different **starting** and **ending widths** as indicated.
	Specify ending width <0.0000>: *.25*	
	Specify next point or [Arc/Close/Halfwidth/Length/Undo/Width]: *@1/2<0*	25. Draw the line segment. Notice the arrowhead effect of your width settings.
	Specify next point or [Arc/Close/Halfwidth/Length/Undo/Width]: *@1/2<0*	26. Continue the polyline. Now AutoCAD uses just the width setting of the endpoint of the last line segment drawn.
	Specify next point or [Arc/Close/Halfwidth/Length/Undo/Width]: *w*	27. Change the **Width** again. Accept the **0.25** default of the **starting width**, but change the **ending width** back to *0*.
	Specify starting width <0.2500>: *[enter]*	
	Specify ending width <0.2500>: *0*	
	Specify next point or [Arc/Close/Halfwidth/Length/Undo/Width]: *@1/2<0*	28. Complete the polyline. Your drawing will look like Figure 9.1.1.28.
	Specify next point or [Arc/Close/Halfwidth/Length/Undo/Width]: *[enter]*	

continued

TOOLS	COMMAND SEQUENCE	STEPS

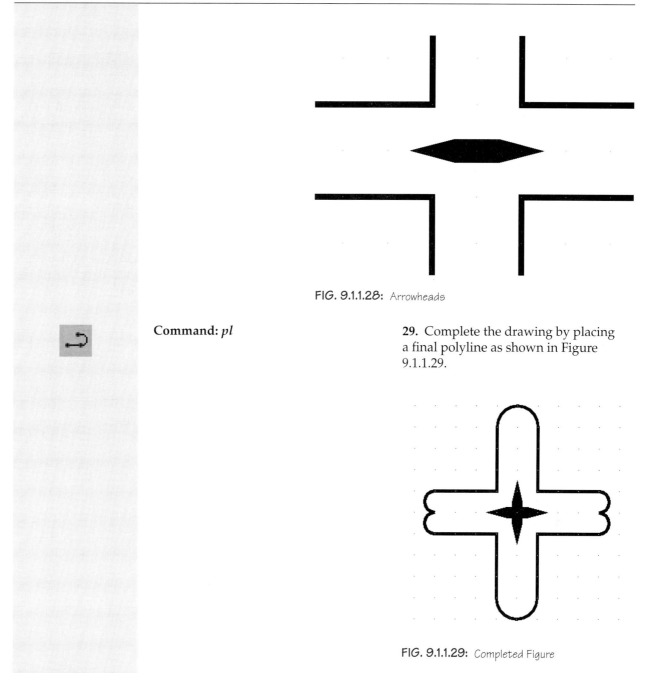

FIG. 9.1.1.28: Arrowheads

	Command: *pl*	29. Complete the drawing by placing a final polyline as shown in Figure 9.1.1.29.

FIG. 9.1.1.29: Completed Figure

	Command: *save*	30. Save the drawing as *MyPline* to the C:\Steps\Lesson09 folder and exit.

Polylines can be a bit frightening at first because of the depth of the command. But do not let too many options prevent you from using one of AutoCAD's more useful tools.

9.2 The Most Useful of the Most Overlooked: AutoCAD's Inquiry Commands

AutoCAD provides four commands—*List*, *Dist*, *Area*, and *ID*—whose simplicity has led to their being almost completely forgotten. We will cover them quickly because they are so simple; but I do recommend them as possible residents of monitor stick 'ems (those tiny bits of paper taped to the side of your monitor with cheater notes).

■ 9.2.1 Tell Me About It: The List Command

One of the problems you may run into on occasion is a modifying command not working exactly as you might expect. The object you are modifying not being what you think it is frequently causes this. For example, you may try to edit a polyline arc only to receive a message from AutoCAD that the object is not a polyline; or you may freeze a layer only to discover that something you thought was on that layer does not freeze. When something unexpected like this happens, do a *List* on the object to see if you can spot the problem. Look at some examples.

> In addition to the command line and toolbar, you can find *all* of the Inquiry commands listed in the **Inquiry** section of the Tools pull-down menu.

9.2.1.1 Do This: Listing an Object's Properties

I. Open the *Samples.dwg* from the C:\Steps\Lesson09 folder. The drawing looks like Figure 9.2.1.1a.

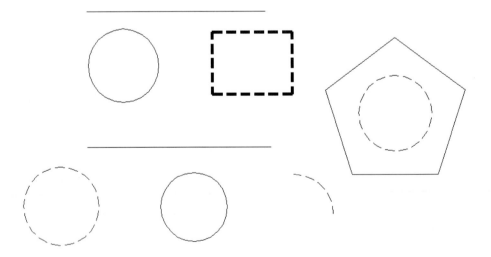

FIG. 9.2.1.1a: *Samples .dwg*

II. Follow these steps.

TOOLS	COMMAND SEQUENCE	STEPS
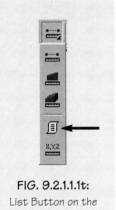 FIG. 9.2.1.1.1t: List Button on the Distance Flyout	**Command:** *ls*	1. Enter the *List* command by typing *list* or *ls* at the command prompt. Or you can go to the **List** button on the Standard toolbar. It is the fourth button down in the **Distance** flyout. It resembles a scroll of paper, as shown in Figure 9.2.1.1.1t. To get to the **List** button, pick and hold (left mouse button) on the **Distance** button. The flyout will appear as shown here. Move your cursor down to the **List** button and release the mouse button.
	Select objects: **Select objects:** *[enter]*	2. Select the upper red line, then hit *enter* to confirm that you are finished selecting. AutoCAD switches to the *text screen* and displays the information in Figure 9.2.1.1.2.

```
          LINE        Layer: "FRED"
                      Space: Model space
              Handle = 26
      from point, X=    2.5424  Y=    7.7686  Z=    0.0000
        to point, X=    7.4140  Y=    7.7686  Z=    0.0000
    Length =    4.8717,  Angle in XY Plane =     0
              Delta X =    4.8717, Delta Y =    0.0000, Delta Z =    0.0000
```

FIG. 9.2.1.1.2: Listed Properties of the Line

Here we see: the type of object (**Line**), its layer (**Fred**), that it was drawn in Model Space, its Handle (for programmers), its beginning and ending point, length and angle.

F2 Key Or X for Exit 		3. Return to the *graphics screen* either by hitting the *F2* key or by picking on the **X** in the upper right corner of the *text screen*.
	Command: *[enter]*	4. Let's repeat the procedure with another object.

continued

TOOLS	COMMAND SEQUENCE	STEPS
	Select objects: **Select objects:** *[enter]*	5. Select the dashed rectangle, then hit *enter*. AutoCAD switches to the text screen and displays the information in Figure 9.2.1.1.5.

```
              LWPOLYLINE   Layer: "0"
                        Space: Model space
                Color: 6 (magenta)      Linetype: "HIDDEN"
                Handle = 28
         Closed
Constant width       0.0625
         area        3.5494
    perimeter        7.6074

       at point  X=    5.9830   Y=    7.2213   Z=    0.0000
       at point  X=    8.1448   Y=    7.2213   Z=    0.0000
       at point  X=    8.1448   Y=    5.5794   Z=    0.0000
       at point  X=    5.9830   Y=    5.5794   Z=    0.0000
```

FIG. 9.2.1.1.5: Listed Properties of the Rectangle

Again, we see the type of object and layer. But since this object is a multisegmented lwpolyline, we also see that: the lwpolyline is **Closed**, its width is **0.0625**, the beginning and ending points of each segment, and the **area** and **perimeter** of the lwpolyline.

Note that the **linetype** and **color** characteristics of this object were not assigned by layer. These characteristics were assigned using the *Color* and *Linetype* methods. As a result, the *List* command also shows **Color** and **Linetype**. It will show these only when they are not defined by layer.

|  | **Command:** *[enter]* | 6. Repeat the *List* command for the rest of the objects in this drawing. (It is best to list objects one at a time.) Notice that *List* provides slightly different information for each. |

■ 9.2.2 How Long or How Far: The Dist Command

Another useful tool helps determine just what its name implies—*distance*. How long is a line or how far is it from here to there?

9.2.2.1 Do This: Determining Distance

I. If you are not still in the *Samples.dwg*, open it. It is in the C:\Steps\Lesson09 folder.

II. Follow these steps.

TOOLS	COMMAND SEQUENCE	STEPS
 FIG. 9.2.2.1.1t: Distance Button on List Flyout	**Command:** *di*	**1.** Enter the *Dist* command by typing *dist* or *di* at the command prompt. Or you can select the **Distance** button from the Standard toolbar. If you are continuing from the previous exercise, AutoCAD will have replaced the **Distance** button on the Standard toolbar with the **List** button. It does this automatically when it uses a *flyout* toolbar. Simply pick and hold the **List** button for the flyout to appear. As indicated in Figure 9.2.2.1.1t, the **Distance** button is the first button.
	Specify first point: _endp of **Specify second point:** _cen of	**2.** Using OSNAPS, pick the **endpoint** of the cyan line. Then pick the **center** of the lower blue circle (see Figure 9.2.2.1.2).
		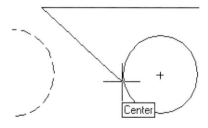 **FIG. 9.2.2.1.2:** From the Endpoint to the Center Point
F2		**3.** AutoCAD returns the information in Figure 9.2.2.1.3 (on the command line—if you cannot see all of the information, hit the *F2* key to toggle to the text screen)

```
Distance = 3.3232,  Angle in XY Plane = 332,  Angle from XY Plane = 0
Delta X = 2.9230,   Delta Y = -1.5811,   Delta Z = 0.0000
```

FIG. 9.2.2.1.3

Here, AutoCAD shows: the true **Distance** (3.3232), the 2-dimensional **Angle in XY Plane** (332), the 3-dimensional **Angle from XY Plane** (0), the distance along the X-plane (or **Delta X = 2.9230**), the distance along the Y-plane (or **Delta Y = −1.5811**), and the distance along the 3-dimensional Z-plane (or **Delta Z = 0**).

■ 9.2.3 Calculating the Area

The *List* command provides the area of closed rectangles, polygons, and circles as you saw in Section 9.2.1. But the boundaries in which we need to determine area are not always closed objects. Sometimes we need an area bounded by simple lines or even multiple objects. For this reason, AutoCAD provides the *Area* command.

The *Area* command sequence is

Command: *area* (or *aa*)
Specify first corner point or [Object/Add/Subtract]: *[select the first corner of the area's boundary]*
Specify next corner point or press ENTER for total: *[this prompt repeats till the user hits enter]*
Specify next corner point or press ENTER for total: *[enter to complete the command]*
Area = 4.0000, Perimeter = 8.0000

The options here are fairly clear.

- The **Specify first corner point** option is the default and simply instructs the user to select the first point on the boundary of the area to be calculated. Auto-CAD follows with **Specify next corner point** prompts until the boundary is defined and the user hits *enter* to complete the command. AutoCAD then shows the **Area** within and **Perimeter** around the boundary.
- The **Object** option allows the user to select an object—a circle, polygon, and so forth—and defines the boundary from the edges of the object.
- **Add** and **Subtract** are ways to keep a running total of several areas or to get the area of a bounded site minus a smaller site—such as the area of a plot of land minus the house sitting on it.

9.2.3.1 Do This: Calculating Area

I. If you are not still in the *Samples.dwg,* open it. It is in the C:\Steps\Lesson09 folder.

II. Follow these steps.

TOOLS	COMMAND SEQUENCE	STEPS
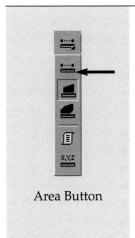 Area Button	**Command:** *aa*	1. Enter the *Area* command by typing *area* or *aa* at the command prompt. Or you can select the **Area** button on the Standard toolbar flyout that contains the **Distance** and **List** buttons.
	Specify first corner point or [Object/Add/Subtract]: *o*	2. Tell AutoCAD you want to use the **Object** option.
	Select objects:	3. Select the circle in the center of the polygon (Figure 9.2.3.1.3).
		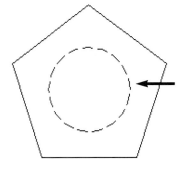
		FIG. 9.2.3.1.3: Select the Circle
	Area = 3.1416, Circumference = 6.2832	AutoCAD returns this information on the command line.
	Command: *[enter]*	4. Repeat the command.

continued

TOOLS	COMMAND SEQUENCE	STEPS
	Specify first corner point or [Object/Add/Subtract]:	5. Select the five points on the polygon (use the endpoint OSNAP). AutoCAD returns this information on the command line.
	Area = 9.5106, Perimeter = 11.7557	
	Command: *[enter]*	6. Repeat the command.
	Specify first corner point or [Object/Add/Subtract]: *a*	7. Tell AutoCAD you want to use the **Add** option.
	Specify first corner point or [Object/Subtract]:	8. AutoCAD prompts again for points or objects. Pick the five points of the polygon as you did in Step 5.
	Specify next corner point or press ENTER for total (ADD mode):	9. After selecting the fifth point, hit *enter* to complete the polygon.
	Area = 9.5106, Perimeter = 11.7557	AutoCAD tells you what the **Area** and **Perimeter** are so far, and the **Total area** defined during this command. It then prompts again as it did previously.
	Total area = 9.5106	
	Specify first corner point or [Object/Subtract]: *o*	10. Type *O* to select the **Object** option.
	(ADD mode) Select objects:	11. Now select the upper blue circle.
	Area = 2.9741, Circumference = 6.1134	AutoCAD tells you the **Area** and **Circumference** of the circle, and adds the area of the circle to the **Total area**. Then it prompts again.
	Total area = 12.4846	
	(ADD mode) Select objects: *[enter]*	12. Hit *enter* to leave the **Add** mode.
	Specify first corner point or [Object/Subtract]: *[enter]*	13. AutoCAD returns you to the initial prompt. Hit *enter* to exit the command.
	Command: *[enter]*	14. Repeat the *Area* command. This time we will subtract one area from another.
	Specify first corner point or [Object/Add/Subtract]: *a*	15. Tell AutoCAD you want to use the **Add** option. We will start by adding the outer boundary; then we will subtract the inner boundary.

continued

TOOLS	COMMAND SEQUENCE	STEPS
	Specify first corner point or [Object/Subtract]:	16. AutoCAD prompts again for points or objects. Pick the five points of the polygon as you did previously.
	Specify next corner point or press ENTER for total (ADD mode): *[enter]*	17. After selecting the fifth point, hit *enter* to complete the polygon.
	Area = 9.5106, Perimeter = 11.7557	
	Total area = 9.5106	
	Specify first corner point or [Object/Subtract]: *s*	18. Now we tell AutoCAD to **Subtract**.
	Specify first corner point or [Object/Add]: *o*	19. We want to subtract an **Object**.
	(SUBTRACT mode) Select objects:	20. Select the circle inside the polygon.
	Area = 3.1416, Circumference = 6.2832	AutoCAD tells you the **Area** and **Circumference** of the circle, then subtracts the area from the **Total area**.
	Total area = 6.3690	This is the **Total area** of the polygon less the area of the circle.
	(SUBTRACT mode) Select objects: *[enter]*	21. Hit *enter* twice to leave the command.
	Specify first corner point or [Object/Add]: *[enter]*	

■ 9.2.4 Identifying Any Point with ID

The last of these simple tools enables the user to identify any point in a drawing. This can prove particularly beneficial to the drafter who works in true coordinates (see the insert in Section 1.10 for an steps of true coordinates) or someone working with the **Ordinate** system. We will discuss the ordinate system in Lesson 15. The command sequence is

Command: *id*
Specify point: *[select a point in the drawing]*
X = 0.0000   Y = 0.0000   Z = 0.0000

As you can see, AutoCAD responds with the X,Y,Z coordinate location of the point selected.

9.2.4.1 Do This: Identifying Coordinates in a Drawing

I. If you are not still in the *Samples.dwg*, open it. It is in the C:\Steps\Lesson09 folder.

II. Follow these steps.

TOOLS	COMMAND SEQUENCE	STEPS
Locate Point Button	**Command:** *id*	1. Enter the **ID** command or go to the **Locate Point** button found on the same Standard toolbar flyout that we have been using.
	Specify point: _cen of	2. Select the center point of the upper blue circle. AutoCAD returns the coordinates locating the center of the circle (Figure 9.2.4.1.2).
	X = 3.5472　　Y = 6.3395　　Z = 0.0000	
	FIG. 9.2.4.1.2: Coordinates of Selected Point	
	Command: *quit*	3. Exit the drawing. Do not save your changes.

9.3 Editing Polylines: The *Pedit* Command

If you thought drawing Polylines was fun, you are in for a real treat now. Remember that second tier of options? The *Pedit* command has one, as well, and several smaller third tiers! Sort of makes you want to go back to the board, does it not? But to encourage you, let me just say that you will probably never need the second (or third) tier—or at least only rarely in the 2-dimensional world.

Two-dimensional polylines are such that it is often easier to erase and redraw than it is to edit. Still, you should be familiar with the *Pedit* command for those benefits that it does provide. These include the ability to change the width of the polyline and to join several polylines into a single object.

> In the 3-dimensional world, it will often be easier to edit a polyline. When you get there, it will be useful if you already know how to add or move a vertex. We will see how to do that here, but you probably will not need it for a while.

The *Pedit* command sequence looks like this:

Command: *Pedit* **(or** *pe***)**
Select polyline: *[select the polyline to edit]*
Enter an option [Open/Join/Width/Edit vertex/Fit/Spline/Decurve/Ltype gen/Undo]: *[tell AutoCAD how you want to edit the polyline]*

This sequence takes us up to the first tier of options. Let's stop here to examine them.

- The **Open** option appears if a closed polyline is selected at the **Select polyline** prompt. Conversely, a **Close** option will appear if an open polyline is selected. The **Open** option will remove the last line segment—the one that closed the polyline. The **Close** option will add a polyline segment between the two open endpoints.
- The **Join** option enables the user to join one polyline to another to form one large polyline. The user must join endpoint to endpoint, and the polylines must touch at the endpoints.
- The **Width** option allows the user to modify the polyline's width.
- **Fit** and **Spline** will soften corners into curves. This was once the tool of choice for drawing contour lines for topographical maps. However, AutoCAD now provides a *Spline* command that was specifically designed for drawing contour lines (more on this in Lesson 12). The difference between **Fit** and **Spline** is that, although **Fit** will create curves that go through each point on the polyline, **Spline**'s curves go through only the first and last point. The rest of the points "pull" the curve but do not insist that the curve touch each point.

> The system variable **Splinetype** controls the amount of curve caused by the **Spline** option. A setting of **5** will cause a more pronounced curve (called a Quadratic B-spline) that is actually tangent to the original polyline. The default setting of **6** causes a softer, less pronounced curve (a Cubic B-spline). There are only the two settings (5 and 6) available.

- **Decurve** removes all curves on the polyline whether put there as **Arcs, Fits,** or **Splines**.
- **Ltype gen** regulates the placement of dashes and spaces in linetypes. Let me make this as simple as possible. Through some fairly complicated mathematics, AutoCAD normally balances dashes and spaces in a line so that the amount of solid line at both ends is the same. If the length of the line does not allow enough room for the dashes and spaces defined by the linetype, none is shown. When turned on, **Ltype gen** calculates the placement of dashes and spaces for the overall polyline. When turned off, the placement is calculated for each individual line segment within the polyline. This will become clearer in our exercise.
- By now you are familiar with the **Undo** option. It undoes the last modification made within the *Pedit* command.

We will look in some detail at the **Edit vertex** option later in this lesson. But let's try an exercise on what we have learned so far.

> In addition to the command line and toolbar, you can find the *Pedit* command by selecting **Object**, then **Polyline**, in the Modify pull-down menu.
>
> The various options of the *Pedit* command are also available in the cursor menu once the *Pedit* command has been entered.

9.3.1 Do This: Working with Simple Polyline Editing Tools

I. Open the *MyPline* drawing you created earlier in this lesson. (*Note:* If this drawing is not available, open the *Pline* drawing in the C:\Steps\Lesson09 folder.)

II. Follow these steps.

TOOLS	COMMAND SEQUENCE	STEPS
 Edit Polyline Button	**Command:** *pe*	1. Enter the *Pedit* command by typing *pedit* or *pe*. Or you may pick the **Edit Polyline** button on the Modify II toolbar.
	Select polyline:	2. Select the outer polyline.
	Enter an option [Open/Join/Width/ Edit vertex/Fit/Spline/Decurve/Ltype gen/Undo]: *o*	3. Let's start with the **Open** option. Type *o* and hit *enter*. Your drawing will look like Figure 9.3.1.3.

continued

TOOLS	COMMAND SEQUENCE	STEPS
		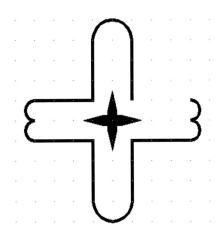 **FIG. 9.3.1.3:** Opened Polyline
	Enter an option [Open/Join/ Width/Edit vertex/Fit/Spline/ Decurve/Ltype gen/Undo]: *c*	**4.** Notice how the prompt has changed from **Open** to **Close**. Type *c* to use the **Close** option. Notice that the figure is closed and looks like it did when you started.
	Enter an option [Open/Join/ Width/Edit vertex/Fit/Spline/ Decurve/Ltype gen/Undo]: *w*	**5.** Type *w* to use the **Width** option.
	Specify new width for all segments: $1/32$	**6.** AutoCAD asks for a new *width*. Enter $1/32$. Notice the difference in the figure. It now looks like Figure 9.3.1.6.

continued

TOOLS	COMMAND SEQUENCE	STEPS
		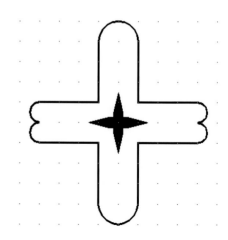 FIG. 9.3.1.6: Changed Width
	Enter an option [Open/Join/ Width/Edit vertex/Fit/Spline/ Decurve/Ltype gen/Undo]: *f*	7. Note where the vertices (corners) are on the figure, then type *f* for the **Fit** option. Your drawing looks like Figure 9.3.1.7.
		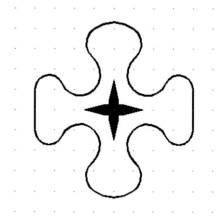 FIG. 9.3.1.7: Fitted Polyline
	Enter an option [Open/Join/ Width/Edit vertex/Fit/Spline/ Decurve/Ltype gen/Undo]: *u*	8. Undo the last modification.

continued

TOOLS	COMMAND SEQUENCE	STEPS
	Enter an option [Open/Join/Width/Edit vertex/Fit/Spline/Decurve/Ltype gen/Undo]: *s*	9. Now try the **Spline** option. Your drawing now looks like Figure 9.3.1.9.

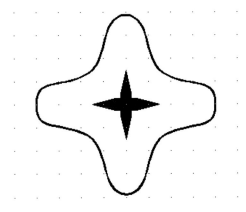

FIG. 9.3.1.9: Splined Polyline

| | Enter an option [Open/Join/Width/Edit vertex/Fit/Spline/Decurve/Ltype gen/Undo]: *d* | 10. **Decurve** the polyline. It looks like Figure 9.3.1.10. |

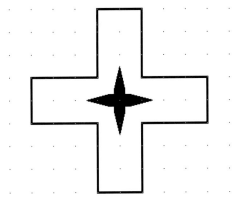

FIG. 9.3.1.10: Decurved Polyline

continued

TOOLS	COMMAND SEQUENCE	STEPS
	Enter an option [Open/Join/Width/Edit vertex/Fit/Spline/Decurve/Ltype gen/Undo]: *u*	11. Undo the last two modifications. The drawing will look as it did in Step 6.
	Enter an option [Open/Join/Width/Edit vertex/Fit/Spline/Decurve/Ltype gen/Undo]: *u*	
	Enter an option [Open/Join/Width/Edit vertex/Fit/Spline/Decurve/Ltype gen/Undo]: *[enter]*	12. Exit the command.
 Properties Button	**Command:** *props*	13. Use *Properties* to change the linetype of the polyline to **Center2**. (Refer to Section 7.1.2 for help loading the linetype, and Section 7.1.5 for help changing it.) Your drawing looks like Figure 9.3.1.13.

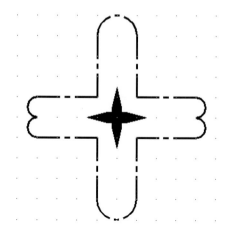

FIG. 9.3.1.13: Center2 Linetype

Notice that there are no dashes and spaces showing in the smaller arcs.

| | **Command:** *pe* | 14. Repeat the *Pedit* command. |
| | **Select polyline:** | 15. Select the same polyline. |

continued

TOOLS	COMMAND SEQUENCE	STEPS
	Enter an option [Open/Join/Width/Edit vertex/Fit/Spline/Decurve/Ltype gen/Undo]: *l*	16. Type *L* for the **Ltype gen** option.
	Enter polyline linetype generation option [ON/OFF] <Off>: *on*	17. Turn it *On*. Notice the change in the drawing (Figure 9.3.1.17)—dashes and spaces now show in the smaller arcs.

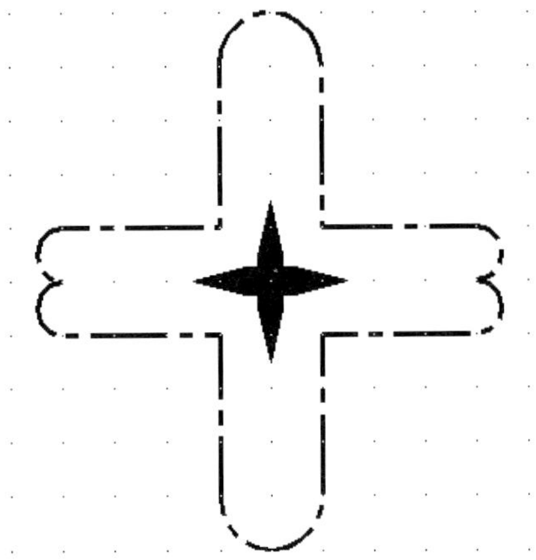

FIG. 9.3.1.17: Ltype Gen Turned On

| | Enter an option [Open/Join/Width/Edit vertex/Fit/Spline/Decurve/Ltype gen/Undo]: *[enter]* | 18. Exit the command. |
|  | Command: *props* | 19. Use *Properties* to return the linetype to **ByLayer**. |

We have just about covered the first tier of options in the *Pedit* command. The only remaining option is the key to the next tier—**Edit vertex**. As a beginning CAD operator in a 2-dimensional world, you will rarely find the need to edit a vertex. But the option comes in quite handy for very complex polylines and 3-dimensional polylines. The **Edit vertex** choice will result in the following list of second-tier options:

Enter a vertex editing option
[Next/Previous/Break/Insert/Move/Regen/Straighten/Tangent/Width/eXit] <N>:

You will notice that, when you enter this level of the *Pedit* command, a small "X" appears on the polyline you are editing. The X is a *locator* to let the user know which vertex is being edited.

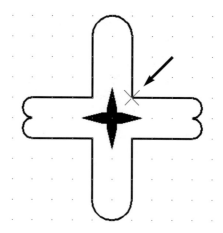

FIG. 9.3.1a: Vertex Locator

- The **Next** (default) and **Previous** options move the locator forward and backward to each vertex around the polyline.
- Use the **Break** option just as you used the *Break* command to remove part of a line.
- **Insert** enables the user to define a new vertex on the polyline and **Straighten** allows the user to remove an existing vertex.

A vertex is the endpoint of a line or curve, or the intersection of two or more lines or curves.

- **Move**, of course, enables the user to move a vertex thus reshaping the polyline.
- In the event that too much editing causes the polyline to display oddly on your screen, you can **Regen** just the polyline while within the editing session. This prevents the user from having to leave the command, regen the drawing, and then return to the *Pedit* command.
- The **Tangent** option allows the user to assign a tangent direction for AutoCAD to use when it **Fits** the polyline (with the **Fit** option on the upper tier).
- The **Width** option on the first tier of choices allowed the user to change the width for the entire polyline. The **Width** option on the second tier allows the user to change the width for a specific segment of the polyline.

The Edit Vertex options appear in the cursor menu when that tier appears on the command line.

Let's look at some of these options.

WWW 9.3.2 Do This: More Complex Polyline Editing Tools

I. If you are not already in the *MyPline* or *Pline* drawing, please open one of them now. Refer to Figure 9.3.2a for this exercise.

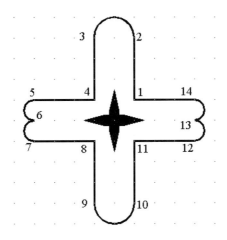

FIG. 9.3.2a: Numbered Vertices

II. Follow these steps.

TOOLS	COMMAND SEQUENCE	STEPS
Edit Polyline Button	Command: *pe*	1. Enter the *Pedit* command by typing *pedit* or *pe*. Or you may pick the **Edit Polyline** button on the Modify II toolbar.
	Select polyline:	2. Select the outer polyline.
	Enter an option [Open/Join/ Width/Edit vertex/Fit/Spline/ Decurve/Ltype gen/Undo]: *e*	3. Enter *e* to access the next tier of options.

continued

TOOLS	COMMAND SEQUENCE	STEPS
	Enter a vertex editing option [Next/Previous/Break/Insert/Move/Regen/Straighten/Tangent/Width/eXit] <N>: *[enter]*	**4.** Reposition the locator to point 2. Notice the default option is **N**. This is the **Next** option, so just hit *enter* to reposition the locator.
	Enter a vertex editing option [Next/Previous/Break/Insert/Move/Regen/Straighten/Tangent/Width/eXit] <N>: *b*	**5.** Select the **Break** option by typing *b*.
	Enter an option [Next/Previous/Go/eXit] <N>: *[enter]*	**6.** Notice that AutoCAD drops to a *third* tier of options. **Next** and **Previous** work as they do in the second tier. **Go** executes the option that dropped you to this level (in this case, the **Break** option). **Exit** to leave this level without executing the level two option. Hit *enter* to accept the **Next** default. The locator will move to point 3.
	Enter an option [Next/Previous/Go/eXit] <N>: *g*	**7.** Type *g* for **Go** and hit *enter* to execute the **Break** option. The segment between points 2 and 3 is removed as seen in Figure 9.3.2.7.

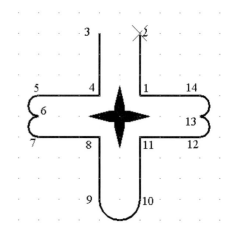

FIG. 9.3.2.7: Broken Polyline

continued

TOOLS	COMMAND SEQUENCE	STEPS
	Enter a vertex editing option [Next/ Previous/Break/Insert/Move/Regen/ Straighten/Tangent/Width/eXit] <N>: *x*	8. Enter *x* to return to the primary tier, then hit *enter* to leave the command.
	Enter an option [Open/ Join/Width/ Edit vertex/Fit/Spline/Decurve/ Ltype gen/Undo]: *[enter]*	
	Command: *[enter]*	9. Repeat Steps 1, 2 and 3.
	Enter a vertex editing option [Next/ Previous/Break/Insert/Move/Regen/ Straighten/Tangent/Width/eXit] <N>: *i*	10. Now let's **Insert** a vertex. Type *i*.
	Specify location for new vertex: _mid of	11. Place the new vertex midway between points 3 and 4 (Figure 9.3.2.11). (Use your OSNAPS!) Notice the locator appears at this new point.

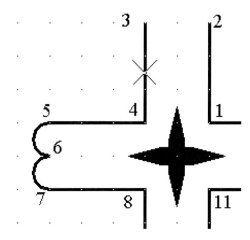

FIG. 9.3.2.11: Add a New Vertex

continued

TOOLS	COMMAND SEQUENCE	STEPS
	Enter a vertex editing option [Next/ Previous/Break/Insert/Move/Regen/ Straighten/Tangent/Width/eXit] <N>: *m* **Specify new location for marked vertex:** *@1<180*	**12.** Now move the new vertex a single unit to the left. Your drawing now looks like Figure 9.3.2.12. 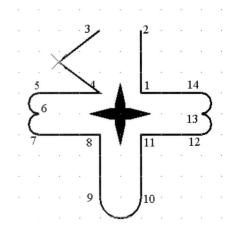 **FIG. 9.3.2.12:** Move the Vertex
	Enter a vertex editing option [Next/ Previous/Break/Insert/Move/Regen/ Straighten/Tangent/Width/eXit] <N>: *p*	**13.** Now we will remove the new vertex. Type *p* to return the locator to point 3.
	Enter a vertex editing option [Next/ Previous/Break/Insert/Move/Regen/ Straighten/Tangent/Width/eXit] <P>: *s*	**14.** Type *s* to straighten the vertex.
	Enter an option [Next/Previous/ Go/eXit] <N>: *[enter]* **Enter an option[Next/Previous/ Go/eXit] <N>:** *[enter]* **Enter an option [Next/Previous/ Go/eXit] <N>:** *g*	**15.** Hit *enter* until the locator moves to point 4. Then type *g* to execute the **Straighten** option. The new vertex disappears and the drawing appears as it did in Step 7.
		16. Return the locator to point 3.
	Enter a vertex editing option [Next/ Previous/Break/Insert/Move/Regen/ Straighten/Tangent/Width/eXit] <P>: *w*	**17.** Now type *w* to change the width of a single line segment.

continued

TOOLS	COMMAND SEQUENCE	STEPS

Specify starting width for next segment <0.0313>: *0*

Specify ending width for next segment <0.0000>: *1/8*

18. Set the starting width to **0** and the ending width to *1/8*. Your drawing will look like Figure 9.3.2.18.

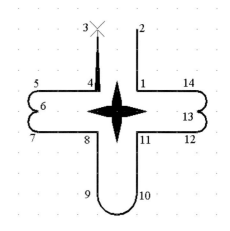

FIG. 9.3.2.18: Change the Width of One Segment

19. Follow Step 8 to exit the *Pedit* command.

Command: *quit*

20. Exit the drawing without saving your changes.

One other aspect of the *Pedit* command is its effect on nonpolyline lines and arcs. If the user selects a line or arc at the **Select polyline:** prompt, AutoCAD will prompt:

Object selected is not a polyline
Do you want to turn it into one? <Y>

This is the conversion method of lines or arcs to polylines. Note, however, that AutoCAD will automatically convert nonpolyline lines or arcs to polylines during execution of the **Join** option.

To convert polylines to lines or arcs, simply use the *Explode* command (or use the **Explode** button on the Modify toolbar). AutoCAD prompts:

Select objects:

The prompt will repeat until the user hits *enter* to confirm completion of the selection set. Be aware, however, that only a polyline can show WYSIWYG width. An exploded polyline becomes a line or arc and loses its width.

9.4 Extra Steps

I have included another Lisp routine—*pe-w.lsp*—in the C:\Steps\Lesson09 folder. This routine includes two commands—**W** and **PLW**—both were designed to help the user change a polyline's width without having to enter the frightening world of the *Pedit* command. **W** will prompt the user for the desired width of a polyline, then prompt the user to select lines, arcs or polylines to change. It saves a bit of time over the *Pedit* method. **PLW** allows the user to select a polyline that has the desired width, then select a polyline to change.

9.5 What Have We Learned?

Items covered in this lesson include:

- *AutoCAD Inquiry Commands:*
  - *List*
  - *Dist*
  - *Area*
  - *ID*

- *Creating and Editing Polyline Commands:*
  - *Pline*
  - *Pedit*
  - *Explode*

This has been a difficult lesson. But polylines are exceptionally powerful tools, and I suppose the nature of the beast requires the multileveled structure of both the *Pline* and *Pedit* commands. Take some time to review the lesson. Repeat the exercises as needed to make yourself comfortable. I can guarantee that polylines will be an important part of your CAD future, so get comfortable with them now.

We also discussed some other important commands. Do you remember what they were? Remember, we discussed that they were so simple they are often overlooked and forgotten. Did you forget them already? I am talking, of course, about **List**, **Dist**, **Area**, and **ID**. Remember to put that cheater sheet on your monitor!

9.6 EXERCISES

1. Start a new drawing from scratch. Set the grid to $1/4"$ and create the image in Figure 9.6.1. Save the drawing as *MyArrows* in the C:\Steps\Lesson09 folder.

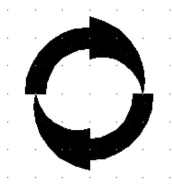

FIG. 9.6.1: Polyline Arrows

2. Set up a new drawing with the following parameters:
   2.1. Lower left limits: 0,0
   2.2. Upper right limits: 16,16
   2.3. Grid: 1
   2.4. Snap: as needed
   2.5. Textsize: $5/8$
   2.6. Text style: Arial font, Bold and Italicized
   2.7. Layers: as created in Lesson 7, Exercise 3
   2.8. Create the backgammon board shown in Figure 9.6.2. Save the drawing as *MyBoard* in the C:\Steps\Lesson09 folder.

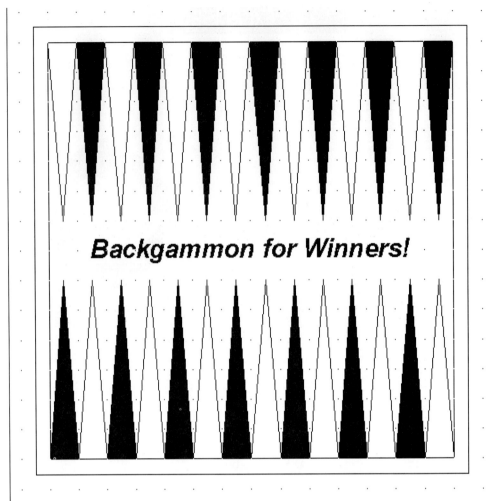

FIG. 9.6.2: Backgammon Board

3. The drawing shown in Figure 9.6.3 is a piping isometric of some pump discharge pipe.
    - **3.1.** Set up the drawing as follows:
        - **3.1.1.** Lower left limits: 0,0
        - **3.1.2.** Upper right limits: 17,11
        - **3.1.3.** Grid: $^1/_2$
        - **3.1.4.** Snap Style: Isometric
        - **3.1.5.** Snap: $^1/_4$
        - **3.1.6.** Textsize: $^1/_4$
        - **3.1.7.** Layers:

LAYER NAME	STATE	COLOR	LINETYPE
0	On	7 (white)	CONTINUOUS
CL	On	7 (white)	CENTER
FITTING	On	6 (magenta)	CONTINUOUS
NOZZLE	On	5 (blue)	CONTINUOUS
PIPE	On	1 (red)	CONTINUOUS
TEXT	On	5 (blue)	CONTINUOUS
VALVE	On	222	CONTINUOUS

    3.2. Details:

        3.2.1. The width of the pipe is $1/16$"

        3.2.2. The width of the valves and fittings is $1/64$"

    3.3. Create the drawing. Save the drawing as *MyPipingIso* in the C:\Steps\Lesson09 folder.

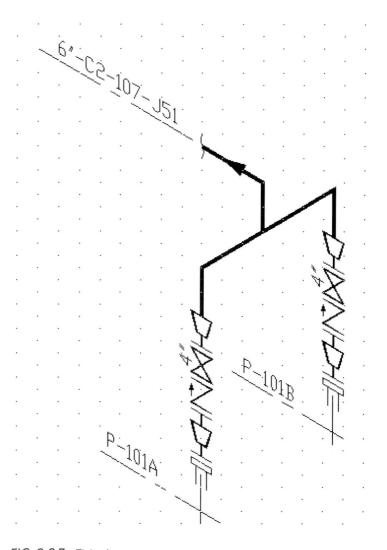

FIG. 9.6.3: Piping Iso

4. Use the same settings as in Exercise 3, but make the polyline width $1/32$ for the pipe and 0 for the valves in the drawings in Figures 9.6.4a and 9.6.4b. The grid is $1/4$". Save the drawings as *MyIso1* and *MyIso2* in the C:\Steps\Lesson09 folder.

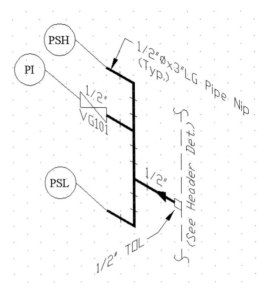

FIG. 9.6.4a: Iso 1

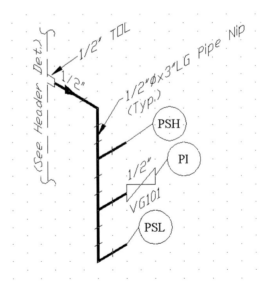

FIG. 9.6.4b: Iso 2

5. Figure 9.6.5 is one of the cards from my computer.
   5.1. Draw it using the following setup:
      5.1.1. Lower left limits: 0,0

5.1.2. Upper right limits: 12,9
5.1.3. Grid: $1/8$
5.1.4. Snap: $1/16$
5.1.5. Layers:

LAYER NAME	STATE	COLOR	LINETYPE
0	On	7 (white)	CONTINUOUS
CARD	On	6 (magenta)	CONTINUOUS
CONNECTION	On	30	CONTINUOUS
MOUNTS	On	5 (blue)	CONTINUOUS
SLIDER	On	7 (white)	CONTINUOUS

5.2. Details:
5.2.1. The heavy polylines are $1/16$" wide.
5.2.2. The lighter polylines are $1/32$" wide.
5.3. Save the drawing as *MyCard* in the C:\Steps\Lesson09 folder.

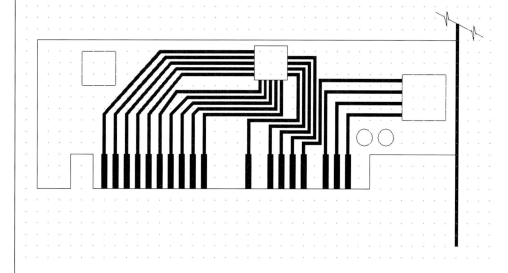

FIG. 9.6.5: *Computer Card*

9.7 REVIEW QUESTIONS

Please write your answers on a separate sheet of paper.

1. Polylines are so named for their ability to be drawn as _____ lines.

2. Capable of containing most of a polyline's information, the _____ is more easily understood and takes up less drawing memory.

3. and 4. Which two polyline options allow the user to draw a polyline with width?

5. When drawing a polyline, what is the first-tier option that repeats until the user hits enter to complete the command?

6. The _____ option of the pline command will present a second-tier of options.

7. (T or F) AutoCAD is *not* case-sensitive.

8. (T or F) By manipulating the width option of the pline command, the user can draw arrowheads.

9. to 12. List AutoCAD's four inquiry commands.

13. and 14. Which two inquiry commands will report the perimeter of a rectangle?

15. Which inquiry command will report the layer on which an object resides?

16. Which inquiry command will you use to identify the coordinate location of the center of an ellipse?

17. What command will you use to determine how far it is from one object to another?

18. It is often easier to _____ a simple polyline than it is to edit it.

19. (T or F) It is possible to join a polyline to a line without losing the line's definition in the database.

20. and 21. Which two options of the pedit command will turn polylines into contour lines?

22. (T or F) The user can create a polyline with different widths at the beginning and ending points.

23. (T or F) The user can modify an existing polyline so that the beginning and ending points have different widths.

24. Which option of the pedit command presents a second-tier of options?

25. Use the _____ command to convert a line to a polyline.

26. Use the _____ command to convert a polyline to a line(s).

Identify the following buttons:

27. 28. 29. 30. 31. 32.

LESSON 10

More Editing Tools

Following this lesson, you will:

➡ Know how to use the advanced Copy Commands:
- **Offset**
- **Array**
- **Mirror**

➡ Know how to use the **Lengthen** Command
➡ Know how to use the **Stretch** Command
➡ Know how to use the **Rotate** Command
➡ Know how to use the **Scale** Command

In Lesson 8, we discussed several modification tools that make computer drafting easier. In this lesson, we will see some more modifying tools also designed to speed and simplify the drafting process. To keep it simple, we will use the same two groupings we began in Lesson 8: *Location and Number* and *Change*. Let's begin where we left off—with the *Location and Number Group*.

10.1 Location and Number

■ 10.1.1 Parallels and Concentrics—The Offset Command

The last modification tools we studied in Lesson 8 were the *Copy* command and the **Multiple** option of the *Copy* command. *Offset* gives us another way to create one or many copies of a single object (line, polyline, circle, arc, etc). The copies will be either parallel (lines) or concentric (circles) to the existing object. The command sequence is

> **Command:** *offset* (or *o*)
> **Specify offset distance or [Through] <Through>:** *[enter]*
> **Select object to offset or <exit>:** *[select the object to offset]*
> **Specify through point:** *[select a point through which the copy should pass]*
> **Select object to offset or <exit>:** *[hit enter to complete the command]*

The only option provided is an important one. The first *Offset* prompt asks the user for an **Offset distance** or defaults to **Through**. **Through** allows the user to select a point in the drawing *through which* the copy will pass. But using the **Offset distance** option allows the user to specify *a perpendicular distance away from* the original object to place the copy. AutoCAD then prompts for the direction in which to place the copy. Let's try these.

> In addition to the command line and toolbar, you will find all of the commands in this lesson in the Modify pull-down menu.

WWW 10.1.1.1 Do This: Using the *Offset* Command

I. Open the *Star+* drawing in the C:\Steps\Lesson10 folder. The drawing looks like Figure 10.1.1.1a.

FIG. 10.1.1.1a: *Star+.dwg*

II. Follow these steps.

TOOLS	COMMAND SEQUENCE	STEPS
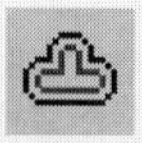 Offset Button	**Command:** *o*	**1.** Enter the *Offset* command by typing *offset* or *o* at the command prompt. Or you may pick the **Offset** button from the Modify toolbar.
	Specify offset distance or [Through] <Through>: *[enter]*	**2.** Hit *enter* to accept the default **Through** option.
	Select object to offset or <exit>:	**3.** Select the horizontal line above the circle.
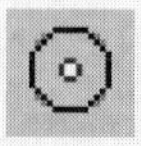	**Specify through point: _cen of**	**4.** Select the center of the circle (use OSNAPS).
	Select object to offset or <exit>: *[enter]*	**5.** Hit *enter* to complete the command. Your drawing looks like Figure 10.1.1.1.5.

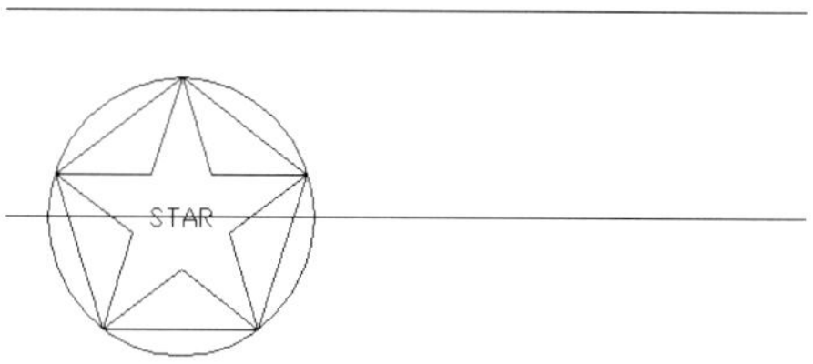

FIG. 10.1.1.1.5: Offset Line Through a Point

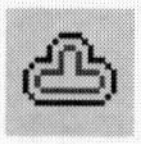	**Command:** *[enter]*	**6.** Repeat the command.
	Specify offset distance or [Through] <Through>: *[enter]*	**7.** Accept the default option.
	Select object to offset or <exit>:	**8.** This time, select the circle around the star.
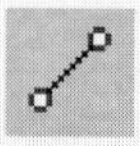	**Specify through point: _endp of**	**9.** Select the endpoint of the upper line.
	Select object to offset or <exit>: *[enter]*	**10.** Hit *enter* to complete the command. Your drawing looks like Figure 10.1.1.1.10.

continued

TOOLS	COMMAND SEQUENCE	STEPS
		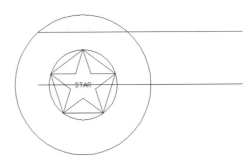

FIG. 10.1.1.1.10: Offset Circle Through a Point

Notice that the new line and circle go through the points you indicated.

Command: *e*

11. *Erase* the new line and circle.

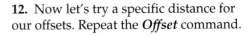

Command: *o*

12. Now let's try a specific distance for our offsets. Repeat the *Offset* command.

Specify offset distance or [Through] <Through>: *.25*

13. Type **.25** to tell AutoCAD to offset each object $1/4$ unit.

Select object to offset or <exit>:

14. Select the upper line and offset it toward the top of the screen.

Specify point on side to offset: *[pick any point above the line]*

Select object to offset or <exit>:

Specify point on side to offset:

15. Select the new line and offset it toward the top of the screen. Repeat this until there are four lines.

Select object to offset or <exit>:

Specify point on side to offset: *[pick any point outside the circle]*

16. While still in the command, select the circle and offset it outward (away from the star). Repeat this until there are four circles. Hit *enter* to complete the command. Your drawing looks like Figure 10.1.1.1.16.

continued

TOOLS	COMMAND SEQUENCE	STEPS

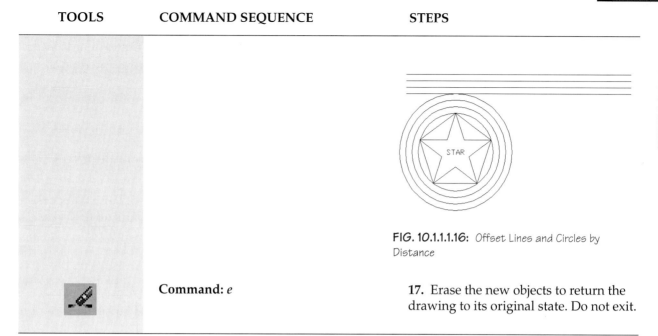

FIG. 10.1.1.1.16: Offset Lines and Circles by Distance

	Command: *e*	17. Erase the new objects to return the drawing to its original state. Do not exit.

> It is very easy to forget that these commands (*Offset*, *Array,* and *Mirror*) are *copy* tools. Many students become confused when the copied objects do not fall into the current layer. Remembering that the entire object (together with all of its properties—including color, linetype and layer) is being copied will help avoid confusion.

Nothing to it, right? *Offset* is a very handy command for being so simple. Its only limitation is that it can only copy a single object at a time. Let's look at something that can handle multiple objects.

■ 10.1.2 Rows, Columns and Circles: The Array Command

Many times we find that not only do we need several copies, but that the copies must be arranged in rows or columns or even circles. Using a normal *Copy* command to do this can evoke groans of tedium from CAD operators. Fortunately, the *Array* command was designed to prevent these groans. (Unfortunately, the *Array* command itself is often tedious.)

> The *Array* command is one of AutoCAD's few unfulfilled cries for a dialog box.

There are actually two command sequences—one for the **Rectangular** option (used to create rows and columns of the object(s) being arrayed) and another for the **Polar** option (used to array the object(s) in a circle or arc). The **Rectangular** sequence looks like this:

Command: *array* (or *ar*)
Select objects: *[select the objects to be arrayed]*
Enter the type of array [Rectangular/Polar] <R>: *[enter]*
Enter the number of rows (---) <1>: *[how many rows do you wish?]*
Enter the number of columns (| | |) <1>: *[how many columns do you wish?]*
Enter the distance between rows or specify unit cell (---): *[how far apart do you want your rows—enter the distance or select two points on the screen?]*
Specify the distance between columns (| | |): *[how far apart do you want your columns?]*
Command:

The only real option given in the *Array* command is whether you want a **Rectangular** or **Polar** array. If the user selects the **Rectangular** option, the remaining prompts simply ask for numbers and spacing.

If the **Polar** option is selected, the remaining sequence looks like this:

Specify center point of array: *[select the center point of the circle around which you want to array the object(s)]*
Enter the number of items in the array: *[how many items do you want?]*
Specify the angle to fill (+=ccw, −=cw) <360>: *[do you want an arc or a full circle?]*
Rotate arrayed objects? [Yes/No] <Y>: *[shall AutoCAD rotate the copies?]*

Notice the line: **Specify the angle to fill (+=ccw, −=cw) <360>:**. The **Angle to fill** request lets the user decide whether to fill a circle or an arc. The **+=ccw, −=cw** figures remind the user that a positive number entered here will yield a *counter-clockwise* (**ccw**) *arc* and a negative number will yield an arc drawn *clockwise* (**cw**).

Let's look at *Array*'s options.

WWW 10.1.2.1 Do This: Using the Array Command

I. Be sure you are still in the *Star+* drawing in the C:\Steps\Lesson10 folder.

II. Erase the horizontal line.

III. Follow these steps.

TOOLS	COMMAND SEQUENCE	STEPS
![Array Button] Array Button	**Command:** *ar*	1. Enter the *Array* command by typing *array* or *ar* at the command prompt. Or you may pick the **Array** button from the Modify toolbar.
	Select objects: 13 found **Select objects:** *[enter]*	2. Select the circle and all its contents. Then hit *enter* to confirm completion of the selection set.
	Rectangular or Polar array (<R>/P): *[enter]*	3. Tell AutoCAD you want a **Rectangular** array.
	Number of rows (---) <1>: *2*	4. Tell AutoCAD you want two rows . . .
	Number of columns (| | |) <1>: *4*	5. . . . and four columns.
	Unit cell or distance between rows (---): *3*	6. Tell AutoCAD to space the rows 3 units apart . . .
	Distance between columns (| | |): *3.5*	7. . . . and the columns 3.5 units apart. Your drawing looks like Figure 10.1.2.1.7.

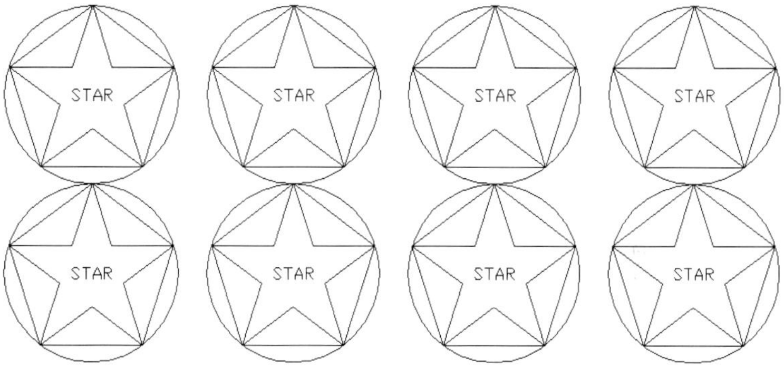

FIG. 10.1.2.1.7: Rectangular Array

![Undo]	**Command:** *u*	8. *Undo* the changes.
![Array Button]	**Command:** *ar*	9. Let's do a **Polar** array. Repeat the *Array* command.
	Select objects: 13 found **Select objects:** *[enter]*	10. Select the same objects.

continued

TOOLS	COMMAND SEQUENCE	STEPS
	Rectangular or Polar array (<R>/P): *p*	11. Tell AutoCAD you want a **Polar** array.
	Base/<Specify center point of array>: *5,5*	12. We will array about a point 3 units above the center of the circle. Type **5,5**.
	Number of items: *5*	13. Let's create five stars ...
	Angle to fill (+=ccw,−=cw) <360>: *[enter]*	14. ... and fill a full circle.
	Rotate objects as they are copied? <Y> *[enter]*	15. Accept the default to **Rotate objects**.
 	**Command:** *z*	16. ***Zoom all*** to better see your changes. Your drawing looks like Figure 10.1.2.1.16.

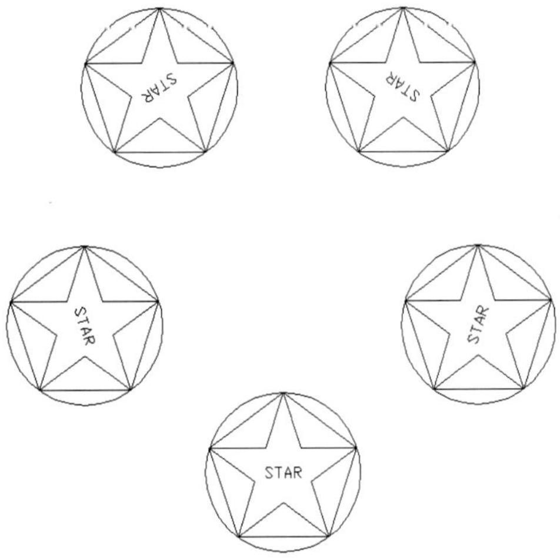

FIG. 10.1.2.1.16: Rotated Polar Array

Notice that the top point of the star always points toward the point we identified as the center of our array. This is because we chose to **Rotate** our objects as they were arrayed.

	Command: *e*	17. ***Erase*** the four new objects.

continued

TOOLS	COMMAND SEQUENCE	STEPS
	Rotate objects as they are copied? <Y> *n*	18. Repeat Steps 9 through 15, but this time tell AutoCAD *not* to **Rotate** the objects. Your drawing looks like Figure 10.1.2.1.18.

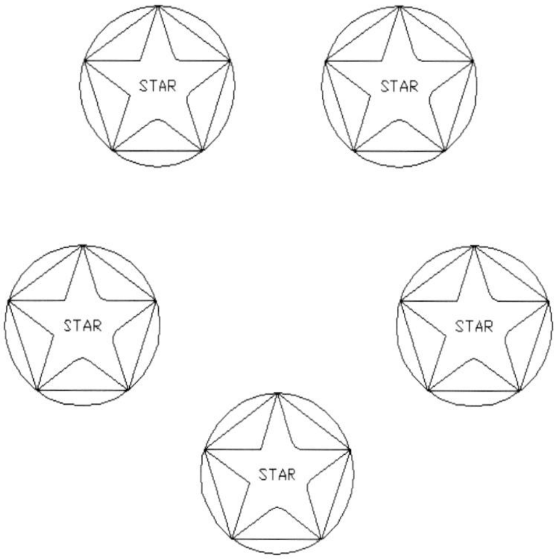

FIG. 10.1.2.1.18: Nonrotated Polar Array

Notice the difference? This time the stars maintain the orientation of the original objects.

	**Command:** *e*	19. *Erase* the new objects. Do not exit the drawing.

■ 10.1.3 Opposite Copies: The Mirror Command

Occasionally, you will run into a situation where you not only need to copy an object, but you need to completely reverse its orientation. This is not a true rotating of the object(s) which would just stand it on its head. What you need is an opposite or a *mirror* image of the original.

The *Mirror* command is one of AutoCAD's simplest. Here is the sequence:

Command: *mirror* **(or** *mi***)**
Select objects: *[select the objects to be mirrored]*
Select objects: *[confirm completion of the selection set]*
Specify first point of mirror line: *[pick the first point of the mirror line]*
Specify second point of mirror line: *[pick the second point of the mirror line]*

Delete source objects? [Yes/No] <N>: *[do you want to keep the original objects?]*
Command:

The only difficulty with this sequence is understanding the mirror line. Refer to Figure 10.1.3a. Consider this: To see a mirror image of an object as it lies on a table, you will lay a mirror along side the object (at a bit of an angle so as to reflect the object). The mirror *line* is the edge of the mirror where it meets the surface of the table. In AutoCAD, your screen is the table. You must define the *line* (the edge of the mirror) by identifying *two points* on it.

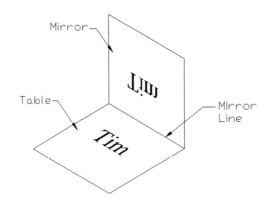

FIG. 10.1.3a: Mirror Line

Let's try it.

10.1.3.1 Do This: Mirroring an Object

I. Be sure you are still in the *Star+* drawing in the C:\Steps\Lesson10 folder.

II. Follow these steps.

TOOLS	COMMAND SEQUENCE	STEPS
Mirror Button	**Command:** *mi*	1. Enter the *Mirror* command by typing *mirror* or *mi*. Or you can pick the **Mirror** button on the Modify toolbar.
	Select objects:	2. Select the circle and all of its contents. Then hit *enter* to confirm completion of the selection set.
	Select objects: *[enter]*	

continued

TOOLS	COMMAND SEQUENCE	STEPS
	First point of mirror line: *7,3* **Second point:**	3. Pick the **First point of mirror line** at *7,3*, then (with the **Ortho *On***) pick a second point to the left (or west) of the first point, as shown in Figure 10.1.3.1.3.

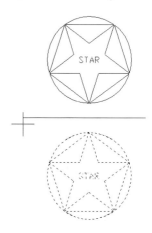

FIG. 10.1.3.1.3: Pick a Point to the Left

Notice the mirror image appearing as you identify the mirror line.

| | **Delete old objects?** <N> *[enter]* | 4. AutoCAD asks if you wish to delete the original. Accept the default—**No**. Your drawing looks like Figure 10.1.3.1.4. |

FIG. 10.1.3.1.4: Mirrored Objects

Notice that everything mirrored *except* the text. See the following insert for more details on this.

> By default, AutoCAD will mirror all objects *except text*. This default will help you avoid the frustration of having to redo all the text in an object just because you needed a mirror image.
>
> If, however, you want the text mirrored as well, you can change the sysvar that controls mirrored text—*Mirrtext*. Here is how:
>
> **Command:** *mirrtext*
>
> **Enter new value for MIRRTEXT <0>:** *1*
>
> Turn on the **Mirrtext** system variable (set it to *1* as indicated), then redo the previous exercise. Notice the difference?

Close the *Star+* drawing without saving your changes. We will return to it shortly.

10.2 More Commands in the Change Group

■ 10.2.1 Two Ways to Change the Length of Lines and Arcs: The Lengthen and Stretch Commands

Prior to the inclusion of the **Lengthen** command in R13, AutoCAD relied on the **Stretch** command to lengthen or shorten one or several lines. The **Stretch** command is still reliable and is really quite simple once the user understands the need to use a *crossing* window to select objects. The command sequence is

> **Command:** *stretch* **(or** *s***)**
>
> **Select objects to stretch by crossing-window or crossing-polygon . . .**
>
> **Select objects: Other corner:** *[note the instruction on the previous line—select the object(s) to be stretched using a* crossing *window]*
>
> **Select objects:** *[enter to confirm completion of the selection set]*
>
> **Specify base point or displacement:** *[pick a base point]*
>
> **Specify second point of displacement:** *[pick a target point]*
>
> **Command:**

The line instructing the user to **Select objects to stretch by crossing-window or crossing-polygon . . .** is easily overlooked as it appears above the actual prompt. But do not overlook its importance. You must select objects using a crossing window (or crossing-*polygon*, which we will discuss in Lesson 11). Unfortunately, AutoCAD does not default to a crossing window, so the user must enter *c* (or use implied windowing) to create a crossing window for object selection.

Notice that the user has the same option provided by the *Copy* and *Move* commands—that is, the user can select a **Base point or displacement**. To review how these work (as well as the **Second point of displacement**), see the discussion in Section 8.2.1.

Let's experiment with the *Stretch* command.

10.2.1.1 Do This: Stretching Objects

I. Reopen the *Star+* drawing in the C:\Steps\Lesson10 folder.

II. Follow these steps.

TOOLS	COMMAND SEQUENCE	STEPS
 (Move icon)	**Command:** *m* **Select objects:** *[select the line]* **Select objects:** *[enter]* **Specify base point or displacement:** *0,3* **Specify second point of displacement or <use first point as displacement>:** *[enter]*	**1.** Move the horizontal line upward (north) about 3 units. Use the **displacement** method.
(Stretch icon) Stretch Button	**Command:** *s* **Select objects to stretch by crossing-window or crossing-polygon ...** **Select objects: Other corner: 8 found**	**2.** Enter the *Stretch* command by typing *stretch* or *s* at the command prompt. Or you can pick the **Stretch** button on the Modify toolbar. **3.** Place a crossing window around the upper part of the star as indicated in Figure 10.2.1.1.3.
		FIG. 10.2.1.1.3: Crossing Window Placement
	Select objects: *[enter]*	**4.** Hit *enter* to complete the selection set.

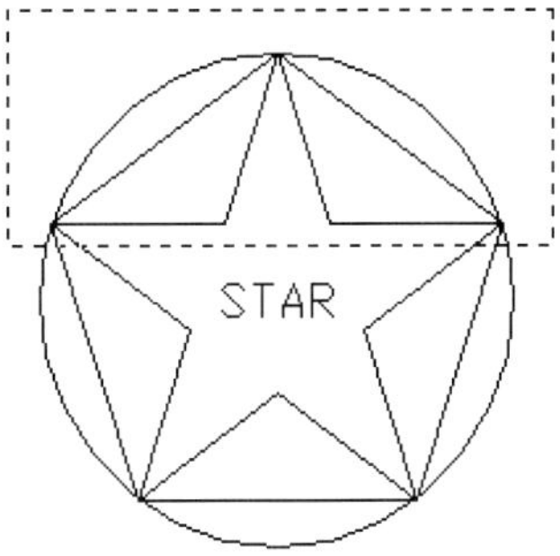

continued

TOOLS	COMMAND SEQUENCE	STEPS
	Specify base point or displacement: *0,2*	**5.** Use the **displacement** option and type a displace of *0,2*.
	Specify second point of displacement: *[enter]*	**6.** Remember, when using the **displacement** option you hit *enter* when prompted for a **Second point**. Your drawing looks like Figure 10.2.1.1.6.
		FIG. 10.2.1.1.6: Stretched Objects
		Notice that the circle did not stretch. It could not stretch and maintain its definition as a circle, so AutoCAD ignored it.
	**Command:** *u*	**7.** *Undo* your changes.

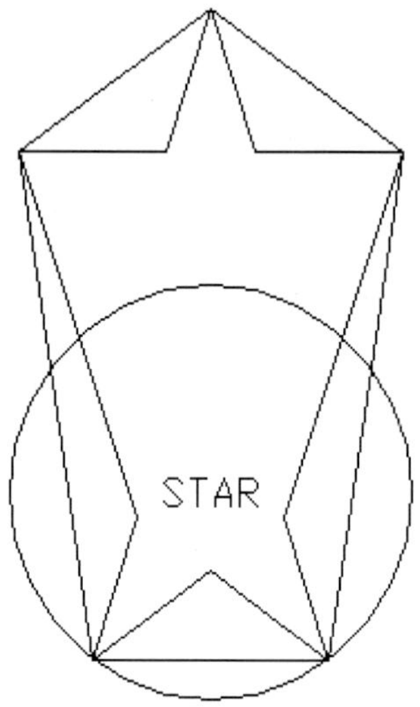

Of course, AutoCAD does not limit the user to one direction or even one object with the *Stretch* command. So in some ways, it still exceeds the newer *Lengthen* command. But the *Lengthen* command has its good points, too. Let's look at the command sequence:

Command: *lengthen* (or *len*)
Select an object or [DElta/Percent/Total/DYnamic]: *[tell AutoCAD what you want to do with the object or select a line or arc; subsequent prompts will change depending on which option you select]*

Current length: X.XXXX *[if you select an object, AutoCAD tells you the current length; if you select an arc, it will tell you the included angle as well]*

Select an object or [DElta/Percent/Total/DYnamic]: *[enter to complete the command]*

Command:

The four options are fairly simple.

- **DElta** (the old Greek word for "change") means *change*. When selected, it prompts:

    Enter delta length or [Angle] <0.0000>: *[enter the amount of change you wish]*
    Select an object to change or [Undo]: *[select a point on the object closest to the end at which you wish to add the length (use a negative number if you wish to shorten the line)]*

    If you select the **Angle** choice of the **DElta** option, you can add to an arc by specifying the angle of the arc to add.

- **Percent** allows the user to change the length of the selected object by percent. AutoCAD prompts:

    Enter percentage length <100.0000>: *[enter the desired change in percents]*

    Note that 100% means no change. More than 100% increases the length of the object and less than 100% decreases the length. Length is added or removed from the end closest to where you select the object.

- **Total** simply means, "How long do you want the object to be?" AutoCAD will add or remove as necessary (again, from the selected end) to make the object as long as the user has specified.

- **DYnamic** allows the user to manually (or *dynamically*) stretch the object to the desired length. AutoCAD prompts:

    Select an object to change or [Undo]:
    Specify new end point:

    The user selects the **object to change** and specifies a **new end point**.

    Let's see what we can do with the *Lengthen* command.

10.2.1.2 Do This: Using the Lengthen Command

I. Be sure you are still in the *Star+* drawing in the C:\Steps\Lesson10 folder.

II. Follow these steps.

TOOLS	COMMAND SEQUENCE	STEPS
	**Command:** *a*	1. Draw an arc as detailed at left.
	Specify start point of arc or [CEnter]: *1,5*	
	Specify second point of arc or [CEnter/ENd]: *c*	
	Specify center point of arc: *1,4*	
	Specify end point of arc or [Angle/chord Length]: *0,4*	
	Command: *z*	2. ***Zoom all.*** Your drawing looks like Figure 10.2.1.2.2.

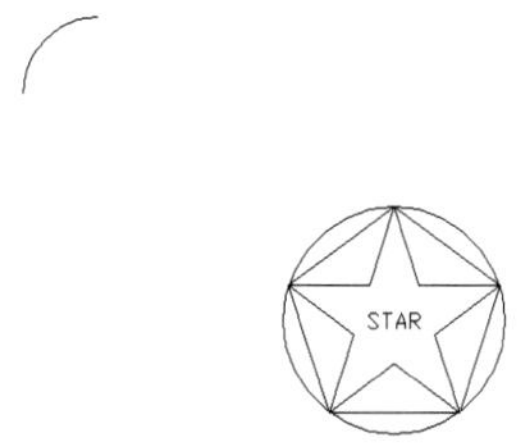

FIG. 10.2.1.2.2: Drawing with Arc

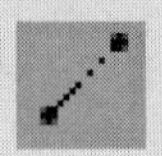 **Lengthen Button**	**Command:** *len*	3. Enter the *Lengthen* command by typing *lengthen* or *len*. Or you can pick the **Lengthen** button on the Modify toolbar.
	Select an object or [DElta/Percent/Total/DYnamic]:	4. Select the arc you drew in Step 1.
	Current length: 1.5708, included angle: 90	5. AutoCAD reports the true length of the arc and the included angle. Hit *enter* to complete the command.
	Select an object or [DElta/Percent/Total/DYnamic]: *[enter]*	
	**Command:** *[enter]*	6. Repeat the command.
	Select an object or [DElta/Percent/Total/DYnamic]: *de*	7. Type *de* to use the **DElta** option to change the length.

continued

TOOLS	COMMAND SEQUENCE	STEPS
	Enter delta length or [Angle] <0.0000>: *1*	8. AutoCAD asks if you wish to make a change by **Angle** or **length**. Since **length** is the default, type *1* to add one unit to the arc.
	Select an object to change or [Undo]:	9. Select the upper end of the arc. AutoCAD adds one unit to the upper end. It now looks like Figure 10.2.1.9.

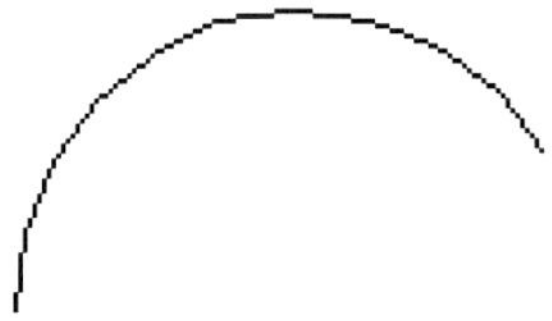

FIG. 10.2.1.2.9: New Arc

	Select an object to change or [Undo]:	10. Now select the left end of the horizontal line. Notice that AutoCAD adds a piece one unit long to the end.
	Select an object to change or [Undo]: *[enter]*	11. Hit *enter* to complete the command.
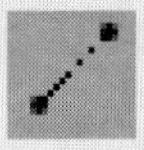	Command: *[enter]*	12. Repeat the command.
	Select an object or [DElta/Percent/ Total/DYnamic]: *p*	13. Let's make an adjustment by the **Percent** method. Type *p* for the **Percent** option.
	Enter percentage length <100.0000>: *50*	14. Let's cut the objects in half. Enter *50* for the **percentage length**.
	Select an object to change or [Undo]: *[select the arc]*	15. Select the lower end of the arc; then select the right end of the line. Hit *enter* to complete the command. The objects now look like Figure 10.2.1.2.15.
	Select an object to change or [Undo]: *[select the line]*	
	Select an object to change or [Undo]: *[enter]*	

continued

TOOLS	COMMAND SEQUENCE	STEPS

FIG. 10.2.1.2.15: 50% of Line and Arc

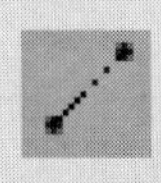

	Command: *[enter]*	**16.** Repeat the command.
	Select an object or [DElta/Percent/Total/DYnamic]: *t*	**17.** This time we will set the **Total** length of arc and line. Type *t* for the **Total** option.
	Specify total length or [Angle] <1.0000>: *4*	**18.** Type *4* to set a total length of four units.
	Select an object to change or [Undo]: *[select the arc]*	**19.** Select the left end of the arc; then select the right side of the line. Notice that **Total** added to the arc but subtracted from the line. Finally, hit *enter* to complete the command. The objects now look like Figure 10.2.1.2.19.
	Select an object to change or [Undo]: *[select the line]*	
	Select an object to change or [Undo]: *[enter]*	

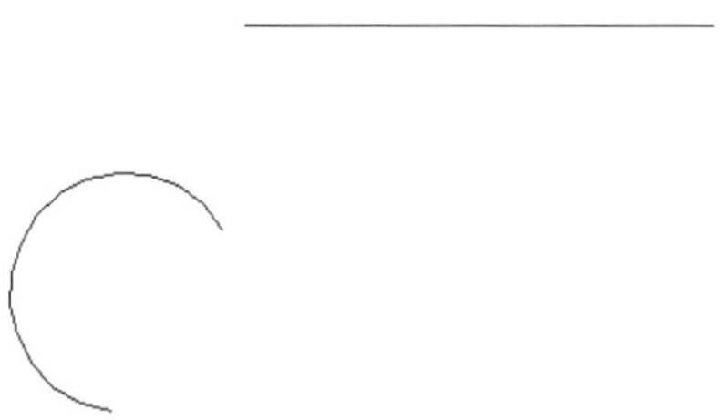

FIG. 10.2.1.2.19: 4" Arc and Line

	Command: *[enter]*	**20.** Repeat the command.
	Select an object or [DElta/Percent/Total/DYnamic]: *dy*	**21.** Using the **DYnamic** option, select the line and watch how it changes as you move the cursor back and forth. Then try it with the arc.
	Command: *quit*	**22.** Close the drawing without saving the changes.

The greatest differences between the *Lengthen* and *Stretch* commands are twofold. First, there is an easy precision allotted by the *Lengthen* command; and, second, the *Stretch* command allows the user to modify more than one object at a time.

I am more prone to use the *Stretch* command partly because, more often than not, I must modify multiple objects. But I use it also out of habit. For such a simple command, it is remarkably versatile and quite useful!

■ 10.2.2 Oh, No! I Drew It Upside Down!: The Rotate Command

Okay. So this will probably never happen to you (then again, you might just be surprised). Still, it is not unusual to find a need to rotate text for a better fit or rotate a piece of equipment or furniture for a more efficient layout. Either way, the *Rotate* command offers a simple solution to problems that, on a drawing board, might cause a redraw.

The *Rotate* command sequence is

Command: *rotate* **(or** *ro***)**
Current positive angle in UCS: ANGDIR=counterclockwise ANGBASE=0
Select objects: *[select the object(s) to rotate]*
Select objects: *[hit enter to confirm the selection set]*
Specify base point: *[select a point around which to rotate]*
Specify rotation angle or [Reference]: *[how much to rotate]*
Command:

■ AutoCAD begins by telling you something about the setup:

- **Current positive angle in UCS:** line refers to 3-dimensional space. You can ignore it for now.
- The **ANGDIR** simply reminds you that angles are measured counterclockwise (unless you changed it during the setup procedure for the drawing)
- **ANGBASE** refers to a system variable that allows the user to change the angle from which AutoCAD begins to measure. For example, if you tell AutoCAD to use a reference angle of 45°, it will add 45 to the value of the **Angbase** sysvar.

■ The default option—**Rotation angle**—simply means, "How much do you want to rotate?" Type in an angle or drag the object on the screen (use ortho to rotate at 90° increments).

■ The other option—**Reference**—prompts again:

Specify the reference angle <0>: *[tell AutoCAD what the current rotation is]*
Specify the new angle: *[tell AutoCAD what you want the rotation to be]*

Use this option when you know what the current rotation angle is and what you want it to be. If you do not know what it is but do know what you want it to be, use this option and select two points on a line that represents the current rotation. AutoCAD will determine the angle from the line you select and rotate to the **new angle**.

This may be confusing. Let's try an example.

10.2.2.1 Do This: Rotating Objects

I. Reopen the *Star+* drawing in the C:\Steps\Lesson10 folder.

II. Follow these steps.

TOOLS	COMMAND SEQUENCE	STEPS
 Rotate Button	**Command:** *ro*	**1.** Enter the *Rotate* command by typing *rotate* or *ro*. Or you can pick the **Rotate** button on the Modify toolbar.
	Current positive angle in UCS: ANGDIR= **counterclockwise ANGBASE=0** **Select objects: 13 found**	**2.** Select the circle and all the objects within it.
	Select objects: *[enter]*	**3.** Hit *enter* to confirm the selection set.
	**Specify base point: _cen of**	**4.** Select the center point of the circle (use OSNAPS).
	Specify rotation angle or [Reference]: *90*	**5.** Tell AutoCAD to rotate the objects 90°. The star will look like Figure 10.2.2.1.5.

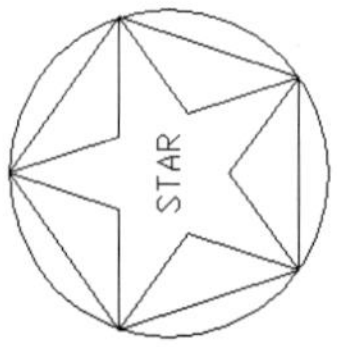

FIG. 10.2.2.15: Rotated Objects

	**Command:** *[enter]*	**6.** Repeat the *Rotate* command.
	Current positive angle in UCS: ANGDIR= **counterclockwise ANGBASE=0** **Select objects: 13 found**	**7.** Select the circle and all the objects within it.

continued

TOOLS	COMMAND SEQUENCE	STEPS
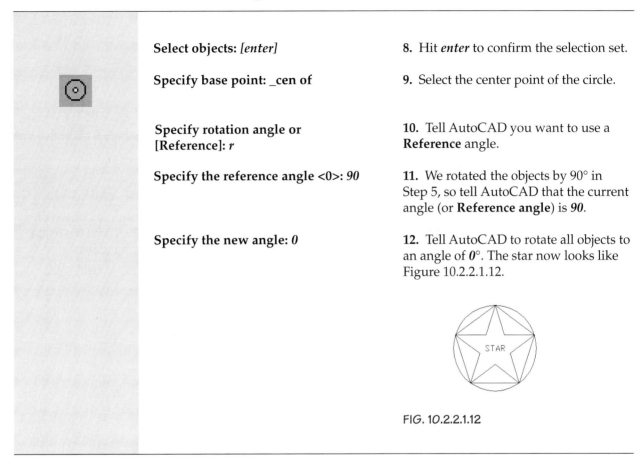	Select objects: *[enter]*	8. Hit *enter* to confirm the selection set.
	Specify base point: _cen of	9. Select the center point of the circle.
	Specify rotation angle or [Reference]: *r*	10. Tell AutoCAD you want to use a **Reference** angle.
	Specify the reference angle <0>: *90*	11. We rotated the objects by 90° in Step 5, so tell AutoCAD that the current angle (or **Reference angle**) is *90*.
	Specify the new angle: *0*	12. Tell AutoCAD to rotate all objects to an angle of *0°*. The star now looks like Figure 10.2.2.1.12.

FIG. 10.2.2.1.12

If only they were all that easy!

■ 10.2.3 "Okay. Give Me Three Just Like It: But Different Sizes": The Scale Command

No, that does not mean you have to draw it two more times. Simply make two copies then *scale* (or *resize*) the copies to meet the customer's requirements. The command sequence is one of the easiest:

Command: *scale* (or *sc*)
Select objects: *[select the object(s) to scale]*
Select objects: *[confirm completion of the selection set]*
Specify base point: *[pick the base point]*
Specify scale factor or [Reference]: *[enter the scale factor]*
Command:

Reference works just like it did in the *Rotate* command. Let's see *Scale* in action.

10.2.3.1 Do This: Scaling Objects

I. Be sure you are still in the *Star+* drawing.

II. Follow these steps.

TOOLS	COMMAND SEQUENCE	STEPS
Scale Button	**Command:** *e*	**1.** Erase the horizontal line.
	Command: *sc*	**2.** Enter the *Scale* command by typing *scale* or *sc*. Or you can pick the **Scale** button on the Modify toolbar.
	Select objects: 13 found	**3.** Select the circle and all the objects inside it.
	Select objects: *[enter]*	**4.** Hit *enter* to confirm the selection set.
	Specify base point: _qua of	**5.** Select the bottom quadrant of the circle as the **Base point**. Notice how the objects change as you move the cursor.
	Specify scale factor or [Reference]: *2*	**6.** Tell AutoCAD you want to scale the objects by a factor of two. The objects look the same but twice as large.
		7. Repeat Steps 2 through 5.
	Specify scale factor or [Reference]: *r*	**8.** Tell AutoCAD you want to use a **Reference** scale.
	Specify reference length <1>: *2*	**9.** We made the objects twice as large in Step 5; so tell AutoCAD the **reference length** is 2.
	Specify new length: *1*	**10.** Tell AutoCAD the **new length** should be 1 (1 where it was 2, or $1/2$). The objects still look the same but are half the size.
	Command: *quit*	**11.** Quit the drawing. Do not save your changes.

10.3 Putting It All Together

We have covered several new modification tools in this lesson using our star for demonstrations. The star is perfect for these demonstrations, but let's try an exercise and use several of these tools to create a drawing.

10.3.1 Do This: A Practical Exercise

I. Open drawing *Grad_cyl* from the C:\Steps\Lesson10 folder. It looks like Figure 10.3.1a.

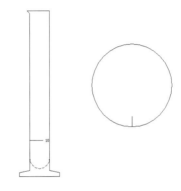

FIG. 10.3.1.1a: Grad_cyl.dwg

II. Follow these steps.

TOOLS	COMMAND SEQUENCE	STEPS
	Command: *z*	1. Use the *Zoom* command to adjust your view to the one shown in Figure 10.3.1.1. 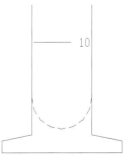 FIG. 10.3.1.1: Zoom In

continued

TOOLS	COMMAND SEQUENCE	STEPS
	Command: *ar* **Select objects:** *[select the line]* **Select objects:** *[enter]* **Enter the type of array [Rectangular/Polar] <R>:** *[enter]* **Enter the number of rows (---) <1>:** *5* **Enter the number of columns (| | |) <1>:** *[enter]* **Enter the distance between rows or specify unit cell (---):** *-.1*	2. Use the *Array* command to make four copies of the red line below the one shown. The spacing should be 0.1" (Figure 10.3.1.2).  **FIG. 10.3.1.2:** Array the Graduation Line
	Command: *len* **Select an object or [DElta/Percent/Total/DYnamic]:** *t* **Specify total length or [Angle] <1.0000)>:** *.75* **Select an object to change or [Undo]:** *[select the right side of the top line]* **Select an object to change or [Undo]:** *[enter]*	3. Use the *Lengthen* command to adjust the length of the top line to **.75"** (Figure 10.3.1.3). 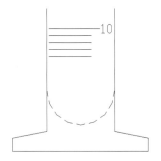 **FIG. 10.3.1.3:** Lengthen the Top Line
	Command: *cp* **Select objects:** *[select the five lines]* **Select objects:** *[enter]* **Specify base point or displacement, or [Multiple]:** *0,¹/₂* **Specify second point of displacement or <use first point as displacement>:** *[enter]*	4. Copy the lines upward $^1/_2$" for a total of 10 lines (Figure 10.3.1.4).

continued

TOOLS	COMMAND SEQUENCE	STEPS

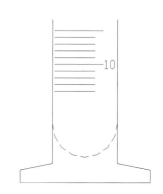

FIG. 10.3.1.4: Copy the Lines

Command: *len*

Select an object or [DElta/Percent/Total/DYnamic]: *t*

Specify total length or [Angle] <0.7500)>: $^{11}/_{16}$

Select an object to change or [Undo]: [*select the right side of the top line*]

Select an object to change or [Undo]: [*enter*]

5. Use the *Lengthen* command to adjust the length of the top line to $^{11}/_{16}$" (Figure 10.3.1.5).

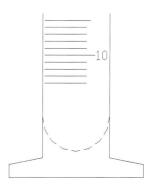

FIG. 10.3.1.5: Shorten the Top Line

Zoom Previous

Command: *z*

Specify corner of window, enter a scale factor (nX or nXP), or

[All/Center/Dynamic/Extents/Previous/Scale/Window] <real time>: *p*

6. Restore the previous view.

continued

TOOLS	COMMAND SEQUENCE	STEPS
	Command: *ar* **Select objects:** *[select the lines and the number]* **Select objects:** *[enter]* **Enter the type of array [Rectangular/Polar] <R>:** *[enter]* **Enter the number of rows (---) <1>:** *5* **Enter the number of columns (I I I) <1>:** *[enter]* **Enter the distance between rows or specify unit cell (---):** *1*	7. Use the *Array* command to place the graduations on the cylinder. (Figure 10.3.1.7). 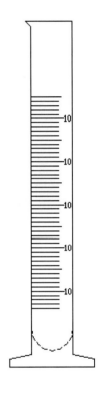 **FIG. 10.3.1.7:** Create the Graduations
	Command: *ed*	8. Use the text editor to change the numbers to read: *10, 20, 30, 40* and *50*. *Erase* the lines above 50 (Figure 10.3.1.8).

continued

TOOLS	COMMAND SEQUENCE	STEPS

FIG. 10.3.1.8: Completed Graduated Cylinder

Your graduated cylinder is complete. Let's draw a speedometer.

Command: *z*

9. Now use the *Zoom* command to change your view to look like Figure 10.3.1.9.

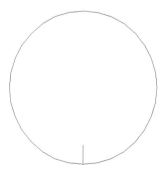

FIG. 10.3.1.9: Zoom In to the Circle

continued

TOOLS	COMMAND SEQUENCE	STEPS
	Command: *o* **Specify offset distance or [Through] <Through>:** *.17* **Select object to offset or <exit>:** *[select the circle]* **Specify point on side to offset:** *[pick a point inside the circle]* **Select object to offset or <exit>:** *[select the inner circle]* **Specify point on side to offset:** *[pick a point inside the circle]* **Select object to offset or <exit>:** *[enter]*	10. Use the *Offset* command to copy the circle inward twice at .17" increments (Figure 10.3.1.10). 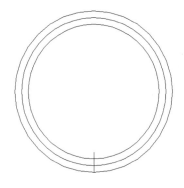 **FIG. 10.3.1.10:** Offset the Circle
	Command: *ar* **Select objects:** *[select the line]* **Select objects:** *[enter]* **Enter the type of array [Rectangular/Polar] <R>:** *p* **Specify center point of array: _cen of** *[pick the center of the circles]* **Enter the number of items in the array:** *10* **Specify the angle to fill (+=ccw, −=cw) <360>:** *36* **Rotate arrayed objects? [Yes/No] <Y>:** *[enter]*	11. Use the *Array* command to make 10 copies of the line along a 36° arc (Figure 10.3.1.11). 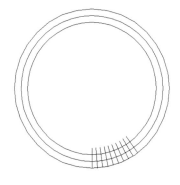 **FIG. 10.3.1.11:** Array the Line
	Command: *tr*	12. Using the inner circles as cutting edges, *trim* the lines, as shown in Figure 10.3.1.12.

continued

TOOLS	COMMAND SEQUENCE	STEPS

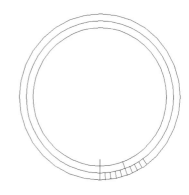

FIG. 10.3.1.12: Trim the Lines

Command: *e*

13. *Erase* the two inner circles.

Command: *ar*

Select objects: *[select the lines]*

Select objects: *[enter]*

Enter the type of array [Rectangular/Polar] <P>: *[enter]*

Specify center point of array: _cen of *[pick the center of the circle]*

Enter the number of items in the array: *10*

Specify the angle to fill (+=ccw, −=cw) <360>: *[enter]*

Rotate arrayed objects? [Yes/No] <Y>: *[enter]*

14. *Array* the lines as shown in Figure 10.3.1.14.

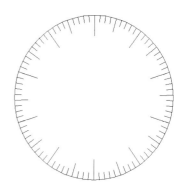

FIG. 10.3.1.14: Array the Lines

Command: *e*

15. *Erase* the bottom set of lines as shown in Figure 10.3.1.15.

continued

TOOLS	COMMAND SEQUENCE	STEPS
		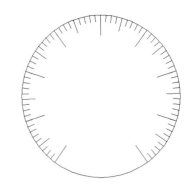 **FIG. 10.3.1.15:** Erase the Bottom **16.** Add text and polylines (use the appropriate layers and a text height of **0.2**) to complete the speedometer. (Hint: Draw the pointer pointing downward using ortho, then *rotate* it.) 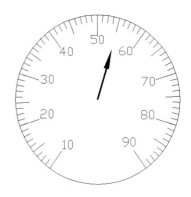 **FIG. 10.3.1.16:** Completed Speedometer
	Command: *qsave*	**17.** Save the drawing.

10.4 Extra Steps

When All Else Fails—Ask For Help.
—My Wife (When I couldn't find the right exit)

We are halfway through the book now. Have you noticed the **Help** pull-down menu yet? Have you tried to use it?

One of the smartest things you (or any student) can do is to know what resources are available to help you when you get into a bind.

Make a list of the resources available to you—begin with this text. Add your instructor (if you are in a classroom environment), the smart guy sitting next to you, and any other classroom references (handouts, books, tests, etc.). Most importantly —*add your willingness to explore and experiment with the software* (never underestimate the power of trial and error as a learning tool).

Now take a look at the **Help** pull-down menu. The first thing you may notice is the number of selections available to you. Let's look at these:

- **AutoCAD Help Topics**: This provides access to a standard Windows Help dialog box. Three tabs provide access to information using a table of contents format, a subject index format and a word search format. I suggest starting with the subject index: Simply enter the subject for which you are searching and pick the **Display** button. If that does not provide a solution, try the word search (**Find**) tab.
- **What's New**: This is for experienced AutoCAD users. It provides update information and it details changes from earlier releases.
- **Learning Assistant**: There should be an extra CD in your AutoCAD package called *Learning Assistant*. This provides tutorials (in some cases, animated tutorials). Put the CD in the computer, then pick this selection.
- **Support Assistant**: This is an "online knowledgebase of technical support information." (You will need Internet access for this.)
- **Autodesk on the Web**: Provides access to autodesk's home page, plug-in store (where you will find things like the Whip—which you will need to view AutoCAD files using your web browser), technical publications, product support, resource guide, and access to a user group (other AutoCAD users who share information over the web).

Now ask yourself, "how can I fail with all this help?!"

I am convinced that we learn more through exploration, experimentation and error than by any other means.

—Anonymous

10.5 What Have We Learned?

Items covered in this lesson include:

- *AutoCAD Modify Commands:*
  - *Offset*
  - *Array*
  - *Mirror*
  - *Lengthen*
  - *Stretch*
  - *Rotate*
  - *Scale*

■ *AutoCAD System Variables:*

- *Mirrtext*
- *Angdir*
- *Angbase*

Have you noticed that the easiest of AutoCAD's tools are invariably modification commands?

With the conclusion of this lesson, you will have learned all of the more common drawing and editing techniques AutoCAD has to offer the 2-dimensional drafter. We will finish Part 2: "Beyond the Basics" with a discussion of some fairly useful drawing tricks and toys. Then we will start our study of some more advanced techniques that you will need to master before venturing into the third dimension (or applying for that CAD position).

10.6 EXERCISES

1. Create a drawing template with the following parameters:
   - **1.1.** Lower left limits: **0,0**
   - **1.2.** Upper right limits: **17,11**
   - **1.3.** Grid: $^1/_2$
   - **1.4.** Snap: $^1/_4$
   - **1.5.** Layers:

1.5.1. LAYER	COLOR	LINETYPE
Obj	red	continuous
Cl	cyan	center2
Hidden	magenta	hidden2
Text	yellow	continuous
Dim	blue	continuous

   - **1.6.** Save the template as *BwLay.dwt* in the C:\Steps\Lesson10 folder.
2. Start a new drawing using the *BwLay.dwt* template.
   - **2.1.** Create the drawing in Figure 10.6.2. You are allowed to draw one box and the outline of the wall. You may use the following commands to help: **Copy, Rotate, Scale** and **Stretch**. The size of each window is 1.75 × the size of the smaller window next to it. Feel free to add layers if you think it necessary.

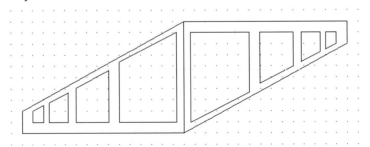

FIG. 10.6.2: Use *Copy, Rotate, Scale* and *Stretch* Commands

3. Open drawing *MyRuler* (created in Lesson 7) or drawing *Ruler*. Create the drawing in Figure 10.6.3. Feel free to add layers if you think it necessary.

   3.1. The circles have a 2" radius. The longer lines in the circles are $1/2$", and the shorter lines are $1/4$".

   3.2. The text in the circles and on the ruler is $1/8$" high.

   3.3. The gradient lines on the ruler are: $3/8$", $1/4$", $1/8$" and $1/16$".

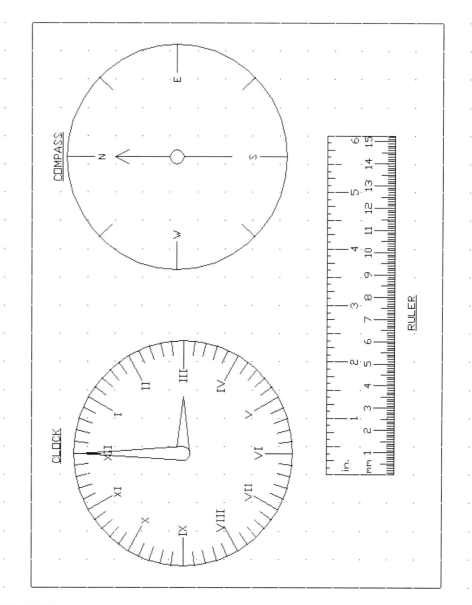

FIG. 10.6.3: *MyRuler.dwg*

4. Using what you have learned, create the Vanguard drawing in Figure 10.6.4. The limits are 0,0 and 9'−6",11'−6"; and the grid is 2".

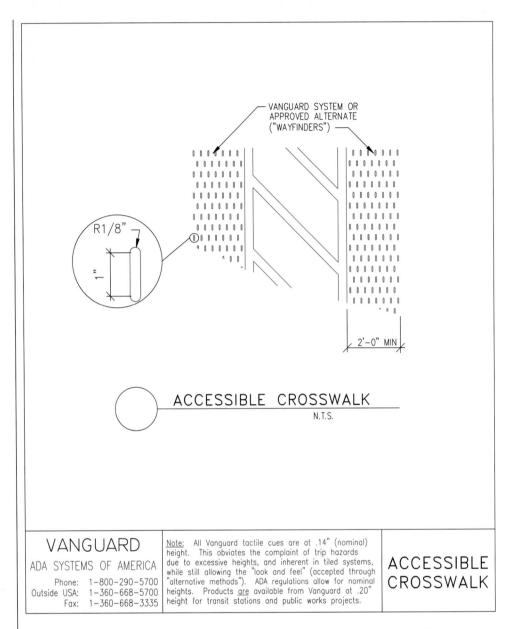

FIG. 10.6.4: Accessible Crosswalk (Special thanks to Tilco Vanguard in Snohomish, Washington, and Jon Julnes for permission to use their drawing.)

5. Now create the drawings in Figures 10.6.5a to 10.6.5d. The Roof Design sheet uses the same border, sheet size, and grid as the previous detail sheets (Lessons 5 and 6). The other drawings will fit on an 11" × 17" sheet of paper.

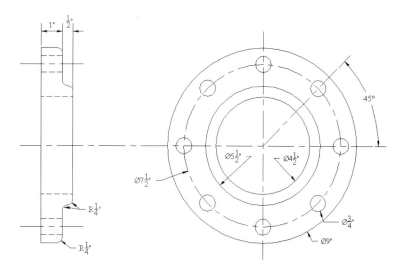

FIG. 10.6.5a: Flange Detail

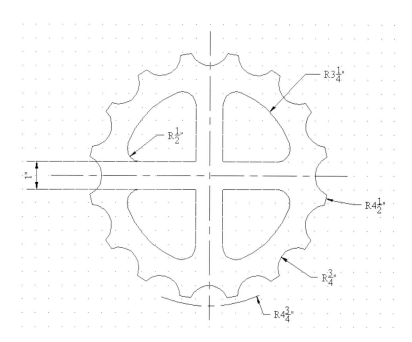

FIG. 10.6.5b: Gear Detail

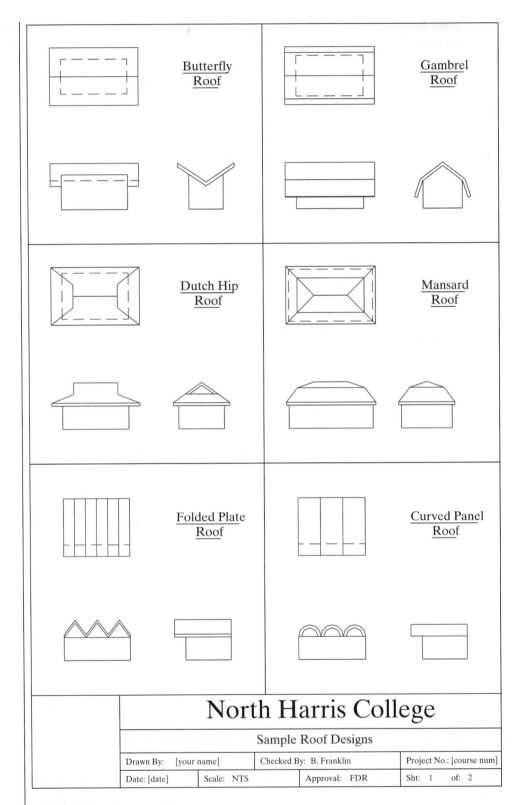

FIG. 10.6.5c: Sample Roof Designs

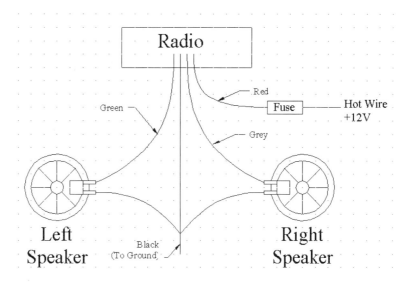

FIG. 10.6.5d: Car Radio Wiring

10.7 REVIEW QUESTIONS

1. to 3. List the three commands that make identical copies of AutoCAD objects.

4. Which command would you use to create parallel or concentric circles?

5. Using the distance option of the offset command enables the user to specify the (linear, perpendicular) distance between the original object and its copy.

6. The _____ command will make a reversed copy of the original objects selected.

7. (T or F) Offset or arrayed copies lie on the layer that was current when the copies were made.

8. (T or F) Offset can only copy a single object at a time.

9. The _____ option of the array command will place objects in a circle or arc.

10. and 11. By default, AutoCAD will mirror all objects except _____ unless the _____ sysvar is set to one.

12. and 13. Use either the _____ or the _____ command to make an object longer.

14. Objects to be stretched must be selected using a _____.

15. and 16. Use the _____ option of the _____ command to make a line half as long as it is.

17. and 18. To make an existing line exactly 5" long, use the _____ option of the _____ command.

19. To lengthen more than one object at a time, use the _____ command.

20. Use the _____ command to adjust the angle of an object.

21. Use the _____ command to adjust the size of an object.

Identify these buttons and the hotkeys that are associated with their commands.

Button							
Command	22.	24.	26.	28.	30.	32.	34.
The hotkey for this comand is:	23.	25.	27.	29.	31.	33.	35.

PART III

Some More Advanced Techniques

This part of our text contains these lessons:
- 11. Some Useful Drawing Tricks
- 12. Guidelines and Splines
- 13. Advanced Lines: Multilines
- 14. Advanced Text: MText
- 15. Basic Dimensioning
- 16. Customizing Dimensions
- 17. Advanced Modfication Techniques

LESSON 11

Some Useful Drawing Tricks

Following this lesson, you will:

➡ Know how to use the **Divide** and **Measure** Commands
➡ Know how to draw with these commands:
  - **Point**
  - **Solid**
  - **Donut**
➡ Be familiar with these Advanced Object Selection Methods:
  - WindowPoly
  - CrossingPoly
  - Fence
  - Last
  - All
  - Previous
  - Add and Remove
➡ Know how to use Quick Select Filters

You have come a very long way since learning the Cartesian Coordinate System so many lessons ago. You have learned the basic 2-dimensional tools for drawing and modifying most objects. You have learned to draw with a precision you probably never dreamt possible on a drawing board. Did you know that this drawing precision enables manufacturers to create products directly from your drawing? It is a system called CAM (Computer Aided Manufacturing). CAD-CAM is one possible direction your career might take if you pursue AutoCAD into the 3-dimensional world.

This lesson allows you to relax a bit and take a look at some of AutoCAD's tricks and toys meant to enhance the productivity of its users. Some of the toys you may never use; some you might rarely use. But all are full of possibilities.

Let's begin.

11.1 Equal or Measured Distances: The *Divide* and *Measure* Commands

Both these commands serve to place markers (nodes) at certain locations on an object. *Divide* places equally spaced nodes along the object. *Measure* places a node at user-set distances along the object. (Note: Neither command actually breaks the object. Rather, both place nodes along the object.)

> We defined a *point*, you will recall, as the place where an X-plane intersects a Y-plane. A *node* is an object that occupies a single point. It serves primarily as an identifier or locator in the drawing. CAD operators frequently use the terms *point* and *node* interchangeably when referring to a single point. It is not that important to remember the difference, but this might help:
>
> I snap to a n**o**de (**o**bject) that occupies a po**i**nt (**i**dea).
> (Wouldn't that make a good test question?!)

The command sequences are very simple. Here is the ***Divide*** command:

Command: *divide* (or *div*)
Select object to divide: *[select the object to divide]*

Enter the number of segments or [Block]: *[tell AutoCAD how many segments you want]*
Command:

- The **Block** option allows the user to place a predefined block rather than a node at the end of each segment. We will study blocks in Lesson 19.

The *Measure* command sequence is

Command: *measure* (or *me*)
Select object to measure: *[select the object to divide]*
Specify length of segment or [Block]: *[tell AutoCAD how far apart to place the nodes]*
Command:

These commands are also available in the Draw pull-down menu. Follow this path:

Draw—Point—Divide (or Measure)

Let's see these commands in action.

WWW 11.1.1 Do This: Dividing & Measuring

I. Open the *mea-div* drawing in the C:\Steps\Lesson11 folder. The drawing looks like Figure 11.1.1a.

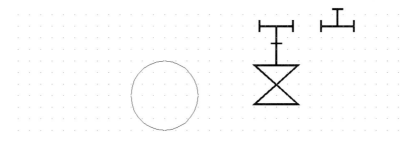

FIG. 11.1.1a: *Mea-div.dwg*

II. Follow these steps.

TOOLS	COMMAND SEQUENCE	STEPS
	**Command:** *z*	1. *Zoom* in around the circle.
	Command: *la*	2. Set **Markers** as the current layer.
	Command: *div*	3. Enter the *Divide* command by typing *divide* or *div* at the command prompt. There is no toolbar button for the *Divide* command.
	Select object to divide:	4. Select the circle.
	Enter the number of segments or [Block]: *5*	5. Tell AutoCAD you want five equal divisions marked off on the circle. Your drawing looks like Figure 11.1.1.5.

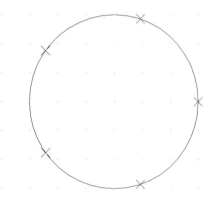

FIG. 11.1.1.5: Divided Circle

 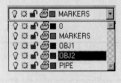	**Command:** *os*	6. Set the Running OSNAP to **Node**. Clear all other settings. 7. Set the current layer to **obj2**.

continued

TOOLS	COMMAND SEQUENCE	STEPS
	Command: *l*	8. Draw lines connecting the nodes so that your drawing looks like Figure 11.1.1.8.
		FIG. 11.1.1.8: Star in Circle
	**Command:** *qsave*	9. Save your drawing, but do not exit.

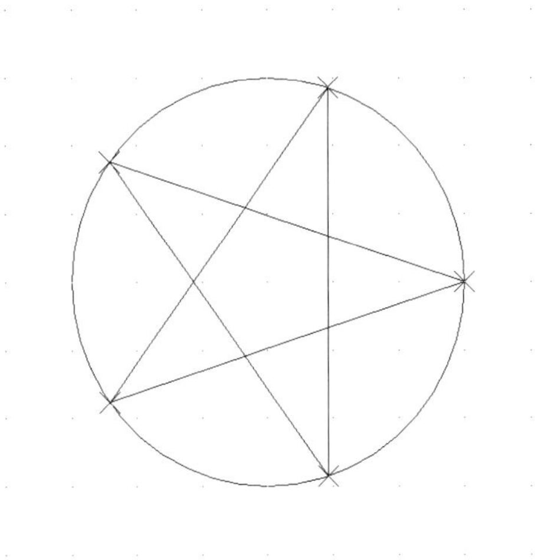

Remember the last star we drew? We used a polygon inside a circle to locate the points. This trick is a bit faster and simpler than the polygon procedure. But the *Divide* command requires a circle whereas the *Polygon* command did not.

III. Let's take a look at the *Measure* command.

TOOLS	COMMAND SEQUENCE	STEPS
	**Command:** *z*	10. *Zoom* out to your previous screen.
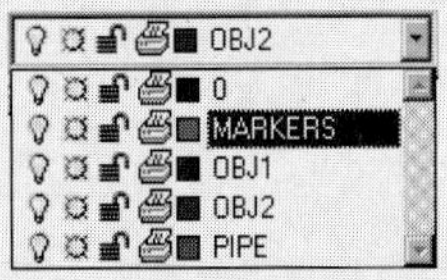	**Command:** *la*	11. Set layer **Markers** as current.
	Command: *me*	12. Enter the *Measure* command by typing *measure* or *me*.

continued

TOOLS	COMMAND SEQUENCE	STEPS
	Select object to measure:	**13.** Select the horizontal line.
	Specify length of segment or [Block]: *6*	**14.** Tell AutoCAD you want the markers placed 6" apart. Your drawing looks like Figure 11.1.1.14.

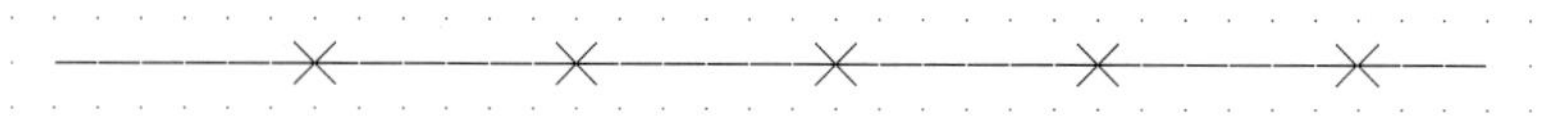

FIG. 11.1.1.14: Measured Line

AutoCAD placed the first mark 6" from the left, then placed marks every 6" thereafter. Notice there is (and will always be) a bit of leftover space at the other end. This space will always be shorter than the specified distance.

| 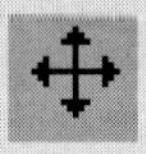 | **Command:** *m* | **15.** Move the *Tee* to the middle node (Figure 11.1.1.15). |

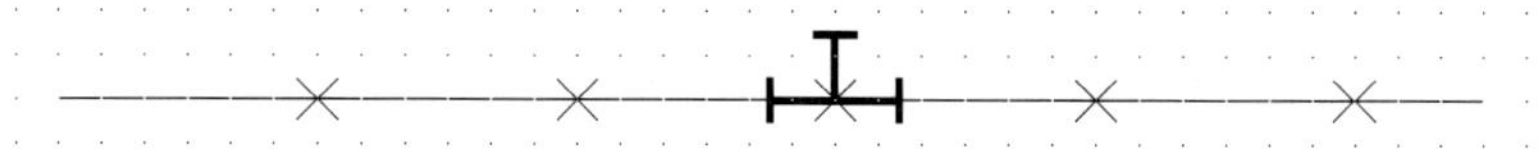

FIG. 11.1.1.15: Move the Tee

| 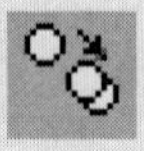 | **Command:** *co* | **16.** Now copy the *Tee and Valve* to each of the remaining nodes (Figure 11.1.1.16). |

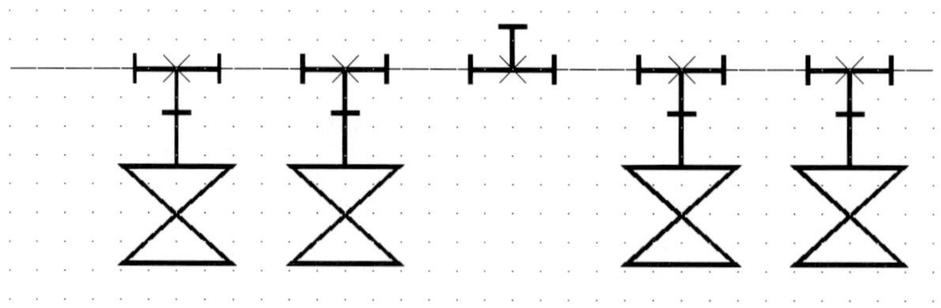

FIG. 11.1.1.16: Copy the Tees and Valves

17. Using the modify tools you know, complete the header drawing, as shown in Figure 11.1.1.17. (*Note:* The pipe is $1/8$" wide.)

continued

TOOLS	COMMAND SEQUENCE	STEPS

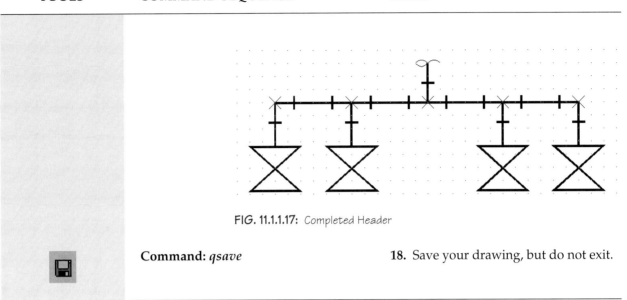

FIG. 11.1.1.17: Completed Header

Command: *qsave*   18. Save your drawing, but do not exit.

As you can see, the *Measure* and *Divide* commands can provide useful assistance in the layout and placement of various objects in your drawing.

11.2 So Where's the Point?

But are these the only tools that make use of those nifty nodes? Of course not. Nodes are a favorite tool of third-party software. I have even included a couple of Lisp routines that make use of nodes. But you can place nodes anywhere you think they might be useful with the *Point* command.

> *Third-party software* refers to any of a myriad of products designed to work within the AutoCAD environment to make life easier for the user. Products are available for most industries.

The *Point* command sequence is

Command: *point* **(or** *po***)**
Current point modes: PDMODE=3 PDSIZE=0'–0" *[AutoCAD tells you how the nodes are currently set to appear—more on this after the exercise]*

Specify a point: *[pick the location]*
Command:

You just cannot get any easier than this one. But remember to always use *OSNAPs* for precise placing of your nodes!

> This command is also available in the Draw pull-down menu. Follow this path:
>
> *Draw—Point—Single Point (or Multiple Points)*

WWW 11.2.1 Do This: Using Points

I. Be sure you are still in the *mea-div* drawing in the C:\Steps\Lesson11 folder.

II. Follow these steps.

TOOLS	COMMAND SEQUENCE	STEPS
 (Zoom tool)	**Command:** *z*	1. **Zoom** in around the circle.
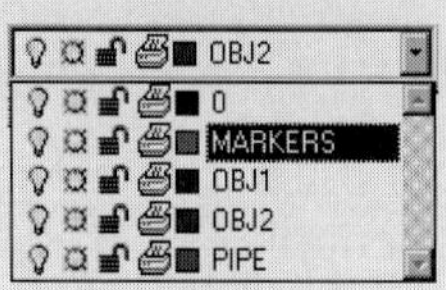 (Layers list with MARKERS selected)	**Command:** *la*	2. Set layer **Markers** as current.
 Point Button	**Command:** *po*	3. Place a node . . .
 (Point style icon)	**Current point modes: PDMODE=3 PDSIZE=0'−0"** **Specify a point: _cen of**	4. . . . in the center of the circle.
 (Erase tool)	**Command:** *e*	5. Erase the circle. Now you can still find the center of the star by using the node (Figure 11.2.1.5).

continued

TOOLS	COMMAND SEQUENCE	STEPS

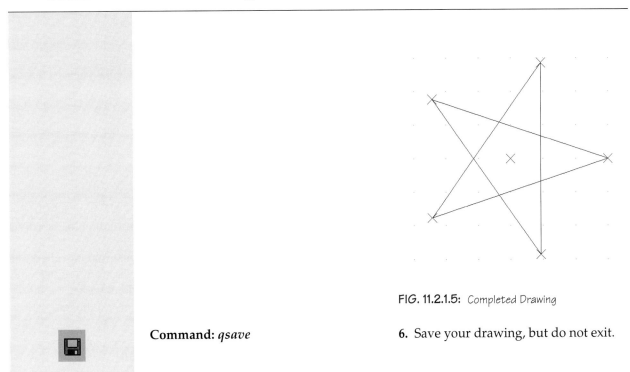

FIG. 11.2.1.5: Completed Drawing

	Command: *qsave*	6. Save your drawing, but do not exit.

We have been using an X to mark our nodes. I set this symbol in the drawing when I created it. But it is not the only (or even the default) symbol available to show a node. To see the various symbols available, or to change the symbol you use, type *DDPType* at the command prompt or select **Point Style** from the Format pull-down menu. AutoCAD provides a dialog box (Figure 11.2a) to help your selection.

To change the node symbol you are using, simply pick the symbol you want to use. Note that the second symbol on the top row is blank. This is an important "symbol" as it clears all the nodes in the drawing without removing them. *You must set the node symbol to blank before plotting a drawing!*

The user can set the point (or node) size **Relative to Screen** or in **Absolute Units**. I recommend the former. AutoCAD will resize the nodes to keep them **Relative** when you alter views (zoom in or out). If they do not automatically resize, *regen* the drawing. **Absolute** nodes can be easily lost if you zoom out too far. I also recommend leaving the **Point Size** at its default 5%. This is large enough to be seen but will not cause your nodes to dominate the screen.

Experiment with the different symbols to see which you prefer. Remember that you must *regen* to see each new setting.

FIG. 11.2a: DDPType Dialog Box

AutoCAD stores the point type you select in a system variable called **PDMode**. If you know the number code of the symbol you want to use, you can set it like this:

> Command: *pdmode*
> Enter new value for PDMODE <0>: *3*
> Command:

11.3 From Outlines to Solids: The *Solid* and *Donut* Commands

Tired of drawing stick figures (outlines)?

The only tool we have seen thus far for drawing objects with width is the polyline. This tool can create most of the objects on which we need to show width. But wait . . . there are others! There are even tools that are easier to use than the polyline! Do you find that hard to believe? Read on!

Two easy tools for showing a solid surface in the 2-dimensional world are the *Solid* command and the *Donut* command. With the *Solid* command, the user fills triangular areas. With the *Donut* command, the user fills round areas.

Let's start with the *Solid* command. The command sequence follows:

> **Command:** *solid* (or *so*)
> **Specify first point:** [*pick the first corner of the area to fill*]
> **Specify second point:** [*pick the second corner of the area to fill*]
> **Specify third point:** [*pick the third corner of the area to fill—if there are four corners or more, select the corner next to the first corner*]
> **Specify fourth point or <exit>:** [*pick the fourth corner*]
> **Specify third point:** [*AutoCAD repeats the third and fourth corner until the user ends the command by hitting enter*]
> **Specify fourth point or <exit>:** [*enter*]
> **Command:**

This command is also available in the Draw pull-down menu. Follow this path:

Draw—Surfaces—2D Solid

Let's try this.

11.3.1 Do This: Creating Solids

I. Be sure you are still in the *mea-div* drawing in the C:\Steps\Lesson11 folder.

II. Follow these steps.

TOOLS	COMMAND SEQUENCE	STEPS
	Command: *la*	1. Set layer **Obj1** current.
	Command: *os*	2. Set the running OSNAP to **Node** and **Intersection**. Clear all other settings.
 Solid Button	Command: *so*	3. Enter the *Solid* command by typing *solid* or *so*. The button for the *Solid* command can be found on the Surfaces toolbar.
	Specify first point: *[select point 1]* Specify second point: *[select point 2]* Specify third point: *[select point 3]* Specify fourth point or <exit>: *[enter]* Specify third point: *[enter]*	4. Select points 1, 2, and 3 as shown in Figure 11.3.1.4. Then hit *enter* twice to exit the command.

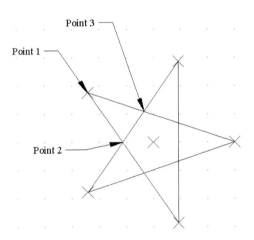

FIG. 11.3.1.4: Select These Points

	Command: *[enter]*	5. Repeat step 4 for each of the points of the star. Your drawing now looks like Figure 11.3.1.5.

continued

TOOLS	COMMAND SEQUENCE	STEPS
	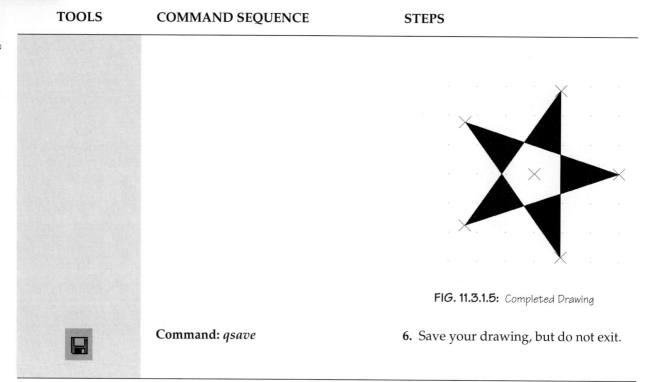 FIG. 11.3.1.5: Completed Drawing	
💾	**Command:** *qsave*	6. Save your drawing, but do not exit.

You see that drawing solids is not difficult. Try drawing a rectangle next to the star, then placing a solid inside. Pick your corners first in clockwise direction. Notice the hourglass shape of the solid (Figure 11.3a).

FIG. 11.3a: Clockwise Solid

Now undo that and create your solid, picking the points in the order shown in Figure 11.3b.

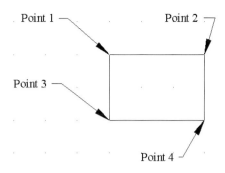

FIG. 11.3b: Use This Order

Notice the difference? You should always reverse Points 3 and 4 from the direction taken from Point 1 to Point 2 in order to create a full solid, as shown in Figure 11.3c.

FIG. 11.3c: Completed Solid

Donuts are polylines and are as easy to draw as solids. But the *Donut* command does ask a couple questions.

Command: *donut* **(or** *do***)**

Specify inside diameter of donut <0'−0 $^1/_2$ ">: *[set the diameter of the donut hole]*

Specify outside diameter of donut <0'−1">: *[set the outer diameter of the donut]*

Specify center of donut or <exit>: *[place the donut]*

Specify center of donut or <exit>: *[enter to complete the command]*

Command:

This command is also available in the Draw pull-down menu. Follow this path:

Draw—Donut

The prompts are self-explanatory, so let's look at donuts in action.

11.3.2 Do This: Using Donuts

I. Be sure you are still in the *mea-div* drawing in the C:\Steps\Lesson11 folder.

II. Follow these steps.

TOOLS	COMMAND SEQUENCE	STEPS
No Button Available	**Command:** *do*	1. Enter the *Donut* command by typing *donut* or *do* at the command prompt. There is no toolbar button for *Donut*.
	Specify inside diameter of donut <0'–0 1/2 ">: *0*	2. Set the **Inside diameter** to *0* . . .
	Specify outside diameter of donut <0'–1">: *2.25*	3. And the **Outside diameter** to 2.25.
	Specify center of donut or <exit>:	4. Place the **center of doughnut** at the **node** in the center of the star.
	Specify center of donut or <exit>: [*enter*]	5. Hit *enter* to complete the command. Your drawing looks like Figure 11.3.2.5.

FIG. 11.3.2.5: Donut Used to Fill Inside the Star

continued

TOOLS	COMMAND SEQUENCE	STEPS
	Command: [*enter*]	6. Let's place some other donuts. Repeat the command.
	Specify inside diameter of donut <0'−0">: .25	7. Make the **inside diameter .25**, and the **outside diameter .5**.
	Specify outside diameter of donut <0'−2 1/4 ">: .5	
	Center of doughnut:	8. Place donuts at the **Nodes** at each of the points of the star. Hit *enter* to complete the command. Your drawing looks like Figure 11.3.2.8. 
	Center of doughnut:	
 Save Button	**Command:** *qsave*	9. Save your drawing, but do not exit.

FIG. 11.3.2.8: Completed Drawing

I must admit that my piping profession offered little opportunity to use solids or donuts. Still, had I been using a computer at the time, these commands would have been quite handy when I was designing those little houses that Santa sits in down at the mall!

Solids and donuts can cause a bit of a slowdown in regeneration time. They can also suck the ink right out of your plotter!

In its infinite wisdom, AutoCAD provided a way to save time and plotter ink. Until you are ready for that final plot, set your **Fillmode** sysvar to *0*, then *regen* the drawing to see the results (see Figure 11.3d).

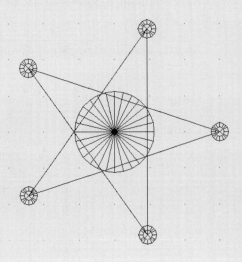

FIG. 11.3d: *Fillmode = 0*

Command: *fillmode*
Enter new value for FILLMODE <1>: *0*
Command: *regen*

AutoCAD hides the filled area of solids and replaces the filled area of donuts with *wireframing*. Set the **Fillmode** back to *1* for the final plot. (Note: The *Fillmode* command also works on polylines.)

11.4 More Object Selection Methods

As our drawings get busier, selecting multiple objects for erasure or modification becomes more difficult. Fortunately, AutoCAD has provided several additional methods to create a selection set. These are **WindowPoly (wp)**, **CrossingPoly (cp)**, **Fence**, **Last**, **All**, **Previous**, **Add,** and **Remove**. Each is quite easily mastered.

- **WindowPoly** and **CrossingPoly** behave like **Window** and **Crossing Window**, except that the user lines out each side of the window. Neither is restricted to the four-sided windows of their nonpoly counterparts.
- **Fence** acts like a single-line crossing window. That is, AutoCAD includes everything crossed by the single-line **Fence** in the selection set. **Fence** is a wonderful tool when it works; unfortunately, it is not always reliable.
- **Last** refers to the last object drawn.

- **Previous** refers to the last active selection set.
- Be careful with the **All** tool. **All** places *all* the thawed objects in the drawing in the selection set. Erasing **All**, then, may empty your drawing if you are not careful.
- **Remove** allows the user to remove objects from the selection set. It is often easier to remove one or two objects from a selected group than it is to individually select multiple objects around the one or two you want to keep.
- **Add** enables the user to put objects into a selection set. Use this when you have made your selection set, removed an object, then decided to *add* something else.

Let's see these in action.

11.4.1 Do This: Selection Practice

I. We will use the *Erase* command for our practice, but any command that presents a **Select objects** prompt can use these tools. Open the *sel-prac* drawing in the C:\Steps\Lesson11 folder. It looks like Figure 11.4.1a.

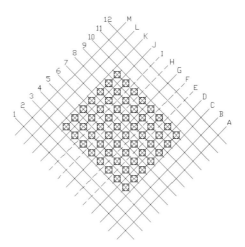

FIG. 11.4.1a: sel-prac.dwg

Notice that the lines exist on different layers with different colors and linetypes. The objects between the lines are nodes.

II. Follow these steps.

TOOLS	COMMAND SEQUENCE	STEPS
	Command: *e*	1. Enter the *Erase* command.

continued

TOOLS	COMMAND SEQUENCE	STEPS
	Select objects: *wp*	2. Tell AutoCAD you wish to select objects using a **WindowPoly** by typing *wp* at the prompt.
	First polygon point:	3. AutoCAD tells you to draw a polygon around the objects you wish to select.
	Specify endpoint of line or [Undo]: *[this lines repeats]*	
	Specify endpoint of line or [Undo]: *[enter]*	Draw a polygon around the brown nodes. (Be sure your OSNAPs are off.) It will continue to prompt until you hit *enter* to say you have completed the set.
	14 found	AutoCAD then tells you how many objects it has found and repeats the **Select objects:** prompt.
	Select objects: *[enter]*	4. Hit *enter* to complete the command. Your drawing looks like Figure 11.4.1.4.

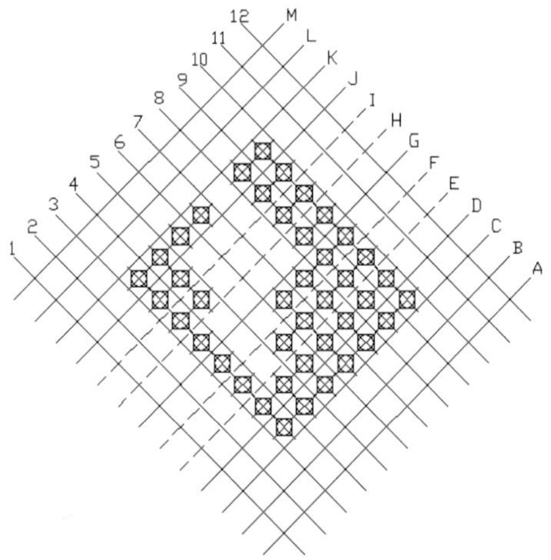

FIG. 11.4.1.4: Erased Using WP

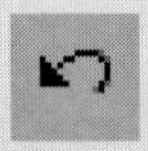	**Command:** *u*	5. Undo the changes.
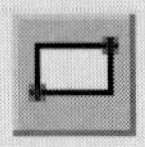	**Command:** *rec*	6. Draw a rectangle in the drawing, as shown in Figure 11.4.1.6.

continued

TOOLS	COMMAND SEQUENCE	STEPS

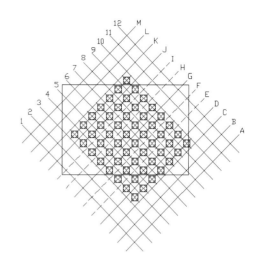

FIG. 11.4.1.6: Draw a Rectangle

Command: *tr*

7. Now enter the *Trim* command.

Current settings: Projection=UCS Edge=None

Select cutting edges ...

Select objects: *l*

1 found

Select objects: *[enter]*

8. Select the rectangle as your cutting edge, but do it by typing *l* for **Last**. AutoCAD selects the last object you drew. Hit *enter* to complete the selection.

Select object to trim or [Project/Edge/Undo]: *f*

First fence point:

Specify endpoint of line or [Undo]: *[this line repeats]*

Specify endpoint of line or [Undo]: *[enter]*

9. Tell AutoCAD you want to use a **Fence** to select the objects to trim. Then draw the fence line around the outside edge of the rectangle (Figure 11.4.1.9). Be sure to cross all the lines. Hit *enter* when done.

continued

TOOLS	COMMAND SEQUENCE	STEPS

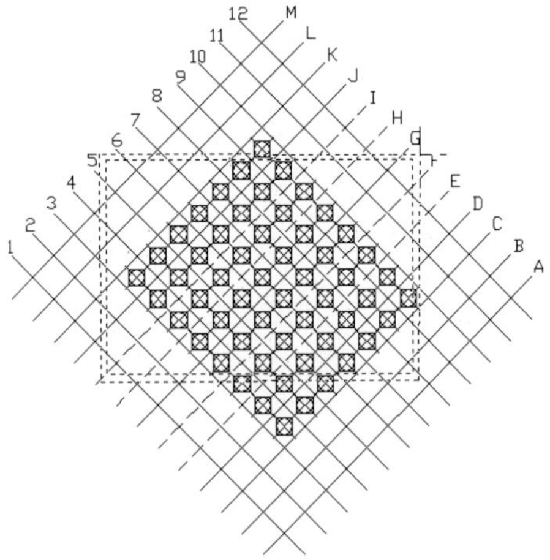

FIG. 11.4.1.9: Fence Location

Select object to trim or [Project/Edge/Undo]: *[enter]*

10. AutoCAD trims the lines, then prompts for more selection. Hit *enter* to confirm completion of the command.

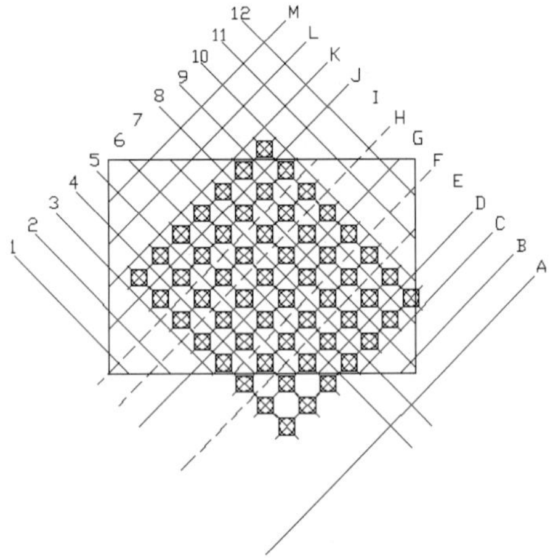

FIG. 11.4.1.10: Trimmed by Fence

Does your drawing look like Figure 11.4.1.10? Notice that not all the lines were trimmed. Remember the **Fence** method is not always reliable.

continued

TOOLS	COMMAND SEQUENCE	STEPS
	Command: *e* **Select objects:** *p* **Select objects:** [*enter*]	**11.** Erase the rectangle now, but tell AutoCAD you want to erase the **Previous** selection set. AutoCAD remembers that you selected the rectangle in your last modification and removes it.
	Command: *u*	**12.** Undo all the changes.
	Command: *e* **Select objects:** *cp*	**13.** This time let's use the **CrossingPoly** method. Enter the *Erase* command and tell AutoCAD to use the **Crossing Polygon** as indicated.
	First polygon point: **Specify endpoint of line or [Undo]:** [*this line repeats*] **Specify endpoint of line or [Undo]:** [*enter*] **81 found**	**14.** Draw a polygon that crosses all the lines but does not touch the numbers (Figure 11.4.1.14).

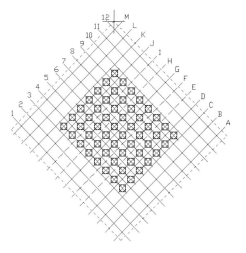

FIG. 11.4.1.14: Cross Each Line

| | **Select objects:** *r* | **15.** Now tell AutoCAD to remove some objects from the selection set. |

continued

TOOLS	COMMAND SEQUENCE	STEPS
	Remove objects: *f* **First fence point:** **Specify endpoint of line or [Undo]:** **Specify endpoint of line or [Undo]:** [*enter*] **5 found, 5 removed**	16. Tell AutoCAD you want to use a **Fence** to select the objects to remove. Place a fence across lines **E** through **I** (Figure 11.4.1.16). 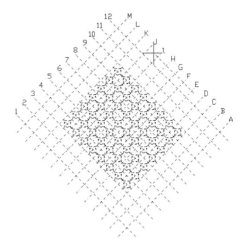 **FIG. 11.4.1.16:** Remove These Lines
	Remove objects: *a* **Select objects: 1 found** **Select objects:** [*enter*]	17. Now tell AutoCAD you want to **Add** one of the lines back into the selection set, then select line **G**. 18. Complete the *Erase* command. Your drawing looks like Figure 11.4.1.18. 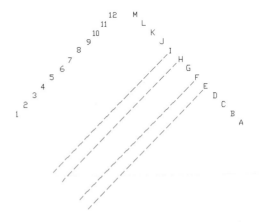 **FIG. 11.4.1.18:** Completed Command

continued

TOOLS	COMMAND SEQUENCE	STEPS
	Command: *u*	19. Undo all the changes.
	Command: *e*	20. This last method is the easiest, but I must again caution you. It erases *everything!* Enter the **Erase** command.
	Select objects: *all* Select objects: [*enter*]	21. At the prompt, type **all** and hit **enter** at the next prompt. See that AutoCAD has removed all the objects in the drawing.
	Command: *quit*	22. *Quit* the drawing without saving your changes.

Well, now you have used all of AutoCAD's selection techniques. What do you think? They sure beat the simple windows we have used up to now, do they not? Certainly, you will almost always use those simple windows and single-object selection boxes in your work. But, occasionally, you will be ever so glad to have learned these advanced selection tricks.

11.5 Object Selection Filters: Quick Select

Selection filters have been available to CAD operators for quite some time. But the introduction of the Quick Select dialog box makes them friendly, convenient, and easy to use!

Use the *QSelect* command to access the dialog box (Figure 11.5a). The dialog box may appear a bit frightening at first; but once mastered, you will find it irreplaceable.

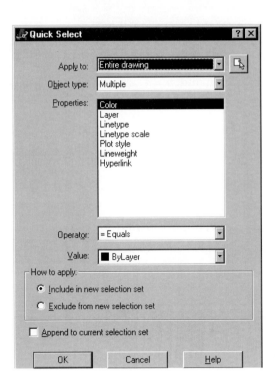

FIG. 11.5a: Quick Select Dialog Box

Let's take a look at it.

- At the top of the box, you find the **Apply to:** control box. Here, AutoCAD allows the user to apply the data listed in the rest of the dialog box to the **Entire drawing** or to the **Current selection** (once a selection has been made).
- Create a selection set to which you can apply the data listed in the dialog box by picking on the **Select objects** button to the right of the **Apply to:** control box. AutoCAD will return you to the graphics area where you can select the objects with which you want to work.
- The **Object type:** control box acts as your first filter. Picking the down arrow will produce a list of the objects currently in use in the drawing (or the selection set). The user can select the type of object with which to work, or **Multiple** if he wishes to apply the data to more than one type of object.
- The **Properties** box allows the user to filter the selection by specific properties.
- The **Operator** and **Value** control boxes work together. The **Operator** box allows the user to set the filter: *equal to, not equal to, less than* or *greater than* the value set in the **Value** box. The properties shown in the **Value** box depend on the property selected in the **Properties** box.
- The **How to apply:** frame allows you to use the filters above to include or exclude objects from the selection set.

It is really not as difficult as it sounds. Let's use filters to repeat what we did in the first part of our last exercise.

11.5.1 Do This: Selection Practice

I. Reopen the *sel-prac* drawing in the C:\Steps\Lesson11 folder.

II. Follow these steps.

TOOLS	COMMAND SEQUENCE	STEPS
No Button Available	Command: *qselect*	1. Open the Quick Select dialog box by typing *qselect* at the command prompt. Or you can pick **Quick Select** from the Tools pull-down menu.
	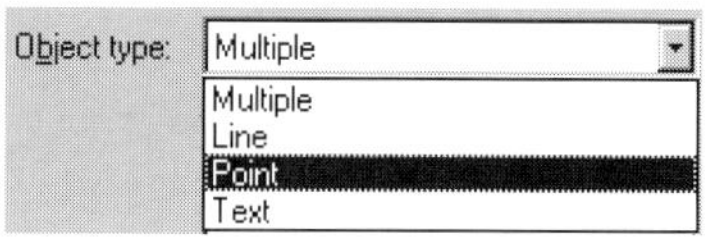	2. Tell AutoCAD you want to select some nodes—pick the down arrow in the **Object type:** control box and pick **Point**.
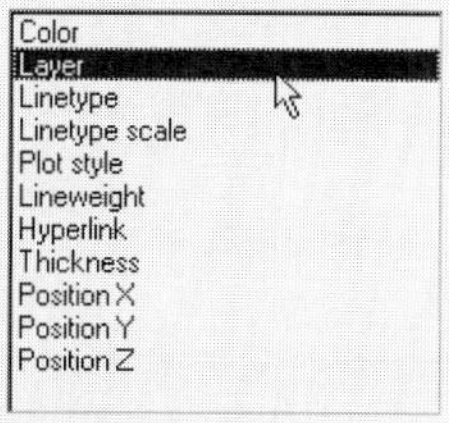		3. We will filter the nodes by layer. Select **Layer** in the **Properties** box.
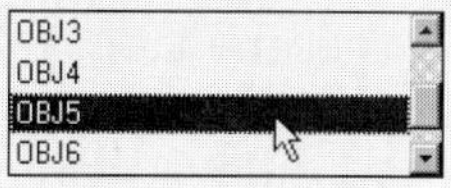		4. We will leave the **Operator:** set to **=Equals** but change the **Value** to the **OBJ5** layer.
		5. Pick the **OK** button to conclude the filter process. Notice that AutoCAD highlights the same nodes we selected in Steps 2, 3, and 4 in our last exercise.
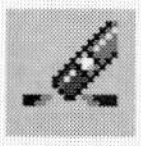	Command: *e*	6. Enter the *Erase* command or pick the **Erase** button on the Modify toolbar. AutoCAD erases the objects (Fig. 11.5.1.6).

continued

TOOLS	COMMAND SEQUENCE	STEPS
		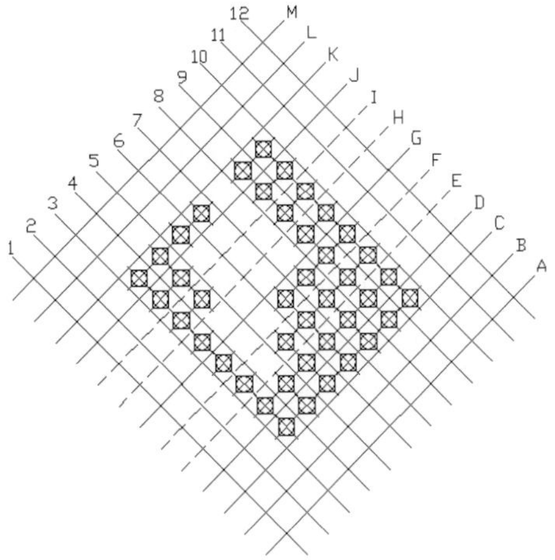 **FIG. 11.5.1.6:** Erased Objects
	Command : *quit*	7. Quit the drawing. Do not save your changes.

I should note that this has been a fairly simple example of what you can do with filters. Filters can (and do) work well in combination with all of the other methods of selection we have discussed. Additionally, once a selection set has been made, you can use filters to modify it—paring it down to exactly what you want. For example, we might select all the nodes on layer **OBJ5** using the color blue and a lineweight of 0.50.

11.6 Extra Steps

I have provided another Lisp routine—this one in the C:\Steps\Lesson11 folder. Simply called *Points.lsp*, this file (when loaded) creates two new commands that you may find useful.

The first command is *PT*. *PT* prompts the user as follows:

PICK A BASE POINT:
HOW FAR: *[enter a distance]*
X, Y, OR P? *[do you want to go along the X-plane, the Y-plane, or polar?]*
ANGLE? *[if polar, AutoCAD will prompt you for the angle]*

PT will create a **Markers** layer and set the node style (**PDMode**) to the *X* we used in our first exercises in this lesson. It then places a node at the requested location. This is a handy tool for helping the user locate a point relative to another point without having to draw guidelines.

The other command is **MW**. This command places a node *midway* between two user-identified points. It prompts:

Command: *mw*
SELECT THE FIRST POINT:
SELECT THE SECOND POINT:

Is that simple enough? Use your OSNAPS to identify the points. This command is ideal for locating the center of rectangles or polyline ellipses.

Start a new drawing and play with these to get comfortable with them.

11.7 What Have We Learned?

Items covered in this lesson include:

- *Commands*
  - *Divide*
  - *Measure*
  - *Point*
  - *Solid*
  - *Donut*
  - *Fillmode*
- *Object Selection Methods*
  - *WindowPoly*
  - *CrossingPoly*
  - *Fence*
  - *Last*
  - *All*
  - *Previous*
  - *Add and Remove*
- *Selection Filters*

How nice that we finish the second part of our text with a short, simple lesson. We have now completed over half the book. (Does that mean we have half-a-mind to use AutoCAD???)

We picked up some neat tricks in Lesson 11. But in your efforts to master so much material, you might easily forget some of the material in this lesson. You probably will not use most of these tools with much frequency; therefore, it will take longer to fully master them. But once mastered, you will probably wish you had done so sooner!

Work through the exercises, and then take some time to relax. The next three lessons are fairly easy to master, so this is a good time to explore and experiment with what you have learned.

11.8 EXERCISES

1. Open *MyRemote* from the C:\Steps\Lesson07 folder (or you can use the *Remote* drawing in the C:\Steps\Lesson11 folder).

    1.1. Using the **Solid** and **Donut** commands you learned in this lesson, fill in the buttons as shown in Figure 11.8.1.

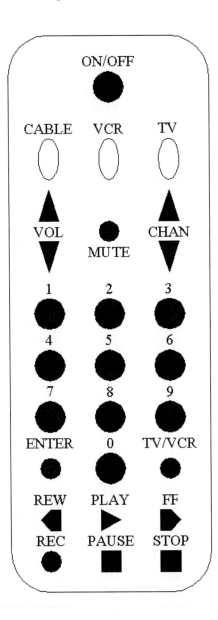

FIG. 11.8.1: Completed Remote

   1.2. Save the drawing as *MyOtherRemote* in the C:\Steps\Lesson11 folder.

2. Set up a new drawing with the following parameters:
   2.1. Limits: [use default limits]
   2.2. Grid: .25
   2.3. Snap: as needed
   2.4. Units: architectural
   2.5. Text Height: .125
   2.6. The following layers:

LAYER NAME	STATE	COLOR	LINETYPE
0	On	white	CONTINUOUS
BUTTON-A	On	cyan	CONTINUOUS
BUTTON-B	On	green	CONTINUOUS
FRAME	On	red	CONTINUOUS
INFRARED	On	magenta	CONTINUOUS
LIGHTS	On	32	CONTINUOUS
SWITCH	On	84	CONTINUOUS
TEXT	On	yellow	CONTINUOUS
VENT	On	white	CONTINUOUS

   2.7. Create the drawing in Figure 11.8.2.

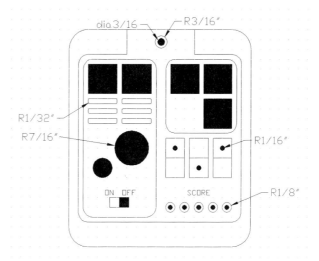

FIG. 11.8.2: Control Panel

   2.8. Save the drawing as *MyControlPanel* in the C:\Steps\Lesson11 folder.
3. This exercise (Figure 11.8.3) can be a lot of fun. Follow these parameters when setting it up:
   3.1. Lower left limits: 0,0
   3.2. Upper right limits: 11'6,7'
   3.3. Grid: 6"

- **3.4.** Snap: as needed
- **3.5.** The wagon is a 30" × 4" solid
- **3.6.** The wheels are donuts with outer diameters of 6" and inner diameters of 4"; the hub is a donut with an outer diameter of 2"
- **3.7.** The pickets are $5\frac{1}{2}$" wide and have a $\frac{1}{2}$" gap between them
- **3.8.** The hedge is made up of arcs
- **3.9.** Use these layers:

LAYER NAME	COLOR	LINETYPE
0	7 (white)	CONTINUOUS
FENCE	6 (magenta)	CONTINUOUS
HEDGE	76	CONTINUOUS
STEEL	5 (blue)	CONTINUOUS
TEXT	7 (white)	CONTINUOUS
WAGON	1 (red)	CONTINUOUS
WHEEL	30	CONTINUOUS

- **3.10.** Save the drawing as *MyYard* in the C:\Steps\Lesson11 folder.

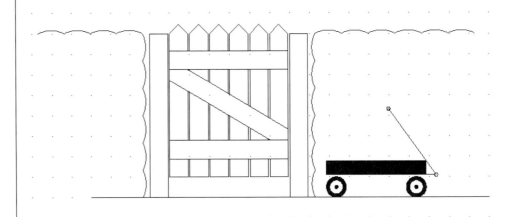

FIG. 11.8.3: Yard

4. Using what you have learned, create the electrical board in Figure 11.8.4. Be sure to use appropriate layers. The grid is $\frac{1}{2}$" and the limits are the default for a new drawing. The polylines are $\frac{1}{16}$" wide and the outer diameter of the donuts is $\frac{3}{16}$".

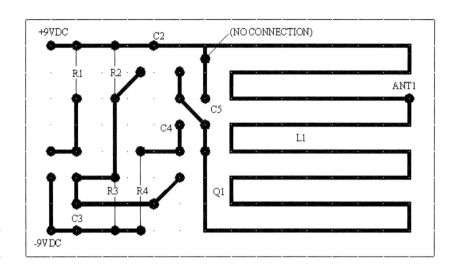

FIG. 11.8.4: Electrical Board

5. Now create the remaining drawings. The grid on the Gable Vent drawing (Figure 11.8.5a) is 3" and the limits are 0,0 and 15',10'. The grid on the Layout for Radiant Electrical Heating System (Figure 11.8.5b) is 6" and the limits are 0,0 and 16'6,15'. Set up your own layers for each drawing.

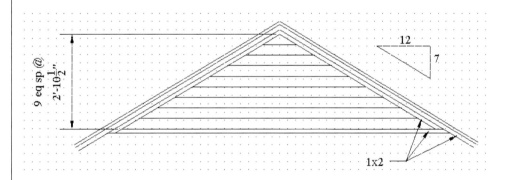

FIG. 11.8.5a: Gable Vent

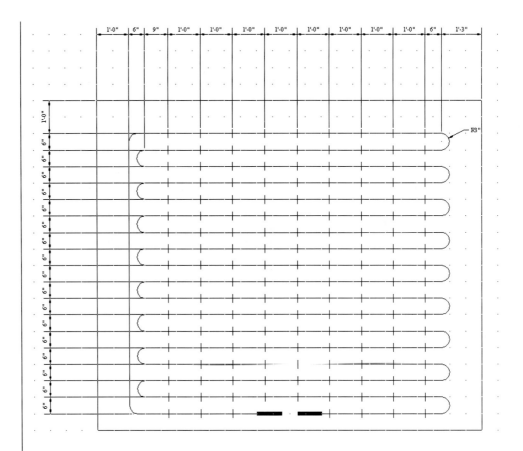

FIG. 11.8.5b: Typical Ceiling Layout for Radiant Electrical Heating System

11.9 REVIEW QUESTIONS

Please write your answers on a separate sheet of paper.

1. What is the name of the discipline that manufactures objects directly from CAD drawings?

2. _____ places nodes at user-set distances along an object.

3. _____ places equally spaced nodes along an object.

4. A _____ is an object that occupies a single point.

5. The user can place nodes anywhere in a drawing using the _____ command.

6. To hide nodes without erasing them or freezing their layer, set the point style to _____.

7. You must _____ the drawing to see any changes to the style of nodes.

8. What is the sysvar that stores the current point type?

9. and 10. What are the two commands used to show a solid surface in the 2-dimensional world?

11. and 12. In order to save ink and regeneration time when using solids or donuts, set the _____ sysvar to _____.

13. When selecting multiple objects, the user may type _____ at the Select objects: prompt to line out each side of a window.

14. When selecting multiple objects, the user may type _____ at the Select objects: prompt to line out each side of a crossing window.

15. A _____ acts like a single-line crossing window.

16. _____ allows the user to remove objects from a selection set.

17. _____ provides a dialog box used to filter objects in a selection set.

LESSON 12

Guidelines and Splines

Following this lesson, you will:
➥ Know how to draw contour lines with the **Spline** Command
➥ Be familiar with the **Splinedit** Command
➥ Know how to create guidelines (or construction lines) using:
  • *XLine*
  • *Ray*

Now we are moving into the more advanced aspects of basic AutoCAD. If you are not comfortable with the first half of our text, it might be a good idea to spend a couple hours going over the less familiar portions before continuing.

Lesson 12 will familiarize you with AutoCAD's version of guidelines —a tool missing through the first twelve releases and welcomed by many when R13 made it available. We will also cover the use of the *Spline* command to draw contour lines (like those found on topographical maps). Prior to R13, the CAD operator used the

Spline option of the *Pedit* command to draw contour lines. This was a tedious and often difficult procedure. But the innovative *Spline* command made contour lines quick and easy.

Continue when you are ready.

12.1 Contour Lines with the *Spline* Command

In Lesson 9, we studied polylines and the *Pedit* command. I am sure you found them as complicated as most people do the first time.

We learned how to make a *spline* (a contour line) from a polyline using the *Pedit* command. Here we will see how we can draw a spline more easily and in a way that shows a contour line as you draw. Additionally, you will see that a spline drawn using the *Spline* command takes up much less memory, and thus reduces the size of your drawing.

The command sequence for the *Spline* command is

Command: *spline* (**or** *spl*)
Specify first point or [Object]: *[pick the starting point]*
Specify next point: *[pick the second point]*
Specify next point or [Close/Fit tolerance] <start tangent>: *[continue selecting points till complete]*
Specify next point or [Close/Fit tolerance] <start tangent>: *[hit enter to complete the spline line]*
Specify start tangent: *[show AutoCAD how you want the beginning of the spline to curve]*
Specify end tangent: *[show AutoCAD how you want the end of the spline to curve]*

By default, the user simply picks points along the spline until complete. But let's look at some of the other options.

- The first option is the easiest. Use the **Object** option to convert a splined polyline into an actual spline. This reduces the drawing size considerably as we will see. (*Note:* To work properly, the polyline must have been converted to a spline using the **Spline** option of the *Pedit* command.)
- The **Close** option works just as it does with the *Line* or *Pline* command. It simply closes the spline.
- The **Fit Tolerance** option tells AutoCAD to draw the spline within a certain distance of the points selected by the user. By default, it is set to zero (meaning, "draw the spline *through* the user selected point"). Personally, I prefer leaving it at the default as it makes my drawing considerably more accurate. However, the user can reset the tolerance as desired.
- The last two options—**start** and **end tangent**—enable the user to control the curve of the spline at the beginning and ending points.

Let's take a look at the *Spline* command.

> This command is also available in the Draw pull-down menu. Follow this path:
>
> *Draw—Spline*
>
> The *Spline* options are also available on the cursor menu once the command has been entered.

12.1 Do This: Working with Splines

I. Open the *splines* drawing in the C:\Steps\Lesson12 folder. The drawing looks like Figure 12.1.1a.

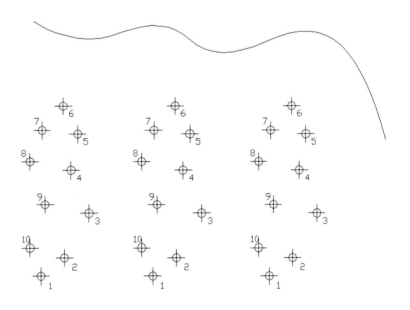

FIG. 12.1.1a: Splines.dwg

II. Follow these steps.

TOOLS	COMMAND SEQUENCE	STEPS
	Command: *li*	1. Do a *List* command on the existing spline. See that it is a polyline (Figure 12.1.1.1). Note, however, that this is actually a *polyline* and not an *lwpolyline*. When it was converted to a spline, AutoCAD automatically converted it to a polyline to hold the required information to define the object.

```
                POLYLINE   Layer: "LAYER3"
                           Space: Model space
                    Handle = FD
           Open spline
starting width      0.0000
  ending width      0.0000
         area      34.0237
       length      22.0284

                VERTEX     Layer: "LAYER3"
                           Space: Model space
                    Handle = 145
           at point, X=    2.0414   Y=  14.3868   Z=   0.0000
starting width      0.0000
  ending width      0.0000
               (Spline control point)

                VERTEX     Layer: "LAYER3"
                           Space: Model space
                    Handle = 14D
           at point, X=    2.0414   Y=  14.3868   Z=   0.0000
starting width      0.0000
Press ENTER to continue:
```

FIG. 12.1.1.1: Listed Polyline Spline

		2. You will have to hit *enter* several times to view all the information attached to the polyline. Or you may hit the **Esc** key on your keyboard to cancel the command.
		3. Hit the **F2** key to return to the graphics screen.
Spline Button	**Command:** *spl*	4. Enter the *Spline* command by typing *spline* or *spl*. Or you may pick the **Spline** button on the Draw toolbar.
	Specify first point or [Object]: *o*	5. Tell AutoCAD you want to convert an **Object** to a spline.

continued

TOOLS	COMMAND SEQUENCE	STEPS
	Select objects to convert to splines	6. Select the polyline spline. Hit *enter* to complete the command.
	Select objects: 1 found	
	Select objects: *[enter]*	
 [list icon]	**Command:** *li*	7. Do another *list* on the spline. Note the difference (see Figure 12.1.1.7). All the information concerning the spline fits on a single screen. There is a related drop in the size of the drawing as well.

```
              SPLINE     Layer:  "LAYER3"
                         Space:  Model space
              Handle = 1B5
                        Length:  22.0422
                         Order:  3
                    Properties:  Planar, Non-Rational, Non-Periodic
              Parametric Range:  Start   0.0000
                                 End     5.0000
       Number of control points: 7
                Control Points:  X =  2.0414  , Y = 14.3868  , Z = 0.0000
                                 X =  4.3990  , Y = 13.0084  , Z = 0.0000
                                 X =  8.7692  , Y = 14.6740  , Z = 0.0000
                                 X = 11.4143  , Y = 12.2618  , Z = 0.0000
                                 X = 16.5321  , Y = 14.5017  , Z = 0.0000
                                 X = 19.2922  , Y = 11.8598  , Z = 0.0000
                                 X = 20.2123  , Y =  8.5287  , Z = 0.0000
```

FIG. 12.1.1.7: Listed Spline

TOOLS	COMMAND SEQUENCE	STEPS
 [spline icon]	**Command:** *spl*	8. Now let's draw a spline. Repeat the *Spline* command.
	Specify first point or [Object]: *[select node 1]*	9. Select nodes 1 to 10 in the left group. (Use OSNAPs.)
	Specify next point: *[select node 2]*	
	Specify next point or [Close/Fit tolerance] <start tangent>: *[this prompt repeats—continue selecting through node 10]*	
	Specify next point or [Close/Fit tolerance] <start tangent>: *c*	10. After selecting node 10, **Close** the spline.
[F8]	**Specify tangent: <Ortho on>**	11. Because you have **Closed** the spline, AutoCAD only prompts once for a tangent. Turn on the **Ortho** tool and pick a point due west. Your drawing looks like Figure 12.1.1.11.

continued

TOOLS	COMMAND SEQUENCE	STEPS

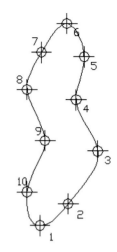

FIG. 12.1.1.11: Drawn Spline

Command: *[enter]*

12. Repeat the *Spline* command.

Specify first point or [Object]: *[select node 1]*

13. Select nodes 1 and 2 in the middle group.

Specify next point: *[select node 2]*

Specify next point or [Close/Fit tolerance] <start tangent>: *f*

14. When AutoCAD prompts like this, type *f* to change the **Fit Tolerance**.

Specify fit tolerance <0.0000>: *1*

15. Set the **tolerance** to *1*.

16. Continue as in Steps 9 through 11. Your drawing looks like Figure 12.1.1.16.

continued

TOOLS	COMMAND SEQUENCE	STEPS
		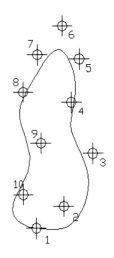
		FIG. 12.1.1.16: Tolerance of 1
		The spline has been drawn within one unit (a **tolerance** of 1) of the nodes selected.
	Command: *[enter]*	**17.** Repeat the *Spline* command.
		18. Repeat Steps 8 and 9 using the third group of nodes.
	Specify next point or [Close/Fit tolerance] <start tangent>: *[enter]*	**19.** Complete the spline, but do not close it. Rather than typing *c* at the last prompt, hit *enter*.
	Specify start tangent: **Specify end tangent:**	**20.** AutoCAD now asks for a **tangent** at the start point and end point. For the **start tangent**, pick a point due east; for the **end tangent**, pick a point due west. Your drawing looks like Figure 12.1.1.20.

continued

TOOLS	COMMAND SEQUENCE	STEPS
	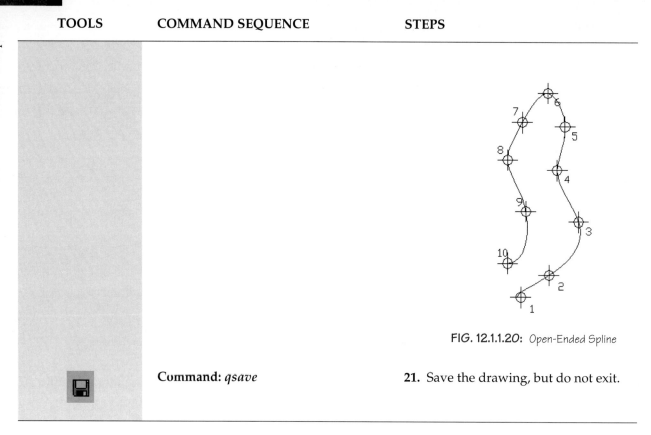 FIG. 12.1.1.20: Open-Ended Spline	
💾	**Command:** *qsave*	21. Save the drawing, but do not exit.

In this exercise, I have provided the location of the spline points; on the job they will be provided by surveyors. Learning to do the survey as well as the drafting might mean more income for the drafter/designer.

We have seen that drawing splines is not difficult. Indeed, you have learned enough in this short exercise to draw some fairly complex topographical maps—if the elevation points are provided. But what about editing the splines?

12.2 Changing Splines: The *Splinedit* Command

Like polylines, it is often easier to erase and redraw simple splines than it is to edit them. However, you will find that some complex drawings require some knowledge of spline editing (particularly in 3-dimensional drawing). Unfortunately, also like polylines, editing is quite a multitiered chore.

Here is the *Splinedit* command sequence:

Command: *splinedit* (or *spe*)
Select spline:
Enter an option [Fit data/Open/Move vertex/Refine/rEverse/Undo]:

Let's look at the options.

- The **Fit Data** option enables the user to edit the *fit points* of the spline. To keep this simple, think of *fit points* on a spline as essentially the same thing as *vertices* on a polyline. This option drops the user into a second tier of options that looks like this:

    [Add/Open/Delete/Move/Purge/Tangents/toLerance/eXit] <eXit>:

    - On this tier, you can **Add**, **Delete,** or **Move** fit points just as you did vertices in the *Pedit* command. You can also **Open** a closed spline or **Close** an opened spline, again just as you did with polyline editing.
    - You can also change **Tangents** and **toLerances** on this level.
    - The **Purge** option removes fit point information from the drawing's database. This makes editing very difficult. Its only real benefit is that it will reduce drawing size. So if you must use it, wait until you have finished the drawing.

- Back on the first tier of *Splinedit* options, we see that **Open** and **Move Vertex** also appear. These are similar to their **Fit Data** counterparts, but the **Move Vertex** option moves a single line whereas the **Fit Data—Move** option moves a curve of the spline. To make this a bit clearer, the **Move Vertex** option may cause a closed spline to open whereas the **Move** option will not. We will see each of these in our next exercise.

- The **Refine** option also presents the user with another tier:

    Enter a refine option [Add control point/Elevate Order/Weight/eXit] <eXit>:

    - The **Add control point** allows the user to add a single control point (points that enable the user to manipulate the spline).
    - **Elevate Order** also causes more control points to appear. But this option adds control points uniformly along the spline. The user determines the number of control points—up to 26. But remember, once added, control points cannot be removed.
    - A control point's **Weight** is similar to tolerance. The **Weight** of a control point controls how much influence, or pull, that point has against the spline. Increasing the **Weight** of a point may cause the spline to pull away from adjacent control points.

- The **Reverse** option switches the start point and endpoint of the spline. On a large spline, the user can hasten an edit using this option; it will enable the user to reach the appropriate control point a bit faster.

Is this starting to feel like the *Pedit* command? Let's try an exercise.

This command is also available in the Modify pull-down menu, and the options are available on the cursor menu once the command has been entered.

12.2.1 Do This: Editing Splines

I. If you are not still in the *splines* drawing, please open it now. It is in the C:\Steps\Lesson12 folder.

II. Follow these steps.

TOOLS	COMMAND SEQUENCE	STEPS
[Layer list showing LAYER3, 0, LAYER1, LAYER2, LAYER3, TEXT]	Command: *la*	1. Freeze **Layer 1** and the **Text** layer. The numbers and nodes disappear from the drawing.
[Splinedit Button icon] Splinedit Button	Command: *spe*	2. Enter the *Splinedit* command by typing *splinedit* or *spe*. Or you may pick the **Edit Spline** button on the Modify II toolbar.
	Select spline:	3. Select the spline on the left. Control points appear as boxes (Figure 12.2.1.3).
		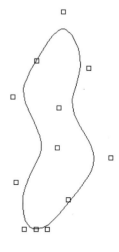 **FIG. 12.2.1.3:** Control Points
	Enter an option [Fit data/Open/Move vertex/Refine/rEverse/Undo]: *o*	4. **Open** the spline (Figure 12.2.1.4). Notice that the **Open** option changes to **Close**.

continued

TOOLS	COMMAND SEQUENCE	STEPS

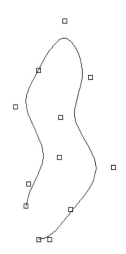

FIG. 12.2.1.4: Opened Spline

	Enter an option [Close/Move vertex/Refine/rEverse/Undo/eXit] <eXit>: *m*	5. **Move** the first **Vertex** . . .
	Specify new location or [Next/ Previous/Select point/eXit] <N>: *@2<180*	. . . two units to the left. Notice the similarity between this tier of options and corresponding options in the *Pedit* command.
	Specify new location or [Next/Previous/Select point/eXit] <N>: *x*	6. **eXit** this tier. Your drawing looks like Figure 12.2.1.6.

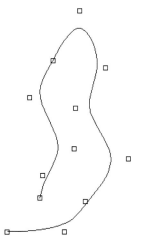

FIG. 12.2.1.6: Moved Vertex

continued

TOOLS	COMMAND SEQUENCE	STEPS
	Enter an option [Close/Move vertex/Refine/rEverse/Undo/eXit] <eXit>: *r*	7. Select the **Refine** option.
	Enter a refine option [Add control point/Elevate order/Weight/eXit] <eXit>: *a*	8. **Add** a single **control point** . . .
	Specify a point on the spline <exit>:	9. Select a point on the spline, as shown in Figure 12.2.1.9.

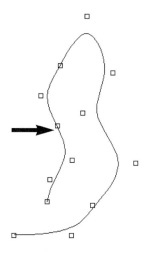

FIG. 12.2.1.9: Select This Location

Notice the new control point.

	Specify a point on the spline <exit>: *[enter]*	10. Hit *enter* to leave this tier.
	Enter a refine option [Add control point/Elevate order/Weight/eXit] <eXit>: *w*	11. Tell AutoCAD you want to change the **Weight** of the new control point.
	Spline is not rational. Will make it so. Enter new weight (current = 1.0000) or [Next/Previous/Select point/eXit] <N>: *[enter]*	12. AutoCAD makes the spline **rational**. (The spline becomes a NURBS—Non-Uniform Rational B-Spline. This is a smoother spline often used in 3-dimensional drawing.) ThenAutoCAD allows you to select the point you want to edit. Hit *enter* till the new control point highlights as seen in Figure 12.2.1.12.

continued

TOOLS	COMMAND SEQUENCE	STEPS
		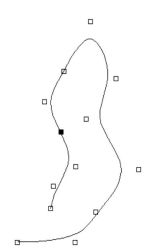 FIG. 12.2.1.12: Highlighted Control Point
	Enter new weight (current = 1.0000) or [Next/Previous/Select point/eXit] <N>: *4*	13. Change the **Weight** of this point to *4*. Notice the change in the spline in relation to the control points next to the one selected (Figure 12.2.1.13).
		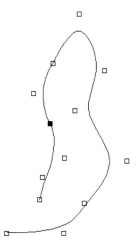 FIG. 12.2.1.13: Changed Weight
	Enter new weight (current = 4.0000) or [Next/Previous/Select point/eXit] <N>: *x*	14. Exit this tier.
	Enter a refine option [Add control point/Elevate order/Weight/eXit] <eXit>: *e*	15. Tell AutoCAD you want to **Elevate** the **Order** of the spline's control points.

continued

TOOLS	COMMAND SEQUENCE	STEPS
	Enter new order <4>: 5	16. Change the **order** from 4 to 5. Notice the increase in the number of control points (Figure 12.2.1.16).

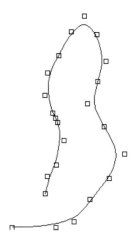

FIG. 12.2.1.16: Additional Control Points via Elevated Order

	Enter a refine option [Add control point/Elevate order/Weight/eXit] <eXit>: *[enter]*	17. **eXit** the tier, then the command. All the control points disappear and your drawing looks like Figure 12.2.1.17.
	Enter an option [Close/Move vertex/Refine/rEverse/Undo/eXit] <eXit>: *[enter]*	

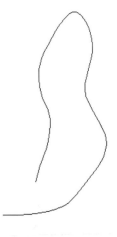

FIG. 12.2.1.17: Completed Spline

| | **Command:** *[enter]* | 18. Repeat the *Splinedit* command. |

continued

TOOLS	COMMAND SEQUENCE	STEPS
	Select spline:	19. Select the middle spline.
	Enter an option [Fit data/Open/ Move vertex/Refine/rEverse/Undo]: *f*	20. Choose the **Fit Data** option.
	Enter a fit data option [Add/Open/ Delete/Move/Purge/Tangents/ toLerance/eXit] <eXit>: *l*	21. Type *l* to tell AutoCAD you want to change the **toLerance** for this spline . . .
	Enter fit tolerance <1.0000>: *0*	22. . . . to *0*. The spline changes to look like Figure 12.2.1.22.

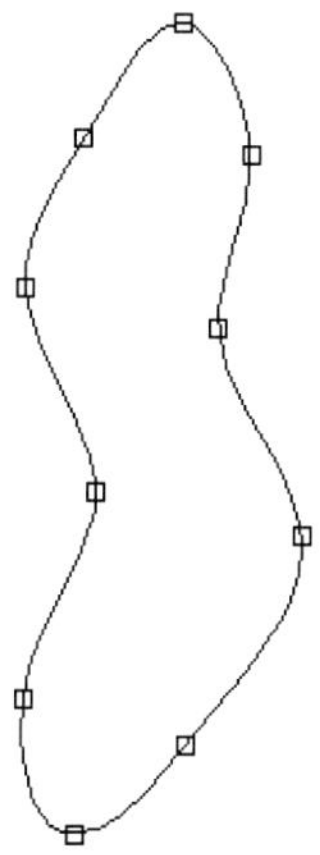

FIG. 12.2.1.22: Tolerance of 1

| | **Enter a fit data option [Add/Open/ Delete/Move/Purge/Tangents/ toLerance/eXit] <eXit>:** *t* | 23. Now tell AutoCAD you want to change the **Tangents**. |
| | **Specify tangent or [System default]:** | 24. Pick a point to the east this time. (Remember that we selected a point to the west when we created the spline.) Your drawing changes to look like Figure 12.2.1.24. |

continued

TOOLS	COMMAND SEQUENCE	STEPS
		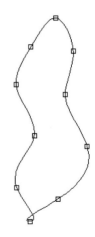

FIG. 12.2.1.24: Adjusted Tangent |
	Enter a fit data option [Add/Open/Delete/Move/Purge/Tangents/toLerance/eXit] <eXit>: *d*	25. Now **Delete** a fit point.
	Specify fit point <exit>:	26. Select the uppermost control point, then hit *enter* to return to the **Fit Data** prompt. Your drawing looks like Figure 12.2.1.26.
	Specify fit point <exit>: *[enter]*	

FIG. 12.2.1.26: Deleted Control Point |
| | **Enter a fit data option [Add/Open/Delete/Move/Purge/Tangents/toLerance/eXit] <eXit>:** *m* | 27. Lastly, let's **Move** a fit point. |
| | **Specify new location or [Next/Previous/Select point/eXit] <N>:** | 28. Move the point indicated, as shown in Figure 12.2.1.28. |

continued

TOOLS	COMMAND SEQUENCE	STEPS
		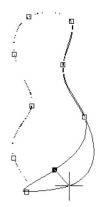
		FIG. 12.2.1.28: Move the Point Like This
	Specify new location or [Next/Previous/Select point/eXit] <N>: *x* **Enter a fit data option** **[Add/Open/Delete/Move/Purge/Tangents/toLerance/eXit] <eXit>:** *[enter]* **Enter an option [Fit data/Open/Move vertex/Refine/rEverse/Undo]:** *[enter]*	**29.** **eXit** this tier of options, then **eXit** the command. Your drawing looks similar to Figure 12.2.1.29. 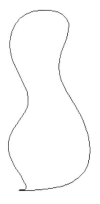 FIG. 12.2.1.29: Completed Spline

In 2-dimensional drafting, simply moving, adding, or deleting fit points is not a problem—especially if your company has some Lisp routines that may shorten the procedures. Indeed, first-tier options are rarely a problem. And in the case of the *Splinedit* command, the **Fit Data** tier is not too difficult. But the complexity of multi-tiered commands, like *Splinedit* and *Pedit*, often presents the user with a choice of erasing and redrawing rather than trying to navigate the editing command itself. Unfortunately, the only way around this is a Lisp routine (as I mentioned) or careful *study* and *familiariziation* with the tools.

12.3 Guidelines

> Beginning with this lesson, we will create a multilesson project. We will continue to have exercises at the end of each lesson; but in a series of <u>Do This</u> sample exercises, we will create the floor plan of a house. We will lay out the drawing in Lesson 12, place the walls in Lesson 13, then add some text in Lesson 14. Finally, in Lesson 16, we will dimension the drawing.

With the advent of Release 13, AutoCAD provided the CAD operator with the equivalent of the *4H* lead. On the board, we used this lead to mark our guidelines—lines used to lay out an area lightly before putting the heavier leads to paper. Precision was important, but the length of the line generally covered an area much larger than was eventually needed. This helped in locating other items on the drawing. Later, we might darken part of the guideline with an *F* or *HB* lead or erase it altogether. The technical term used for these guidelines was *construction line*.

Create a construction line in AutoCAD with the **XLine** or **Ray** command. In fact, the terms *xline* and *construction line* are used to mean the same thing. The only real difference between an xline and a ray is that an xline is infinite in both directions, whereas a ray is infinite in only one direction. In fact, the **Ray** command, with no options, is often overlooked in favor of the more versatile **XLine**.

Of course, we can convert the xline or ray into a drawing object just as we could darken our construction line with a heavier lead. Simply trim away what you do not need and the xline/ray becomes an ordinary line!

Let's look at each. The command sequence for **XLine** is

Command: *xline* **(or** *xl***)**

Specify a point or [Hor/Ver/Ang/Bisect/Offset]: *[pick a point on the proposed line]*

Specify through point: *[pick a second point on the proposed line]*

Specify through point: *[AutoCAD will continue to place xlines through the initial selected point and any second point the user selects—hit enter to complete the command]*

Command:

- By default, AutoCAD requires the user to select two points on the xline. It then draws an infinite line (a line that continues infinitely in both directions) through those two points. The user may, however, select the **Hor**, **Ver**, or **Ang** options to force AutoCAD to draw horizontal, vertical, or angular xlines. Then the user need only pick a single point on the xline. **Hor** and **Ver** need no further input from the user except the location. **Ang** will prompt the user for the desired angle:

Enter angle of xline (0) or [Reference]:

The user can enter the desired angle via the keyboard, or use the **Reference** option to get an angle from an existing object.

- The **Bisect** option allows the user to find the bisector of an existing angle. It prompts:

Specify angle vertex point: *[select the corner of the angle]*
Specify angle start point: *[select any point on one of the lines forming the angle]*
Specify angle end point: *[select any point on the other line]*
Specify angle end point: *[enter to complete the command]*

Once you have used the **Bisect** option, you will never want to go back to the old compass method!

- The last option—**Offset**—is peculiar. It replaces the *XLine* command prompt with the *Offset* command prompt. It then behaves like the *Offset* command. I recommend ignoring this option in favor of the actual *Offset* command.

> An interesting point about xlines and rays: although they are infinite in length, neither affects a *zoom extents* (zooming to include *all* the objects in a drawing). XLines and rays will print; but otherwise, AutoCAD treats them as background images.

The *Ray* command sequence is much easier:

Command: *ray*
Specify start point: *[pick the start point]*
Specify through point: *[pick a point through which the ray will pass]*
Specify through point: *[continue picking through points or hit enter to complete the command]*

> These commands are also available in the Draw pull-down menu. Follow this path:
>
> *Draw—Construction Line (or Ray)*
>
> The options for each are also available on the cursor menu once the command has been entered.

Let's begin our floor plan.

WWW 12.3.1 Do This: Construction Lines

I. Open the *flr-pln12* drawing in the C:\Steps\Lesson12 folder. This drawing has been set up to create a floor plan on a $1/4" = 1'-0"$ scale, on a C-size sheet of paper.

II. Follow these steps.

TOOLS	COMMAND SEQUENCE	STEPS
 Construction Line Button		1. Be sure **Const** is the current layer. This is the layer on which we will draw our construction lines.
	Command: *xl*	2. Enter the *XLine* command by typing *xline* or *xl*. Or you may pick the **Construction Line** button on the Draw toolbar.
	XLINE Specify a point or [Hor/Ver/Ang/Bisect/Offset]: *v*	3. Tell AutoCAD you want to draw a vertical xline by typing *v*.
	Specify through point: *1′,0*	4. Place two vertical xlines as indicated.
	Specify through point: *87′,0*	
	Specify through point: *[enter]*	5. Hit *enter* to complete the command.
	Command: *[enter]*	6. Repeat the command.
	XLINE Specify a point or [Hor/Ver/Ang/Bisect/Offset]: *h*	7. This time, let's draw horizontal xlines . . .
	Specify through point: *0,1′*	8. . . . through these points.
	Specify through point: *0,67′*	
	Specify through point: *[enter]*	9. Hit *enter* to complete the command. Your drawing looks like Figure 12.3.1.9.

continued

TOOLS	COMMAND SEQUENCE	STEPS

FIG. 12.3.1.9: Border Outline

Command: *o*

Specify offset distance or [Through] <Through>: *[enter]*

Select object to offset or <exit>: *[pick the bottom line]*

Specify through point: *@87<90*

Select object to offset or <exit>: *[pick the bottom line]*

Specify through point: *@18<90*

Select object to offset or <exit>: *[pick the bottom line]*

Specify through point: *@36<90*

Select object to offset or <exit>: *[pick the bottom line]*

Specify through point: *@57<90*

Select object to offset or <exit>: *[enter]*

10. Use the *Offset* command to offset the bottom line upward as indicated.

continued

TOOLS	COMMAND SEQUENCE	STEPS
	Command: *[enter]* **Specify offset distance or [Through] <Through>:** *[enter]* **Select object to offset or <exit>:** *[pick the right line]* **Specify through point:** *@216<180* **Select object to offset or <exit>:** *[pick the right line]* **Specify through point:** *@48<180* **Select object to offset or <exit>:** *[pick the right line]* **Specify through point:** *@132<180* **Select object to offset or <exit>:** *[enter]*	**11.** Now offset the right vertical line as indicated. The lower right corner of your drawing now looks like Figure 12.3.1.11. *(Note:* Your grid may be toggled off if it bothers you.) 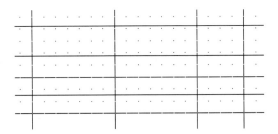 **FIG. 12.3.1.11:** Title Block Outline
	Command: *tr*	**12.** Use the *Trim* command to clean up the area to form the title block shown in Figure 12.3.1.12 (zoom as needed to ease your view). 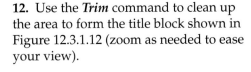 **FIG. 12.3.1.12:** Completed Title Block
 	Command: *tr*	**13.** Finish trimming the border. Then change all the lines from the **Const** layer to the **Border** layer. Your drawing now looks like Figure 12.3.1.13.

continued

TOOLS	COMMAND SEQUENCE	STEPS

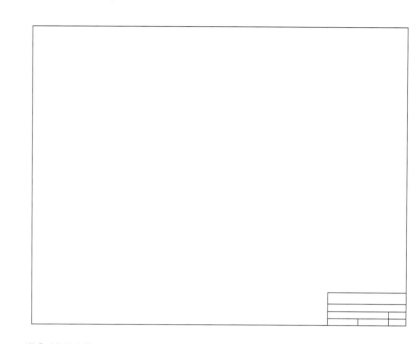

FIG. 12.3.1.13: Completed Border

Command: *qsave*

14. Remember to *save* occasionally.

Command: *xl*

15. Now let's locate the walls of our house. Repeat the *XLine* command.

Specify a point or [Hor/Ver/Ang/Bisect/Offset]: *v*

16. Place three **Vertical** xlines as indicated (Figure 12.3.1.16).

Specify through point: *10',0*

Specify through point: *33'4,0*

Specify through point: *60',0*

Specify through point: *[enter]*

continued

TOOLS	COMMAND SEQUENCE	STEPS

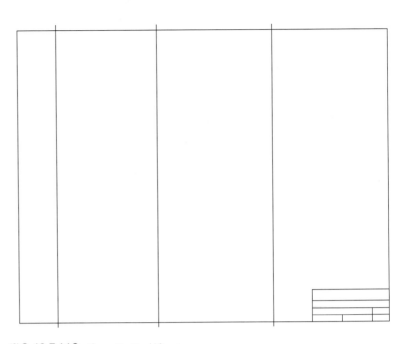

FIG. 12.3.1.16: Three Vertical Xlines

Command: *[enter]*

Specify a point or [Hor/Ver/Ang/Bisect/Offset]: *h*

Specify through point: *0,10'*

Specify through point: *0,33'*

Specify through point: *0,60'*

Specify through point: *[enter]*

17. Place three **Horizontal** xlines as indicated (Figure 12.3.1.17).

continued

TOOLS	COMMAND SEQUENCE	STEPS

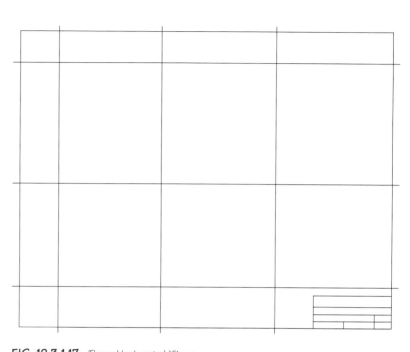

FIG. 12.3.1.17: Three Horizontal Xlines

Command: *tr*

18. Trim away some of the excess lines so that your drawing looks like Figure 12.3.1.18.

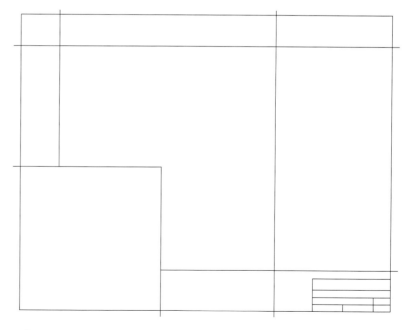

FIG. 12.3.1.18: Trim Some Lines

continued

TOOLS	COMMAND SEQUENCE	STEPS
	**Command:** *la*	**19.** Now let's locate some inner walls. Make layer **Const2** current.
	Command: *xl*	**20.** Create **Vertical** xlines at these coordinates: **26'8,0**; **40'4,0**; **43'10,0** and **47'10,0**; create **Horizontal** xlines at these coordinates: **0,19'10**; **0,23'4**; **0,28'10**; **0,39'10**; **0,50'**. Now your drawing looks like Figure 12.3.1.20.

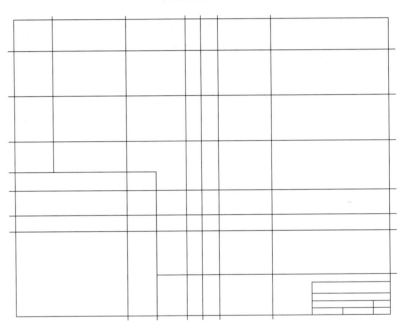

FIG. 12.3.1.20: Additional Xlines

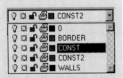	**Command:** *la*	**21.** Make the **Const** layer current.
	Command: *ray* **Specify start point:** *33'4,31'* **Specify through point:** *[pick a point to the left]* **Specify through point:** *[enter]*	**22.** Let's add a bay window. With the ortho toggled **On**, draw a ray from the point indicated due west.

continued

TOOLS	COMMAND SEQUENCE	STEPS
	Command: *[enter]*	**23.** Offset the leftmost vertical line to the right by 3' and again by 12'. The house now looks like Figure 12.3.1.23.

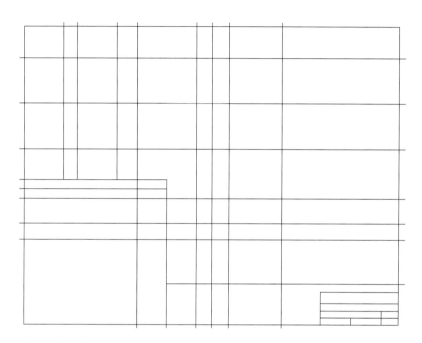

FIG. 12.3.1.23: Offset Xlines

	Command: *xl* **Specify a point or [Hor/Ver/Ang/Bisect/Offset]:** *a* **Enter angle of xline (0) or [Reference]:** *135* **Specify through point:** *_int of* **Specify through point:** *[enter]*	**24.** Use the angle method to place an xline for the left wall of the bay window (Figure 12.3.1.24). (Use your OSNAPs to hit the intersection indicated.)

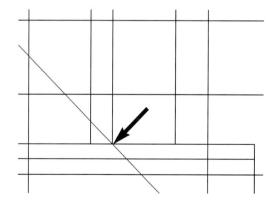

FIG. 12.3.1.24: Left Wall of the Bay Window

continued

TOOLS	COMMAND SEQUENCE	STEPS
	Command: *xl* **Specify a point or [Hor/Ver/Ang/Bisect/Offset]:** *b* **Specify angle vertex point:** *[select the intersection of the Vertex shown]* **Specify angle start point:** *[select a point nearest to Point 1]* **Specify angle end point:** *[select a point nearest to Point 2]* **Specify angle end point:** *[enter]*	25. Use the **Bisect** option to place the other wall. Refer to Figure 12.3.1.25.

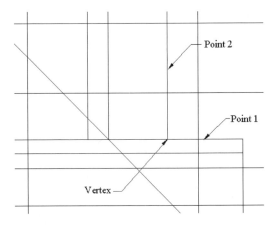

FIG. 12.3.1.25: Reference

	Command: *tr*	26. Use the *Erase* and *Trim* commands to complete the drawing. It will look like Figure 12.3.1.26.

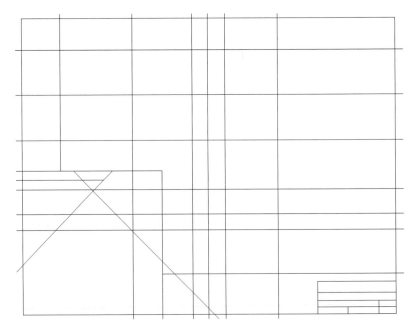

FIG. 12.3.1.26: Completed Layout

	Command: *saveas*	27. Save the drawing as *MyFlr-Pln13* in the C:\Steps\Lesson13 folder.

What do you think? It does not look much like a floor plan, does it?

Consider how a pencil layout might appear at this stage of production. All we have done is place some guidelines (construction lines) to help us locate our walls. We will save the drawing for now and use it in our next lesson when we discuss the *MLine* (multiline) command.

12.4 Extra Steps

In the first half of this lesson, we discussed how to draw contour lines. Contour lines generally represent changes in elevation on topographical or site maps. Go to your local library, check the encyclopedia or ask your employer for some samples of topographical or marine drawings to help familiarize yourself with these fascinating tools.

12.5 What Have We Learned?

Items covered in this lesson include:

■ *Commands*
- *Spline*
- *Splinedit*
- *Xline*
- *Ray*

We have covered some interesting tools in this lesson. Splines will prove themselves surprisingly versatile in the fields where they are used. Construction lines, while not as critical to computer drafting as they are to board work, are nonetheless quite handy in most disciplines. By the time we have finished the next few lessons, this will become apparent.

In our next lesson, we will add walls to the *MyFlr-Pln13* drawing. After that (in Lessons 14 and 16), we will add some notes and dimensions. By the time we complete this section of the book, you will have accomplished quite a drawing!

continued

12.6 EXERCISES

1. Open *topography.dwg* from the C:\Steps\Lesson12 folder.

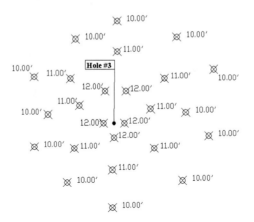

FIG. 12.6.1: *Topography.dwg*

We shot several elevations during a survey of an area golf course. The elevations for Hole 3 are shown.

Using the ***Spline*** command, draw contour lines connecting like elevations. This will afford you a topographical view of the area.

2. Open *Piping plan 12.dwg* in the C:\Steps\Lesson12 folder. The drawing looks like Figure 12.6.2a.

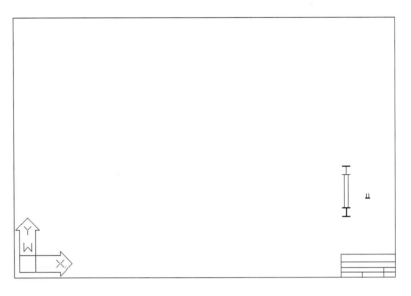

FIG. 12.6.2a: *Piping plan 12.dwg*

2.1. On the **Const** layer, create horizontal constructions lines at these coordinates:

2.1.1. 0,18'; 0,28'; 0,48'; 0,39'6; 0,16'

2.2. Create vertical constructions at these coordinates:

2.2.1. 12'6,0; 27'6,0; 57'6,0; 42'6,0; 50',0; 72'6,0

2.3. Create a 45° angled construction line at 50',48'.

Your drawing looks like Figure 12.6.2b (without the arrow).

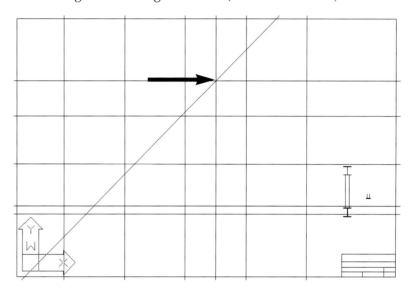

FIG. 12.6.2b: Drawn Construction Lines

2.4. On the **Equip** layer, draw a 15'-diameter circle at the intersection indicated by the arrow in Figure 12.6.2b.

2.5. Using modification commands, position the ibeams, pipe supports, and nozzle as indicated in Figure 12.6.2c.

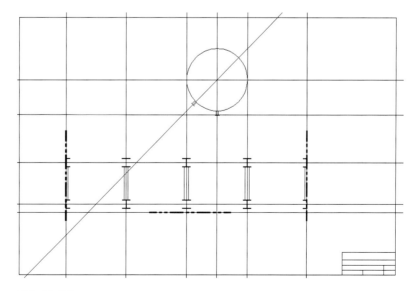

FIG. 12.6.2c: Copy the Ibeams, Pipe Supports, and Nozzle

2.6. Add the matchlines as shown in Figure 12.6.2c (use 3" wide polylines on the **Matchlin** layer). Trim and change the layer of the construction lines to **Center**. Your drawing will look like Figure 12.6.2d.

2.7. Save the drawing as *MyPipingPlan13* in the C:\Steps\Lesson13 folder.

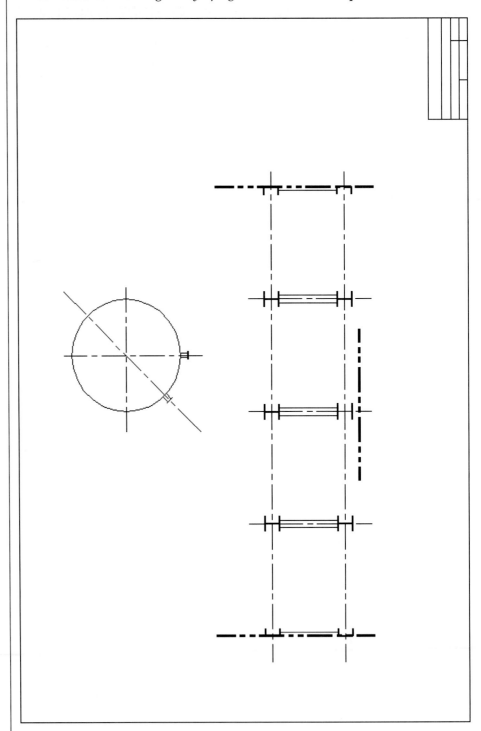

FIG. 12.6.2d: Completed Drawing

3. Start a new drawing from scratch. Save it as *MyContPanel.dwg* in the C:\Steps\Lesson12 folder.

   3.1. Accept the default limits, but set the grid to $1/2$ and the snap to $1/4$. Create layers as needed during the drawing session, but begin with the basic *const*, *object*, *text*, *centerline*, and *hidden* layers with appropriate colors and linetypes.

   3.2. Using the grid as a guide, draw construction lines on the *const* layer, as shown in Figure 12.6.3a.

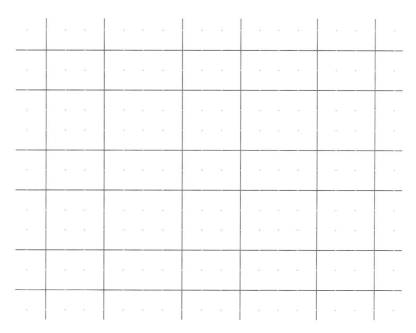

FIG. 12.6.3a: Construction Lines

   3.3. On the *object layer* draw rectangles, as shown in Figure 12.6.3b.

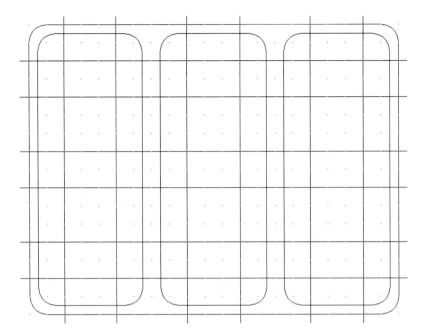

FIG. 12.6.3b: Rectangle with ½" Fillets

3.4. Now draw the buttons as shown in Figure 12.6.3c. Use the grid to guide you as you draw the inner ellipses. Offset those $1/16$" outward for the outer ellipses. Put the buttons on their own layer.

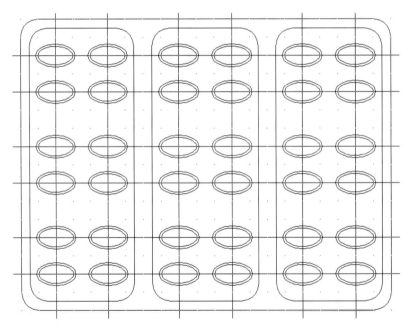

FIG. 12.6.3c: Buttons

3.5. Now complete the drawing as shown in Figure 12.6.3d.

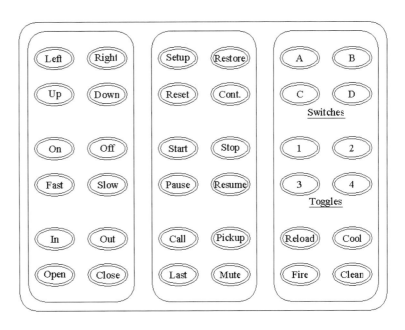

FIG. 12.6.3d: Completed Drawing

4. Start a new drawing from scratch as you did in Exercise 3. Save the drawing as *MyHolder2.dwg* in the C:\Steps\Lesson12 folder.

   4.1. On the *const* layer, create the construction lines shown in Figure 12.6.4a.

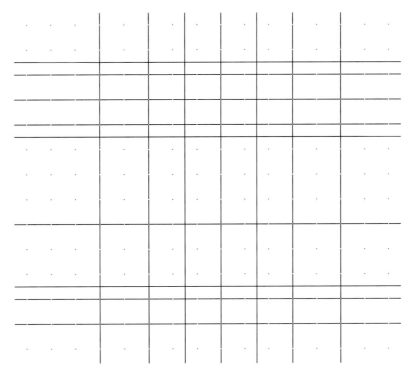

FIG. 12.6.4a: Construction Lines

4.2. On the *object* layer, create the circles shown in Figure 12.6.4b. (The smaller circles are $3/8$" dia. And the inner circle is offset $1/16$ from one next to it.)

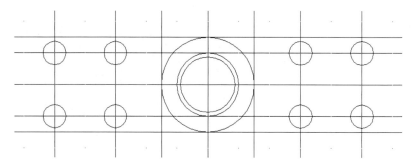

FIG. 12.6.4b: Circles

4.3. Snapping to the points indicated in Figure 12.6.4c, draw a spline as shown.
4.4. Draw the notches and add construction lines as indicated in Figure 12.6.4d.

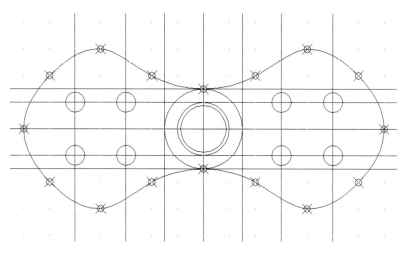

FIG. 12.6.4c: Spline

4.5. Complete the drawing (see Figure 12.6.4e).

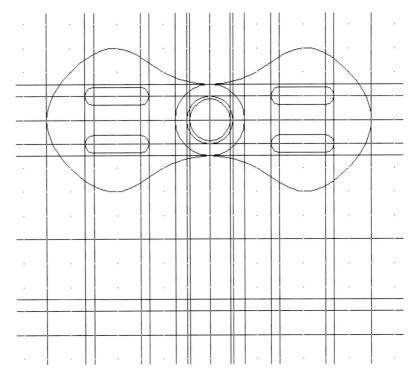

FIG. 12.6.4d: Notches and Additional Construction Lines

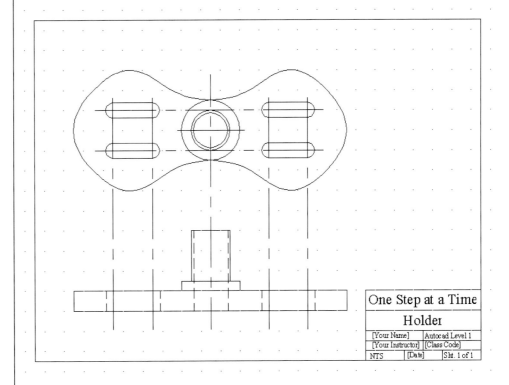

FIG. 12.6.4e: Completed Drawing

5. Using what you have learned, create the drawings shown in Figures 12.6.5a to 12.6.5c. Do not draw the hatching. (Special thanks to Randy Behounek at Sarpy County Surveyors Office in Papillion, Nebraska, for permission to use these drawings.)

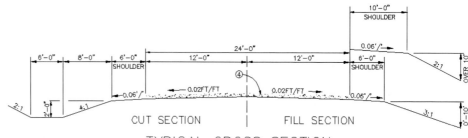

FIG. 12.6.5a

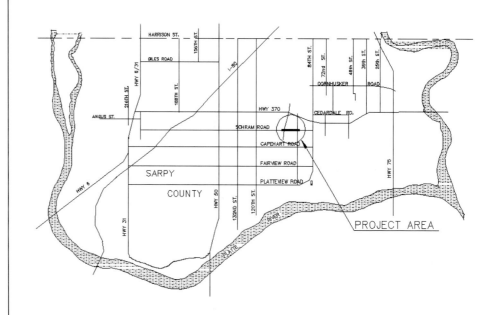

FIG. 12.6.5b: Sarpy and Project Area Map for Exercise

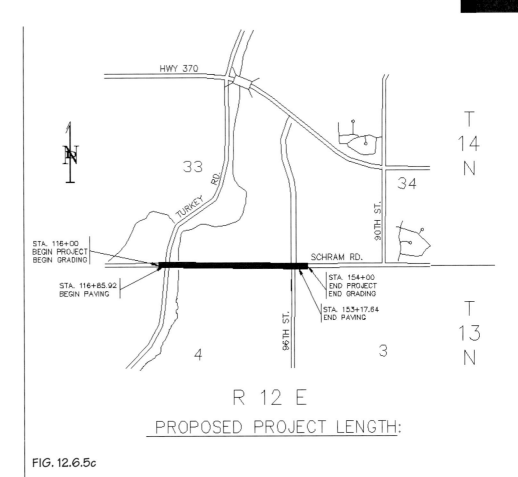

FIG. 12.6.5c

12.7 REVIEW QUESTIONS

Please answer these questions on a separate sheet of paper.

1. Which takes up less drawing memory, a spline or a splined polyline?

2. Use the _____ option of the spline command to convert a splined polyline into a spline.

3. The _____ option tells AutoCAD to draw the spline within a certain distance of the points selected by the user.

4. Normally, the points used to define a contour line (spline) will be provided by a _____.

5. The _____ on a spline are essentially the same thing as vertices on a polyline.

6. Which option must I select on the first tier of splinedit options to find the tier with the purge option?

7. If the user wants to add a control point to a spline, he must first select which option of the splinedit command?

8. A control point's _____ determines how much influence that point has against the spline.

9. A smoother spline often used in 3-dimensional drawing is called a _____.

10. and 12. Use either the _____ or the _____ command to create a construction line in AutoCAD.

13. Of the two previous commands, which is infinite in *both* directions?

14. (T or F) Because a construction line is infinite in both directions, use of constructions disables the zoom extents command.

15. Use the (xline, ray) command to bisect an existing angle.

LESSON 13

Advanced Lines: Multilines

Following this lesson, you will:
➥ Know how to draw several lines at once with the **MLine** Command
➥ Be able to create multiline styles with the **MLStyle** Command
➥ Be able to edit multilines with the **MLEdit** Command

Back in grade school, when I had been naughty (and got caught), I was punished by being made to write "penance" sentences—"I will be good in school," one hundred times. Oh! The degradations of childhood!

But in my childish attempts to cut corners (there were always ways to cut corners), I would tape four pencils together. Then I only had to "be good in school" twenty-five times!

Apparently, someone at AutoDESK learned a similar childhood lesson. The result was the *MLine* command. The *MLine* command does just what taping four pencils together did—it enables the user to create more than one line at a time.

AutoCAD's multilines actually involve three commands: *MLine*, *MLEdit*, and *MLStyle*. **The first actually draws the lines; the second enables the user to edit, or change, the lines; the third enables the user to define the lines.**

In Lesson 13, we will look at each.

13.1 Many at Once: AutoCAD's Multilines and the *MLine* Command

The multiline saga began before R13. Someone created a Lisp routine that would draw two lines at once. It was called *DLine* (for DoubleLine) and was enjoyed by CAD users for some time. In the piping industry, however, we needed to draw three lines at once—the outer walls of a pipe and a centerline. So we adjusted the Lisp routine to add an additional line. I am sure our counterparts in architecture did the same thing to add brick ledges around their houses.

All this got back to AutoCAD, and in R13 they included the *MLine* command. This is an easy-to-use tool designed to enhance the efficiency of multiline drawing. The command sequence is

Command: *mline* **(or** *ml***)**
Current Settings: Justification = Top, Scale = 1.00, Style = STANDARD
Specify start point or [Justification/Scale/Style]: *[pick the start point]*
Specify next point: *[pick the next point]*
Specify next point or [Undo]: *[either continue picking points or hit enter to end the command]*
Command:

As you can see, the actual command sequence is not very different from drawing any other line. AutoCAD prompts **Specify start point** and then repeats **Specify next point** until you have completed the line. There are, however, some simple options the user must set.

- AutoCAD makes three choices available when the user selects the **Justification** options:

  Enter justification type [Top/Zero/Bottom]<top>:

  These involve where AutoCAD places the lines in relation to the user-identified point. (See Figures 13.1a to 13.1c) to see how they work.

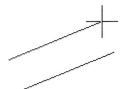
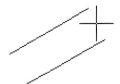
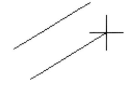

FIG. 13.1a: Top Justification

FIG. 13.1b: Zero Justification

FIG. 13.1c: Bottom Justification

This may seem a bit strange, but it really does have its uses. For example, most pipe is drawn using centerlines. For a piper, then, the **Zero** justification would be easiest to use. But for an architect who dimensions to outer walls, the **Top** or **Bottom** justification may be just the ticket. Of course, the architect can switch to the **Zero** option when adding interior walls!

- The **Scale** option enables the user to insert a multiline at any width. For example, a piper creates a multiline style that has three lines—two outer walls and a centerline. The outer walls are one unit apart according to the style's definition. The piper needs a 3" diameter pipe, so he inserts the multiline at a **Scale** of 3.5 (the outer diameter, or OD, of the pipe). Similarly, the architect may insert a basic two-line wall at a **Scale** of 4 or 5.5 to cover different building materials for walls.

- The **STyle** option enables the user to tell AutoCAD which multiline style is needed for this particular multiline. The default is **Standard**, which is a basic two-line multiline with a separation of 1 unit between the lines. We will look at creating multiline styles in Section 13.2.

Let's draw a simple multiline using the **Standard** style.

The *MLine* command is also available in the Draw pull-down menu. The options are available on the cursor menu once the command has been entered.

13.1.1 Do This: Creating Multilines

I. Open a new drawing from scratch.

II. Follow these steps.

TOOLS	COMMAND SEQUENCE	STEPS
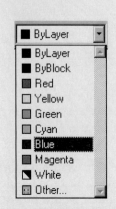	**Command:** *col*	1. Set the current color to Blue.
 Multiline Button	**Command:** *ml*	2. Enter the *MLine* command by typing *mline* or *ml*. Or you can pick the **Multiline** button on the Draw toolbar.
	Current settings: Justification = Top, Scale = 1.00, Style = STANDARD **Specify start point or [Justification/Scale/STyle]:** *1,1* **Specify next point:** *1,6* **Specify next point or [Undo]:** *5,6* **Specify next point or [Close/Undo]:** *5,1* **Specify next point or [Close/Undo]:** *[enter]*	3. Accept the defaults and draw a simple multiline as indicated. Your drawing looks like Figure 13.1.1.3. 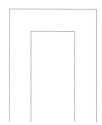 **FIG. 13.1.1.3:** Multiline Notice that the color of the multiline does not reflect the current color setting. This is because multiline colors are set according to the Multiline Style definition (Section 13.2).

continued

TOOLS	COMMAND SEQUENCE	STEPS
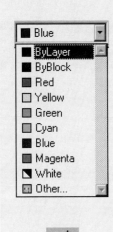	Command: *col*	4. Set the current color back to **Bylayer**.
	Command: *ml*	5. Repeat the *MLine* command.
	Current settings: Justification = Top, Scale = 1.00, Style = STANDARD	6. Tell AutoCAD you want to change the **Justification** . . .
	Specify start point or [Justification/Scale/STyle]: *j*	
	Enter justification type [Top/Zero/Bottom] <top>: *z*	7. . . . to **Zero**.
	Current settings: Justification = Zero, Scale = 1.00, Style = STANDARD	8. Repeat Step 3 using the same coordinates. You are now locating the multiline from the center. Your drawing with both multilines looks like Figure 13.1.1.8.

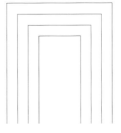

FIG. 13.1.1.8: Multilines with Different Justifications

	Command: *[enter]*	9. Repeat the *MLine* command.

continued

TOOLS	COMMAND SEQUENCE	STEPS
	Current settings: Justification = Zero, Scale = 1.00, Style = STANDARD **Specify start point or [Justification/Scale/STyle]:** *s*	10. This time we will change the **Scale** . . .
	Enter mline scale <1.00>: *2*	11. . . . to 2.
	Current settings: Justification = Zero, Scale = 2.00, Style = STANDARD **Specify start point or [Justification/Scale/STyle]:** *7,1* **Specify next point:** *7,5* **Specify next point or [Undo]:** *12,5* **Specify next point or [Close/Undo]:** *12,1* **Specify next point or [Close/Undo]:** *[enter]*	12. Draw the multiline as indicated. Your drawing looks like Figure 13.1.1.12. **FIG. 13.1.1.12:** Scaled Multiline
	Command: *quit*	13. Exit the drawing without saving.

So you see how easy it is to draw multilines. But we have only used AutoCAD's **Standard Style**. Let's try something different. Let's define another **Style**.

13.2 Options, Options, Options: The MLStyle Command

We define multiline styles using the *MLStyle* command. It can be a fairly complex task, but fortunately, AutoCAD provides dialog boxes to help. Upon receiving the *MLStyle* command, AutoCAD presents the Multiline Styles dialog box (Figure 13.2a).

FIG. 13.2a: Multiline Styles Dialog Box

The first time I attempted to create a multiline style, I became quite confused. So let me give you a procedure to make it easy.

Multiline Style Creation

1. **Name** your new style.
2. **Add** your new style to the current drawing.
3. Set the **Element Properties** for your new style.
4. Set the **Multiline Properties** (if required) for your new style.
5. If you want to make the style available to other drawings, **Save** it to a file.
6. **OK** your work.

The button we did not cover in our creation procedure was the **Load** button. This button enables the user to load a multiline style from a file (the same file to which you **Saved** your style in Step 5). We will see it in Exercise 13.2.1.

Each of these steps involves a different dialog box. And the best way to get familiar with each is to create a new style and explain the dialog boxes as we go. Let's begin.

The *MLStyle* command is also available in the Format pull-down menu. Follow this path:

Format—Multiline Style . . .

13.2.1 Do This: Defining Multilines with MLStyle

I. Open a new drawing from scratch. We will follow our procedure to create a basic style to be used in drawing pipe.

II. Follow these steps.

TOOLS	COMMAND SEQUENCE	STEPS
No Button Available	**Command:** *mlstyle*	**1.** Enter the *MLStyle* command. AutoCAD will present the Multiline Styles dialog box (Figure 13.2a).
	 Name: PIPE	**2.** Place the name of your new style (*Pipe*) in the **Name** box.
 Add		**3.** Pick the **Add** button to add the new style to the drawing. The name is added and made current (Figure 13.2.1.3).
		Multiline Style Current: PIPE Name: PIPE Description: Load... Save... Add Rename
		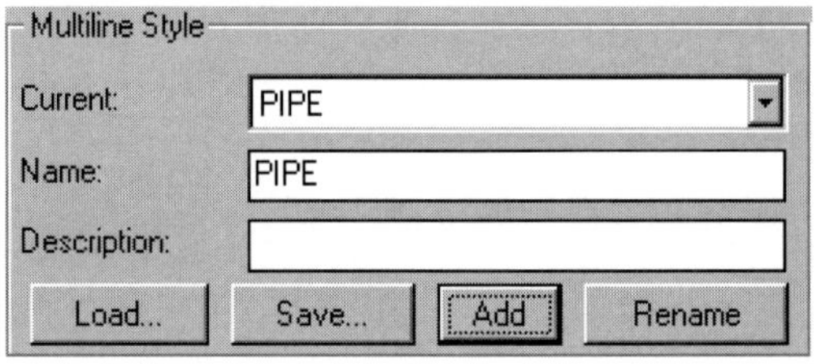 **FIG. 13.2.1.3:** Added Style
 Element Properties ...		**4.** Pick the **Element Properties** button. The Element Properties dialog box appears (Figure 13.2.1.4). This is where we will tell AutoCAD how many and what type of lines to include in our style.

continued

TOOLS	COMMAND SEQUENCE	STEPS

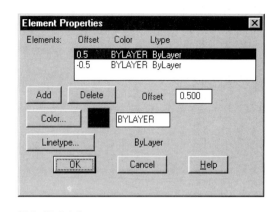

FIG. 13.2.1.4: Element Properties Dialog Box

5. Let's start by adding a line. Pick the **Add** button. A line definition appears in the **Elements** box at the top. The offset distance for the new line is given as **0.0**. Here you also see that the existing lines' offsets have been defined as 0.5 unit above and 0.5 unit below the center. This gives our style an overall size of 1.0. We can change the **Offset**, **Color**, and **Linetype** using the buttons provided.

6. Let's change the linetype of our new line. Highlight the new line in the Elements box.

7. Pick the **Linetype** button. The Select Linetype dialog box appears. You are familiar with this box from Lesson 7. Follow the procedures you learned in that lesson to load the **Center** linetype and set it as the linetype for your new definition. The Elements box now shows **Center** as the linetype for the middle line (Figure 13.2.1.7).

FIG. 13.2.1.7: Center Linetype

continued

TOOLS	COMMAND SEQUENCE	STEPS

8. To help differentiate between the pipe walls and the centerline, let's change the color of our new line as well. Pick the **Color** button. Again, you are familiar with the Select Color dialog box from Lesson 7. Set the color to **Blue**. The Element Properties dialog box now looks like Figure 13.2.1.8.

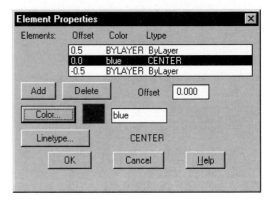

FIG. 13.2.1.8: Completed Element Properties

9. Pick the **OK** button to return to the Multiline Styles dialog box. Notice that an image of the current (pipe) style is shown in the middle of the box (Figure 13.2.1.9).

FIG. 13.2.1.9: Demo of Style

Notice also that the centerline does not appear as a centerline. This is a bit of a flaw in the program. This box will not show linetypes, only colors and the number of lines.

10. Pick the **Save** button to save the multiline style to a file so you can use it in this lesson's exercise. AutoCAD will present a standard Save File type dialog box. Save the file to the C:\Steps\Lesson13 folder as *MyMLines.mln* (AutoCAD will put the *.mln* extension on the filename automatically).

continued

TOOLS	COMMAND SEQUENCE	STEPS
OK		11. Pick the **OK** button here to complete the command.
		12. You can use the mline now. But we will look next at how to load it into another drawing.
		Exit AutoCAD. Do not save the changes to this drawing. Then reenter and start a new drawing from scratch.
(layers icon)	**Command:** *la*	13. Create a layer called *Pipe* that is **Cyan** with a **Continuous** linetype. Make this layer current.
	Command: *mlstyle*	14. Open the Multiline Styles dialog box.
Load...		15. Pick the **Load** button. AutoCAD presents the Load Multiline Styles dialog box (Figure 13.2.1.15).

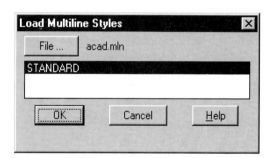

FIG. 13.2.1.15: Load Multiline Styles Dialog Box

File...		16. Pick the **File** button. AutoCAD presents the standard Open File dialog box. Open *MyMLines* from the C:\Steps\Lesson13 folder. (*Note:* If that file is not available, open *MLines* instead.)
OK		17. The Load Multiline Styles dialog box now shows the *MyMLines* file contents. Be sure **Pipe** is highlighted (Figure 13.2.1.17) then pick the **OK** button.

continued

TOOLS	COMMAND SEQUENCE	STEPS
		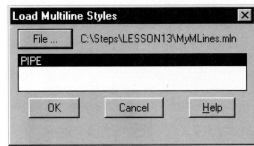 **FIG. 13.2.1.17:** Load the Pipe Style
		18. Notice that **Pipe** is now the current multiline style. Pick **OK** to leave the command.
	Command: *ml*	19. Now let's draw two multilines. Enter the *Mline* command by typing *mline* or *ml* at the command prompt. Or you may pick the **Multiline** button on the Draw toolbar.
	Current settings: Justification = Top, Scale = 1.00, Style = PIPE **Specify start point or [Justification/Scale/STyle]:** *s* **Enter mline scale <1.00>:** *1.375*	20. Set the scale for the OD of a 1" pipe as indicated.
	Current settings: Justification = Top, Scale = 1.38, Style = PIPE **Specify start point or [Justification/Scale/STyle]:** *6,1* **Specify next point:** *6,7* **Specify next point or [Undo]:** *[enter]*	21. Draw the multiline as indicated. Your drawing looks like Figure 13.2.1.21. 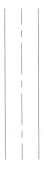 **FIG. 13.2.1.21:** Multiline Using the Pipe Style

continued

TOOLS	COMMAND SEQUENCE	STEPS
		Notice that the outer walls (whose color defaulted to **ByLayer** in the multiline style's definition) assume the characteristics of the current layer. But the centerline's **Color** and **Linetype** were defined separately in the multiline style's definition and reflect the characteristics assigned there.
	Current settings: Justification = Top, Scale = 1.38, Style = PIPE **Specify start point or [Justification/Scale/STyle]:** *3,5* **Specify next point:** *11,5* **Specify next point or [Undo]:** *[enter]*	22. Draw another multiline crossing the last. Your drawing looks like Figure 13.2.1.22. **FIG. 13.2.1.22:** Completed Drawing
💾	**Command:** *save*	23. Save this drawing as *MyMlines* to the C:\Steps\Lesson13 folder.

Did you notice that we skipped Step 4 in our Multiline Style Creation procedure? This step is for creating a bit of razzle-dazzle in your multiline.

Let's take a look at what the **Multiline Properties . . .** button can do.

13.2.2 Do This: More on Multiline Styles

I. Open a new drawing from scratch. Follow the last <u>Do This:</u> Exercise 13.2.1 to create a multiline style called *Demo*. *Demo* should have seven lines. Use these offsets: **0.75, 0.50, 0.25, 0.0, −0.25, −0.50**, and **−0.75**. Leave the color and linetype for each set to **ByLayer**.

II. Save this multiline style in the *MyMlines.mln* file in the C:\Steps\Lesson13 folder.

III. Follow these steps.

TOOLS	COMMAND SEQUENCE	STEPS
![Multiline Properties...]		**1.** Pick the **Multiline Properties . . .** button. AutoCAD presents the Multiline Properties dialog box (Figure 13.2.2.1).

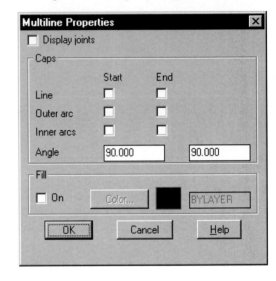

FIG. 13.2.2.1: Multiline Properties Dialog Box

Let's take a look at each of these options.

2. Put a check in the **Display joints** check box. Then pick **OK** to return to the Multiline Styles dialog box. Notice the line in the middle of the demo box (Figure 13.2.2.2). This indicates that AutoCAD will put a line in the multiline joints.

FIG. 13.2.2.2: Demo Box Showing Joints

3. Pick the **OK** button to complete the command.

continued

TOOLS	COMMAND SEQUENCE	STEPS

Command: *ml*

4. Enter the *MLine* command by typing *mline* or *ml*. Or you can pick the **Multiline** button on the Draw toolbar.

Current settings: Justification = Top, Scale = 1.00, Style = DEMO

Specify start point or [Justification/Scale/STyle]: *1,1*

Specify next point: *1,6*

Specify next point or [Undo]: *5,6*

Specify next point or [Close/Undo]: *5,1*

Specify next point or [Close/Undo]: *[enter]*

5. Accept the defaults and draw a multiline as indicated. Your drawing looks like Figure 13.2.2.5. Notice the joints.

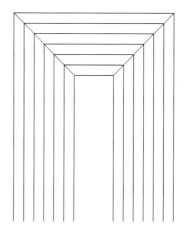

FIG. 13.2.2.5: Multiline with Joints

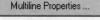

Command: *mlstyle*

6. Repeat the *MLStyle* command. **Add** a new style called *Demo1*. The new style is created from the properties of the current (*Demo*) style. Pick the **Multiline Properties...** button.

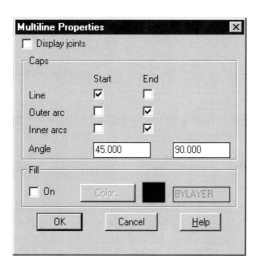

7. Make these adjustments: Remove the check from the **Display joints** check box, add a check to the **Line** check box under the **Start** column, and to the **Outer** and **Inner arcs** check boxes under the **End** column, and change the **Angle** under the **Start** column to *45°*.

continued

TOOLS	COMMAND SEQUENCE	STEPS
 [OK button]		8. Pick the **OK** button to return to the Multiline Styles dialog box (the image in the demo box looks like Figure 13.2.2.8) and again to complete the command. **FIG. 13.2.2.8:** Demo of Multiline with Lined and Arced Ends
[Erase icon]	Command: *e*	9. Erase the last multiline you drew.
[Multiline icon]	Command: *ml* Current settings: Justification = Top, Scale = 1.00, Style = DEMO1 Specify start point or [Justification/Scale/STyle]: 2,7 Specify next point: 9,7 Specify next point or [Undo]: 9,2 Specify next point or [Close/Undo]: *[enter]*	10. Draw a new multiline as indicated. Your drawing looks like Figure 13.2.2.10. **FIG. 13.2.2.10:** Multiline with Lined and Arced Ends Notice the angle of the start point is 45°. You can also see the line at the start point and the inner and outer arcs at the endpoint as you set in Step 7.
 [Multiline Properties button]	Command: *mlstyle*	11. Repeat the *MLStyle* command. **Add** a new style called *Demo2* then pick the **Multiline Properties ...** button.

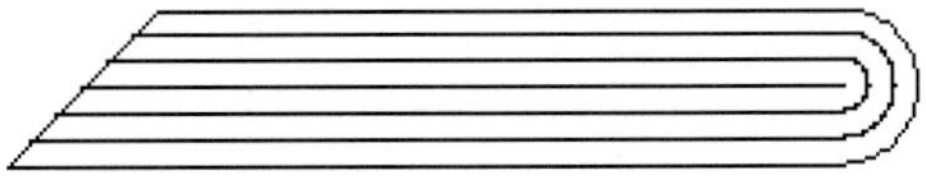

continued

TOOLS	COMMAND SEQUENCE	STEPS
	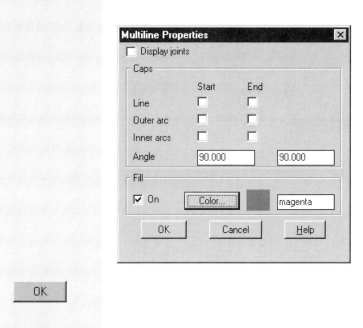	12. Set the properties as follows: remove all the checks, set the **Start Angle** to *90°*, put a check in the **On** check box in the **Fill** section of the dialog box. Set the color to *magenta*.
		13. Pick the **OK** buttons to complete the command.
	Command: *ml* **Current settings: Justification = Top, Scale = 1.00, Style = DEMO2** **Specify start point or [Justification/Scale/STyle]:** *1,3* **Specify next point:** *5,3* **Specify next point or [Undo]:** *[enter]* **Command:** *quit*	14. Draw a new multiline as indicated. The multiline is filled (Figure 13.2.2.14). 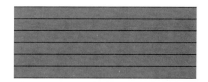 **FIG. 13.2.2.14:** Filled Multiline 15. Exit the drawing without saving.

As you have seen, you can do a lot of things with multilines. You will find them particularly useful in two fields—Piping and Architecture. But they are also quite useful in laying out multilane (or single-lane) streets.

You should, however, be aware of the limitations of multilines. The worst of these is the inability of the user to set different layers for each of the individual lines. Because of this, centerlines cannot be manipulated separately from outer lines. Another limitation is the need for a specific tool for modifying multilines. This tool is called *MLEdit* and is the subject of the next section of our lesson.

13.3 Editing Multilines: The MLEdit Command

As I mentioned, editing multilines requires a special tool. Oh, some of AutoCAD's more basic modifying tools—such as *Copy* and *Move*—work on multilines, but the tools you might find most useful—*Trim* and *Extend*—do not. This leaves the user with two options—either *explode* the multiline (into individual objects as you would a polyline) or use the *MLEdit* command.

Exploding the multiline is not a bad idea. In fact, once it has been drawn, there is very little reason to maintain its multiline definition except that, as a multiline, it can be manipulated as a single object.

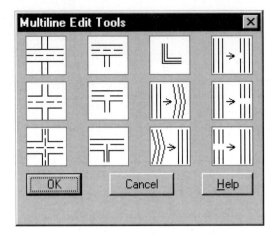

FIG. 13.3a: Multiline Editing Tools Dialog Box

But let's look at the *MLEdit* command. The command calls the Multiline Edit Tools dialog box (Figure 13.3a). Selecting any of the images in the dialog box will cause a cheater line to appear in the bottom left corner telling the user what the selection does. The possibilities include three types of crosses and tees, corner creation, adding or removing a multiline vertex, breaking all or part of a multiline, and welding a broken multiline.

Let's see these in action.

The *MLEdit* command is also available in the Modify pull-down menu.

13.3.1 Do This: Editing Multilines

I. Open the *MyMlines* file you created in the C:\Steps\Lesson13 folder. (If this file is unavailable, open *Mlines* in the same folder.) The drawing looks like Figure 13.3.1a.

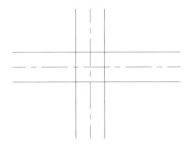

FIG. 13.3.1a: *MyMlines.dwg*

II. Follow these steps.

TOOLS	COMMAND SEQUENCE	STEPS
 Edit Multiline Button Closed Cross	**Command:** *mledit*	1. Enter the *MLEdit* command by typing *mledit* at the command. There is no hotkey, but you can pick the **Edit Multiline** button on the Modify II toolbar. 2. Select the **Closed Cross** button and then pick **OK**. (Or, you can double-click on the button.)
	Select first mline:	3. Pick the lower part of the vertical multiline . . .

continued

TOOLS	COMMAND SEQUENCE	STEPS
	Select second mline:	4. . . . then the right part of the horizontal line. Hit *enter* to complete the command. Your drawing looks like Figure 13.3.1.4.
	Select mline or [Undo]: *[enter]*	

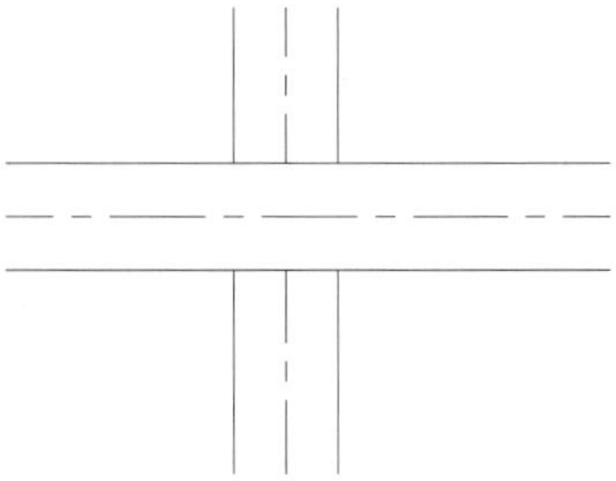

FIG. 13.3.1.4: Closed Cross

Notice that the line you selected first is the one broken.

TOOLS	COMMAND SEQUENCE	STEPS
	**Command:** *u*	5. Undo the last command.
	**Command:** *mledit*	6. Repeat the *MLEdit* command.
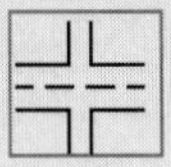 Open Cross		7. Select the **Open cross** button then pick the **OK** button.
	Select first mline:	8. Pick the lower part of the vertical multiline . . .

continued

TOOLS	COMMAND SEQUENCE	STEPS
	Select second mline:	**9.** . . . then the right part of the horizontal line. Hit *enter* to complete the command. Your drawing looks like Figure 13.3.1.9.
	Select mline or [Undo]: *[enter]*	
		FIG. 13.3.1.9: Open Cross
		Notice the centerline of the second selected multiline does not break.
 ↺	**Command:** *u*	**10.** Undo the last command.
	**Command:** *mledit*	**11.** Repeat the *MLEdit* command.
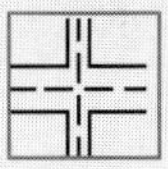 Merged Cross		**12.** Select the **Merged Cross** button then pick the **OK** button.
	Select first mline:	**13.** Pick the lower part of the vertical multiline . . .

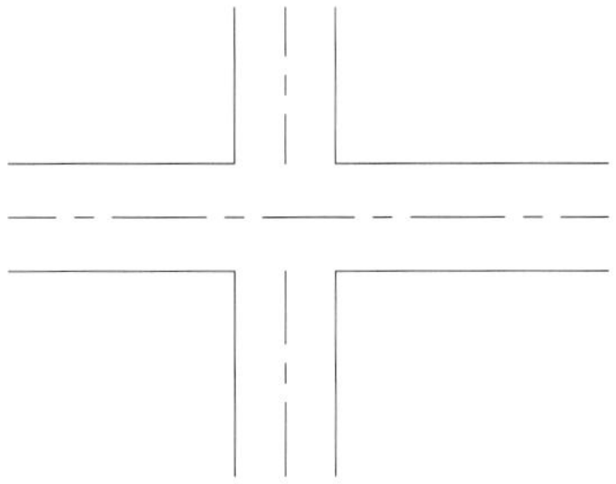

continued

TOOLS	COMMAND SEQUENCE	STEPS
	Select second mline:	**14.** ... then the right part of the horizontal line. Hit *enter* to complete the command. Your drawing looks like Figure 13.3.1.14.
	Select mline or [Undo]: *[enter]*	
		FIG. 13.3.1.14: Merged Cross
		Notice that, this time, neither centerline breaks.
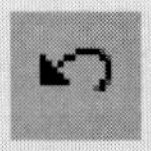	**Command:** *u*	**15.** Undo the last command.
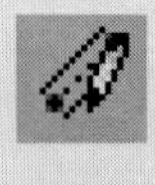	**Command:** *mledit*	**16.** Repeat the *MLEdit* command.
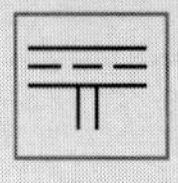 Closed Tee		**17.** Select the **Closed Tee** button and then pick the **OK** button.
	Select first mline:	**18.** Pick the lower part of the vertical multiline ...
	Select second mline:	**19.** ... then the right part of the horizontal line. Hit *enter* to complete the command. Your drawing looks like Figure 13.3.1.19.
	Select mline or [Undo]: *[enter]*	
		FIG. 13.3.1.19: Closed Tee

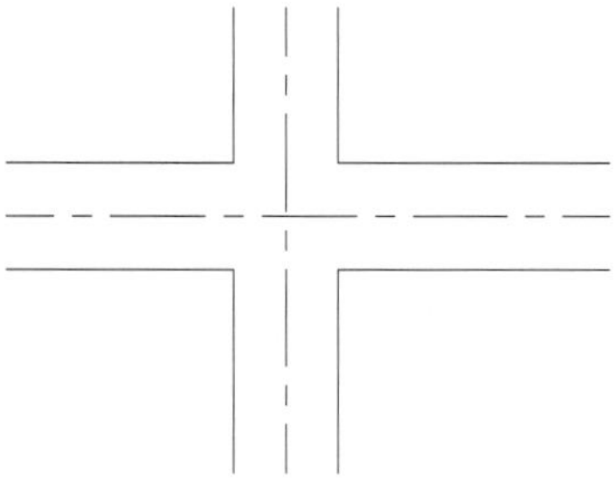

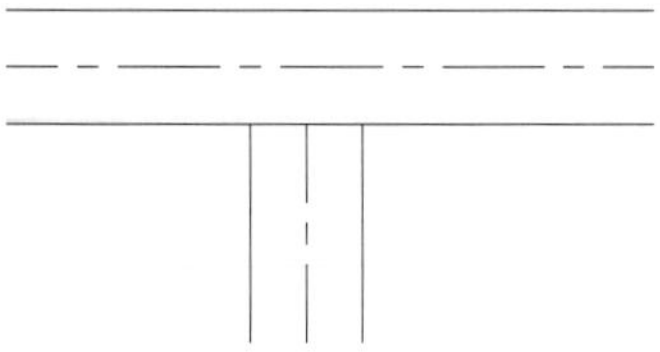

continued

TOOLS	COMMAND SEQUENCE	STEPS
	**Command:** *u*	20. Undo the last command.
	**Command:** *mledit*	21. Repeat the *MLEdit* command.
 Open Tee		22. Select the **Open Tee** button then pick the **OK** button.
	Select first mline:	23. Pick the lower part of the vertical multiline . . .
	Select second mline: **Select mline or [Undo]:** *[enter]*	24. . . . then the right part of the horizontal line. Hit *enter* to complete the command. Your drawing looks like Figure 13.3.1.24.

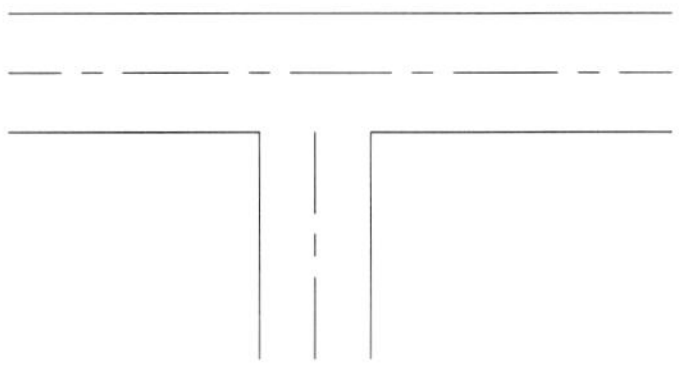

FIG. 13.3.1.24: Open Tee

Notice that the horizontal centerline does not break.

	**Command:** *u*	25. Undo the last command.
	**Command:** *mledit*	26. Repeat the *MLEdit* command.
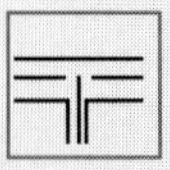 Merged Tee		27. Select the **Merged Tee** button and then pick the **OK** button.
	Select first mline:	28. Pick the lower part of the vertical multiline . . .

continued

TOOLS	COMMAND SEQUENCE	STEPS
	Select second mline: **Select mline or [Undo]:** *[enter]*	**29.** . . . then the right part of the horizontal line. Hit *enter* to complete the command. Your drawing looks like Figure 13.3.1.29.

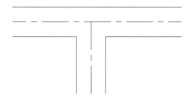

FIG. 13.3.1.29: Merged Tee

Notice that the centerlines form a tee as well.

TOOLS	COMMAND SEQUENCE	STEPS
	Command: *u*	**30.** Undo the last command.
	Command: *mledit*	**31.** Repeat the *MLEdit* command.
Corner Joint		**32.** Select the **Corner Joint** button, then pick the **OK** button.
	Select first mline:	**33.** Pick the lower part of the vertical multiline . . .
	Select second mline: **Select mline or [Undo]:** *[enter]*	**34.** . . . then the right part of the horizontal line. Hit *enter* to complete the command. Your drawing looks like Figure 13.3.1.34.

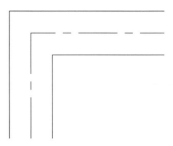

FIG. 13.3.1.34: Corner Joint

continued

TOOLS	COMMAND SEQUENCE	STEPS
	Command: *u*	35. Undo the last command.
	Command: *mledit*	36. Repeat the *MLEdit* command.
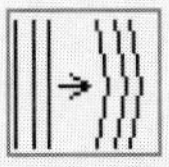 Add Vertex		37. Select the **Add Vertex** button and then pick the **OK** button.
	Select second mline: _mid of Select mline or [Undo]: *[enter]*	38. Pick the midpoint of the horizontal line. Hit *enter* to complete the command.
	Command: *s* Select objects to stretch by crossing-window or crossing-polygon . . . Select objects: Other corner: 2 found Select objects: Base point or displacement: *1,0* Second point of displacement: *[enter]*	39. Stretch the middle of the horizontal line upward one unit. Your drawing looks like Figure 13.3.1.39.

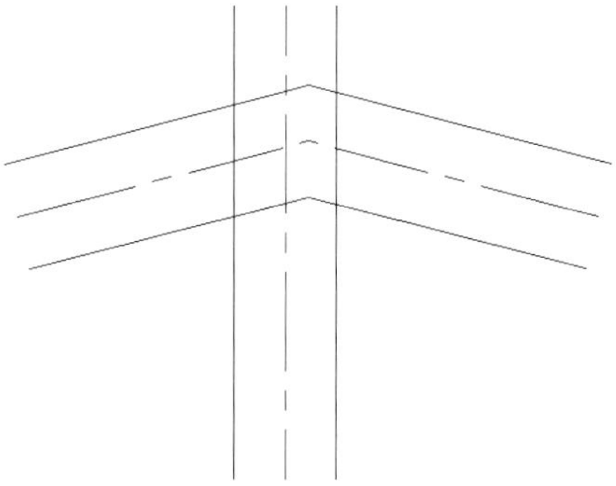

FIG. 13.1.39: Added Vertex

(*Note:* The **Stretch** procedure was to help you see the new vertex.)

	Command: *mledit*	40. Repeat the *MLEdit* command.
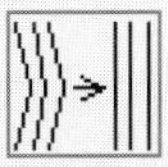 Delete Vertex		41. Select the **Delete Vertex** button and then pick the **OK** button.
	Select mline: Select mline or [Undo]:	42. Select the new vertex on the horizontal multiline. Hit *enter* to complete the command. Your drawing looks like Figure 13.3.1.42.

continued

TOOLS	COMMAND SEQUENCE	STEPS
		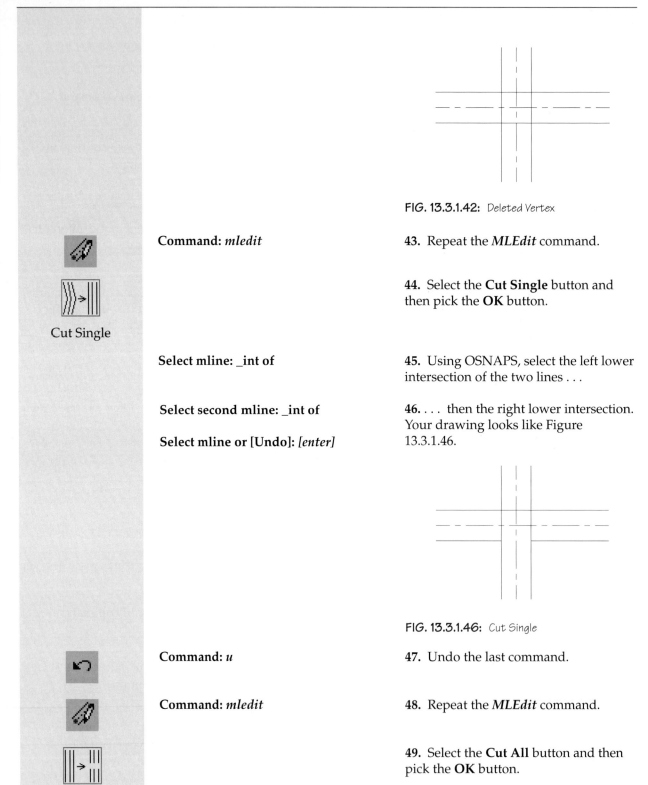
		FIG. 13.3.1.42: Deleted Vertex
	Command: *mledit*	43. Repeat the *MLEdit* command.
Cut Single		44. Select the **Cut Single** button and then pick the **OK** button.
	Select mline: _int of	45. Using OSNAPS, select the left lower intersection of the two lines . . .
	Select second mline: _int of	46. . . . then the right lower intersection. Your drawing looks like Figure 13.3.1.46.
	Select mline or [Undo]: *[enter]*	
		FIG. 13.3.1.46: Cut Single
	Command: *u*	47. Undo the last command.
	Command: *mledit*	48. Repeat the *MLEdit* command.
Cut All		49. Select the **Cut All** button and then pick the **OK** button.

continued

TOOLS	COMMAND SEQUENCE	STEPS
	Select mline: _int of	50. Using OSNAPS, select the left lower intersection of the two lines . . .
	Select second mline: _int of	51. . . . then the right lower intersection. Your drawing looks like Figure 13.3.1.51.
	Select mline or [Undo]: *[enter]*	

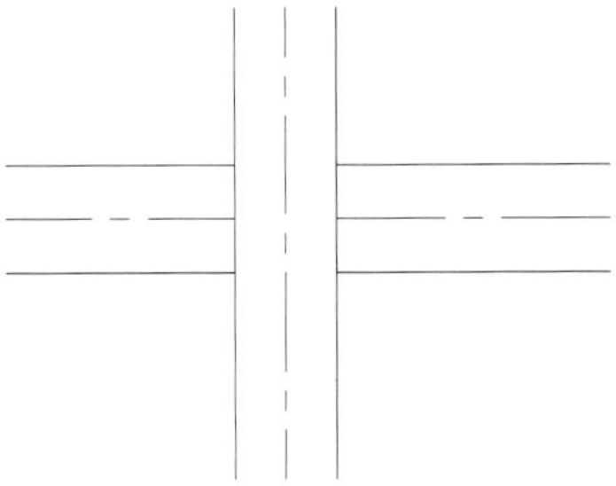

FIG. 13.3.1.51: Cut All

	**Command:** *mledit*	52. Repeat the *MLEdit* command.
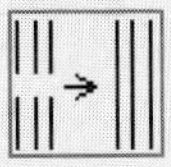 Weld All		53. Select the **Weld All** button and then pick the **OK** button.
	Select mline:	54. Select the two endpoints of the multiline you just broke. Your drawing looks like Figure 13.3.1.54.
	Select second point:	
	Select mline or [Undo]: *[enter]*	

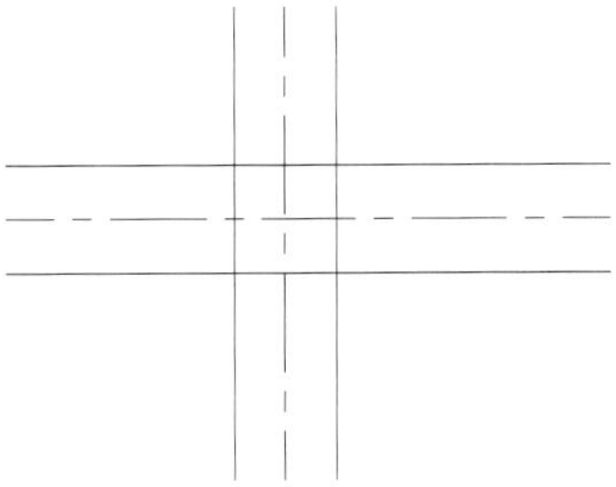

FIG. 13.3.1.54: Weld All

	Command: *quit*	55. Quit the drawing without saving the changes.

13.4 The Project

Now that you have had some practice in drawing and editing multilines, what do you think? Perhaps it is difficult to see the full benefits of these marvelous tools without some actual experience. Let's go back to our floor plan and add some walls!

13.4.1 Do This: Add Some Walls

I. Open the *MyFlr-Pln 13* file you created in the C:\Steps\Lesson13 folder. (If this file is unavailable, open *FlrPln13* in the same folder.) The drawing looks like Figure 13.4.1a.

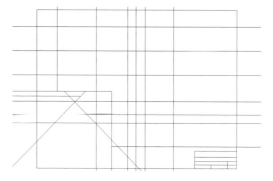

FIG. 13.4.1a: *MyFlr-Pln 13.dwg*

II. Follow these steps (do *not* explode the multilines!).

TOOLS	COMMAND SEQUENCE	STEPS
	**Command:** *la*	1. Set **Walls** as the current layer.
	Command: *ml*	2. Enter the **MLine** command by typing *mline* or *ml* at the command prompt. Or you may select the **Multiline** button on the Draw toolbar.
		3. Draw a $5^1/_2$" wide multiline to show the outer walls. Use the construction lines to guide you. (*Hints:* Set the multiline scale to **5.5** and use **Top** justification.) Your drawing looks like Figure 13.4.1.3.

continued

TOOLS	COMMAND SEQUENCE	STEPS

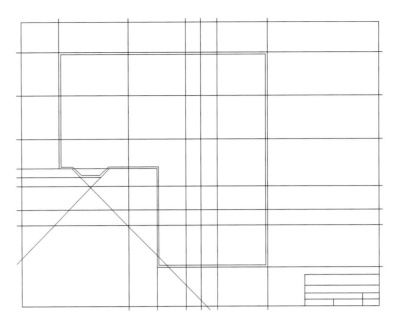

FIG. 13.4.1.3: Outer Walls

Command: *ml*

4. Draw 4" wide multilines to show the inner walls. Use the construction lines to guide you. (*Hints:* Set the multiline scale to **4** and use **Zero** justification.) Your drawing looks like Figure 13.4.1.4.

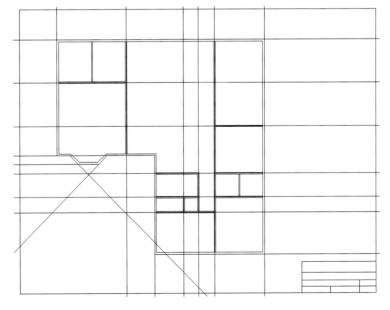

FIG. 13.4.1.4: All Walls

continued

TOOLS	COMMAND SEQUENCE	STEPS
		5. Erase the guidelines.
Cut All	**Command:** *mledit*	**6.** Cut the walls for doors and windows, as shown in Figure 13.4.1.6.
		All door openings are 36" except the smaller closet doors, which are 30". Window openings are 42". Door and window sizes and locations are provided in Figure 13.4.1.6.
		There are many possible ways to locate the openings. One suggestion follows (I will locate and size the left window on the back wall):
		6a. Load the *Points* Lisp routine found in the C:\Steps\Lesson11 folder.
	Command: *pt*	**6b.** Enter the command *pt* at the command line (this command has been defined by the Lisp routine you just loaded).
	PICK A BASE POINT	**6c.** AutoCAD asks you to **PICK A BASE POINT:**. Pick the left endpoint of the top wall.
	HOW FAR?: *12'7–³/₈*	**6d.** AutoCAD asks **HOW FAR:**. Enter the dimension given (*12'7-³/₈*). (Be sure your Running OSNAPs are off.)
	X, Y, OR P: *x*	**6e.** AutoCAD asks **X, Y, OR P:** (or do you want to go along the X-Plane, the Y-Plane, or in a Polar direction). Type *X*. AutoCAD places a node at that location.
	Command: *cp*	**6f.** We see by the number shown at this location that this is a 42" window. Copy the node 21" to the left and 21" to the right.
	Command: *mle*	**6g.** Using the multiline editing tools, break the wall using the nodes placed in Step 6f as guides. Then freeze the **Markers** layer.

continued

TOOLS	COMMAND SEQUENCE	STEPS

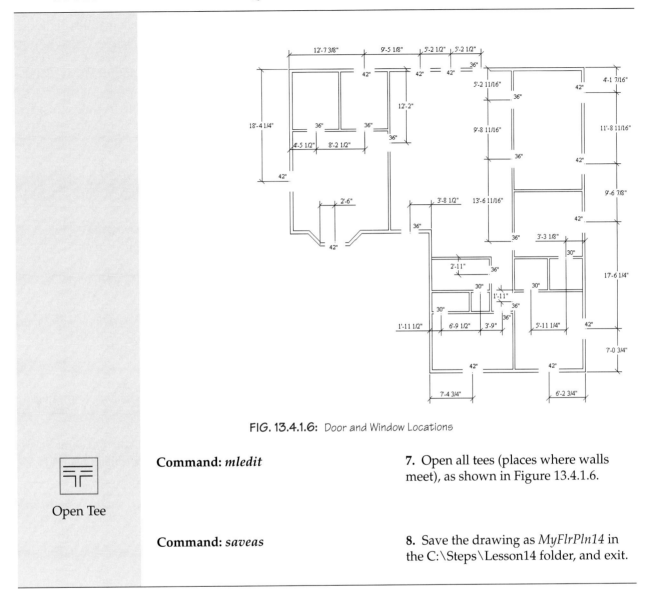

FIG. 13.4.1.6: Door and Window Locations

Open Tee	**Command:** *mledit*	7. Open all tees (places where walls meet), as shown in Figure 13.4.1.6.
	Command: *saveas*	8. Save the drawing as *MyFlrPln14* in the C:\Steps\Lesson14 folder, and exit.

Well, that was fun! Does your drawing look more like a floor plan now? Are you beginning to feel like a CAD operator?

13.5 Extra Steps

- Go to a new subdivision in your area and pick up some sample floor plans from the sales office. Practice drawing them on the computer. You do not have to be exact on the dimensions (you will only get rough dimensions on the plans anyway).

▌ I know it takes a while to define styles for multilines, but now is a good time to establish a policy of never doing anything twice that can be done once and placed in a template (remember templates?). Anything that requires definition should be defined and put into a template to save time later. But does that mean you will have thousands of templates from which to choose when creating a new drawing? Heavens, no! You should have a handful of templates—each with a lot of information.

Create some new templates that you might use on the job. Include units, grid and snap settings, limits, borders, and multiline definitions. But do not relax—there will be more to add to your templates later.

13.6 What Have We Learned?

Items covered in this lesson include:

▌ *Commands*
- *MLine*
- *MLStyle*
- *MLEdit*

This has been a fun lesson—and not nearly as difficult as its length might indicate.

I understand that you may be a bit frustrated that the last exercise did not go quite as fast as you might like. My instructions were not as detailed as some earlier exercises. I am beginning to count on you to make a necessary transition.

> Many instructors rely heavily on the step-by-step approach. Others refuse to use it at all. They argue that, in the end, all the student learns is how to follow steps. So they rely on student self-reliance to accomplish the necessary tasks.
>
> We are using a combination of the step-by-step approach and self-reliance. I lead you at first, relying more and more on what you learn as we go.

After practicing the additional floor plans in the preceding Extra Steps section, you probably noticed that your speed increased as your comfort with multilines increased. Eventually, multilines will prove themselves to be a timesaving tool in your arsenal.

In our next lesson, we will add some notes to our floor plan.

13.7 EXERCISES

1. Open the *FLRPLN-HVAC.dwg* file in the C:\Steps\Lesson13 folder. Add the ductwork shown in Figure 13.7.1. to the floor plan. Do not add the detail—it is there as a guide only.

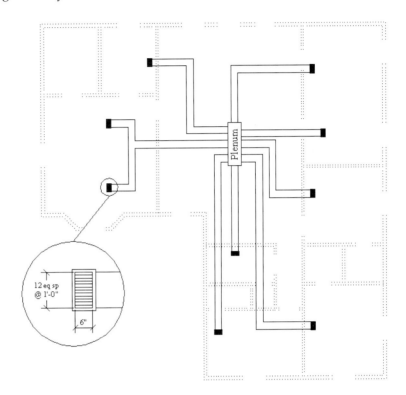

FIG. 13.7.1: *Flrpln-HVAC.dwg*

2. Create the cabinet face shown in Figure 13.7.2. The limits are 0,0 and 6',4'-6". The toe board is a solid.

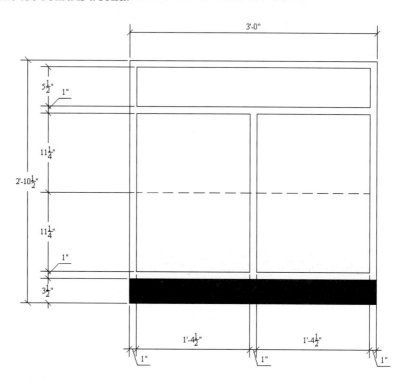

FIG. 13.7.2: Cabinet Face

3. Now try the piping spool drawing in Figure 13.7.3d. This is the drawing a welder will use to actually create sections of pipe. Use the details in Figures 13.7.3a to 13.7.3c as a guide. Can you determine the limits yourself? The drawing will be plotted on a 17" × 11" sheet of paper. Save the drawing as *MySpool* in the C:\Steps\Lesson13 folder.

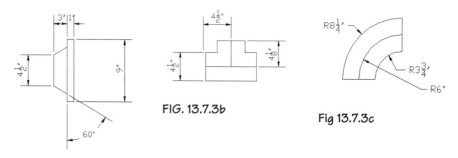

FIG. 13.7.3b

Fig 13.7.3c

FIG. 13.7.3a

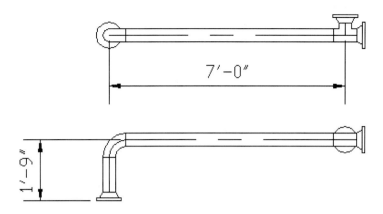

FIG. 3.7.3d: Pipe Spool

4. Setup a new drawing from scratch. Create the drawing in Figure 13.7.4b.

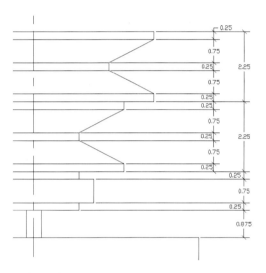

FIG. 13.7.4a: Dimension Details

- Use a 1=1 scale on a 36" × 24" sheet of paper.
- Set the grid to 1; set the snap as needed.
- Create layers as needed
- Use multilines whenever possible.
- The dimensions detail (Figure 13.7.4a) should help—but do not draw the dimensions.)
- Save the drawing as *MyMotorAssmbly.dwg* in the C:\Steps\Lesson13 folder.

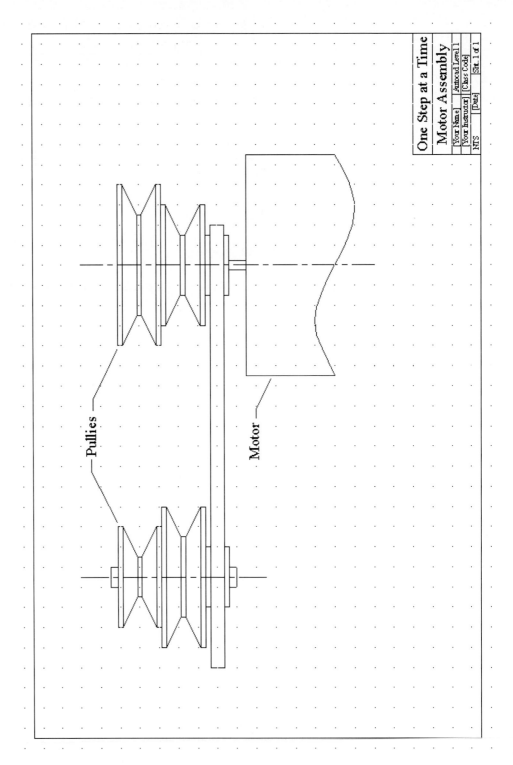

FIG. 13.7.4b: Motor Assembly

5. Start a new drawing from scratch. Create the drawing in Figure 13.7.5.
   - Use a $1/2$" grid on a standard 11" × 8.5" sheet of paper.
   - No scale is required.
   - Use the Times New Roman font.
   - Save the drawing as *MyDirections* in the C:\Steps\Lesson13 folder.

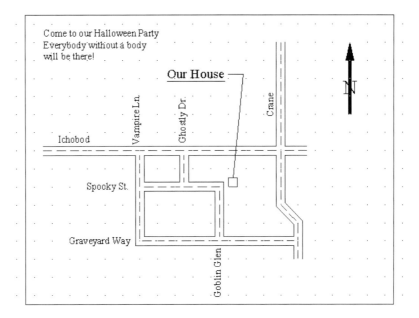

FIG. 13.7.5: *MyDirections.dwg*

6. Create the vessel drawing in Figure 13.7.6. The grid is 1" and the limits are 0,0 and 72",56". Save the drawing as *MyVessel* in the C:\Steps\Lesson13 folder.

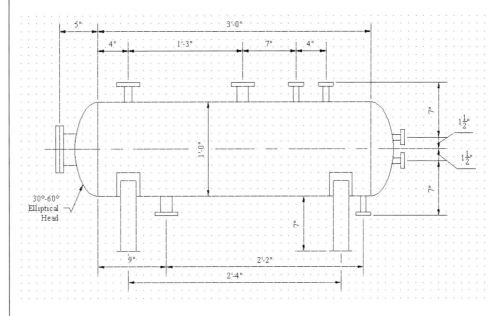

FIG. 13.7.6: *MyVessel.dwg*

7. Open the *MyPipingPlan13* drawing you created in Lesson 12 (in the C:\Steps\Lesson13 folder). (If this is not available, open *Piping Plan 13* in the same folder.) Refer to Figure 12.6.2f in Lesson 12 to see what it looks like.

   7.1. Add the pipe shown in Figure 13.7.7f. These notes might help:

   7.1.1. 8" pipe has an OD (outer diameter) of 8.625"

   7.1.2. 6" pipe has an OD of 6.625"

   7.1.3. 4" pipe has an OD of 4.5"

   7.1.4. The pipe coming from the vessel is 6"

   7.2. Use multilines whenever possible—**Load** the *Pipe* multiline style from the *MyMLines* style file in the C:\Steps\Lesson13 folder. (Use *MLines.mnl* if the *MyMLines.mnl* file is not available.)

   7.3. Some useful dimensions are shown in Figures 13.7.7a to 13.7.7e.

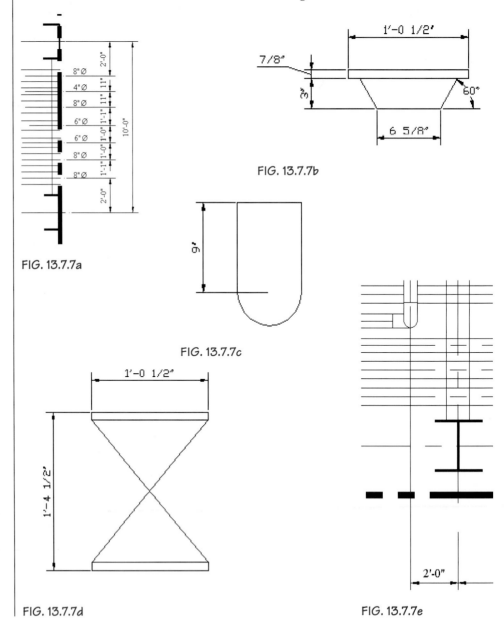

FIG. 13.7.7a

FIG. 13.7.7b

FIG. 13.7.7c

FIG. 13.7.7d

FIG. 13.7.7e

7.4. Save the drawing as *MyPipPln14* in the C:\Steps\Lesson14 folder.

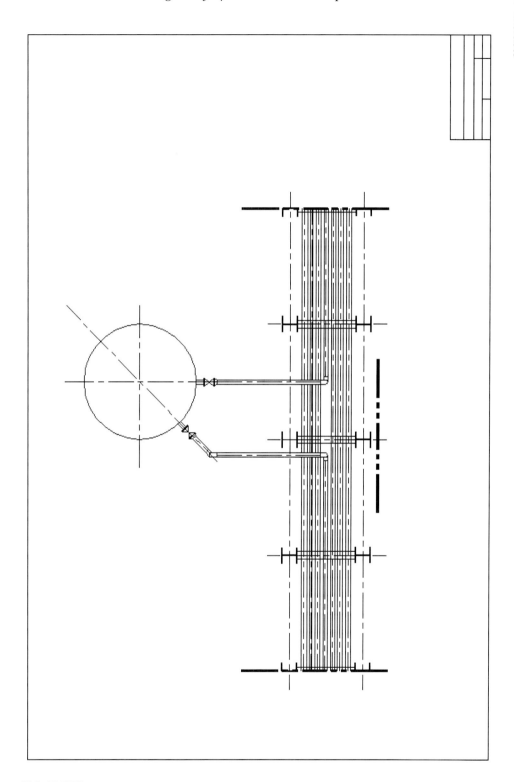

FIG. 13.7.7f: MyPipPln14

13.8 REVIEW QUESTIONS

Please write your answers on a separate sheet of paper.

1. Which mline option enables the user to draw multilines from a point centered between the two lines?

2. Must the user create a multiline style for each size of wall in an architectural drawing?

3. If not, what option is available to avoid creating several styles for different size walls?

4. The _____ option allows the user to switch between different multiline styles.

5. Multiline colors are set according to the _____ definition.

6. Define multiline styles using the _____ command.

7. to 12. Fill in the blank

Multiline Style Creation Procedure

a. _____ your new style.

b. _____ your new style to the current drawing.

c. Set the _____ for your new style.

d. Set the _____ for your new style.

e. If you want to make the style available to other drawings, _____ it to a file.

f. _____ your work.

13. Tell AutoCAD how many and what type of lines to include in your style in the _____ dialog box.

14. The extension for a file containing multiline style definitions is _____.

15. A line appearing down the middle of the multiline in the demo box indicates that AutoCAD will put a line in the multiline _____.

16. (T or F) The user cannot set different layers for individual lines in a multiline.

17. The tool intended for use in editing multilines is the _____ command.

18. Another option used to edit multilines is to _____ the multiline then use standard modifying tools.

19. In editing multilines, the (first, second) line selected is the one broken.

LESSON 14

Advanced Text: *MText*

Following this lesson, you will:

➥ Know how to create and edit paragraph (multiline) text: The **MText** Command

➥ Be familiar with AutoCAD's **Spell** Checker

➥ Know how to Find and Replace text in a drawing

In Lesson 5, we learned about *Text*, text *styles* and the *QText* command. All of this information will remain quite useful—indeed, *Text* will probably remain the primary method for creating call-outs on your drawings.

But using *Text* to create notes—that long list of construction information down the side of many drawings—can prove difficult. Editing a long list of notes can prove downright aggravating, especially if you must add or remove lines!

When I was on the boards, we often sweet-talked the project secretary into typing our notes. We then photocopied them onto sticky transparencies that we attached to our drawings. This worked well until revisions required additional notes or changing existing notes. Then we had to peel the sticky transparency from the drawing without damaging the paper.

With the introduction of the *MText* command, we have the benefits of the project secretary's typing abilities without the hassles of sticky transparencies. The secretary can type the notes using a computer's word processor (MS Word, WordPerfect, etc.), then give us the file to import into our drawings.

Of course, we can do the typing ourselves if the keyboard does not intimidate us. We can even share the secretary's dictionary to keep our spelling accurate (or at least consistent with the rest of the project). But why not take advantage of what is available—in many cases, this is a secretary's superior typing ability.

In this lesson, we will see how to create, import, and edit notes in an AutoCAD drawing using the Multiline Text Editor.

14.1 AutoCAD's Word Processor: The Multiline Text Editor

AutoCAD took great pains to make their word processor look and behave like other word processors designed for the Windows environment. Obviously, they could not incorporate the complete workings of an MSWord or WordPerfect; but the AutoCAD word processor looks and works very much like WordPad—the simple word processor that ships with Windows. So if you are familiar with the more common word processors on the market today, this lesson will be a snap!

> A *word processor* is the computer equivalent of a typewriter. It is what you use to type letters, notes, résumés and even AutoCAD books. There are several on the market, but the most common are Microsoft's Word and Novell's WordPerfect.

To access the Multiline Text Editor, follow this command sequence:

Command: *mtext* (or *t* or *mt*)
Current text style: STANDARD. Text height: 0.2000

Specify first corner: *[you will place a border around the area where you want the text to go; pick a corner of that border here]*

Specify opposite corner or [Height/Justify/Line Spacing/Rotation/Style/Width]: *[pick the other corner of the border]*

AutoCAD first lets you know the style and text height you are using. Then it prompts the user to place a border around the area in which to place the text. The **opposite corner** of the border is a bit tricky. There are a few options from which to choose—some refer to the bordered area, others refer to the text inside the border, and still others refer to both. Let's look at these options.

- The **Height** option allows the user to set the text height. It prompts the user:

  Specify height <0.2000>:

- The **Justify** option allows the user to justify text within the border. The options are the same as the *Text* justification options, except that **Fit** and **Align** are missing.

  The **Justify** option also controls how the multiline text object is placed in the drawing. For example, using the default **TL** (top left) option, AutoCAD anchors the text at the topmost and leftmost corner of the user-defined border. If the text entered is too large for the border, AutoCAD will automatically expand the border downward. Using the **BC** (bottom center) option causes the anchor point to be placed at the bottom and center of the border. Expansion of the border then occurs upward. Other justifications affect expansion in a similar manner.

- The **Line Spacing option**, new to AutoCAD 2000, allows the user the opportunity to control the spacing between lines of text.
- The **Rotation** option allows the user to control the rotation of the text just as it does within the *Text* command. It also controls the direction of expansion of the user-defined text border. For example, a downward-expanding border will actually expand toward the bottom of the text—or toward the right of the drawing when using a 90° text rotation (Figures 14.1a and 14.1b).
- The next option is simple enough. **Style** allows the user to specify the text style AutoCAD will use in the text box. Refer to Lesson 5 for more on *Style*.
- The **Width** option allows the user to define the width of the text border more precisely. This option is more in keeping with the precision priority of CAD use.

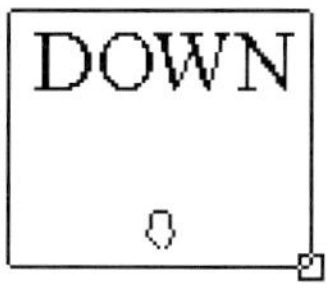

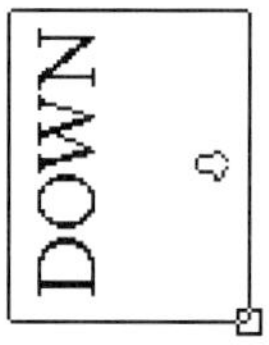

The user can reset each of these options from within the editor.

On completion of the text border, the Multiline Text Editor dialog box appears (Figure 14.1c). This resizable dialog box closely resembles that of a word processor and may be familiar to the experienced computer user.

FIG. 14.1a
0° Expansion

FIG. 14.1b
90° Expansion

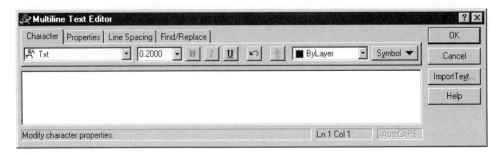

FIG. 14.1c: Multiline Text Editor

Let's take a look at the Multiline Text Editor.

▌ At the bottom of the dialog box is the status line, which contains useful information for the CAD operator.

- The left corner of the status line gives a brief description of the currently selected tab.
- The right corner provides a toggle for the AutoCAPS tool. When toggled **On**, AutoCAPS will convert any new text (typed or imported) into uppercase letters.
- To the left of the AutoCAPs toggle is the **Ln/Col** box that shows the location of the cursor in the text window. *Ln* refers to the number of the line down from the top, and *col* refers to the number of spaces over from the beginning (left end) of the line.

▌ The right side of the dialog box has four buttons.

- The top two buttons —**OK** and **Cancel**—are standard to most dialog boxes. **OK** sends the text entered in the dialog box to the drawing. **Cancel**, of course, cancels the command.
- The **Import Text** . . . button presents Windows' standard Open dialog box. The user can insert any text file or file saved in Rich Text Format (*.rtf). (This is how you will insert the file the project secretary types for you.)

A file created in most word processors (including WordPad) can be saved in simple Text (*.txt*) format. This format strips away any bold, italics, and underline formatting that may have been used in creating the file. Or the file can be saved in Rich Text Format (*.rtf*). This preserves the basic formatting.

Obviously, the user should opt for the *.rtf* format (if it is available) when importing text from outside AutoCAD. Whenever possible, then, the project secretary should save the file accordingly. Unfortunately, it will not always be possible. Therefore, in our exercises, we will use the more common *.txt* format when importing files so that we can study AutoCAD's methods of formatting text.

- The **Help** button calls AutoCAD's Command Reference dialog box—opened to the Multiline Text Editor page.

▌ You will notice that there are four tabs in the Multiline Text Editor.

- By default, the **Character** tab appears on top. The **Character** tab allows the user the opportunity to select the font independently of the current style. In fact, styles can be mixed within the multiline text, allowing the user more independence in creating the final appearance of the text.

  Depending on the font selected, the **Character** tab also gives the user access to the more common tools of emphasis in word processors—**Bold**, **Italicized**, or **Underlined** words. The user may also make use of color changes within the text for emphasis.

  Notice the **a/b** button on this tab. The user may highlight an area—anything that uses the front slash (/)—and pick this button to stack the text. This comes in handy when the user wants to stack fractions. (If the user enters fractions into the text—as in $^3/_8$—AutoCAD will prompt with a dialog box allowing the user the opportunity to automatically stack fractions.)

- The **Symbol** button will present the user with a simple way to insert the special characters we discussed back in Lesson 5. Additionally, the **Other** option (presented by the **Symbol** button) presents the Character Map dialog box (Figure 14.1d). This dialog box shows the user all of the specific symbols—letters, numbers, oddball shapes, and doodles available in a specific font. The lower right corner of the dialog box tells the user the keystrokes needed to insert the symbol when the font is current. We will see this in our exercises.

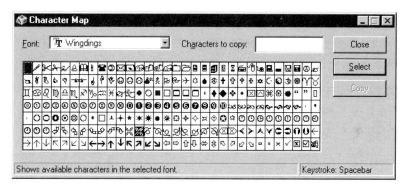

FIG. 14.1d: The Character Map (here showing the special characters available in the Wingdings font that ships with Windows)

▌ The next tab of the Multiline Text Editor dialog box is the **Properties** tab. When the user selects this, AutoCAD presents the options shown in Figure 14.1e.

FIG. 14.1e: Properties Tab of Multiline Text Editor

Using these control boxes, the user may set (or reset) any of the corresponding options first presented at the *MText* command prompt.

- Next to the **Properties** tab, you will find the **Line Spacing** tab (Figure 14.1f).

FIG. 14.1f: Line Spacing Tab of the Multiline Text Editor

The user can set the line spacing **At Least** or **Exactly** equal to typing standard spacings of 1, 1.5, or 2 using the control boxes here.

- The last tab is the **Find/Replace** tab (Figure 14.1g).

FIG. 14.1g: Find/Replace Tab of the Multiline Text Editor

The user can type a word, symbol, or group of words in the **Find** box, pick on the **Find** button (Binoculars), and be taken to the next occurrence of that word/group/symbol in the Multiline Text Editor. If the user types another word in the **Replace with** box, then picks the **Replace** button (A/B), AutoCAD will replace all occurrences (prompting for each) of the **Find** word(s) with the **Replace with** words. The user may specify that the word search be specific for **Case** (Capitals) and **Whole Word** (no plurals or parts of words) by putting a check in the appropriate box.

We will look at these tools in our exercise. But first, if you are not familiar with word processors, please study the chart in <u>Appendix D</u>. This chart shows the keystrokes for maneuvering through and manipulating the text in the Multiline Text Editor. Familiarity with these keystrokes will save a considerable amount of time in Multiline Text Editing.

Now let's put some text into our floor plan.

> The *Mtext* command is also available in the Draw pull-down menu. Follow this path:
>
> *Draw—Text—Multiline Text . . .*

14.1.1 Do This: Using Multiline Text

I. Open the *MyFlrPln14* drawing in the C:\Steps\Lesson14 folder. If this is not available, open the *FlrPln14* drawing in the same folder. The drawing looks like Figure 14.1.1a.

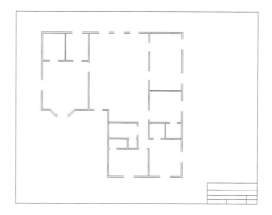

FIG. 14.1.1a: *FlrPln14.dwg*

II. Follow these steps.

TOOLS	COMMAND SEQUENCE	STEPS
	Command: *st*	1. Create a new text style called *Times*. Use the *Times New Roman* font. **Apply** it to the current drawing. (Refer to Lesson 5 if you need help.)
Multiline Text Button	Command: *la*	2. Create a layer called *Text*. Assign the color *yellow* to it and set it current.
	Command: *t*	3. Enter the *MText* command by typing *mtext, mt,* or *t* at the command prompt. Or you can pick the **Multiline Text** button on the Draw toolbar.
	Current text style: TIMES Text height: 3/16" **Specify first corner:** *69',62'*	4. AutoCAD displays the **Current text style** and **Text height**, then prompts for the **first corner** of the text border. Enter the coordinates shown.
	Specify opposite corner or [Height/Justify/Line spacing/ Rotation/Style/Width]: *w*	5. Tell AutoCAD you want to specify the **Width** of the text border . . .

continued

TOOLS	COMMAND SEQUENCE	STEPS
	Specify width: *15'*	**6.** . . . then set the **Width** to *15'*. AutoCAD presents the Multiline Text Editor dialog box with the *Times* style current (Figure 14.1.1.6).

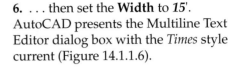

FIG. 14.1.1.6: Multiline Text Editor

7. Set the font size (text height) to 12". (The scale factor of a $1/4$" drawing is 48. The desired text height at plot time is $1/4$". $1/4$" × 48 = 12")

Bold Button

8. Pick the **Bold** button. Notice that it appears depressed when active.

9. Type *"The "*

10. Pick the **Italics** button.

Italics Button

11. Type *"Tara II "*

12. Pick the **Italics** button again.

13. Type *"Floor Plan"*, then hit *enter* to start a new line.

14. Pick the **OK** button. The upper right corner of your drawing looks like Figure 14.1.1.14.

continued

TOOLS	COMMAND SEQUENCE	STEPS

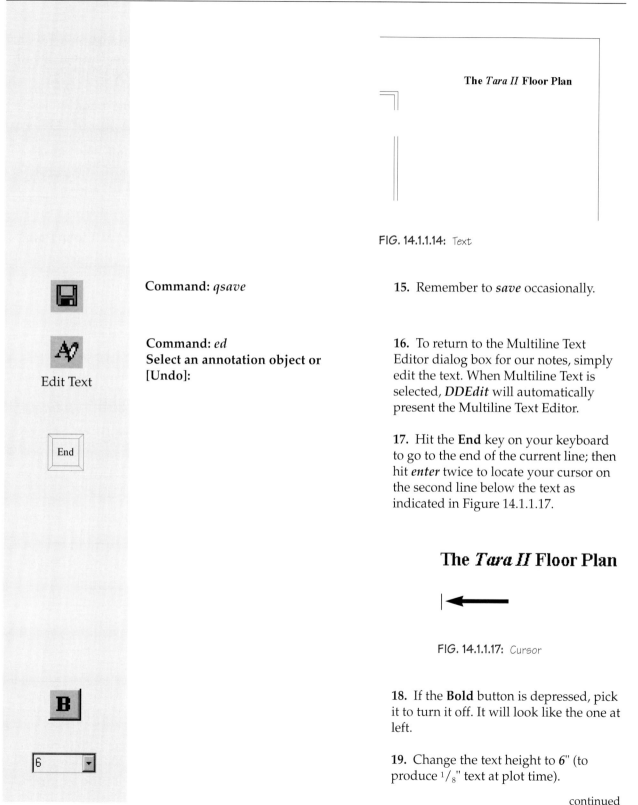

FIG. 14.1.1.14: Text

Command: *qsave*

15. Remember to *save* occasionally.

Command: *ed*
Select an annotation object or [Undo]:

16. To return to the Multiline Text Editor dialog box for our notes, simply edit the text. When Multiline Text is selected, ***DDEdit*** will automatically present the Multiline Text Editor.

17. Hit the **End** key on your keyboard to go to the end of the current line; then hit *enter* twice to locate your cursor on the second line below the text as indicated in Figure 14.1.1.17.

FIG. 14.1.1.17: Cursor

18. If the **Bold** button is depressed, pick it to turn it off. It will look like the one at left.

19. Change the text height to **6"** (to produce $1/8$" text at plot time).

continued

TOOLS	COMMAND SEQUENCE	STEPS
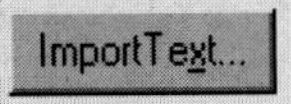 [ImportText...]		20. Now let's import some notes created by our summer intern. Pick the **Import text...** button. AutoCAD displays Windows' standard Open dialog box. Open the *Notes.txt* file found in the C:\Steps\Lesson13 folder.
 [OK]	Select an annotation object or [Undo]: *[enter]*	21. Pick the **OK** button, then hit *enter* to complete the command. Your text looks like Figure 14.1.1.21.

The *Tara II* Floor Plan

NOTES:
SQUARE FEET:
BEDROOMS:
BATHS:
HVAC: Entire central heating and cooling system or heat pump ready to operate
PLUMBING: Builder to hook-up to MUD utilities
ELECTRICAL: Builder to instal range, range hood, bathroom exhaust fan, clothes dryer hook-up, and cable television and telephone jacks
INSULATION: To be fiberglass; R14 min. in all walls and R24 min in ceiling.
WALLS: Interior walls to be gipsum with builder to tape and float; Buyer to select paint from contractor options
DOORS & WINDOWS: Interior door to be hollow, exterior doors to be solid wood; windows to be sun-blocked; builder to provide storm doors
CABINETS: Buyer to select from builder options
FLOORING: Buyer to select fom builder options

FIG. 14.1.1.21: Completed Exercise

TOOLS	COMMAND SEQUENCE	STEPS
 [save icon]	Command: *qsave*	22. Save the drawing, but do not exit.

We have seen how easy it is to create Multiline Text both by keyboard entry and by importation. We created our text using the *Times* style we set up earlier; but we could just as easily have used AutoCAD's default *Standard* style and changed fonts in the Multiline Text Editor itself.

We have also seen that we will use the Multiline Text Editor both to create and edit Multiline Text. Next, we will look at some of our options for changing, or editing, our text.

The *DDEdit* command is also available in the Modify pull-down menu. Follow this path:

> *Modify—Text . . .*

WWW 14.1.2 Do This: Editing Multiline Text

I. Be sure your are in the *MyFlrPln14* drawing in the C:\Steps\Lesson14 folder. If not, please open it now.

II. Follow these steps.

TOOLS	COMMAND SEQUENCE	STEPS
	Command: *ed*	1. Return to the Multiline Text Editor dialog box. (Select the existing text.)
		2. Locate the cursor to the left of the word, "NOTES".
 Right Arrow		3. Hold down the **Shift** key and hit the **Right arrow** key six times. (Or you can hold down the **Shift** key and hit the **End** key to select the whole line.) The word and following colon are selected.
  Bold & Underline		4. Pick the **Bold** and **Underline** buttons.
 Home Key Down Arrow		5. Hit the **Home** key followed by the **Down arrow** key. The text looks like Figure 14.1.2.5. ### The *Tara II* Floor Plan <u>**NOTES:**</u> SQUARE FEET: BEDROOMS: BATHS: **FIG. 14.1.2.5:** Edited Text

continued

TOOLS	COMMAND SEQUENCE	STEPS

Left Arrow

End Key

Bold Button

Down Arrow

Bold Button

Down Arrow

Bold Button

6. Hold down the **Shift** key and hit the down arrow three times. The next three lines are selected. Pick the **Bold** button.

7. Hit the **Left arrow**, the **End** key, then the **Bold** button. Your cursor is at the end of the **SQUARE FEET:** line.

8. Type *1950*.

9. Hit the **Down arrow** once, then the **Bold** button. Your cursor is now at the end of the **BEDROOMS:** line.

10. Type *Four*.

11. Hit the **Down arrow** once, then the **Bold** button. Your cursor is now at the end of the **BATHS:** line.

12. Type *Two*.

continued

TOOLS	COMMAND SEQUENCE	STEPS
[OK]	**Select an annotation object or [Undo]:** *[enter]*	13. Pick the **OK** button, then hit *enter* to complete the command. The text looks like Figure 14.1.2.13.

The *Tara II* Floor Plan

NOTES:
SQUARE FEET: 1950
BEDROOMS: Four
BATHS: Two
HVAC: Entire central heating and cooling system or heat pump ready to operate
PLUMBING: Builder to hook-up to MUD utilities

FIG. 14.1.2.13: More Edited Text

TOOLS	COMMAND SEQUENCE	STEPS
[disk icon]	**Command:** *qsave*	14. Remember to save often.
[A✓ icon]	**Command:** *ed*	15. Repeat Step 1.
		16. Let's look at some other ways to select text. Move the cursor to the beginning of the **HVAC:** line.
[→]		17. Hold down the **Shift** and **Control** keys while hitting the **Right arrow** twice. This is faster but requires more fingers.
[B]		18. Embolden the text.
[→]		19. Another way to select text is this: double click on the word **PLUMBING**. See that the word is selected but not the colon. Hold down the **Shift** key and hit the **Right arrow** key. Now the colon is selected as well.
[B]		20. Embolden the text.
		21. Embolden the rest of the capitalized words (include the colons): ELECTRICAL, INSULATION, WALLS, DOORS & WINDOWS, CABINETS, and FLOORING.

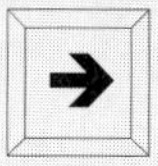

continued

TOOLS	COMMAND SEQUENCE	STEPS
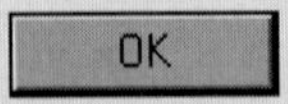 [OK]	Select an annotation object or [Undo]: *[enter]*	22. Pick the **OK** button then hit *enter* to complete the command. The multiline text looks like Figure 14.1.2.22.

The *Tara II* Floor Plan

NOTES:
SQUARE FEET: 1950
BEDROOMS: Four
BATHS: Two
HVAC: Entire central heating and cooling system or heat pump ready to operate
PLUMBING: Builder to hook-up to MUD utilities
ELECTRICAL: Builder to instal range, range hood, bathroom exhaust fan, clothes dryer hook-up, and cable television and telephone jacks
INSULATION: To be fiberglass; R14 min in all walls and R24 min in ceiling.
WALLS: Interior walls to be gipsum with builder to tape and float; Buyer to select paint from contractor options
DOORS & WINDOWS: Interior door to be hollow, exterior doors to be solid wood; windows to be sun-blocked; builder to provide storm doors
CABINETS: Buyer to select from builder options
FLOORING: Buyer to select fom builder options

FIG. 14.1.2.22: Completed Mtext

	Command: *qsave*	23. Save your drawing, but do not exit.

Note that you can also select large pieces of text by placing the cursor on one end of the text to select, holding down the left mouse button and dragging to the other end.

Now you have seen the ease with which you can edit multiline text. What do you think? Let's see what else the Multiline Text Editor can do.

14.1.3 Do This: Editing with the Cursor Menu

I. Be sure you are in the *MyFlrPln14* drawing in the C:\Steps\Lesson14 folder. If not, please open it now.

II. Follow these steps.

TOOLS	COMMAND SEQUENCE	STEPS
	**Command:** *ed*	**1.** Return to the Multiline Text Editor dialog box. (Select the existing text.)
		2. We want to move a section of text from one location to another. Let's take the **HVAC** section and place it just above the **CABINETS** section. First select the **HVAC** section (Figure 14.1.3.2) using one of the techniques mentioned in exercise 14.1.2.
		BATHS: Two **HVAC:** Entire central heating and cooling system or heat pump ready to operate **PLUMBING:** Builder to hook-up to MUD FIG. **14.1.3.2:** Select This Text
		3. Right-click anywhere in the text area of the Multiline Text Editor dialog box. AutoCAD will present a cursor menu. Select **Cut** on that menu.
		The highlighted text disappears. (If necessary, hit the **Delete** key on the keyboard to get rid of the empty line.)
		4. Move your cursor just to the left of the word **CABINETS:**.
		5. Right-click anywhere in the text area. Notice that **Cut** and **Copy** are grayed-out. These options are not available unless some text has been selected. Pick **Paste** on the menu. The **HVAC** text appears. (if the **CABINETS:** text is right behind it, hit the *enter* key to position it where it belongs.)

continued

TOOLS	COMMAND SEQUENCE	STEPS
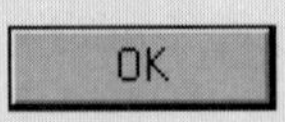 [OK]	Select an annotation object or [Undo]: *[enter]*	6. Pick the **OK** button and then hit *enter* to complete the command. The multiline text looks like Figure 14.1.3.6.

The *Tara II* Floor Plan

NOTES:
SQUARE FEET: 1950
BEDROOMS: Four
BATHS: Two
PLUMBING: Builder to hook-up to MUD utilities
ELECTRICAL: Builder to instal range, range hood, bathroom exhaust fan, clothes dryer hook-up, and cable television and telephone jacks
INSULATION: To be fiberglass; R14 min. in all walls and R24 min in ceiling.
WALLS: Interior walls to be gipsum with builder to tape and float; Buyer to select paint from contractor options
DOORS & WINDOWS: Interior door to be hollow; exterior doors to be solid wood; windows to be sun-blocked; builder to provide storm doors
HVAC: Entire central heating and cooling system or heat pump ready to operate
CABINETS: Buyer to select from builder options
FLOORING: Buyer to select fom builder options

FIG. 14.1.3.6: Completed Editing

TOOLS	COMMAND SEQUENCE	STEPS
 [disk icon]	Command: *qsave*	7. Save your drawing, but do not exit.

> The same cursor-menu technique will work for copying text rather than moving (cutting and pasting) it. You could also use the ***Ctrl + C*** then ***Ctrl + V*** method to copy and paste or the ***Ctrl + X*** then ***Ctrl + V*** method to cut and paste. This is actually a Windows redundancy carried into AutoCAD. It seems that nobody wants to tie users to a single approach to doing something!

Next, we will look at the **Search and Replace** capabilities of the Multiline Text Editor. You will really like this. The **Search** tool helps the user find specific text. In a large group of notes, this is a real timesaver. The **Replace** tool allows the user to replace that text with something new.

WWW 14.1.4 Do This: Search and Replace Text

I. Be sure you are in the *MyFlrPln14* drawing in the C:\Steps\Lesson14 folder. If not, please open it now.

II. Follow these steps.

TOOLS	COMMAND SEQUENCE	STEPS
(Find/Replace icon) Find/Replace	**Command:** *ed*	**1.** Return to the Multiline Text Editor dialog box. (Select the existing text.)
		2. Pick the **Find/Replace** tab. AutoCAD presents the options shown in Figure 14.1.4.2.

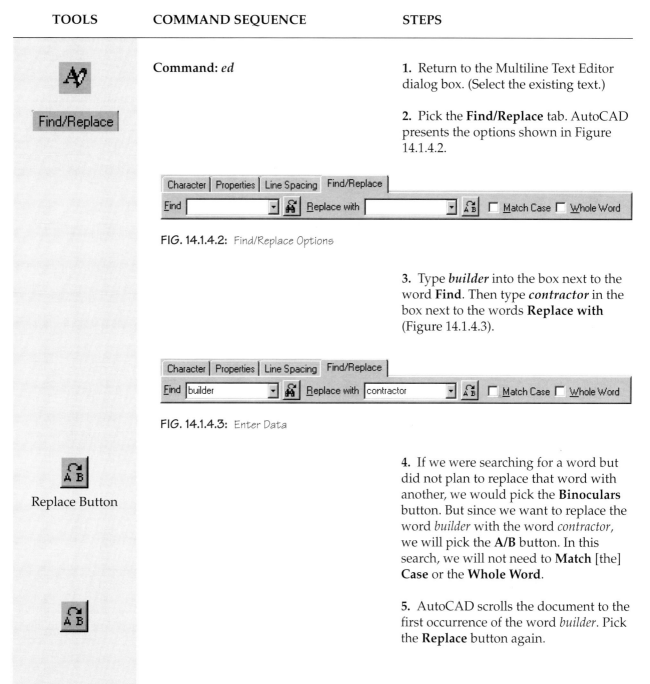

FIG. 14.1.4.2: Find/Replace Options

3. Type *builder* into the box next to the word **Find**. Then type *contractor* in the box next to the words **Replace with** (Figure 14.1.4.3).

FIG. 14.1.4.3: Enter Data

(A/B icon) Replace Button		**4.** If we were searching for a word but did not plan to replace that word with another, we would pick the **Binoculars** button. But since we want to replace the word *builder* with the word *contractor*, we will pick the **A/B** button. In this search, we will not need to **Match** [the] **Case** or the **Whole Word**.
(A/B icon)		**5.** AutoCAD scrolls the document to the first occurrence of the word *builder*. Pick the **Replace** button again.

continued

TOOLS	COMMAND SEQUENCE	STEPS
		6. Repeat Step 4 for each occurrence of the word *builder*. When AutoCAD can find no more occurrences, the status line of the Multiline Editor dialog box will display the message in Figure 14.1.4.5.
		No occurrences of builder found.
		FIG. 14.1.4.5: No More Occurrences Found
[OK]	**Select an annotation object or [Undo]:** *[enter]*	7. Pick the **OK** button and then hit *enter* to complete the command. The multiline text looks like Figure 14.1.4.7.
		The *Tara II* Floor Plan
		NOTES: SQUARE FEET: 1950 BEDROOMS: Four BATHS: Two PLUMBING: contractor to hook-up to MUD utilities ELECTRICAL: contractor to instal range, range hood, bathroom exhaust fan, clothes dryer hook-up, and cable television and telephone jacks INSULATION: To be fiberglass; R14 min. in all walls and R.24 min in ceiling. WALLS: Interior walls to be gipsum with contractor to tape and float; Buyer to select paint from contractor options DOORS & WINDOWS: Interior door to be hollow, exterior doors to be solid wood; windows to be sun-blocked; contractor to provide storm doors HVAC: Entire central heating and cooling system or heat pump ready to operate CABINETS: Buyer to select from contractor options FLOORING: Buyer to select fom contractor options
		FIG. 14.1.4.7: Completed Text Replacement
[save icon]	**Command:** *qsave*	8. Save your drawing, but do not exit.

Before we finish with the Multiline Text Editor, let's take a quick look at that **Symbol** button. As I mentioned earlier, this comes in handy when you want to insert something a bit out of the ordinary.

WWW 14.1.5 Do This: Symbol Insertion

I. Be sure you are in the *MyFlrPln14* drawing in the C:\Steps\Lesson14 folder. If not, please open it now.

II. Follow these steps.

TOOLS	COMMAND SEQUENCE	STEPS
	**Command:** *ed*	1. Return to the Multiline Text Editor dialog box. (Select the existing text.) 2. Move your cursor to the blank area just below **The *Tara II* Floor Plan**. 3. Pick the **Symbol** button. 4. Notice the options. Most of these are familiar to you from our Lesson 5 study of the *Text* command. (The Non-breaking Space will insert a space that will not allow the words on either side to be split between lines.) Select the **Other** . . . option. AutoCAD presents the Character Map, as shown in Figure 14.1d (p. 509).
	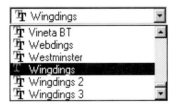	5. Pick the down arrow in the **Font** control box. Scroll till you find the *Wingdings* font; select it.
 Happy Face		6. Now select the happy face you will find in the second row, eleventh column. Notice that the lower right corner of the dialog box gives you the key (*J*) that will produce this character when using the *Wingdings* font.
		7. Pick the **Close** button. AutoCAD returns to the Multiline Text Editor.

continued

TOOLS	COMMAND SEQUENCE	STEPS
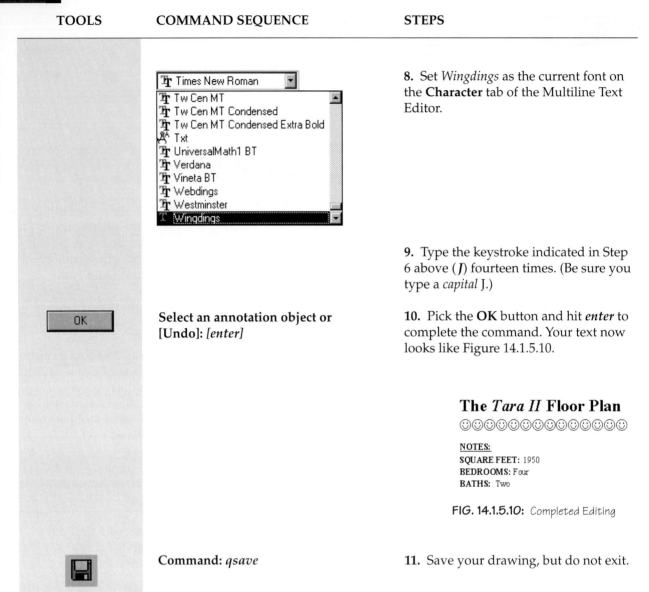		8. Set *Wingdings* as the current font on the **Character** tab of the Multiline Text Editor.

9. Type the keystroke indicated in Step 6 above (*J*) fourteen times. (Be sure you type a *capital* J.) |
| | Select an annotation object or [Undo]: *[enter]* | 10. Pick the **OK** button and hit *enter* to complete the command. Your text now looks like Figure 14.1.5.10.

The *Tara II* Floor Plan
☺☺☺☺☺☺☺☺☺☺☺☺☺☺
NOTES:
SQUARE FEET: 1950
BEDROOMS: Four
BATHS: Two

FIG. 14.1.5.10: Completed Editing |
| | Command: *qsave* | 11. Save your drawing, but do not exit. |

14.2 Okay I Typed It, but I Don't Know If It's Right!: AutoCAD's *Spell* Command

The next aspect of text we must consider is AutoCAD's **Spell Checker**. This remarkably simple tool can provide an inestimable service to those of us who never made it to the national spelling bee. It works in much the same way that spell checkers work in the major word processors.

The command sequence is simply:

Command: *spell* (**or** *sp*)
Select objects: *[select the text object or objects]*
Select objects: *[enter]*

Rather than select individual pieces of text or multiline text, the user can type *all* at the **Select objects:** prompt. AutoCAD will then spell-check the entire drawing.

After the user selects some text, AutoCAD presents the Check Spelling dialog box (Figure 14.2a). This box offers the same options as most spell checkers, but it also gives the user the opportunity to change dictionaries. This can prove quite valuable to the project. It means that the project lead can assign a single person to control the project dictionary (usually that wonderful typist—the project secretary). That person creates the custom dictionary using a word processor. The secretary must then save the dictionary with a *.cus* extension. AutoCAD needs this extension for recognition but your word processor should have no problem reading the file as well. (See the Extra Steps in Section 14.4 for a procedure to set up a custom dictionary.)

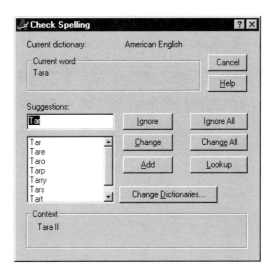

FIG. 14.2a: Check Spelling Dialog Box

- If AutoCAD finds a word that is not in its dictionary, it presents the word in the **Current word** frame. It makes a few suggestions as to what word it thinks you may be trying to spell in the **Suggestions** list box with the word it thinks that you are most likely trying to spell highlighted in the text box.
- The user may pick a button to **Ignore** this word or **Ignore All** (ignore it throughout this checking session), **Change** the word or **Change All** (change it every time it occurs in the selected text), or **Add** it to the dictionary.
- The user can also type a word into the **Suggestions** text box and pick the **Lookup** button to have AutoCAD check its spelling.

- If the user wants to use the project dictionary, the **Change Dictionaries** ... button provides that opportunity.

Let's take a look at AutoCAD's spell checker in action.

> The *Spell* command is also available in the Tools pull-down menu. Follow this path:
>
> *Tools—Spelling*

14.2.1 Do This: Checking Your Spelling

I. Be sure your are in the *MyFlrPln14* drawing in the C:\Steps\Lesson14 folder. If not, please open it now.

II. Follow these steps.

TOOLS	COMMAND SEQUENCE	STEPS
No Button Available	**Command:** *sp*	1. Enter the *Spell* command by typing *spell* or *sp* at the command prompt.
	Select objects:	2. The spell checker works on all types of text in the drawing. Select the Multiline Text we have been using.
	Select objects: *[enter]*	3. Confirm your selection. AutoCAD presents the Check Spelling dialog box (Figure 14.2a). It identifies **Tara** as being misspelled.
 Add		4. We know that **Tara** is spelled correctly and will probably occur again somewhere in the project documents, so let's add it to the dictionary. Pick the **Add** button.

continued

TOOLS	COMMAND SEQUENCE	STEPS

5. (Note: AutoCAD may indicate that our special font—JJJJJ, etc.—is a misspelling. If so, pick the **Ignore** button.) AutoCAD looks for the next misspelling and presents the word **hook-up**. It shows that the correct spelling should be **hookup**. Our summer intern thought the first spelling was correct and might have used it more than once in our text, so let's pick the **Change All** button to be sure we fix every occurrence.

6. Next AutoCAD finds **instal**. This was probably a typo so let's just pick the **Change** button.

7. Change **hook-up,** to **hookup,** (apparently, the comma prevented the earlier fix).

8. Now AutoCAD finds the abbreviation **min**. Normally, we would like to put a period after an abbreviation, but the project standard is to omit the period. So we will **Ignore** this prompt.

9. Change **gipsum** to **gypsum**.

10. **Ignore All** occurrences of the hyphenated words **sun-blocked**, and the abbreviation **HVAC:**. (Normally, we would **Add** HVAC; but in this instance, the colon would go with it.)

11. It looks like we omitted the **r** in the word **from**. Scroll down the list box to find the word **from** at the bottom of the list. Pick on it and notice that it appears in the text box above.

12. Pick on the **Change** button to fix the word.

continued

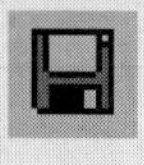

TOOLS	COMMAND SEQUENCE	STEPS
		13. Now AutoCAD presents a message box telling you that it has completed the spell check (Figure 14.2.1.13).
		FIG. 14.2.1.13: Message Box
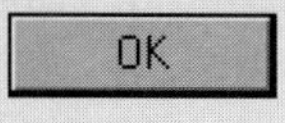 OK		Pick the **OK** button.
		14. Your text now looks like Figure 14.2.1.14.
		FIG. 14.2.1.14: Completed Exercise
💾	**Command:** *qsave*	**15.** Save your drawing, but do not exit.

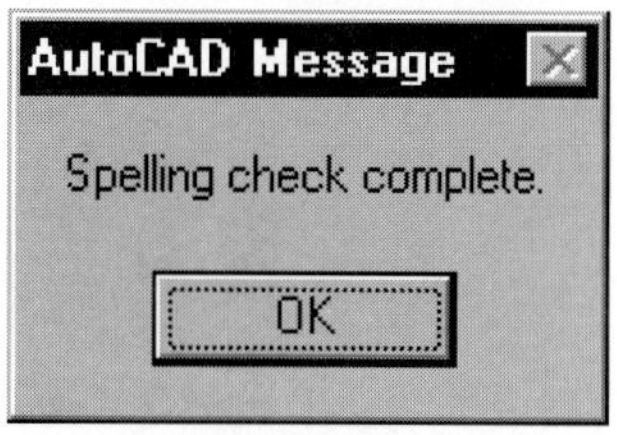

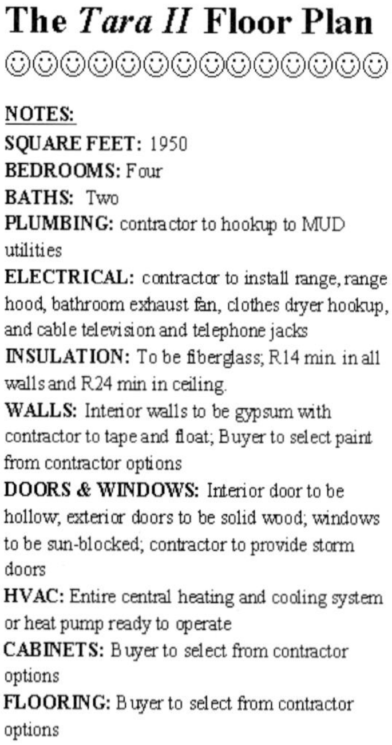

The *Tara II* Floor Plan
☺☺☺☺☺☺☺☺☺☺☺☺

NOTES:
SQUARE FEET: 1950
BEDROOMS: Four
BATHS: Two
PLUMBING: contractor to hookup to MUD utilities
ELECTRICAL: contractor to install range, range hood, bathroom exhaust fan, clothes dryer hookup, and cable television and telephone jacks
INSULATION: To be fiberglass; R14 min in all walls and R24 min in ceiling.
WALLS: Interior walls to be gypsum with contractor to tape and float; Buyer to select paint from contractor options
DOORS & WINDOWS: Interior door to be hollow, exterior doors to be solid wood; windows to be sun-blocked; contractor to provide storm doors
HVAC: Entire central heating and cooling system or heat pump ready to operate
CABINETS: Buyer to select from contractor options
FLOORING: Buyer to select from contractor options

Now complete the text as seen in Figure 14.2b as follows:

- Place all text on the **Text** layer.
- Text heights are: 12", 9", and 6".
- Put your own information into the title block (school name, your name, date, etc.).
- When you have finished, save the drawing as *MyFlrPln16* in the C:\Steps\Lesson16 folder.

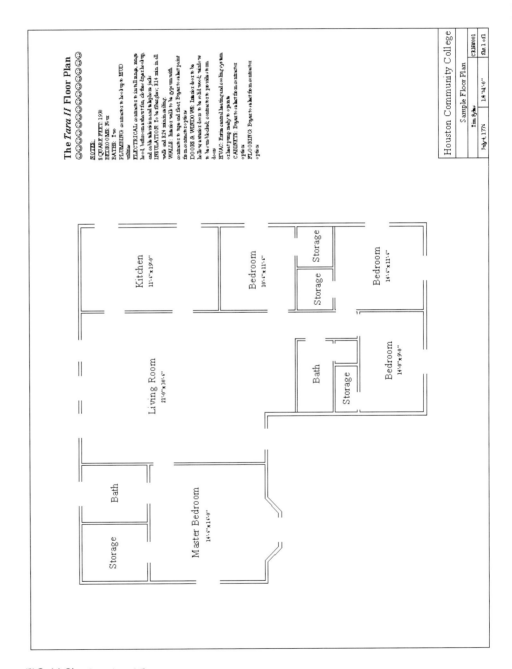

FIG. 14.2b: Completed Text

14.3 Find and Replace

A new feature of AutoCAD 2000 is the ability to perform **Search and Replace** on text, regardless of whether or not it was placed as text or mtext. The Multiline Text Editor is not required and the search can be done on the entire drawing.

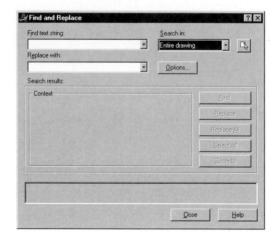

FIG. 14.3a: Find and Replace Dialog Box

The command *Find* is the same to find text or to find and replace text. It presents the Find and Replace dialog box in Figure 14.3a.

Let's take a look at it.

- Place the text you wish to locate in the **Find text string:** control box. Text that has been previously entered can be selected using the down arrow.
- Place the new text (the text with which you wish to replace the found text) in the **Replace with:** control box. Again, text that has been previously entered can be selected using the down arrow.
- Use the **Search in:** control box or the **Select objects** button (next to the control box) to define your search area (select where you want to search).
- The **Options**... button presents the Find and Replace Options dialog box (Figure 14.3b). Here the user can filter the selection set to include (or not include) certain types of text. A great benefit is the inclusion of the **Block Attribute Value** option that allows the user to find and replace attribute values (more on attributes in Lesson 20).

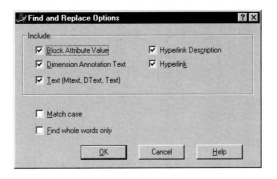

FIG. 14.3b: Find and Replace Options Dialog Box

The user can also tell AutoCAD to **Match case** (match capital letters in the text) or **Find whole words only** in the search.

- The **Search results:** frame presents the same buttons you found in the spell checker with the addition of a **Zoom to** button. I am sure you guessed that this button will cause AutoCAD to zoom in to the located text.

Let's try the **Find and Replace** tool.

14.3.1 Do This: Find and Replace without the Multiline Text Editor

I. Be sure you are in the *MyFlrPln14* drawing in the C:\Steps\Lesson14 folder. If not, please open it now.

II. Zoom all.

III. Follow these steps.

TOOLS	COMMAND SEQUENCE	STEPS
Find and Replace Button	Command: *find*	1. Enter the *Find* command by typing *find* at the command prompt. Or you can pick the **Find and Replace** button on the Standard toolbar.
	Find text string: Storage Replace with: Closet	2. Let's replace all the **Storage** call-outs with the word **Closet**.

continued

TOOLS	COMMAND SEQUENCE	STEPS
		3. Tell AutoCAD to **Replace All** occurrences of the word **storage**. AutoCAD responds in the message box, as shown in Figure 14.3.1.3.

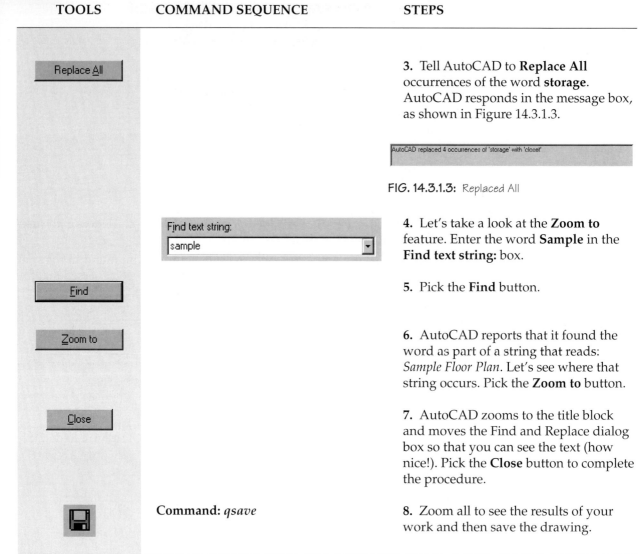

FIG. 14.3.1.3: Replaced All

4. Let's take a look at the **Zoom to** feature. Enter the word **Sample** in the **Find text string:** box.

5. Pick the **Find** button.

6. AutoCAD reports that it found the word as part of a string that reads: *Sample Floor Plan*. Let's see where that string occurs. Pick the **Zoom to** button.

7. AutoCAD zooms to the title block and moves the Find and Replace dialog box so that you can see the text (how nice!). Pick the **Close** button to complete the procedure.

Command: *qsave*

8. Zoom all to see the results of your work and then save the drawing.

14.4 Extra Steps

You can impress the boss by being the only one on the contract capable of setting up the project to use a shared dictionary. Here is how to do it. (If you do not have access to MS Word 8.0, go to Step 2.)

1. First create the dictionary in Word.
    - **1.1.** Go to the **Tools** pull-down menu in Microsoft Word. Select the **Spelling and Grammar . . .** option.
    - **1.2.** Pick the **Options** button. Word presents the Spelling and Grammar dialog box.
    - **1.3.** Pick the **Dictionaries** button. Word presents the Custom Dictionaries dialog box.

1.4. Pick the **New** button. Word presents the Create Custom Dictionary dialog box.

1.5. Next to the words **Save as type:**, pick the down arrow. Set the box to **All Files (*.*)**.

1.6. Save the dictionary as *MyCustom.cus* (you will need to include the extension) to the C:\Steps\Lesson14 folder (or to a folder on the office network to share it properly). (MSWord will create the new dictionary using all the words in its default dictionary. But the user can add to or modify the new dictionary as desired.)

1.7. Word returns to the Custom Dictionaries dialog box. Be sure there is a check beside **MyCustom** so Word will use it. Pick the **OK** button.

1.8. Word returns to the Spelling and Grammar dialog box. Set the **Custom Dictionary:** to *MyCustom*.

1.9. Pick the **OK** button.

1.10. Word returns to the Spelling and Grammar dialog box. Pick the **X** in the upper right corner of the dialog box to exit.

2. Next you must tell AutoCAD to use the *MyCustom* dictionary.

   2.1. On the Check Spelling dialog box you encountered in Section 14.2, pick the **Change Dictionaries...** button. AutoCAD presents the Change Dictionaries dialog box.

   2.2. Pick the **Browse** button. AutoCAD presents a Windows standard Open dialog box (this one is called Select Custom Dictionary). Open the *MyCustom.cus* file in the C:\Steps\Lesson14 folder (or the file you created in Step 1; or the *Custom.cus* file if neither is available).

   2.3. AutoCAD returns to the Change Dictionaries dialog box. Pick the **Apply & Close** button.

   2.4. AutoCAD returns to the Check Spelling dialog box. You may now continue with the spell check or exit the dialog box.

This little trick is guaranteed to impress the boss and save quite a bit of project time. Good luck!

14.5 What Have We Learned?

Items covered in this lesson include:

- *Multiline text creation and editing*
- *AutoCAD's spell checker*
- *How to find and replace text*
- *Commands:*
  - *Mtext*
  - *Spell*
  - *Find*

If you were already familiar with word processors, this has probably been a fairly easy—if not downright boring—lesson. The only new thing for you would have been the Extra Steps part of the lesson in which we learned to share dictionaries.

But if you were not familiar with word processors, this has probably been one of the easier—almost fun—lessons.

The decision to use **MText** or *Text* will not always be an easy one. *Text* is less complicated (once the styles have been created); **MText** has more capabilities. I would suggest that any time it is a toss up, go with the easier *Text* (assuming, of course, that the style you need has already been created). AutoCAD, however, tends to favor the **MText** command. So it is really a question of preference. Do what is easier for you.

Take a break so that you will be fresh when you start the next lesson. In Lesson 15, we will look at dimensioning!

14.6 EXERCISES

1. Open the *MyHolder2.dwg* file you created in Lesson 12. It is in the C:\Steps\Lesson12 folder. (Or you can use the *Holder14.dwg* in the C:\Steps\Lesson14 folder.) Create the notes shown in Figure 14.6.1.
   - **1.1.** Use the Times New Roman font.
   - **1.2.** Use a text height of $1/8$".
   - **1.3.** The width of the text box is 3.25".
   - **1.4.** Save the drawing as *MyHolder14.dwg* in the C:\Steps\Lesson14 folder.

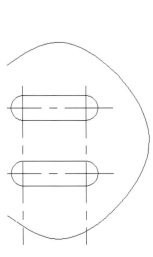

FIG. 14.6.1: *MyHolder14.dwg*

2. Open the *MyDirections.dwg* file you created in Lesson 13. It is in the C:\Steps\Lesson13 folder. (Or you can use the *Directions2.dwg* file in the C:\Steps\Lesson14 folder.) Create the notes shown in Figure 14.62.

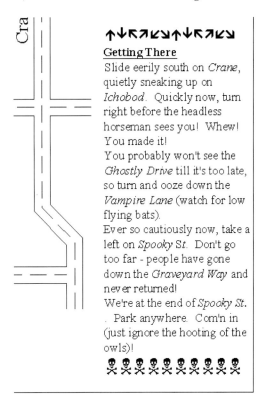

FIG. 14.6.2: *MyDirections.dwg*

2.1. Find the arrows at top in the Wingdings font. Make them 0.15" in height.
2.2. Find the images at the bottom in the Wingdings font. Make them 0.25" in height.
2.3. All other text uses the Times New Roman font at a height of 0.125".
2.4. The width of the text box is 2.25".
2.5. Save the drawing as *MyDirections2.dwg* in the C:\Steps\Lesson13 folder.

3. Add the text shown to the *MyVessel* drawing (Figure 14.6.3) you created in Lesson 13. It is in the C:\Steps\Lesson13 folder.

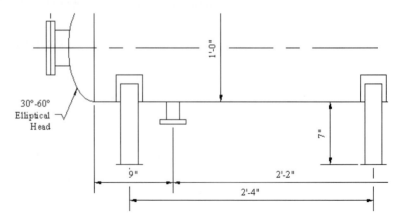

NOTES:
1. Design Temp: 550°
2. Design Pressure: 200PSIG
3. 1/8" Corrosion Allowance for shell and heads
4. Fireproofing by client.
5. Bolt holes to straddle N/S centerline.
6. All nozzles are 300# except where noted.
7. Paint per client spec. 1003254789SPN

FIG. 14.6.3: *MyVessel.dwg*

4. Open the *Star Soot Trap* drawing in the C:\Steps\Lesson14 folder and add the text shown in Figure 14.6.4.

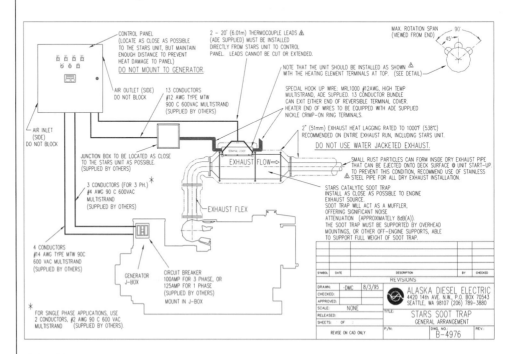

FIG. 14.6.4: *Star Soot Trap.dwg* (Special thanks to Alaska Diesel Electric of Seattle, Washington, and Mr. Lee Pangilinan for allowing the use of this drawing.)

5. Add the text and notes to the piping plan on which we have been working. See the completed drawing in Figure 14.6.5g. The file is *My Piping Plan 14.dwg* in the C:\Steps\Lesson14 folder. If that file is not available, use the *Piping Plan 14.dwg* file located in the same folder.

   5.1. When plotted, the text heights should be $1/4$", $3/16$", and $1/8$". The drawing will plot on a $3/8$" = 1'−0" scale. All text uses the *Times* style.

   5.2. The images in Figures 14.6.5a to 14.6.5f will help you see the text better.

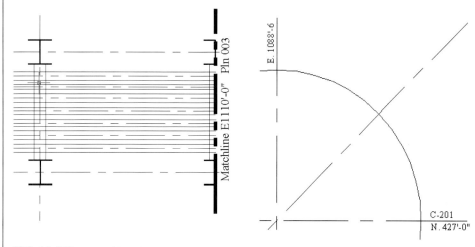

FIG. 14.6.5a: Matchline Callout

FIG. 14.6.5b: Vessel Coordinates

NOTES:
1. High and low point vents and drains required on all pipe.
2. See drawings *SPC 4001* through *SPC 40009* for Piping Support Standards.
3. See drawings *INS2022* through *INS2025* for Instrumentation Installation Details.
4. **Hold drawing for Stress Analysis.**
5. See customer spec *TXCO502331-103a/b* for construction specifications.
6. HPFS is 101'-0"

FIG. 14.6.5c: Notes. Insert the *Piping Notes.txt* found in the C:\Steps\Lesson 14 folder. Then modify it to look like this.

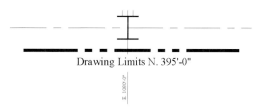

FIG. 14.6.5d: Drawing Limits Callout

8"-C2-109-J51
4"-C2-108-J51
8"-CWS-102-JPH52
6"-CWR-102-JPH52
6"-MOU-110-J52
8"-C3-105-J51
8"-C3-106-J51
Matchline E1050'-0" Pln 001

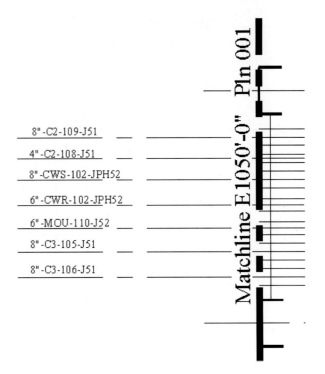

FIG. 14.6.5e: Line Number Callouts

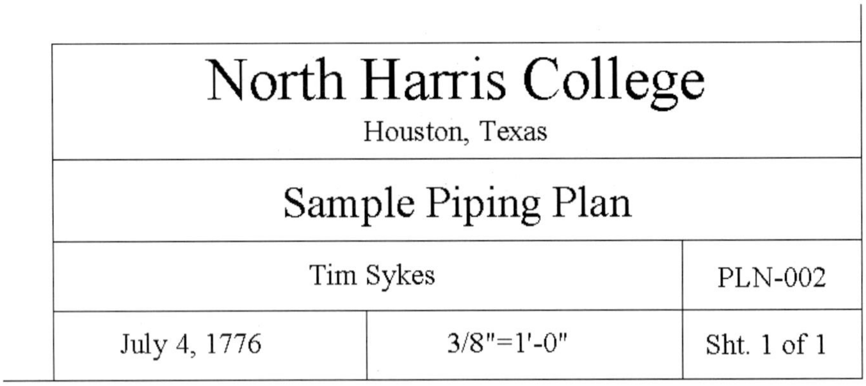

North Harris College
Houston, Texas
Sample Piping Plan
Tim Sykes
PLN-002
July 4, 1776
3/8"=1'-0"
Sht. 1 of 1

FIG. 14.6.5f: Title Block

14.6 Exercises

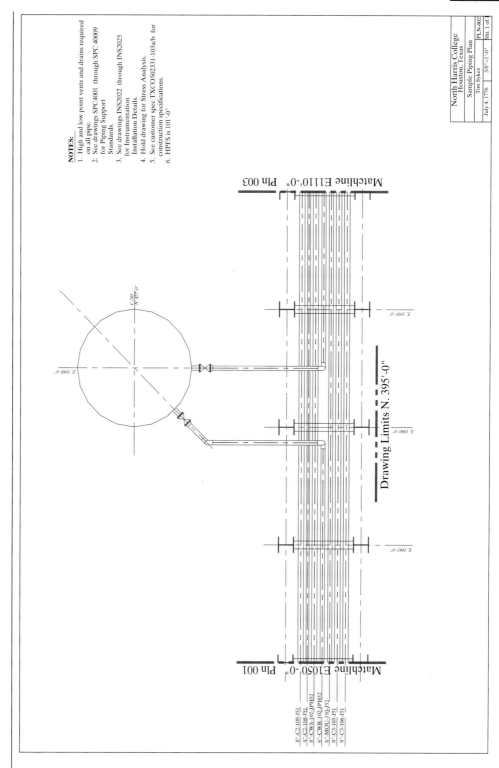

FIG. 14.6.5g: Completed Drawing

14.7 REVIEW QUESTIONS

Please write your answers on a separate sheet of paper.

1. A _____ is the computer equivalent of a typewriter.

2. The AutoCAD word processor looks and works very much like _____—the simple word processor that ships with Windows.

3. To access AutoCAD's Multiline Text Editor, enter the _____ command.

4. To anchor text at the bottom center of the multiline text border, I would first select the _____ option in the mtext command, then use the BC option.

5. If the text entered is too large for the default border, AutoCAD will automatically expand the border _____.

6. When using a 90° text rotation, the multiline text border will expand in which direction?

7. (T or F) The user can reset each of the mtext options from within the Multiline Text Editor.

8. and 9. What are the two types of files—listed by file extension—that can be imported into AutoCAD's Multiline Text Editor?

10. to 13. List the four ways the user can emphasize text in the Multiline Text Editor.

14. Use the _____ button on the Multiline Text Editor to insert special characters.

15. Pick the _____ tab on the Multiline Text Editor to change the justification of the text.

16. Pick the _____ button on the Find/Replace tab of the Multiline Text Editor to replace one word with another.

17. When selecting objects for AutoCAD's Spell Checker, type _____ to tell AutoCAD to spell check the entire drawing.

18. To share the project dictionary, it should be given a _____ extension.

LESSON 15

Basic Dimensioning

Following this lesson, you will:
➥ Know how to create Dimensions in AutoCAD using a host of dimensioning commands
➥ Know the difference between Associative and Normal Dimensions
➥ Know how to edit dimensions using AutoCAD's **Dimedit** and **Dimtedit** commands
➥ Know how to dimension an isometric

Creating dimensions on a drawing board requires a certain amount of expertise, a bit of patience, a calculator, and a good eye. If the dimension does not fit between dimension lines just right, the drafter can always scrunch it in a bit. If the drafter is experienced, he or she can often do this without the results being flagrantly noticeable. The board drafter can also cheat on a dimension fairly easily. Indeed, the company for which I worked in the early eighties maintained a 3" plus or minus tolerance for dimensioning.

AutoCAD dimensioning also requires a certain amount of expertise. But in contrast to board work, the CAD operator needs a lot of patience, no calculator (AutoCAD will perform the calculations), and AutoCAD's "eye" for precision. If a dimension does not fit between dimension lines just right, some (often) complex maneuvers are required to reposition it. The user can cheat on a dimension fairly easily, but it involves overriding AutoCAD's precision and is rarely a good idea.

AutoCAD's dimensioning tools come in three types: dimension *Creation*, dimension *Editing*, and dimension *Customization*.

- In the dimension Creation category, AutoCAD provides several different dimensioning commands and a leader command. Use these to actually draw the dimensions.

- The Editing category includes two dimension-editing commands. Use these to change, reposition, or reorient the dimension.

- The Customization category is one that will allow the user to create dimension styles. Creating dimension styles involves several tabs of a dialog box for each style you create. You will use these tabs to set at least 60 dimension variables (dimvars).

But as complicated as dimensioning is, it is not as complicated as it seems! We will cover the Creation and Editing categories in this lesson. But we will save Customization for Lesson 16.

Let's get right to it.

15.1 First, Some Terminology

If you are an experienced drafter, you may already be familiar with dimension terminology. But I have provided the drawing in Figure 15.1a for the novice (and as a review for those who have been drafting for so long that dimensioning has become second nature—even though the textbook terminology has been long forgotten).

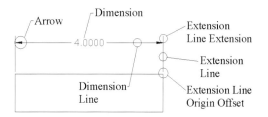

FIG. 15.1a: Dimension Terminology

The dimension shown in this drawing, and those used throughout this lesson, use AutoCAD defaults for everything from location of the *Dimension* to the size of the *Extension Line Origin Offset*. You can see that AutoCAD uses Arrows by default rather than the slashes more often seen in architectural drafting. AutoCAD also defaults to decimal units (regardless of the units the drawing has been set up to use), a 0.18" text height, a 0.18" *Extension Line Extension* and *Arrow* length, and an *Extension Line Origin Offset* of 0.0625".

A new term with which you should become familiar is *Associative*. Associative Dimensions are tied to the objects being dimensioned. As the user stretches those objects, the dimension will change with the object. (Notice the new dimension as the box is stretched in Figure 15.1b.) Additionally, when Associative Dimensioning is active, AutoCAD uses true dimensions and does not prompt the user for verification.

FIG. 15.1b: Associative Dimensions Change with the Object

But perhaps the most important thing to remember about Associative Dimensioning is that, like polylines and splines, Associative Dimensions behave as a single object. That is, a single pick will select the dimension for erasure or modification. The problem with this lies in the way the single dimension object responds to modification commands like **Break** and **Trim**. Simply put, the user cannot break or trim the extension lines to allow room for text. The experienced CAD operator, like the experienced drafter, knows that this is often necessary in a crowded drawing.

AutoCAD begins with Associative Dimensioning activated by default.

To turn Associative Dimensioning **Off**, enter the following at the command prompt:

Command: *dimaso*
Enter new value for DIMASO <On>: *off*

Another name for nonassociative dimensioning is *Normal* Dimensioning. Unlike Associative Dimensioning, AutoCAD creates Normal Dimensions as separate objects (lines, arrows, text, etc.). Therefore, the user can break or trim the extension lines as needed. Normal Dimensioning requires the user to manually update the dimension when he edits the objects dimensioned. (In other words, normal dimensions will not automatically update as the object is stretched.) Additionally, although Normal Dimensioning reads the distance between the extension lines just as Associative Dimensioning does, AutoCAD will prompt the user with the true dimension and give the user the opportunity to override it with another number.

See the following chart for a comparison of Associative and Normal Dimensions.

MODIFICATION	ASSOCIATIVE DIMENSION	NORMAL DIMENSION
Dimension will automatically update when stretched	Yes	No
Extension Lines can be trimmed or broken to allow room for text in a crowded drawing	No	Yes
Prompts the user to allow the true dimension to be overridden	No	Yes

The question of which type of dimension to use will usually be decided between the CAD coordinator for the job (the guru) and the project manager. The CAD operator should check with one of them before entering the wrong type of dimension since, although it can be tedious to convert from Associative to Normal, you cannot convert the other way at all!

To convert from Associative to Normal, *explode* the dimension (just as you exploded a polyline into lines). You will then find it necessary to change the layer of the separate objects to the dimension layer and the color of the objects to **ByLayer**.

15.2 Dimension Creation—Dimension Commands

15.2.1 Linear Dimensioning

One of the first things I tell my students when we study dimensions is to use the Dimension toolbar whenever possible. AutoCAD's dimension commands can be difficult to remember, but the toolbar has pictures to show what each does. I suggest activating the Dimension toolbar and docking it on the top of the screen (see Lesson 1 if you need help).

The workhorse of AutoCAD dimensioning is the *Dimlinear* command. It will draw all the straight horizontal and vertical dimensions. The command sequence is

> **Command:** *dimlinear* (or *dimlin* or *dli*)
> **Specify first extension line origin or** <select object>: *[select the first dimension point]*
> **Specify second extension line origin:** *[select the second dimension point]*
> **Specify dimension line location or [Mtext/Text/Angle/Horizontal/Vertical/Rotated]:** *[locate the dimension line]*
> **Dimension text** = X.XXXX *[AutoCAD reports the dimension value]*
> **Command:**

- The first option AutoCAD presents is easily missed. AutoCAD prompts for the **First extension line origin**. Most users stop there, select the **First** and **Second extension line origins** and proceed with the dimension. But the second part of that first prompt reads **or** <**select object**>. This option allows the user to simply select an object to dimension. If the user presses enter and selects an object, AutoCAD determines where the object's endpoints lie and places the extension lines accordingly.

The next set of options occurs after the extension line origins are located.

- AutoCAD will determine automatically if the dimension should be **Horizontal** or **Vertical** by the location of the crosshairs. However, the user may override AutoCAD's determination by typing **H** or **V**.
- The **Text** and **MText** options allow the user to override AutoCAD's automatic determination of the actual dimension.

    The dimension will not automatically update if the user overrides here. I do not recommend doing this as it defeats the purpose of Associative Dimensioning. (*Note:* The **MText** prompt will not appear if Associative Dimensioning is turned **Off**.)
- The **Angle** option allows the user to rotate the dimension text for a better fit. This handy tool is far too often overlooked (Figure 15.2.1a).

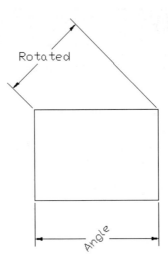

FIG. 15.2.1a: Angled versus Rotated Dimensions

- The **Rotated** option allows the user to measure a dimension at an angle to the object other than 90°. (See Figure 15.2.1a for a comparison of the **Angle** and **Rotated** options.)

All of the commands in this lesson can be found in the **Dimension** pull-down menu.

Let's look at the *Dimlinear* command.

15.2.1.1 Do This: Linear Dimensioning

I. Open the *Ex-Linear.dwg* file in the C:\Steps\Lesson15 folder. The drawing looks like Figure 15.2.1.1a.

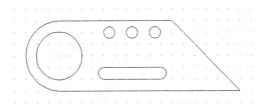

FIG. 15.2.1.1a: *Ex-Linear.dwg*

II. Follow these steps.

TOOLS	COMMAND SEQUENCE	STEPS
 Linear Dimension Button	Command: *dli*	1. Enter the *Dimlinear* command by typing *dimlinear*, *dimlin*, or *dli* at the command prompt. Or you can pick the **Linear Dimension** button on the Dimension toolbar.
	Specify first extension line origin or \<select object\>: _cen of	2. Pick the Center of the circle at point 1 (Figure 15.2.1.1.2) . . .
		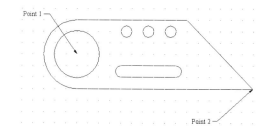 **FIG. 15.2.1.1.2:** Extension Line Origins
	Specify second extension line origin: _endp of	3. . . . then the endpoint at point 2 (Figure 15.2.1.1.2).
	Specify dimension line location or [Mtext/Text/Angle/Horizontal/Vertical/Rotated]: Dimension text = 8.0000	4. Locate the dimension line four grid points below the bottom line of the object (Figure 15.2.1.1.4).
		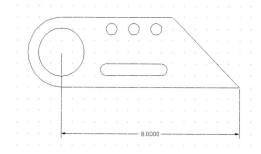 **FIG. 15.2.1.1.4:** Linear Dimension

continued

TOOLS	COMMAND SEQUENCE	STEPS
	Command: *[enter]*	5. Repeat the *Dimlinear* command.
	Specify first extension line origin or <select object>: *[enter]*	6. Let's use the **select object** option. Hit *enter*.
	Select object to dimension:	7. Select the angled line on the right end of the object.
	Specify dimension line location or [Mtext/Text/Angle/Horizontal/Vertical/Rotated]:	8. Pick a point four grid points to the right (Figure 15.2.1.1.8).
	Dimension text = 3.0000	

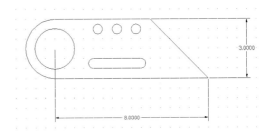

FIG. 15.2.1.1.8: Completed Dimesioning

	Command: *saveas*	9. Save the drawing, but do not exit.

■ 15.2.2 Dimensioning Angles

Our next dimension command is *Dimangular*. You will use this to dimension angles as well as for angular dimensions on circles and arcs. The command sequence for dimensioning an arc or circle is

> **Command:** *dimangular* (or *dimang* or *dan*)
> **Select arc, circle, line, or <specify vertex>:** *[show AutoCAD what you want to dimension or hit enter to select the vertex first]*
> **Specify second angle endpoint:** *[this prompt appears only if a circle is being dimensioned]*

Specify dimension arc line location or [Mtext/Text/Angle]: *[locate the dimension line]*
Dimension text = *[AutoCAD reports and places the dimension]*
Command:

If the user selects a **line** at the first prompt, AutoCAD prompts:

Select second line:

The user must select a line that forms an angle with the first selected line, then locate the dimension as prompted.
If the user opts to select the vertex first, AutoCAD prompts:

Specify angle vertex: *[select the vertex]*
Specify first angle endpoint: *[select a point on the first line]*
Specify second angle endpoint: *[select a point on the second line]*

The *MText*, *Text*, and *Angle* options are the same as in the *Dimlinear* command. Let's look at the *Dimangular* command.

15.2.2.1 Do This: Dimensioning Angles

I. Be sure you are in the *Ex-Linear.dwg* file in the C:\Steps\Lesson15 folder. If not, open it now.

II. Follow these steps.

TOOLS	COMMAND SEQUENCE	STEPS
 Angular Dimension Button	**Command:** *dan*	1. Enter the *Dimangular* command by typing *dimangular*, *dimang*, or *dan* at the command prompt. Or you can pick the **Angular Dimension** button on the Dimension toolbar.
	Select arc, circle, line, or <specify vertex>:	2. Select the arc on the left end of the object.

continued

TOOLS	COMMAND SEQUENCE	STEPS
	Specify dimension arc line location or [Mtext/Text/Angle]: **Dimension text = 180**	3. Place the dimension, at the end of the bottom linear dimension, as shown in Figure 15.2.2.1.3.

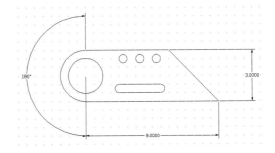

FIG. 15.2.2.1.3: Arc Dimension

	Command: *[enter]*	4. Now let's dimension the angle at the other end of the object. Repeat the command.
	Select arc, circle, line, or <specify vertex>:	5. Select the bottom line on the object . . .
	Select second line:	6. . . . then the angled line.
	Specify dimension arc line location or [Mtext/Text/Angle]: **Dimension text = 45**	7. Notice that the angle changes as you change the mouse position. Place the dimension line at the bottom arrow of the vertical dimension (Figure 15.2.2.1.7).

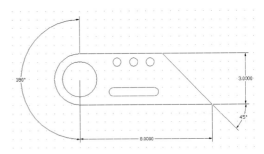

FIG. 15.2.2.1.7: Dimensioned Angle

continued

TOOLS	COMMAND SEQUENCE	STEPS
🖫	**Command:** *qsave*	**8.** Save the drawing, but do not exit.

15.2.3 Dimensioning Radii and Diameters

There is very little difference between dimensioning radii and dimensioning diameters. In fact, the command sequences are identical:

> **Command:** *dimradius* (or *dimrad* or *dra*)
> [OR *dimdiameter* (or *dimdia* or *ddi*)]
> **Select arc or circle:** *[show AutoCAD what you want to dimension]*
> **Dimension text** = *[AutoCAD reports the dimension]*
> **Specify dimension line location or [Mtext/Text/Angle]:** *[locate the dimension]*
> **Command:**

AutoCAD will automatically place either a diameter symbol (ø) or an **R** in front of the dimension to indicate what has been dimensioned.

WWW 15.2.3.1 Do This: Dimensioning Diameters and Radii

I. Be sure you are in the *Ex-Linear.dwg* drawing file in the C:\Steps\Lesson15 folder. If not, open it now.

II. Follow these steps.

TOOLS	COMMAND SEQUENCE	STEPS
🧽	**Command:** *e*	**1.** Erase the 180° arc dimension.
⌖ Radius Dimension Button	**Command:** *dra*	**2.** Enter the *Dimradius* command by typing *dimradius*, *dimrad*, or *dra* at the command prompt. Or you can pick the **Radius Dimension** button on the Dimension toolbar.
	Select arc or circle:	**3.** Select the arc at the left end of the object.

continued

TOOLS	COMMAND SEQUENCE	STEPS
	Dimension text = 1.5000 **Specify dimension line location or [Mtext/Text/Angle]:**	**4.** Locate the dimension as shown in Figure 15.2.3.1.4. 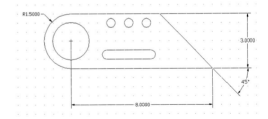 **FIG. 15.2.3.1.4:** Radius Dimension Notice that AutoCAD places the dimension and a *Center Mark* locating the center of the arc being dimensioned.
 Diameter Dimension Button	**Command:** *ddi*	**5.** Enter the *Dimdiameter* command by typing *dimdiameter*, *dimdia*, or *ddi* at the command prompt. Or you can pick the **Diameter Dimension** button on the Dimension toolbar.
	Select arc or circle:	**6.** Select the large circle to the left inside the object.
	Dimension text = 2.0616 **Specify dimension line location or [Mtext/Text/Angle]:**	**7.** Place the dimension as shown in Figure 15.2.3.1.7. 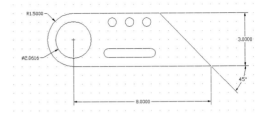 **FIG. 15.2.3.1.7:** Diameter Dimension
	Command: *qsave*	**8.** Save the drawing, but do not exit.

■ 15.2.4 Dimension Strings

It is often preferable to *string* dimensions. This makes it easier for the contractor to find and read dimensions, and it enhances the overall appearance of the drawing.

To string dimensions, we can repeat the ***Dimlinear*** command over and over, or we can begin our string with the ***Dimlinear*** command and then follow it with the ***Dimcontinue*** command. The ***Dimcontinue*** command will place dimensions along the same line begun by the ***Dimlinear*** command. With each selection, the second extension line from the previous dimension is used as the first extension line for the continued string. Some of the many benefits of this approach include:

- Not overwriting the extension line. Too many lines atop one another may cause plotting problems later.
- The user does not have to locate the dimension line with each new selection. AutoCAD simply continues along the previous string.
- The command automatically repeats until the user stops it. This saves the time required to repeat the command.

The command sequence is

Command: *dimcontinue* (or *dimcont* or *dco*)

Specify a second extension line origin or [Undo/Select] <Select>: *[simply select the origin of the second extension line—AutoCAD assumes the first extension line origin to be the second extension line of the last dimension entered]*

Dimension text = *[AutoCAD reports the dimension]*

Specify a second extension line origin or [Undo/Select] <Select>: *[AutoCAD repeats the command until the user hits enter to exit]*

Select continued dimension: *[when the user hits enter at the first prompt, AutoCAD will provide the opportunity to select a different dimension from which to continue; hitting enter at this prompt will exit the command]*

15.2.4.1 Do This: String Dimensions

I. Be sure you are in the *Ex-Linear.dwg* drawing file in the C:\Steps\Lesson15 folder. If not, open it now.

II. Follow these steps.

TOOLS	COMMAND SEQUENCE	STEPS
	Command: *dli*	1. Create a linear dimension between the center of the larger circle and the center of the leftmost arc on the slot (Figure 15.2.4.1.1). Be sure to select the circle first.

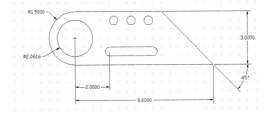

FIG. 15.2.4.1.1: Linear Dimension

 Continue Dimension Button	**Command:** *dco*	2. Enter the *Dimcontinue* command by typing *dimcontinue*, *dimcont*, or *dco* at the command prompt. Or you can pick the **Continue Dimension** button on the Dimension toolbar.
	Specify a second extension line origin or [Undo/Select] <Select>: _cen of	3. Select the center of the arc on the other end of the slot.
	Dimension text = 2.5000	4. Hit *enter* twice to complete the command. Your drawing looks like Figure 15.2.4.1.4.
	Specify a second extension line origin or [Undo/Select] <Select>: *[enter]*	
	Select continued dimension: *[enter]*	

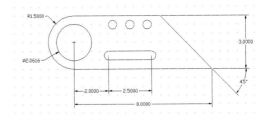

FIG. 15.2.4.1.4: Continued Linear Dimension

continued

TOOLS	COMMAND SEQUENCE	STEPS
	Command: *dli* **Specify first extension line origin or <select object>:** _cen of *[middle small circle]* **Specify second extension line origin:** _cen of *[left small circle]* **Specify dimension line location or [Mtext/Text/Angle/Horizontal/ Vertical/Rotated]:** *a* **Specify angle of dimension text:** *60* **Specify dimension line location or [Mtext/Text/Angle/Horizontal/ Vertical/Rotated]:** **Dimension text = 1.0000**	**5.** Now let's use what we know to dimension the upper circles. Put a linear dimension between the two left smaller circles (pick the middle one first). **6. Angle** the dimension at 60°. Your drawing looks like Figure 15.2.4.1.6. 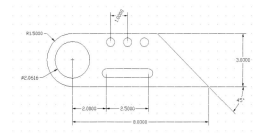 **FIG. 15.2.4.1.6:** Angled Linear Dimension
	Command: *dco* **Specify a second extension line origin or [Undo/ Select] <Select>:** _cen of **Dimension text = 2.2500** **Specify a second extension line origin or [Undo/Select] <Select>:** *[enter]* **Select continued dimension:** **Specify a second extension line origin or [Undo/Select] <Select>:** _cen of	**7.** Continue the dimension to the center of the large circle. **8.** We want to continue the dimension in the other direction now, so hit *enter* to go to the **Select continued dimension:** prompt. **9.** Select the right extension line (above the smaller middle circle). **10.** Select the center of the smaller right circle.

continued

TOOLS	COMMAND SEQUENCE	STEPS
	Dimension text = 1.0000	**11.** Exit the command. Your drawing looks like Figure 15.2.4.1.11.
	Specify a second extension line origin or [Undo/Select] <Select>: *[enter]*	
	Select continued dimension: *[enter]*	

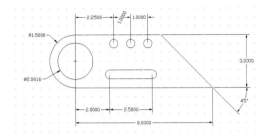

FIG. 15.2.4.1.11: Continued Dimensions

12. Now complete the dimensioning as shown in Figure 15.2.4.1.12.

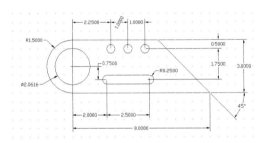

FIG. 15.2.4.1.12: Completed Drawing

	Command: *qsave*	**14.** Save the drawing and exit.

■ 15.2.5 Aligning Dimensions

Nonlinear objects often require nonlinear dimensioning. This usually means that the dimension must be aligned with the object for clarity.

The ***Dimaligned*** command works almost like the ***Dimlinear*** command. The difference is that the aligned dimension parallels the first and second *extension line origins.* The command sequence is

> **Command:** *dimaligned* (or *dimali* or *dal*)
> **Specify first extension line origin or <select object>:** *[select the first dimension point]*

Specify second extension line origin: *[select the second dimension point]*
Specify dimension line location or
[Mtext/Text/Angle]: *[locate the dimension line]*
Dimension text = *[AutoCAD reports and places the dimension]*
Command:

The options are identical to those used in the *Dimlinear* command.

WWW 15.2.5.1 Do This: Aligned Dimensions

I. Open the *Ex-base.dwg* drawing file in the C:\Steps\Lesson15 folder. The drawing looks like Figure 15.2.5.1a.

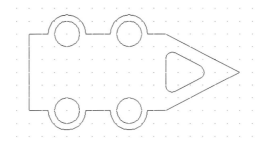

FIG. 15.2.5.1a: *Ex-Base.dwg*

II. Follow these steps.

TOOLS	COMMAND SEQUENCE	STEPS
Aligned Dimension Button	**Command:** *dal*	1. Enter the *Dimaligned* command by typing *dimaligned*, *dimali*, or *dal* at the command prompt. Or you can pick the **Aligned Dimension** button on the Dimension toolbar.
	Specify first extension line origin or <select object>: *[enter]*	2. Hit *enter* to select the object to dimension.
	Select object to dimension:	3. Select the upper angled line on the right end of the object.

continued

TOOLS	COMMAND SEQUENCE	STEPS
	Specify dimension line location or [Mtext/Text/Angle]: **Dimension text = 3.3541**	4. Locate the dimension line—three grid points above the object (Figure 15.2.5.1.4).

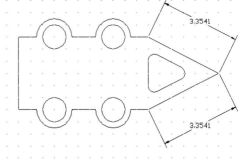

FIG. 15.2.5.1.4: Aligned Dimension

	Command: *[enter]*	5. Repeat the command.
	Specify first extension line origin or <select object>: *[enter]* **Select object to dimension:** **Specify dimension line location or [Mtext/Text/Angle]:** **Dimension text = 3.3541**	6. Place a dimension for the lower angled line, as shown in Figure 15.2.5.1.6.

FIG. 15.2.5.1.6: Completed Dimensions

	Command: *qsave*	7. Save the drawing, but do not exit.

15.2.6 Baseline Dimensions

Like the *Dimcontinue* command, *Dimbaseline* works from an original linear dimension. But where continued dimensions were based on the *second* extension line origin of the last or selected linear dimension, the baseline dimension starts from the same *first* extension line origin of the last or selected linear dimension. The command sequence parallels that of the *Dimcontinue* command.

> **Command:** *dimbaseline* (or *dimbase* or *dba*)
>
> **Specify a second extension line origin or [Undo/Select] <Select>:** *[select the origin of the second extension line—AutoCAD assumes the first extension line origin to be the first extension line of the last dimension entered]*
>
> **Dimension text** = *[AutoCAD reports and inserts the dimension]*
>
> **Specify a second extension line origin or [Undo/Select] <Select>:** *[AutoCAD repeats the command until the user hits enter to exit]*
>
> **Select base dimension:** *[when the user hits enter at the first prompt, AutoCAD will provide the opportunity to select a different dimension from which to base the baseline dimensions; hitting enter at this prompt will exit the command]*

Let's try a baseline dimension.

15.2.6.1 Do This: Baseline Dimensions

I. Be sure you are in the *Ex-base.dwg* drawing file in the C:\Steps\Lesson15 folder. If not, open it now.

II. Follow these steps.

TOOLS	COMMAND SEQUENCE	STEPS
	Command: *dli*	1. Draw a linear dimension, as shown in Figure 15.2.6.1.1. Be sure to select the extension line origin on the right first. 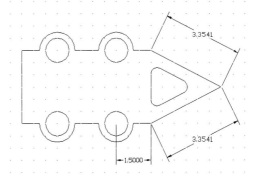 **FIG. 15.2.6.1.1:** Linear Dimension

continued

TOOLS	COMMAND SEQUENCE	STEPS
 Baseline Dimension Button	**Command:** *dba*	**2.** Enter the *Dimbaseline* command by typing *dimbaseline*, *dimbase*, or *dba* at the command prompt. Or you can pick the **Baseline Dimension** button on the Dimension toolbar.
	Specify a second extension line origin or [Undo/Select] <Select>: **Dimension text = 4.0000**	**3.** Pick the center of the next bolt hole to the left. AutoCAD reports and places the dimension.
	Specify a second extension line origin or [Undo/Select] <Select>: **Dimension text = 5.5000**	**4.** Pick the lower left endpoint of the object.
	Specify a second extension line origin or [Undo/Select] <Select>: *[enter]* **Select base dimension:** *[enter]*	**5.** Hit *enter* twice to exit the command. Your drawing looks like Figure 15.2.6.1.5.

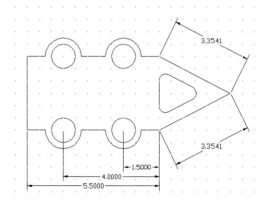

FIG. 15.2.6.1.5: Baseline Dimensions

	Command: *qsave*	**6.** Save the drawing, but do not exit.

■ 15.2.7 Placing Leaders

Use the *Leader* command to place simple call-outs that require an arrow or line to reference an object.

Contrary to the way we learned to place call-outs in board drafting, CAD leaders should be placed *before* the call-out text. Here is the sequence:

Command: *leader* (or *lead* or *le*)
Specify first leader point, or [Settings]<Settings>: *[pick the place you want the tip of the arrow]*
Specify next point: *[pick the outer end of the first leader line]*
Specify next point: *[you may continue to draw the leader line, or hit enter to begin the call out]*
Enter first line of annotation text <Mtext>: *[if your call-out requires more than one line, you may continue to type; or you may hit enter to exit the command]*

There are several options associated with the *Leader* command. Luckily, AutoCAD has placed them in a user-friendly dialog box (Figure 15.2.7a).

FIG. 15.2.7a: Leader Settings Dialog Box

Access the dialog box via the **Settings** option at the first leader prompt (just hit *enter*). Notice the three tabs. We will look at each beginning with the default—**Annotation**.

■ The **Annotation** tab is divided into three frames:

- Use the **Annotation Type** frame to determine what to place after the leader:
  - **Mtext** is simply the standard text entry.
  - **Copy an Object** allows the user to copy text or an object from one part of the drawing to the end of the leader line.
  - **Tolerance** will be discussed in more detail in Lesson 16.
  - **Block Reference** allows the user to place a block at the end of the leader line. (More on blocks in Lessons 19 and 20.)
  - **None**, of course, allows the user to create a leader without text. Use this option when the call-out has already been placed on the drawing (as you did on the board).

- The **Mtext options** frame provides three options:
  - **Prompt for width** causes AutoCAD to prompt the user for the width of the mtext entry (much like the *MText* command does).
  - **Always left justify** can be a nuisance. By default, AutoCAD will determine the text justification depending on which way the leader line goes.
  - **Frame text** will place a frame around the leader text.
- The last frame—**Annotation Reuse**—allows the user to reuse the text entered behind the leader. Leave this set to **None** unless you plan to enter the same call-out several times.

▌ The **Leader Lines & Arrows** tab (Figure 15.2.7b) presents four frames of options.

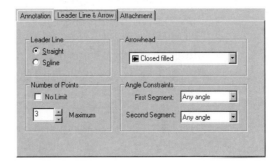

FIG. 15.2.7b: Leader Lines & Arrows Tab

- The **Leader Line** frame allows the user to use the standard (**Straight**) leader line or a fancier **Spline** line.
- The **Arrowhead** frame contains a control box from which the user may select the type of arrowhead (or no arrowhead at all) to use on the leader.
- **Number of Points** refers to the number of points allowed to define the leader line. The default (3) means that the user can draw a leader with two lines. A check in the check box removes all limitations on the number of lines your leader can have.
- **Apply Constraints** defines how the leader may be drawn. I prefer the default **Any angle** for the first segment; but then I require the second segment to be **Horizontal** to lead into the text.

▌ The last tab—**Attachment**—allows the user to position the text in relation to the leader line.

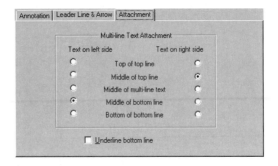

FIG. 15.2.7c: Attachment Tab

The number of options for the leader command reflects the importance of leaders in drafting. No one preference can be universal—although the defaults come fairly close. So the user should become familiar with the Leader Settings dialog box.

> The *Leader* command in AutoCAD 2000 has incorporated the more primitive command from earlier releases and the *QLeader* (or Quick Leader) command that accompanied R14. You will hear it referred to by both names in AutoCAD 2000.

Let's draw some leaders.

WWW 15.2.7.1 Do This: Leaders

I. Be sure you are in the *Ex-base.dwg* drawing file in the C:\Steps\Lesson15 folder. If not, open it now.

II. Follow these steps.

TOOLS	COMMAND SEQUENCE	STEPS
Quick Leader Button	**Command:** *le*	1. Enter the *Leader* command by typing *leader*, *lead*, or *le* at the command prompt. Or you can pick the **Quick Leader** button on the Dimension toolbar.
	Specify first leader point, or [Settings]<Settings>: *[enter]*	2. Hit *enter* to access the Leader Settings dialog box.
Leader Line: ○ Straight ● Spline		3. On the **Leader Line & Arrow** tab, change the leader line to use a spline.
OK		4. Pick the **OK** button to exit the dialog box.

continued

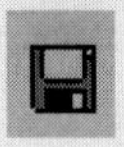

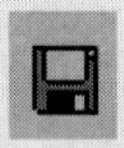

TOOLS	COMMAND SEQUENCE	STEPS
	Specify first leader point, or [Settings]<Settings>:	5. Place the leader, as shown in Figure 15.2.7.1.5.
	Specify next point:	
	Specify next point: *[enter]*	

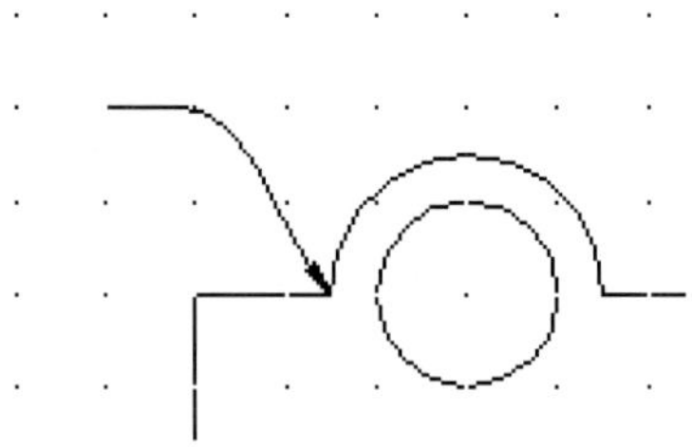

FIG. 15.2.7.1.5: Place the Leader

	Enter first line of annotation text <Mtext>: *No Fillet*	6. Type in the text indicated in Figure 15.2.7.1.6, then exit the command.
	Enter next line of annotation text: *[enter]*	

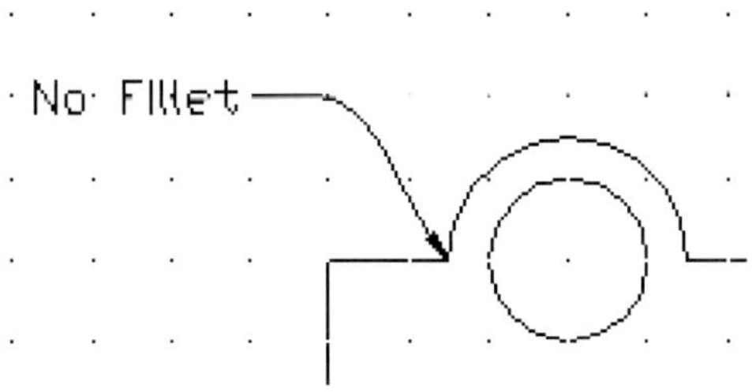

FIG. 15.2.7.1.6: Leader Text

💾	**Command:** *qsave*	7. Save the drawing, but do not exit.
		8. Complete the dimensions, as shown in Figure 15.2.1.7.8.

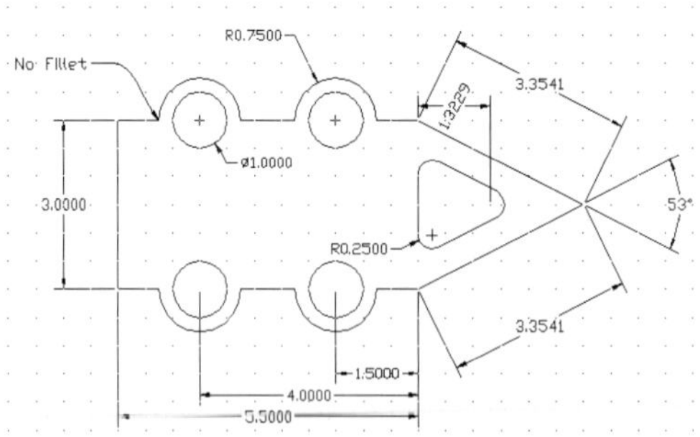

FIG. 15.2.7.1.8: Completed Drawing

💾	**Command:** *qsave*	9. Save the drawing.

15.2.8 Ordinate Dimensions

Ordinate dimensioning is a valuable tool in some types of manufacturing. *Dimordinate* places a *dimension* that is actually a distance from the 0,0 coordinate of the drawing. To be useful then, 0,0 must be located somewhere on the object itself (usually the lower left corner).

The command sequence for the *Dimordinate* command is

Command: *_dimordinate* (or *dimord* or *dor*)
Specify feature location: *[select the feature to be dimensioned]*
Specify leader endpoint or [Xdatum/Ydatum/Mtext/Text/Angle]: *[the user can specify the X or Y distance from 0,0 or identify the datum by mouse movement]*
Dimension text = *[AutoCAD reports and places the distance]*

- As indicated, the user can specify an **Xdatum** (X-distance from 0,0) or a **Ydatum** (Y-distance from 0,0) by typing the appropriate letter for that option.
- If the user prefers to place **Text** or **MText** at the ordinate location, those options are also available.
- The **Angle** option behaves just it does on other dimension tools—that is, the user may define the angle of the dimension text.

Let's try an exercise.

15.2.8.1 Do This: Ordinate Dimensioning

I. Open the *ex-ordinate.dwg* drawing file in the C:\Steps\Lesson15 folder. The drawing looks like Figure 15.2.8.1a.

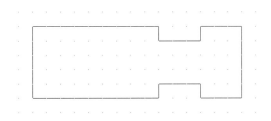

FIG. 15.2.8.1a: *Ex-Ordinate.dwg*

(Note that the limits on this drawing have been placed so that the lower left corner of the object drawn is at coordinate 0,0.)

II. Follow these steps.

TOOLS	COMMAND SEQUENCE	STEPS
 Ordinate Dimension Button	**Command:** *dor*	**1.** Enter the *Dimordinate* command by typing *dimordinate*, *dimord*, or *dor* at the command prompt. Or you can pick the **Ordinate Dimension** button on the Dimension toolbar.
	Specify feature location:	**2.** Select the left side of the lower indentation (Figure 15.2.8.1.3).
	Specify leader endpoint or [Xdatum/Ydatum/Mtext/Text/Angle]: **Dimension text = 4.5000**	**3.** Pick a point two grid marks due south of the point selected in Step 2. Your drawing looks like Figure 15.2.8.1.3.

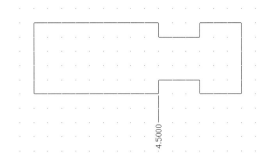

FIG. 15.2.8.1.3: Ordinate Dimension

	Command: *[enter]*	**4.** This time specify the datum you will use. Repeat the *dimordinate* command.
	Specify feature location:	**5.** Select the lower right corner of the object.
	Specify leader endpoint or [Xdatum/Ydatum/Mtext/Text/Angle]: *x*	**6.** Type *x* to tell AutoCAD you want an **XDatum**.
	Specify leader endpoint or [Xdatum/Ydatum/Mtext/Text/Angle]: **Dimension text = 7.5000**	**7.** Select a point two grid marks south and two grid marks west (Figure 15.2.1.8.11).
	Command: *[enter]*	**8.** Repeat the command.
	Specify feature location:	**9.** Select the same point selected in Step 6.

continued

TOOLS	COMMAND SEQUENCE	STEPS
	Specify leader endpoint or [Xdatum/ Ydatum/Mtext/Text/Angle]: *y*	10. Type *y* to tell AutoCAD that you want a **YDatum** this time.
	Specify leader endpoint or [Xdatum/ Ydatum/Mtext/Text/Angle]: **Dimension text = 0.0000**	11. Select a point two grid marks east and two grid marks north. Your drawing looks like Figure 15.2.1.8.11.

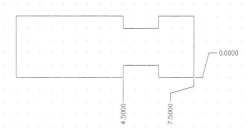

FIG. 15.2.8.1.11: X and Y Datum Dimensions

12. Complete the drawing, as shown in Figure 15.2.8.1.12.

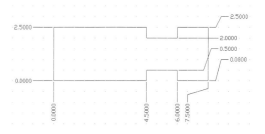

FIG. 15.2.8.1.12: Completed Drawing

	Command: *qsave*	13. Save the drawing.

15.3 And Now the Easy Way: Quick Dimensioning (QDim)

AutoCAD 2000 includes one of the niftiest new tools to come along in quite a while —Quick Dimensioning (*QDim*). With *QDim*, the user can create a string of dimensions using two picks to select the objects (with a window), one click to confirm the selection and one pick to locate the dimension. That is only four mouse clicks!

The command sequence looks like this:

Command: *qdim*
Select geometry to dimension: *[use a window or crossing window to select the objects to dimension]*
Select geometry to dimension: *[hit enter to complete the selection process]*
Specify dimension line position, or [Continuous/Staggered/Baseline/Ordinate/Radius/Diameter/datumPoint/Edit] <Continuous>: *[tell AutoCAD where to place the dimension]*
Command:

Most of the options are fairly obvious—**Continuous** for a continuous string, **Baseline** for a baseline string, **Ordinate** for ordinate dimensions, and so forth. But let's look at the last two.

- **datumPoint** allows the user to define a selected point as the 0,0 coordinate of the dimension string (very handy when ordinate dimensioning);
- **Edit** prompts:

Indicate dimension point to remove, or [Add/eXit] <eXit>:

Using the **Edit** option then, the user can add or remove Extension Line Origins to the selection set.

Let's give this nifty new tool a try.

15.3.1 Do This: Quick Dimensions

I. Open the *ex-linear2.dwg* file in the C:\Steps\Lesson15 folder. The drawing looks just like the *ex-linear.dwg* file we used in Exercise 15.2.1.1.

II. Follow these steps.

TOOLS	COMMAND SEQUENCE	STEPS
Quick Dimension Button	Command: *qdim*	1. Enter the **QDim** command or pick the **Quick Dimension** button on the Dimension toolbar.

TOOLS	COMMAND SEQUENCE	STEPS
	Select geometry to dimension:	**2.** Use a crossing window to select the bottom line, the slot, and the large circle. Do not select anything else (see Figure 15.3.1.2).

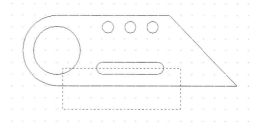

FIG. 15.3.1.2: Selection Window

	Select geometry to dimension: *[enter]*	**3.** Hit *enter* to confirm the selection set.
	Specify dimension line position, or [Continuous/Staggered/Baseline/ Ordinate/Radius/Diameter/ DatumPoint/Edit] <Continuous>:	**4.** Locate the dimension below the object (Figure 15.3.1.4).

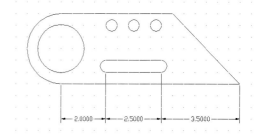

FIG. 15.3.1.4: Quick Dimensions

	Command: *[enter]*	**5.** Now let's try a baseline dimension. Repeat the command.

continued

TOOLS	COMMAND SEQUENCE	STEPS
	Select geometry to dimension:	**6.** Use a crossing window to select the upper line and the four circles, as shown in Figure 15.3.1.6.

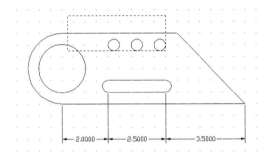

FIG. 15.3.1.6: Select the Circles

	Select geometry to dimension: *[enter]*	**7.** Hit *enter* to confirm the selection set.
	[Continuous/Staggered/Baseline/ Ordinate/Radius/Diameter/ datumPoint/Edit] **<Baseline>:** *b*	**8.** Tell AutoCAD you want to create a **Baseline** dimension.
	Specify dimension line position, or **[Continuous/Staggered/Baseline/ Ordinate/Radius/Diameter/ datumPoint/Edit]** **<Baseline>:**	**9.** Locate the dimensions, as shown in Figure 15.3.1.9.

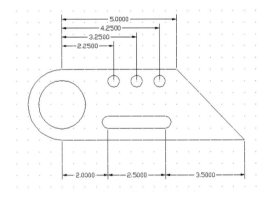

FIG. 15.3.1.9: Completed Dimensions

	Command: *quit*	**10.** Exit the drawing without saving.

Does that not make you wonder why we ever did it any other way?

Still, by now, you may have noticed that dimensions do not always appear just the way we think they should. And sometimes (God, forbid), the dimension has to change to accommodate some new design information. Our next section will show you how to modify dimensions.

15.4 Dimension Editing: The *Dimedit* and *DimTedit* Commands

AutoCAD provides two commands for editing *Associative* dimensions. These commands show a bit of overlap in their functions (redundancy again), but the user will generally change or rotate the text or change the angle of the extension lines with the *Dimedit* command. With the *DimTedit* command, the user can change the position of the text.

> Although AutoCAD provides tools for editing Associative dimensions, no special tools are required for Normal dimensions. Use basic modification commands on dimensions as you would on any other objects. Edit the text using the text editing command (*DDEdit*).

■ 15.4.1 Positioning the Dimension: The *DimTedit* Command

The *DimTedit* command sequence looks like this:

Command: *dimtedit* (**or** *dimted*)

Select dimension: *[pick the dimension to reposition]*

Specify new location for dimension text or [Left/Right/Center/Home/Angle]: *[pick the new position or select an option]*

The user can change the position of the dimension text dynamically by dragging it with the mouse, or by using AutoCAD's default **Left**, **Right,** or **Center** position. (*Note*: The **Left** or **Right** option redefines the *Home* position of the text next to the left or right arrow.) The **Angle** option allows the user to rotate the dimension text, and the **Home** option returns the dimension text to its original position (Home) and rotation.

> I know the term Text Edit is confusing since the *DimTedit* command is not used to edit the text but rather to edit the *position* of the text. *Dimedit* is actually used to edit the dimension text. But these are AutoCAD's terms, so we will just have to get used to them.

15.4.1.1 Do This: Repositioning Dimension Text

I. Reopen the *ex-base.dwg* file in the C:\Steps\Lesson15 folder. If this file has not been completed, open the *ex-base2.dwg* in the same folder.

II. Follow these steps.

TOOLS	COMMAND SEQUENCE	STEPS
 Dimension Text Edit Button	**Command:** *dimted*	1. Enter the **DimTedit** command by typing **dimtedit** or **dimted** at the command prompt. Or you can pick the **Dimension Text Edit** button on the Dimension toolbar.
	Select dimension:	2. Select the dimension below the right end of the object, as indicated in Figure 15.4.1.1.2.
		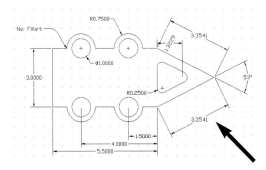 FIG. 15.4.1.1.2: Select This Dimension
	Specify new location for dimension text or [Left/Right/Center/Home/Angle]: *[l or r]*	3. Experiment with repositioning the text using the **Left** and **Right** options. Repeat this step until you are comfortable with these options.
	Command: *u*	4. Use the **U** command to return the text to its original position.
	Command: *dimted* **Select dimension:**	5. Repeat the **Dimtedit** command and select the same dimension.

continued

TOOLS	COMMAND SEQUENCE	STEPS
	Specify new location for dimension text or [Left/Right/Center/Home/Angle]:	6. Notice that the dimension moves dynamically as you move the crosshairs. Experiment with repositioning the text manually. Repeat this step until you are comfortable with this method.
	**Command:** *[enter]* **Select dimension:**	7. Repeat the command and select the same dimension.
	Specify new location for dimension text or [Left/Right/Center/Home/Angle]: *h*	8. Use the **Home** option to return the text to its original position.
	**Command:** *[enter]* **Select dimension:**	9. Repeat the command and select the same dimension.
	Specify new location for dimension text or [Left/Right/Center/Home/Angle]: *a*	10. Tell AutoCAD you want to change the angle of the text . . .
	Specify angle for dimension text: 27	11. . . . to 27°. It now looks like Figure 15.4.1.1.11.

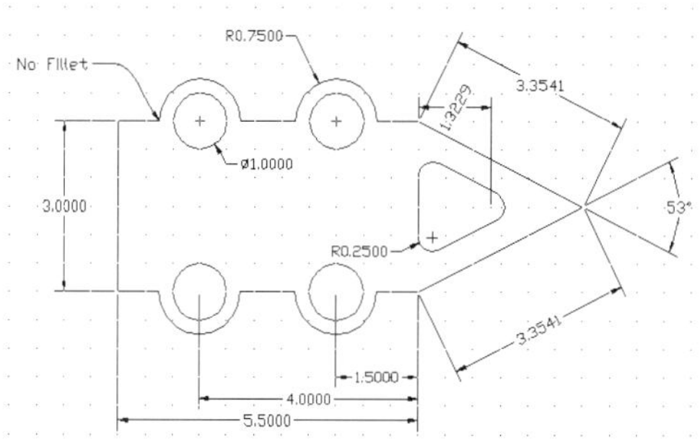

FIG. 15.4.1.1.11: Angled Dimension

| | **Command:** *qsave* | 12. Save the drawing. |

15.4.2 Changing the Dimension Text: The Dimedit Command

The *Dimedit* command sequence is

> **Command:** *dimedit* (or *dimed* or *ded*)
> **Enter type of dimension editing [Home/New/Rotate/Oblique] <Home>:** *[hit enter to accept the default or tell AutoCAD what you want to do]*
> **Select objects:** *[select the object(s) to change]*

Notice that the *Dimedit* command uses a default (<**Home**>) for the initial prompt. Notice also that the option must be selected *before* the dimension *objects*.

- Most of the options are simple. In fact, half the options are redundancies. **Home** does exactly what it did in the *DimTedit* command. **Rotate** does what the **Angle** option did.
- The **New** option presents the Multiline Text Editor with these symbols in the editing window: <>. Delete these symbols before typing the desired text. When the user picks the **OK** button, AutoCAD will prompt to **Select objects:**. It will then replace the dimension text with what was typed in the Multiline Text Editor. But be aware that once the dimension text has been changed, it will no longer automatically adjust when the object/dimension is modified.
- You will find the **Oblique** option quite valuable when dimensioning isometrics. To dimension an isometric, you will use the *Dimaligned* command, then reposition the dimension in the correct isometric plane using this tool. (You will see more on this in Section 15.5.)

15.4.2 Do This: Changing Dimension Text

I. Be sure you are still in the *ex-base.dwg* (or the *ex-base2.dwg*) file in the C:\Steps\Lesson15 folder. If not, open it now.

II. Follow these steps.

TOOLS	COMMAND SEQUENCE	STEPS
 Dimension Edit Button	**Command:** *ded*	1. Enter the *Dimedit* command by typing *dimedit*, *dimed* or *ded* at the command prompt. Or you can pick the **Dimension Edit** button on the Dimension toolbar.

continued

TOOLS	COMMAND SEQUENCE	STEPS
	Enter type of dimension editing [Home/New/Rotate/Oblique] <Home>: *n*	2. Tell AutoCAD you want to edit a dimension (select the **New** option).
		3. AutoCAD displays the Multiline Text Editor with two brackets showing (<>). Delete these brackets and type *4.125*.
[OK]		4. Pick the **OK** button.
	Select objects: 1 found **Select objects:** *[enter]*	5. Select the 4.0000 dimension indicated in Figure 15.4.2.1.5. Then complete the command.

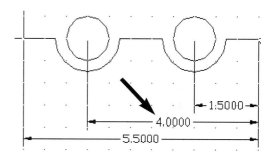

FIG. 15.4.2.1.5: Select This Dimension

| | **Command:** *s*

Select objects to stretch by crossing-window or crossing-polygon ...

Select objects: Specify opposite corner: 15 found

Select objects: *[enter]*

Specify base point or displacement:

Specify second point of displacement: *@1<180* | 6. AutoCAD replaces the dimension text with the new text. Now try stretching the object one unit to the left, as shown in Figure 15.4.2.1.6. |

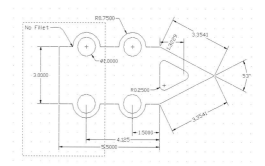

FIG. 15.4.2.1.6: Stretch These Objects

continued

TOOLS	COMMAND SEQUENCE	STEPS
		7. Notice (in Figure 15.4.2.1.7) that AutoCAD automatically updated the 5.5000 dimension but the 4.125 dimension did not change. 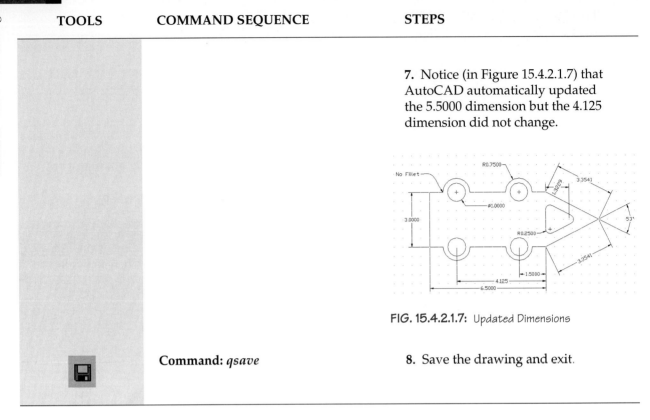 FIG. 15.4.2.1.7: Updated Dimensions
	Command: *qsave*	8. Save the drawing and exit.

15.5 Isometric Dimensioning

It might relieve you to know that there is no actual isometric dimension command!

Dimension your isometrics using the ***Dimaligned*** command. Then you must edit the aligned dimension using the **Oblique** option of the ***Dimedit*** command to adjust the aligned angle to the appropriate isometric plane.

Sound complicated? It is not as complicated as it is tedious. Let's see how it works.

15.5.1 Do This: Dimensioning Isometrics

I. Open the *Iosdim.dwg* file in the C:\Steps\Lesson15 folder. The drawing looks like Figure 15.5.1a.

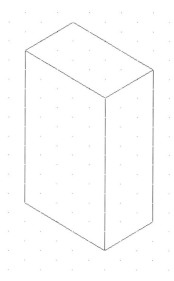

FIG. 15.5.1a: *Isodim.dwg*

II. Follow these steps.

TOOLS	COMMAND SEQUENCE	STEPS
	Command: *dal*	1. Use the *Dimaligned* command to place the dimensions shown in Figure 15.5.1.1. 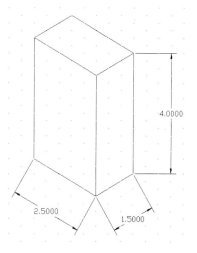 **FIG. 15.5.1.1:** Place These Dimensions

continued

TOOLS	COMMAND SEQUENCE	STEPS
	Command: *ded*	2. Repeat the *Dimedit* command.
	Enter type of dimension editing [Home/New/Rotate/Oblique] <Home>: *o*	3. Tell AutoCAD you want to use the **Oblique** option.
	Select objects: 1 found	4. Select the 4.0000 vertical dimension.
	Select objects: *[enter]*	
	Enter obliquing angle (press ENTER for none): *30*	5. Tell AutoCAD to angle the extension lines at 30°. Your drawing looks like Figure 15.5.1.5.

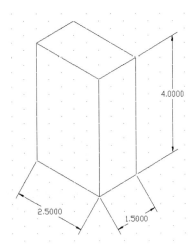

FIG. 15.5.1.5: Dimension *Obliqued* to 30°

continued

TOOLS	COMMAND SEQUENCE	STEPS
	Command: *[enter]* **Enter type of dimension editing [Home/New/Rotate/Oblique] <Home>:** *o* **Select objects:** **Select objects:** *[enter]* **Enter obliquing angle (press ENTER for none):** *270*	6. Now set the obliquing angle of the lower two dimensions to 270°. Your drawing looks like Figure 15.5.1.6. 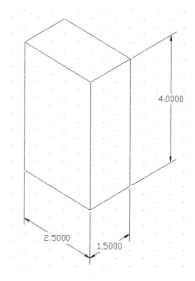 **FIG. 15.5.1.6:** Isometric Dimensions
	Command: *qsave*	7. Save the drawing and exit.

To change the text to an isometric format, create appropriate text styles (Lesson 5) then use the *Props* command (Lesson 7) to change the style of the dimension text.

15.6 Extra Steps

We have used the command prompt and the toolbars throughout this lesson. Take some time to familiarize yourself with the **Dimension** pull-down menu. This one has the best chance of surviving the assaults of programmers because of the necessity of simplifying the dimensioning tools. You will see that the tools available on the pull-down are the same as those provided on the toolbar. The **Style** and **Update** options correspond to commands that we will discuss in Lesson 16.

15.7 What Have We Learned?

Items covered in this lesson include:

- *AutoCAD Dimension Terminology*
- *AutoCAD Dimension Commands:*
  - *Dimlinear*
  - *Dimangular*
  - *Dimradius*
  - *Dimdiameter*
  - *Dimcontinue*
  - *Dimaligned*
  - *Dimbaseline*
  - *Leader*
  - *Dimordinate*
  - *QDim*
  - *Dimedit*
  - *DimTedit*

You are now familiar with the various methods of creating and editing AutoCAD's dimensions. For the most part, you can place dimensions as you would on any board. But if you spend much time with dimensioning, you will discover some shortfalls in your understanding. This is because you are only halfway through your study of AutoCAD's dimensioning tools.

In Lesson 16, we will discuss customizing your dimensions. This seems difficult, but it is where you will learn:

- to use architectural units instead of decimals
- how to change arrow styles and text sizes
- how to have AutoCAD insert tolerances automatically
- and much, much more!

15.8 EXERCISES

1. Dimension drawings *Iso1* and *Iso2* in the C:\Steps\Lesson15 folder. They will look like Figures 15.8.1a and 15.8.1b when completed.

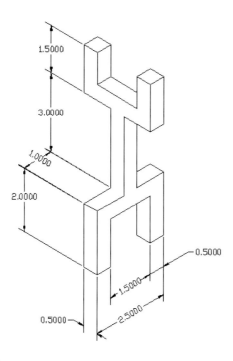

FIG. 15.8.1a: Completed Iso1.dwg

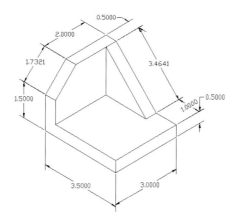

FIG. 15.8.1b: Completed Iso2.dwg

2. Dimension the *Brake.dwg* file in the C:\Steps\Lesson15 folder.
   Remember:
   - the ***Dimaso*** command toggles Associative and Normal dimensioning
   - the ***Explode*** command converts an Associative Dimension to a Normal Dimension.
   - dimensions do not always fall where you want them, so remember your editing commands

The drawing will look like Figure 15.8.2 when complete:

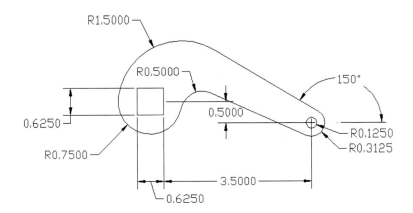

FIG. 15.8.2: Brake.dwg

3. Dimension the *Pulley-slider.dwg* file in the C:\Steps\Lesson15 folder. (*Hints*: Remember the text tricks in Lesson 5.) The drawing will look like Figure 15.8.3 when completed.

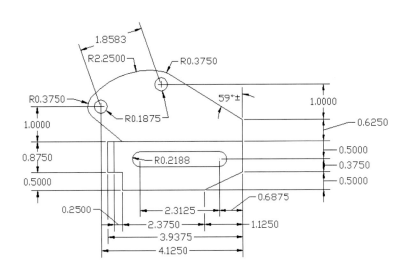

FIG. 15.8.3: Pulley-slider.dwg

4. Add the dimensions shown to the *MyBracket* drawing you created in Lesson 8 (or you can use the *Bracket15.dwg* file found in the C:\Steps\Lesson15 folder). Your drawing will look like Figure 15.8.4 when completed.

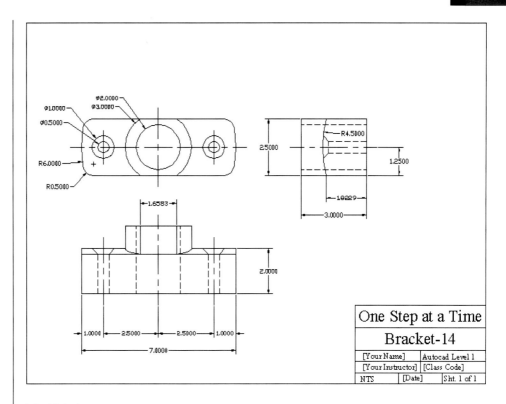

FIG. 15.8.4: *MyBracket.dwg*

5. Add the dimensions shown to the *MyGear* drawing you created in Lesson 10 (or you can use the *Gear15.dwg* file found in the C:\Steps\Lesson15 folder). Your drawing will look like Figure 15.8.5 when completed.

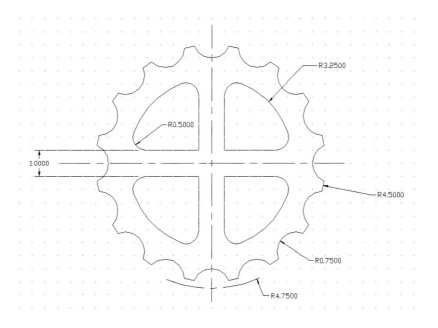

FIG. 15.8.5: *MyGear.dwg*

15.9 REVIEW QUESTIONS

Write your answers on a separate sheet of paper.

1. to 6. Identify.

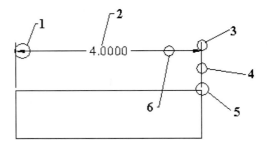

FIG. 15.9.1: Identify

7. AutoCAD defaults to _____ units when dimensioning regardless of the units the drawing has been set up to use.

8. and 9. _____ dimensions are tied to the objects being dimensioned and _____ dimensions are not.

10. (T or F) The user cannot break or trim extension lines on a normal dimension.

11. What is the system variable (command) that controls associative/normal dimensioning?

12. Use the _____ command to convert associative dimensions to normal dimension.

13. to 20. List the eight dimension commands.

21. When using the **Press Enter to Select** option in dimensioning, (the user, AutoCAD) determines the endpoints of the object being dimensioned.

22. If allowed, AutoCAD will determine the location of the linear dimension by the location of the _____.

23. The _____ option of the dimlinear command allows the user to rotate the dimension text for a better fit.

24. Use the _____ command to dimension angles, circles and arcs.

25. (T or F) When dimensioning diameters or radii, it is necessary for the user to enter the diameter symbol (by holding down the control key while typing D) or R.

26. Use the _____ command to create a continuous string of dimensions.

27. and 28. Whereas dimcontinue works off of the (first, last) extension line origin, dimbaseline works off of the (first, last).

29. Use the _____ command to place simple call-outs that require an arrow or line to reference an object.

30. _____ places a dimension that is a distance from the 0,0 coordinate of the drawing.

31. In editing associative dimensions, the user will generally change or rotate the text or change the angle of the extension lines with the _____ command.

32. The user can change the position of associative dimension text using the _____ command.

33. To edit the dimension text in a normal dimension, use the _____ command.

34. and 35. When dimensioning isometrics, you must use the _____ option of the _____ command to orient the dimension in the correct isometric plane.

36. Use the _____ command to place dimensions on an isometric drawing.

Identify these buttons.

37. **38.** **39.** **40.** **41.**

42. **43.** **44.** **45.** **46.**

LESSON 16

Customizing Dimensions

Following this lesson, you will:
➥ Know how to create and use Dimension Styles
➥ Know how to remove unused Dimension Styles: The *Purge* Command

Every design industry has its own preferences about such things as units, dimension arrows, text sizes, tolerances, and so forth. If all industries had to accept decimal dimensions (with four decimal places), AutoCAD would probably lose its business fairly quickly—especially from architectural and petrochemical designers who rely on the ever-present feet and inches of architectural drafting. Indeed, even metric users might wash their hands of taking millimeters to four decimal places.

In this lesson, we will look at changing the way AutoCAD creates dimensions to fit different industrial standards.

16.1 Creating Dimension Styles: The *DDim* Command

> We will use dialog boxes for our customization. AutoCAD will store each of our settings in a DIMension VARiable (Dimvar). The user can adjust each dimvar manually at the command prompt (as we did prior to R13). But take my word for it, only a programmer would want to customize dimensions this way! Still, for those who are into that sort of thing, I have included a list of AutoCAD's dimvars with their default settings in Appendix E.

Each drawing can incorporate several dimension styles, but there is rarely need for more than one. AutoCAD's clever use of the *Family* method of setup enables the user to create overall dimension settings (**Parent** settings) for the drawing with different settings for each of its six children: **Linear**, **Radial**, **Diameter**, **Ordinal**, **Angular**, and **Leader**. This is a remarkable accomplishment when compared with other CAD systems or even earlier releases of AutoCAD. The downside of this, of course, is that the user must navigate the Dimension Style Manager to create the proper settings for the parent and each child. But remember, set it up once and save it to a template, then you do not have to do it again!

Let's look at the first step. Open a new drawing from scratch and follow along on the screen.

Access the Dimension Style Manager by typing *DDim* at the command prompt or picking the **Dimension Style** button (Figure 16.1a) on the Dimension toolbar. Or you can pick **Style...** from the **Dimension** pull-down menu. AutoCAD presents the Dimension Style Manager shown in Figure 16.1b. Let's take a look at this.

FIG. 16.1a: Dimension Style Button

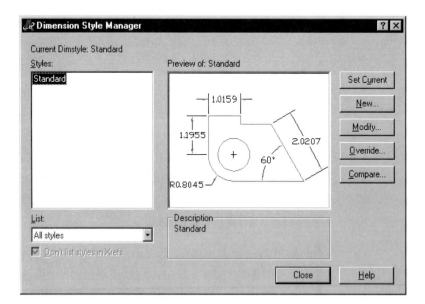

FIG. 16.1b: Dimension Style Manager

- In the **Styles:** list box, AutoCAD will display a list of all of the styles currently defined in this drawing. The left-justified names are the Parent styles. The children of that parent will be listed below it and slightly indented.
- The **List:** control box controls what the user will see in the **Styles:** list box. Here the user can tell AutoCAD to list **All styles** or just the **Styles in use** in the current drawing.
- AutoCAD shows the current dimension setup in the **Preview of:** display box. Following the words "Preview of," AutoCAD will list the dimension style being displayed. A written description of the dimstyle may appear in the **Description** frame below the display box.
- The five buttons down the right side of the Dimension Style Manager allow the user to **Set Current** the dimstyle highlighted in the **Styles:** list box, create a **New** or **Modify** an existing dimstyle, **Override** the settings in the current dimstyle, or **Compare** two dimstyles to find the differences. Let's look at these.
  - You must set a dimension style as current before you can use it (just as you did with layers). You can do that by picking the **Set Current** button here. But it is easier to use the Dim Style control box on the Dimension toolbar (as we will see in our exercises).
  - The **New . . .** and **Modify . . .** buttons both present a tabbed dialog box used to set up the dimstyle (more on these in a moment).
  - Overriding a dimension style is rarely a good idea. But selecting the **Override . . .** button will present the same tabbed dialog box as the previous buttons. The difference is that **Override . . .** will not use the changes to redefine the dimstyle (although you can use the changes in your dimensioning). (More on **Override** in Section 16.3.2.)
  - The **Compare . . .** button presents a dialog box that allows the user to compare the differences between two dimstyles.

Now that you have some idea of what the Dimension Style Manager looks like, we need to get into the specifics of exactly how to define or create a new dimension style.

How do you think you would start a *New* style? Of course, you would pick the **New . . .** button. When you do, AutoCAD will present the **New Dimension Style** dialog box (Figure 16.1c). Let's look at our options.

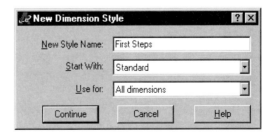

FIG. 16.1c: New Dimension Style Dialog Box

- Name your dimstyle in the **New Style Name:** text box (here we have called our new style *First Steps*).

- In the **Start With:** control box, you can select to use the settings from any of the existing dimstyles as the basis for your new dimstyle.
- The **Use for:** control box is the key to the parent/child relationship of AutoCAD's dimension styles. By default, the settings you make for your new style will be used for **All dimensions** you create. However, you can create separate settings for linear, angular, radius, diameter, ordinate, or leader dimensions (the children).

Let's start a new dimension style.

16.1.1 Do This: Create a new Dimstyle

I. Open the *Drill-gizmo-16.dwg* file in the C:\Steps\Lesson16 folder. The drawing looks like Figure 16.1.1a.

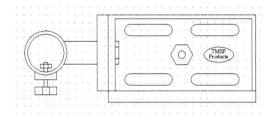

FIG. 16.1.1a: *Drill-gizmo-16.dwg*

II. Follow these steps.

TOOLS	COMMAND SEQUENCE	STEPS
	Command: *ddim*	1. Enter the *Dimension Styles* command by typing *ddim* at the command prompt. Or you can pick the **Dimension Style** button on the Dimension toolbar. AutoCAD presents the Dimension Style Manager (Figure 16.1b p. 590).
New...		2. Start a **New ...** dimstyle.

continued

TOOLS	COMMAND SEQUENCE	STEPS
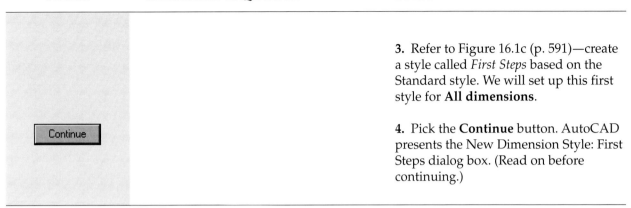		3. Refer to Figure 16.1c (p. 591)—create a style called *First Steps* based on the Standard style. We will set up this first style for **All dimensions**. 4. Pick the **Continue** button. AutoCAD presents the New Dimension Style: First Steps dialog box. (Read on before continuing.)

Once you have completed the selections in the New Dimension Style dialog box, you will **Continue** to the tabbed **New Dimension Style: First Steps** dialog box (where First Steps is the name of your new style), as shown in Figure 16.1d. Here you will actually define the settings.

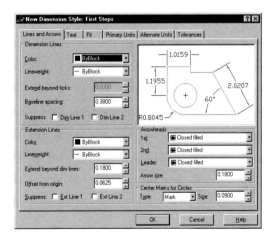

FIG. 16.1d: New Dimension Style Settings Tabs

Let's take a look at each tab.

▌ The first tab—**Lines and Arrows**—contains five frames. You will define the physical appearance of your dimensions in these frames.
- Preview your setup in the upper right **Preview** frame.
- Use the first frame to set up the **Dimension Line**.
  ⇒ The **Color** and **Lineweight** control boxes allow the user to set the color and lineweight of the dimension. The default for each is **ByBlock**.
  ⇒ The **Extension beyond ticks** control box remains gray unless the arrowheads have been set to tick marks (we will look at **Arrowheads** in a moment). If tick

marks have been selected, you can tell AutoCAD how far to *extend* the dimension line beyond the extension line (refer to Figure 16.1e) by typing a distance here.

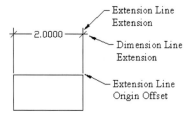

FIG. 16.1e: Extension Line Extension

- **Baseline spacing** refers to the distance between Baseline dimensions. It is probably best to leave this setting at its default unless your industry uses a different standard.

> The user should set all the sizes in a dimension style to the actual size required on the final plot. The scale factor of the dimension objects will be set on the **Fit** tab.

- The check boxes at the bottom of the frame allow the user to **Suppress** (not draw) the dimension line of the 1st or 2nd side of the dimension. This can be useful in a crowded area.
- The **Extension Lines** frame is very similar to the **Dimension Line** frame.
  - The **Suppress** options work on extension lines as they did on dimension lines. Use these to avoid placing one extension line atop another when stringing dimensions. It will prevent your plotter from drawing too many lines in one place (and possibly digging a hole in the paper). I do not recommend this; most papers can handle two lines. If you are placing more than that in one place, rethink your drawing strategy.
  - **Color** and **Lineweight** work the same as they did for dimension lines. I recommend the same setting for dimension lines and extension lines for continuity.
  - The **Extend beyond dim lines** setting determines the distance the extension lines continue beyond the dimension line (the **Extension Line Extension** as seen in Figure 16.1e).

- ⇒ The **Offset from origin** setting determines the distance away from the origin the extension line will begin (see the **Extension Line Origin Offset** in Figure 16.1e).

* The user will determine the size and style of the arrowheads in the **Arrowheads** frame.
  - ⇒ AutoCAD provides sample images for each of the **1st**, **2nd** and **Leader** arrowheads in the control boxes. AutoCAD provides 20 selections as well as options to use a user-define block or no arrowhead at all (**None**).
  - ⇒ Just below the control boxes is a size box in which the user can adjust the size of the arrowhead.

* The **Center marks for circles** frame allows the user to select what symbol, if any, should mark the center of a circle or arc when dimensioned. It also allows the user to change the size of the center marks or the extension of the centerlines outside the circle.

Let's continue our exercise.

16.1.2 Do This: Setting up Lines and Arrows

I. Continue the previous exercise. Follow these steps.

TOOLS	COMMAND SEQUENCE	STEPS
Text	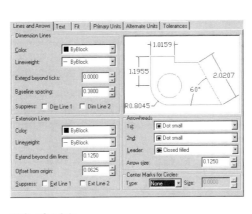 FIG. 16.1.2.1t	1. (Refer to Figure 16.1.2.1t.) (a) Accept the default settings in the **Dimension Lines** frame. (b) Set the extension lines to **Extend** 0.125" **beyond the dim lines**. (c) Use **Dot small** for the 1st and 2nd arrowheads for the dimensions but **Closed filled** arrowheads for the leaders. (d) Make the arrowheads *0.125"* (e) Do not use **Center marks for Circles**. 2. Pick the **Text** tab. (Read on before continuing.)

- The **Text** tab (Figure 16.1f) contains four frames for defining how to use text in your dimensions.

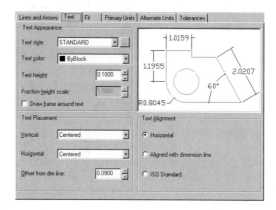

FIG. 16.1f: The Text Tab

- Use the **Preview** frame to see what your dimension looks like as you create it.
- The **Text Appearance** frame contains settings for the appearance of the text.
  - Set the style of text for the dimensions using the **Text style:** control box. If you have not yet created the style, use the button to the right of the control box to access the Text Style dialog box.
  - Set the text color and height using the appropriate boxes.
  - When using fractional or architectural units, you can set the size of the fractions using the **Fraction height scale:** number box.
  - Place a frame around the dimension text by checking the **Draw frame around text** box.
- The **Text Placement** frame allows the user to control where the text will be placed in relation to the dimension line.
- In the **Text Alignment** frame, the user determines whether to place dimension text **Horizontal**, **Aligned with the dimension line**, or in the **Iso Standard** mode. (Iso Standard means that the text will be aligned with the dimension line if it falls between the extension lines or horizontally if it does not fit between the extension lines.)

Let's set up our text.

WWW 16.1.3 Do This: Setting up Dimension Text

I. Continue the previous exercise. Follow these steps.

TOOLS	COMMAND SEQUENCE	STEPS
	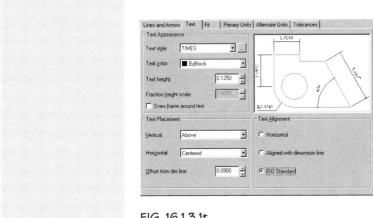 FIG. 16.1.3.1t	1. [Refer to Figure 16.1.3.1t.] (a) Use the **Times** text style (it has already been set up) and a text height of 0.125". (b) Place the text above the line vertically but centered horizontally. (c) Use the ISO Standard alignment mode.
Fit		2. Pick the **Fit** tab. (Read on before continuing.)

■ The **Fit** tab (Figure 16.1g) has five frames to help fine-tune your dimension style.

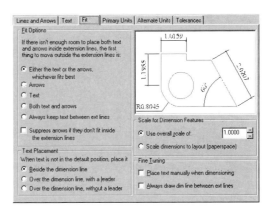

FIG. 16.1g: Fit Tab

- The **Fit Options** frame allows the user to tell AutoCAD what to do when the dimension parts do not fit between the extension lines.

- Move text and/or arrowheads from inside to outside the extension lines.
- Do not place arrowheads on the dimension lines.

- In the **Text Placement** frame, the user tells AutoCAD *where* to place the text if it does not fit between the extension lines—beside the dimension line or over the dimension line (with or without a leader).
- The **Scale for Dimension Features** frame contains some critical settings. Scaling the dimension objects is quite a bit easier than scaling text. The user should set dimension size according to the desired plot size. Simply identify the scale factor using the **overall scale** size box. AutoCAD will do the rest! That is, where the user *manually* scales text (Lesson 4), AutoCAD will *automatically* scale dimension objects.

    (Ignore the paperspace option for now. We will cover paperspace in our advanced text.)
- The **Fine Tuning** frame allows the user to manually place text or to force dimension lines between extension lines (even when nothing else will fit).

Let's continue our setup.

16.1.4 Do This: Setting up the Fit Tab

I. Continue the previous exercise. Follow these steps.

TOOLS	COMMAND SEQUENCE	STEPS
	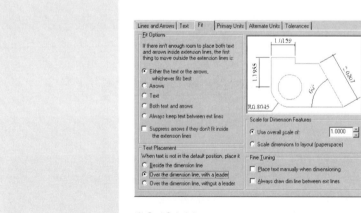 FIG. 16.1.4.1t	1. (Refer to Figure 16.1.4.1t.) (a) Allow AutoCAD to determine what to do if the text and arrows do not fit—leave the bullet in the first option of the **Fit Options** frame. (b) When text is not in the default position, place it over the dimension with a leader (the **Text Placement** frame). (c) Accept the rest of the defaults. 2. Pick the **Primary Units** tab. (Read on before continuing.)
Primary Units		

Dimension units operate independently of the drawing units and must be set separately.

▌ The **Primary Units** (Figure 16.1h) and the **Alternate Units** tabs allow the user to set the dimension units separately from the drawing units. The **Primary Units** tab has two large frames to set up.

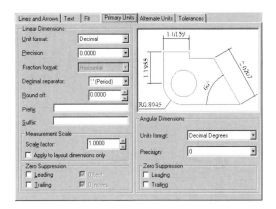

FIG. 16.1h: Primary Units Tab

- The Linear Dimension frame has several options.
  - It begins with the actual **Unit format:** option. Select the type of units to use in the control box.
  - Use the **Precision** control box to set the precision of your dimensions (how many decimal places or the fractional denominator to use).
  - When using Fractional or Architectural units, you can make your fractions stack horizontally or vertically, or you can choose not to stack them at all. These options will be available in the **Fraction format:** control box.
  - Use the **Decimal separator:** control box to separate decimals from whole numbers using a period (as done in the United States), a comma (as done in Europe), or with a space.
  - Use the **Round off:** number box to round off your dimensions.
  - The **Prefix** and **Suffix** boxes are provided to allow the user to place leading or trailing text with your dimensions (like "mm" marks).
  - You should probably leave the Measurement Scale settings at their defaults. The **Scale factor:** number box tells AutoCAD how to scale dimensions. Simply put, if the setting is *1.0000* (its default), dimensions reported are the same as the true distance between extension line origins. AutoCAD will take any other setting, multiply it by the true distance and use the results as the dimension text. This is handy if you are using various details—drawn at different scales—on your drawing.

- **Zero Suppression** enables the user to dimension without unnecessary zeros. See the examples in Figures 16.1i to 16.1l.

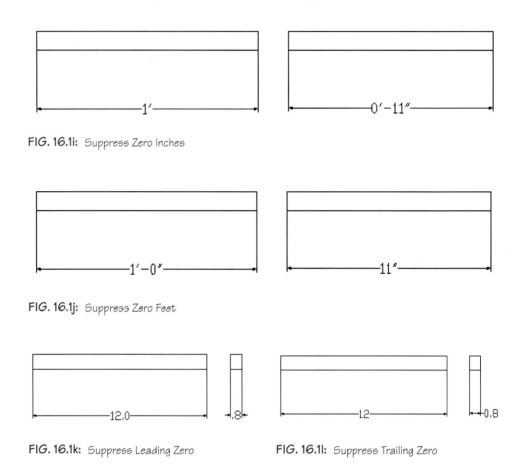

FIG. 16.1i: Suppress Zero Inches

FIG. 16.1j: Suppress Zero Feet

FIG. 16.1k: Suppress Leading Zero

FIG. 16.1l: Suppress Trailing Zero

- The **Primary Units** tab allows the user to set up **Angular Dimensions** as well. The **Alternate Units** tab does not allow the setup of angular dimensions but does provide a frame to allow the user to determine the placement of alternate dimensions.

  Alternate units are useful when a company in the United States is working on a European or Asian project that must show metrics as well as feet and inches. A dimension may look something like the one in Figure 16.1m.

  Notice that AutoCAD places the alternate units in brackets after the primary units (by default). Also by default, AutoCAD sets the alternate to a scale of 25.4—or *millimeters*.

FIG. 16.1m
Alternate Units

Let's continue our setup.

16.1.5 Do This: Setting up the Units

I. Continue the previous exercise. Follow these steps.

TOOLS	COMMAND SEQUENCE	STEPS
	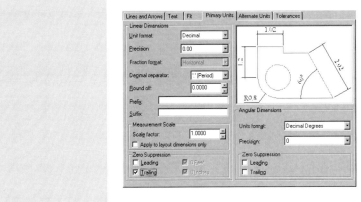 FIG. 16.1.5.1t	**1.** (Refer to Figure 16.1.5.1t.) **(a)** Accept the decimal format, but set the precision to two decimal places. **(b)** Suppress trailing zeros.
Alternate Units		**2.** Pick the **Tolerances** tab (we will not use alternate units in this exercise). (Read on before continuing.)

■ The last tab we must examine is the **Tolerance** tab (Figure 16.1n). Here the user sets up any tolerances that may be required in the dimensioning. Again, we have two frames.

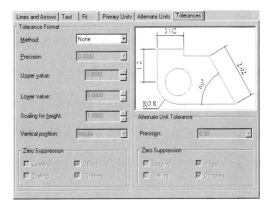

FIG. 16.1n: Tolerance Tab

- The first frame allows the user to set up the **Tolerance Format**.
  ➠ See the examples in Figures 16.1o to 16.1t for **Method:** and **Vertical position:** values.

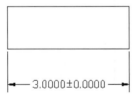

FIG. 16.1o: Symmetrical Tolerance

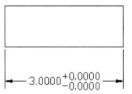

FIG. 16.1p: Deviation Tolerance (Middle Justification)

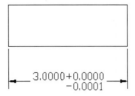

FIG. 16.1q: Deviation Tolerance (Top Justification)

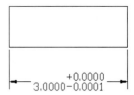

FIG. 16.1r: Deviation Tolerance (Bottom Justification)

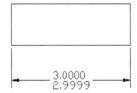

FIG. 16.1s: Limits Tolerance

FIG. 16.1t: Basic Tolerance

  ➠ **Precision** works just as it did in the **Units** tabs.
  ➠ **Upper** and **Lower value:** identify the value of the tolerances.
  ➠ **Scaling for height:** allows the user to set a separate text height (in proportion to the height set on the **Text** tab) for the tolerances.
  ➠ **Zero Suppression** works as it did on the **Units** tabs.
- **Alternate Unit Tolerance** controls tolerances for alternate units.

Let's complete our setup.

16.1.6 Do This: Setting up Tolerances

I. Continue the previous exercise. Follow these steps.

TOOLS	COMMAND SEQUENCE	STEPS
	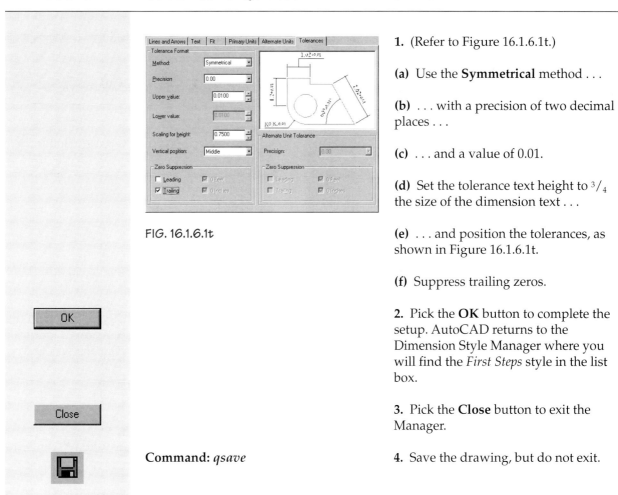 FIG. 16.1.6.1t	1. (Refer to Figure 16.1.6.1t.) (a) Use the **Symmetrical** method . . . (b) . . . with a precision of two decimal places . . . (c) . . . and a value of 0.01. (d) Set the tolerance text height to $3/4$ the size of the dimension text . . . (e) . . . and position the tolerances, as shown in Figure 16.1.6.1t. (f) Suppress trailing zeros.
[OK]		2. Pick the **OK** button to complete the setup. AutoCAD returns to the Dimension Style Manager where you will find the *First Steps* style in the list box.
[Close]		3. Pick the **Close** button to exit the Manager.
[disk icon]	Command: *qsave*	4. Save the drawing, but do not exit.

Remember, the user can set each of the values you have studied in these last several pages *separately* for each member of the dimension **Family**! This can prove quite handy when, for example, you want to place the dimension text *above* the dimension line, but you want to place radial text at the *end* of a leader. Or you may want AutoCAD to automatically locate the dimension text for linear dimensions but allow you to manually locate it for angular dimensions.

Let's set up a quick child for radial dimensions.

16.1.7 Do This: Setting up a Child

I. Be sure you are still in the *drill-gizmo-16.dwg* in the C:\Steps\Lesson16 folder. If not, open it now.

II. Follow these steps.

TOOLS	COMMAND SEQUENCE	STEPS
	Command: *ddim*	1. Reopen the Dimension Style Manager.
First Steps / STANDARD		2. Select the **First Steps** style in the list box.
New...		3. Pick the **New** button to create a new child of the First Steps style.
	New Dimension Style — New Style Name: First Steps: Diameter; Start With: First Steps; Use for: Radius dimensions	4. Use the **Use for:** control box to set the child for **Radius dimensions**. Notice that the **New Style Name:** box is grayed and unavailable for child styles.
Continue		5. Pick the **Continue** button to continue.
	Arrowheads — 1st: Closed filled; 2nd: Closed filled; Leader: Closed filled; Arrow size: 0.1250	6. On the **Lines and Arrows** tab, change the arrowheads to **Closed filled**. Remember that this will only affect radial dimensions!
	Text Placement — Vertical: Centered; Horizontal: Centered; Offset from dim line: 0.0900	7. On the **Text** tab, set the vertical placement of the radial text as centered on the leader line …

continued

TOOLS	COMMAND SEQUENCE	STEPS
[Text Alignment dialog: Horizontal selected, Aligned with dimension line, ISO Standard]		8. ... and set the text alignment to **Horizontal**. This will prevent radial dimensions appearing at oddball angles.
[OK button]		9. Pick the **OK** button to complete the setup. AutoCAD returns to the Dimension Style Manager. Notice the **Radial** child in the **Styles** list box.
[Close button]		10. Pick the **Close** button to complete the command.
[Save icon]	**Command:** *qsave*	11. Save the drawing, but do not exit.

> Obviously, there is a tremendous amount of information to absorb if you want to master dimension styles. But unless you are determined to become the contract guru, it may not be necessary to memorize these dialog boxes. AutoCAD provides for everyone, even the casual user!
>
> Your best approach to dimension styles may be to simply follow this text to set up the dimension style you prefer or the one you want to make standard for your contract. But be smart! Set up your dimension styles as part of your *templates*. Then you can forget them. (Another smart move will be to write down each setting as you go—computers crash and information gets lost!)
>
> And remember—you can copy the dimstyle from one drawing to another using the AutoCAD Design Center (Lesson 7).

Let's try some advanced dimensioning.

16.1.8 Do This: Dimension the Drawing

I. Be sure you are still in the *Drill-gizmo-16.dwg* file in the C:\Steps\Lesson16 folder.

II. Dimension the drawing. It will look like Figure 16.1.8.1 when you have finished.

TOOLS	COMMAND SEQUENCE	STEPS

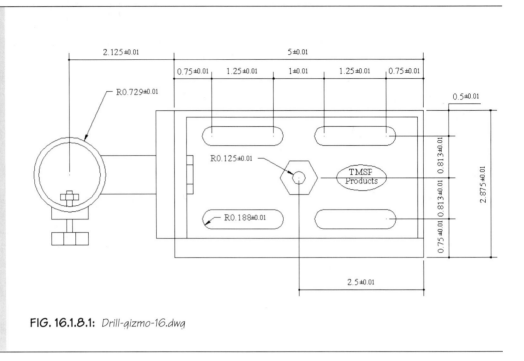

FIG. 16.1.8.1: *Drill-gizmo-16.dwg*

16.2 Try One

Now that you have some understanding of what it takes to create dimension styles and dimension a drawing, let's dimension our floor plan. As we are three-quarters of the way through our text, I am going to let you do as much as you can on your own. Refer back to the appropriate portion of the text as needed.

16.2.1 Do This: Dimension the Floor Plan

I. Open your *MyFlrPln16.dwg* file from the C:\Steps\Lesson16 folder. If that one is not available, open *flr-pln16.dwg* drawing file in the same folder.

II. Follow these steps.

TOOLS	COMMAND SEQUENCE	STEPS
	Command: *la*	1. Create a dimension layer called *dim*. Assign the color **Yellow** to your new layer and make it current.

continued

TOOLS	COMMAND SEQUENCE	STEPS
	Command: *ddim*	2. Open the Dimension Style Manager.
	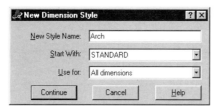	3. Pick the **New** button and create a new style called *Arch*. **Continue** to the New Dimension Style: Arch dialog box.
	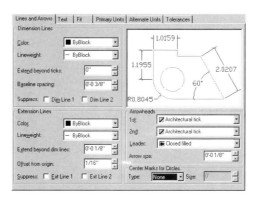 FIG. 16.2.1.4t	4. Set up the **Lines and Arrows** tab, as shown in Figure 16.2.1.4t.
	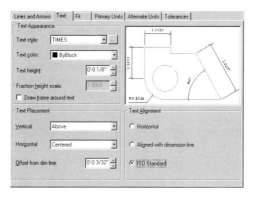 FIG. 16.2.1.5t	5. Set up the **Text** tab, as shown in Figure 16.2.1.5t.

continued

TOOLS	COMMAND SEQUENCE	STEPS
	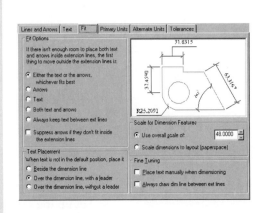 FIG. 16.2.1.6t	6. Set up the **Fit** tab, as shown in Figure 16.2.1.6t.
	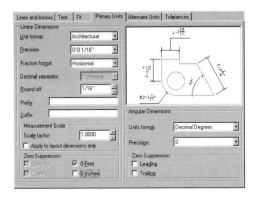 FIG. 16.2.1.7t	7. Set up the **Primary Units** tab as shown in Figure 16.2.1.7t.
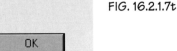		8. We will not need to do any setup on the **Alternate Units** or **Tolerances** tabs. Pick the **OK** button to continue ...
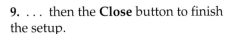		9. ... then the **Close** button to finish the setup.
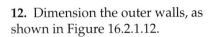		10. Use the **Dim Style** control box on the Dimension toolbar to set the **arch** dimstyle current.
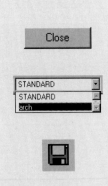	**Command:** *qsave*	11. Save the drawing, but do not exit.
		12. Dimension the outer walls, as shown in Figure 16.2.1.12.

continued

TOOLS	COMMAND SEQUENCE	STEPS

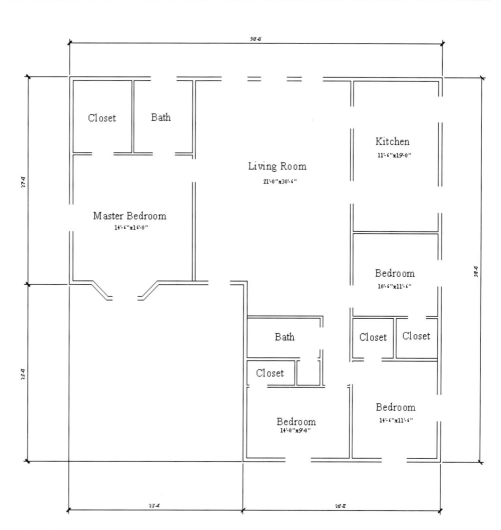

FIG. 16.2.1.12: Dimension the Outer Walls

13. Now dimension the rest of the drawing. Notice that inner walls and door and window openings are dimensioned to their centers. There are a couple of ways to do this:

(a) use the **From** OSNAP, or

(b) load the *Points* Lisp routine in C:\Steps\Lesson11 folder, and use the **MW** command it gives you to place a node *midway* between two points. Locate the midway point for the doors and windows and use that node for dimensioning.

continued

TOOLS	COMMAND SEQUENCE	STEPS
		Your completed drawing looks like Figure 16.2.1.13.

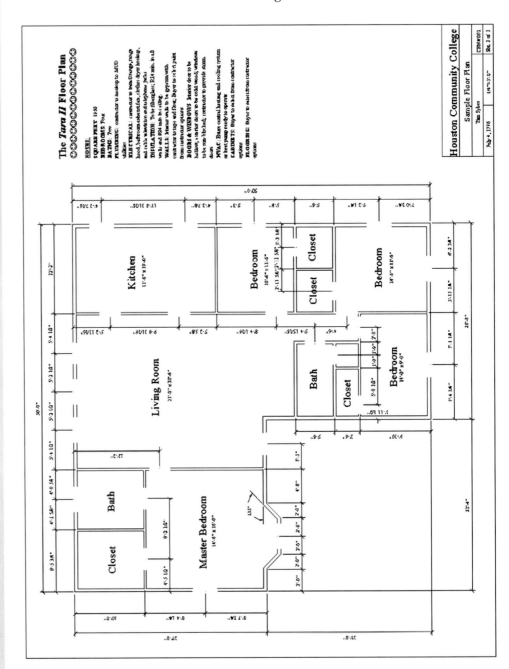

FIG. 16.2.1.13: Completed Drawing

continued

TOOLS	COMMAND SEQUENCE	STEPS
	Command: *saveas*	14. Save the drawing as *MyFlrPln19* to the C:\Steps\Lesson19 folder.

16.3 Simple Repairs

Setting up dimensions is no easy chore. Fortunately, AutoCAD provides some tools for accidents, boo boos, uh ohs, and #$%^&!

> Occasionally, you may find it necessary to rid a drawing of an unwanted dimension style. AutoCAD provides a command to do just that—the *Purge* command. It works like this:
>
> **Command:** *purge*
> **Enter type of unused objects to purge**
> **[Blocks/Dimstyles/LAyers/LTypes/Plotstyles/SHapes/STyles/ Mlinestyles/All]:** *d*
> **Enter name(s) to purge <*>:** *baddims*
> **Verify each name to be purged? [Yes/No] <Y>:** *[enter]*
> **Purge dimension style "baddims"? <N>** *y*
>
> As you can see, you can also rid the drawing of unwanted blocks, layers, linetypes, plotstyles and much more by using the *Purge* command. But most of these provide **Delete** buttons in their dialog boxes.
>
> (*Note*: For a dimstyle (or any of the other options) to be removed, it *must* be unused.)

■ 16.3.1 Experimenting with Dim-Update

To change a dimension to reflect current dimstyle settings, use the dimension *Update* command. Follow this sequence:

> **Command:** *dim*
> **Dim:** *update*
> **Select objects:** *[select the dimension you want to change]*
> **Dim:** *exit*

Of course, it will be easier to select the **Dimension Update** button on the Dimension toolbar.

> You will notice in the preceding command sequence, that AutoCAD provides a dimension prompt. You may issue dimension commands at the command line using this prompt as we did in previous releases, but I do not recommend it. It is quite tedious compared to the Dimension toolbar.

■ 16.3.2 Overriding Dimensions

You can use the **Override ...** button on the Dimension Style Manager to set up dimension variables that differ from the style's settings. But if you use dimension overrides, be aware that as soon as you set a different style as current, the override settings disappear.

I advise against doing this; after all, if you need different settings once, you might need them again. It is best to set up a new style.

A better approach might be to use the **Object Property Manager** (Lesson 7) to adjust your dimensions' variables once they have been placed.

16.4 Extra Steps

Take a few minutes to go over the dimension variables (dimvars) in Appendix E. The user can set each of these at the command line simply by typing the dimvar and a value.

Then take a look at the command line approach to dimension styles. It looks like this:

Command: *dimstyle*
Current dimension style: arch *[AutoCAD tells you which style is current]*
Enter a dimension style option
[Save/Restore/STatus/Variables/Apply/?] <Restore>: *[hit enter to make a different style current]*
Current dimension style: arch
Enter a dimension style name, [?] or <select dimension>: *[type the name of the dimension style you wish to make current]*
Command:

- Use the **Save** option to create a new style based on the current dimvars. Use the **Restore** option (the default) to switch from one style to another.
- The **Status** option will show you the current settings of the dimvars.

- Using the **Variables** option, the user can select a dimension and read the values of the dimvars used to create that dimension.
- The **Apply** option will change user-selected dimensions so they use the current dimvars.

This should make you appreciate the *DDim* command and the Dimension Style Manager!

16.5 What Have We Learned?

Items covered in this lesson include:

- *Defining Dimensions*
- *Commands:*
  - *Ddim*
  - *Update*
  - *Purge*

It is important to take the time now to get comfortable with dimensioning. After all, drafting without dimensioning would not be of much use to anyone. Dimensioning itself does not have to be difficult if the user takes the time to get familiar with the toolbar. But dimvars and dimension styles can easily overwhelm the AutoCAD novice.

If you do not feel comfortable with dimensioning, go back to Lesson 15 and do 15 and 16 again. Then say to yourself, "I have met the challenge and am wiser for it!"

16.6 EXERCISES

1. Open the *drillguide.dwg* file in the C:\Steps\Lesson16 folder. Create an appropriate dimension style to dimension the image as indicated in Figure 16.6.1.
   1.1. Use small dot arrowheads but no center marks.
   1.2. Use an overall scale of 1.
   1.3. Use decimal units accurate to three decimal places (suppress trailing zeros).
   1.4. Use standard $1/8$" text.
   1.5. Allow a tolerance deviation of 1° on all angles, and a precision of zero decimal places.
   1.6. All dimensions should be above the dimension line except radii, which should be centered on the leader.

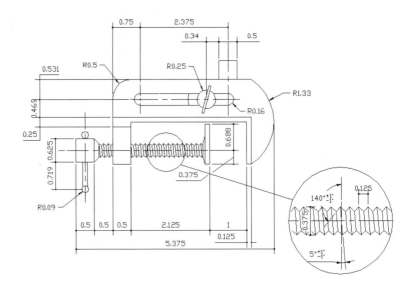

FIG. 16.6.1: *drillguide.dwg*

2. Open the *My Piping Plan 16.dwg* file in the C:\Steps\Lesson16 folder. If this is not available, open the *Piping Plan 16.dwg* file in the same folder. Create an appropriate dimension style to dimension the image as indicated in Figure 16.6.2.

   2.1. To determine the overall scale, refer to the scale factor in the Drawing Scales Chart (Appendix A). You will plot this drawing at $^3/_8$" = 1'−0" on a D-Size sheet of paper.

   2.2. Use architectural units with a $^1/_{16}$" precision.

   2.3. Suppress zero feet.

   2.4. Use the Times text style and a $^1/_8$" text height.

   2.5. Center the dimension text above the dimension line.

   2.6. Use **Closed Filled** Arrowheads.

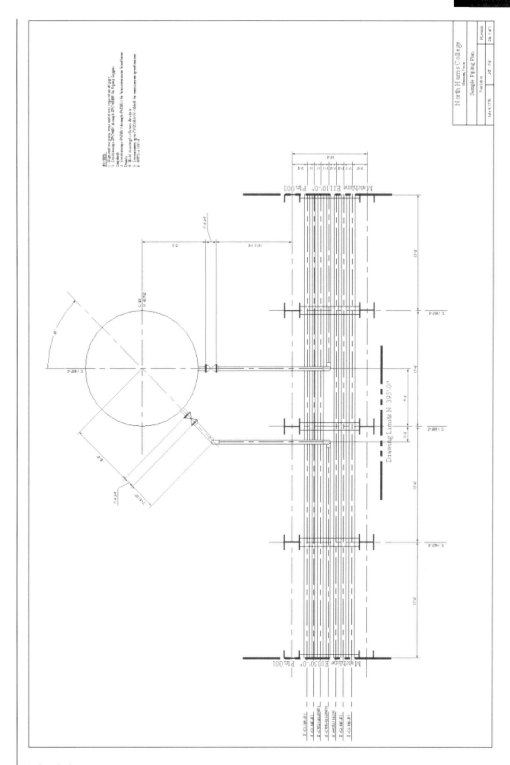

FIG. 16.6.2: Piping Plan 16.dwg

3. Now let's dimension some of the other drawings we have created.
   3.1. Open the *Block Guide.dwg* file you created in Lesson 7 (in the C:\Steps\Lesson07 folder). (Or you can use the *Block Guide 16.dwg* file in the C:\Steps\Lesson16 folder.) Dimension the drawing as shown in Figure 16.6.3.1. (*Hint*: The dimension text size is $3/16$" and the arrow size is $1/8$".)
      3.1.1. Save the drawing as *MyBlockGuide16.dwg* in the C:\Steps\Lesson16 folder.

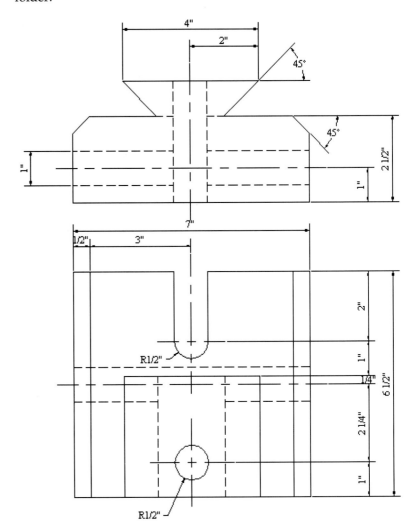

FIG. 16.6.3.1: *Block Guide 16.dwg*

   3.2. Open the *MyHolder.dwg* you created in Lesson 6 (in the C:\Steps\Lesson06 folder). (Or you can use the *Holder 16.dwg* file in the C:\Steps\Lesson16 folder.) Dimension the drawing as shown in Figure 16.6.3.2 (*Hint*: The dimension text size is $3/16$" and the arrow size is $1/8$".)

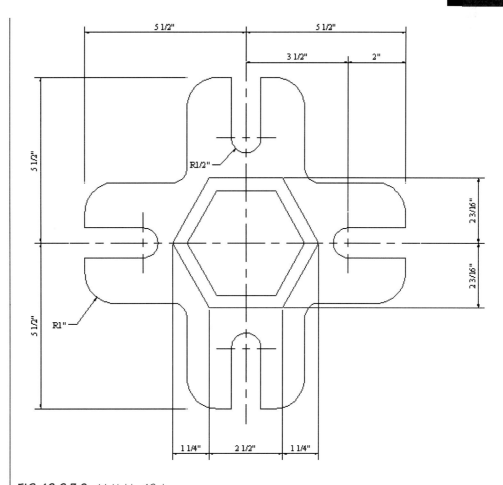

FIG. 16.6.3.2: *MyHolder 16.dwg*

3.2.1. Save the drawing as *MyHolder16.dwg* in the C:\Steps\Lesson16 folder.

3.3. Open the *Bracket.dwg* file you created in Lesson 8 (in the C:\Steps\Lesson08 folder). (Or you can use the *Bracket 16.dwg* file in the C:\Steps\Lesson16 folder.) Dimension the drawing as shown in Figure 16.6.3.3 (*Hint*: The dimension text size is $3/16$" and the arrow heads are $1/8$" obliques.)

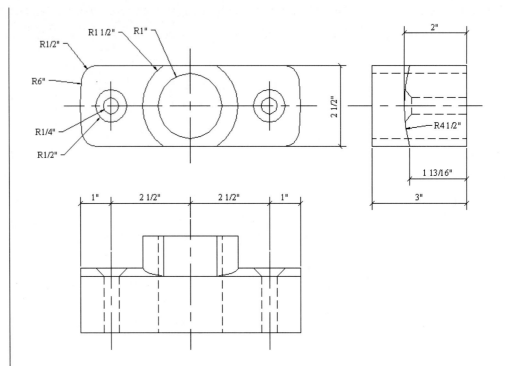

FIG. 16.6.3.3: *MyBracket 16.dwg*

    3.3.1. Save the drawing as *MyBracket16.dwg* in the C:\Steps\Lesson16 folder.
4. Open the *Soot Trap* drawing in the C:\Steps\Lesson16 folder and add the dimensions shown in Figure 16.6.4.

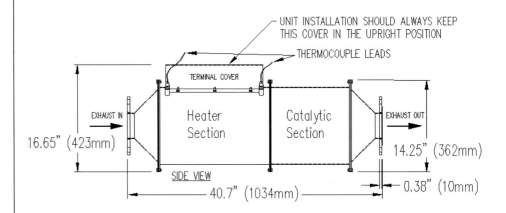

FIG. 16.6.4: *Soot Trap.dwg*

16.7 REVIEW QUESTIONS

Write your answers on a separate sheet of paper.

1. Another name for a dimension variable is _____.

2. Access the Dimension Styles Manager with the _____ command.

3. (T or F) The Dimension Style Manager is not the only way to create dimension styles, but they are the easiest.

4. Overall dimension settings are called _____ settings.

5. to 10. List the six children available to a parent dimension style.

11. The user can set the size of arrowheads and extension line origin offset on the _____ tab of the Dimension Style Manager.

12. The overall scale of dimensions should be set to the _____ for the drawing.

13. Dimension text, which is not horizontal, will be placed _____ to the dimension line.

14. _____ determines where along the dimension line AutoCAD will place the dimension text.

15. (T or F) Once units have been set for the drawing (using DDUnits or AutoCAD's setup wizard), it is not necessary to set up units for dimensions.

16. (T or F) Careful consideration should be given to the size of a working copy of a drawing before setting dimensioning to use stacked fractions.

17. and 18. To dimension various details—drawn at various scales—on a drawing, set the scale of the detail in relation to the rest of the drawing (for example, 2 × or .5 ×) in the _____ box of the _____ tab of the Dimension Style Manager.

19. (T or F) The only text style available in dimensioning is AutoCAD's standard style.

20. Use the _____ command to rid a drawing of unwanted dimension styles.

21. and 22. To read the dimension variables used to create a particular dimension, use the _____ option of the _____ command.

LESSON 17

Advanced Modification Techniques

Following this lesson, you will:

➡ Know how to use alternate selection techniques found in the Selections Tab of the Options dialog box, including:
- Noun/Verb Selection
- Shift to Add
- Press and Drag
- Implied Windowing

➡ Know how to use Grips commands to modify objects
- **Stretch**
- **Move**
- **Rotate**
- **Mirror**
- **Scale**

After all that effort to master dimensioning, why not take some time for something different (and much easier than the last two lessons)? Lesson 17 will deal with some nifty techniques that, although not often used by anyone other than well-trained CAD operators, can enhance your drawing speed even beyond the techniques you have already learned!

We will begin with some fairly simple techniques that you may have (indeed, *probably* have) stumbled across already. (Have you noticed that selecting an object *before* picking the Erase button produces the same result as if you had picked the button first?) Knowing how to use these techniques can help you draw faster; knowing how to turn them off if you do not want to use them can save your sanity!

We will spend most of the lesson discovering *Grips*. These are the culprits responsible for those little blue squares that dot an object when you select it at the Command: prompt instead of a Select objects: prompt. But mastering grips will make you the envy of any CAD environment.

First, however, let's take a look at some advanced Object Selection methods.

17.1 Object Selection Settings

> It will be easier to understand **Noun/Verb Selection** (one of the Selection Settings) and **Grips** if they are not both active at once. Let's begin our study of the Selection Settings by turning *off* **Grips**. Follow this procedure:
>
> **Command:** *grips*
>
> **New value for GRIPS <1>:** *0*
>
> We will reactivate them when we are finished with this section of the text.

There are several different Selection Settings designed to enhance your use of AutoCAD, each with a different sysvar that the user can set at the command line. But AutoCAD also provides a tab on the Options dialog box that might be easier to use than half a dozen system variables.

The dialog box with the **Selections** tab on top can be seen in Figure 17.1a. (*Caution*: Avoid changes to the other tabs at this time.) We will concentrate on the Selection Modes Frame for now. Let's start at the top.

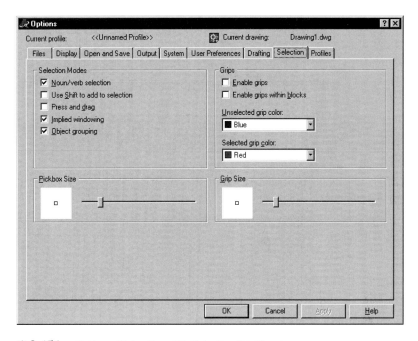

FIG. 17.1a: Options Dialog Box with Selection Tab Showing

■ The "normal" approach to object modification in AutoCAD is to issue a command (like *Erase*), then select the object(s) of that command. This approach can be called *Verb/Noun* as the command is invariably a verb and the object is a noun. Placing a check in the box next to **Noun/Verb Selection** (as the default does) means that AutoCAD will allow the user to select the object of the command first, then enter the command itself. The sequence (using the *Erase* command) looks like this:

Command: *[select an object(s)]*
Command: *e*

The sysvar for **Noun/Verb Selection** is *Pickfirst*. Its default setting is *1*.

17.1.1 Do This: Noun/Verb Selection

I. Open the *grips.dwg* file in the C:\Steps\Lesson17 folder. The drawing looks like Figure 17.1.1a.

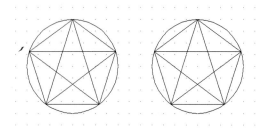

FIG. 17.1.1a: *Grips.dwg*

II. Be sure Grips are turned off as detailed in the insert on the previous page.

III. Follow these steps.

TOOLS	COMMAND SEQUENCE	STEPS
 Named Views Button	**Command:** *v*	1. Restore the *Selection* view. It looks like Figure 17.1.1.1. 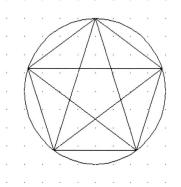 **FIG. 17.1.1.1:** *Selection View*
No Button Available	**Command:** *op*	2. Enter the *Options* command by typing *options* or *op*. AutoCAD presents the Options dialog box shown in Figure 17.1a. (Pick the **Selections** tab to put it on top.)

continued

TOOLS	COMMAND SEQUENCE	STEPS
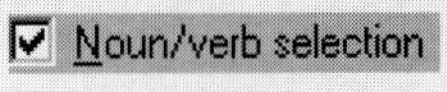 ☑ Noun/verb selection		**3.** Verify that there is a check in the **Noun/Verb Selection** check box.
 OK		**4.** Pick the **OK** button to return to the drawing.
	Command:	**5.** Select the two lines that form the topmost point of the star (Figure 17.1.1.5).

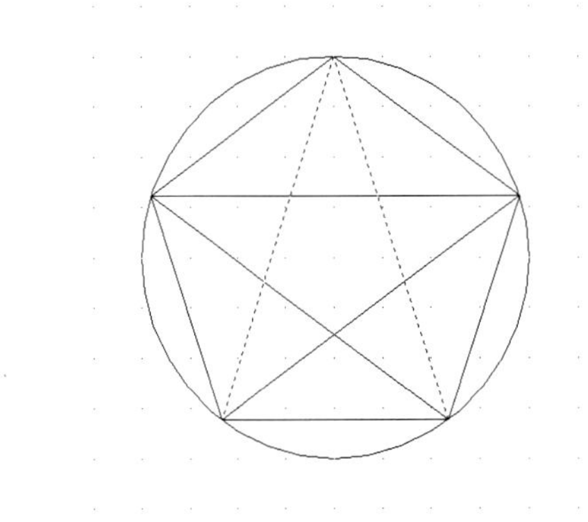

FIG. 17.1.1.5: Select These Two Lines

| | **Command:** *e* | **6.** Enter the *Erase* command. Notice that the lines are erased (Figure 17.1.1.6). |

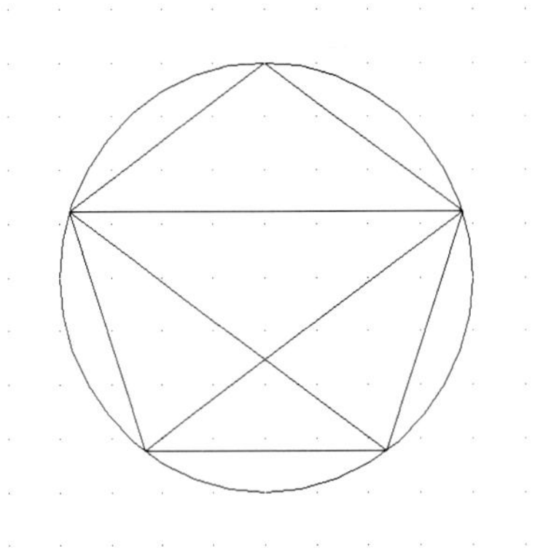

FIG. 17.1.1.6: Erased Lines

|  | **Command:** *qsave* | **7.** Save the drawing, but do not exit. |

■ In many Windows programs, selecting more than a single file or object requires the user to hold down the **Shift** key. AutoCAD makes an allowance for users accustomed to this approach in the next option—**Use Shift to Add**. If this box contains a check, selecting a second object will remove all previous objects from the selection set unless the user holds down the **Shift** key while selecting.

The sysvar for **Use Shift to Add** is *Pickadd*. Its default setting is *1*.

17.1.2 Do This: Using Shift to Add

I. Be sure you are still in the *grips.dwg* file in the C:\Steps\Lesson17 folder. If not, open it now.

II. Follow these steps.

TOOLS	COMMAND SEQUENCE	STEPS
No Button Available	**Command:** *op*	**1.** Enter the *Options* command by typing *options* or *op*. AutoCAD presents the Options dialog box. Pick the **Selections** tab.
	☑ Use **S**hift to add to selection	**2.** Place a check in the **Use Shift to Add** check box.
OK		**3.** Pick the **OK** button to return to the drawing.
	Command: *e*	**4.** Enter the *Erase* command.
	Select objects:	**5.** Select the horizontal line and then the two angled lines. Notice that as you select an object, the previously selected object is removed from the selection set (it is no longer highlighted).
	Select objects:	**6.** Now select the horizontal line. Then, while holding down the **Shift** key, select the angled lines. Notice that they all highlight.

continued

TOOLS	COMMAND SEQUENCE	STEPS
	Select objects: [*enter*]	**7.** Complete the command. The view now looks like Figure 17.1.2.7.

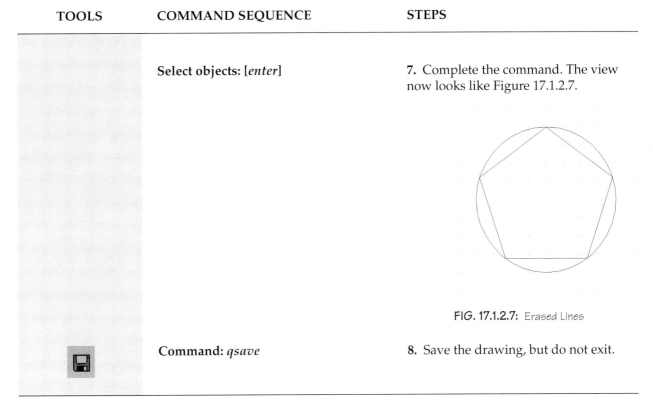

FIG. 17.1.2.7: Erased Lines

	Command: *qsave*	**8.** Save the drawing, but do not exit.

- Again, in many Windows programs, placing a window around objects requires the user to hold down the mouse button between corner selections. AutoCAD does not require this, but it allows for the user who has become accustomed to it. A check in the **Press and Drag** box will change window creation to the following procedure:

  1. Pick the first corner of the window.
  2. Hold down the mouse button while positioning the crosshairs at the opposite corner.
  3. Release the mouse button.

The sysvar for **Press and Drag** is *Pickdrag*. Its default setting is *0*.

17.1.3 Do This: Using Press and Drag

I. Be sure you are still in the *grips.dwg* file in the C:\Steps\Lesson17 folder. If not, open it now.

II. Follow these steps.

TOOLS	COMMAND SEQUENCE	STEPS
No Button Available	**Command:** *op*	**1.** Enter the *Options* command by typing *options* or *op*. AutoCAD presents the Options dialog box. Pick the **Selections** tab. **2.** Remove the check from the **Use Shift to Add** check box. Add a check in the **Press and Drag** check box. **3.** Pick the **OK** button to return to the drawing.
	**Command:** *e*	**4.** Enter the *Erase* command.
	Select objects:	**5.** Try to begin an implied window as you normally would—by selecting a point around the grid mark at coordinate 6,5. Notice that you lose the window as soon as you release the mouse button.
	Select objects:	**6.** Try it again. But this time hold down the mouse button as you move the lower right corner of the window to the grid mark around coordinate 11,0. Now you get the window. **7.** Release the mouse button to complete the selection window.
	Select objects: [*enter*]	**8.** Hit *enter* to complete the command. The view is now empty.
	Command: *z*	**9.** Zoom to the limits of the drawing, which now looks like Figure 17.1.3.9.

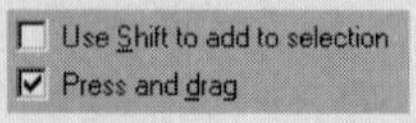

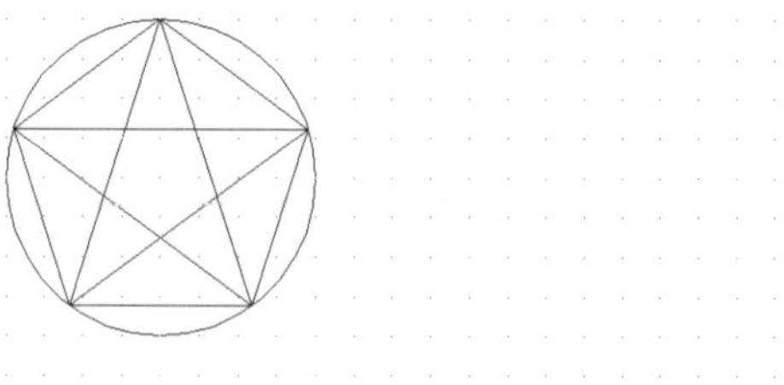

FIG. 17.1.3.9: Erased Objects

continued

TOOLS	COMMAND SEQUENCE	STEPS
	Command: *qsave*	10. Save the drawing, but do not exit.

- We have been using **Implied Windowing** since Lesson 2. This is what enables us to pick an open area of the drawing to automatically begin a selection window. It is *On* by default. Removing the check will turn it *Off*.
  The sysvar for **Implied Windowing** is *Pickauto*. Its default setting is *1*.

- A check in the **Object Grouping** option box means that AutoCAD will recognize grouped objects as a single object. We will learn more about grouping in Lesson 19.
  The sysvar for **Object Grouping** is *Pickstyle*. Its default setting is *1*. **CTRL + A** will also toggle object grouping *On* and *Off*.

- Below the **Selection Modes** frame, you will find the **Pickbox Size** frame. Here the user can adjust the size of the box used to select objects in a drawing. The slider bar makes adjustment easy as the user can watch the sample selection box resize as it is adjusted.
  The sysvar for **Pickbox Size** is *Pickbox*. I prefer a setting of *4* or *5* (an allowance for my aging eyes—many experienced users prefer *3*), but adjust it until you are happy.

17.2　A Whole New Ball Game!: Editing with Grips

I have always approached grips with reluctance in my basic AutoCAD classes. It is not that they are not remarkable tools; it is just that most students in a basic class are still trying to master the "normal" modifying tools. These are fairly simple and straightforward—if you want to copy something, you type *copy*. AutoCAD then leads you through the necessary steps by prompts at the command line. Grips, on the other hand, are more intuitive. That is, they require that the user *know* how to move from one step to the next and that the user know where the specific desired modification tool is located in the Grips prompts.

Exactly what are grips?

Grips are control points located on all objects, blocks, and groups. We will look at blocks and groups in Lesson 19, but we can gain an understanding of grips using simple lines, polylines and circles.

> Before we begin our experiments with grips, we must reactivate them. Follow this sequence:
>
> **Command:** *grips*
> **New value for GRIPS <0>:** *1*

> It will also help if we deactivate **Noun/Verb Selection**. It is not mandatory; they will work together. But **Noun/Verb Selection** might confuse your study of Grips, so let's turn it *Off*. You can use the Options dialog box just covered or you can follow this sequence:
>
> **Command:** *pickfirst*
>
> **New value for PICKFIRST <1>:** *0*

Grips are assigned to specific locations on specific objects. That is, all lines and arcs will have grips at the endpoints and midpoint. All circles will have grips at the center point and quadrants. And all polylines will have grips at the vertices.

There are five basic modification commands available using grips: *Stretch*, *Move*, *Rotate*, *Scale*, and *Mirror*. *Copy* is also available as an option of each of the primary Grip commands.

To access the commands, first select the object(s) you wish to modify. (Do this without entering a command.) The blue grips that display for each of the objects are called *unselected* grips. Picking on a grip will cause it to change (as shown in Figure 17.2a) to a filled, red, *selected* grip. Your crosshairs will automatically snap to a grip regardless of the **Snap** setting. You must *select* a grip to modify the object. (*******To* disable [clear] *grips, hit the* Escape *key twice or pick* **Deselect all** *on the cursor menu*.)

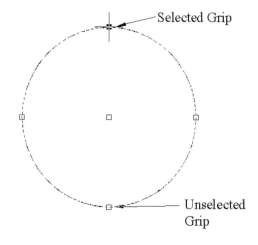

FIG. 17.2a: Grips

> The user can control all aspects of the Grips display using the Options dialog box (Figure 17.1a). Here the user can set up different grips colors and sizes (not recommended), enable/disable grips in the drawing (the dialog box method of setting the **Grips** system variable) or enable/disable grips within blocks (more on blocks in Lesson 19).

Note that you can select multiple objects using a window or crossing window just as you have always done. Selecting multiple objects simply means that the grips for several objects will display at once. To remove an object from a selection set, hold down the shift key and select the object to remove. Its grips will remain showing, but the object will no longer be highlighted and will not be affected by the editing procedure.

Once a grip has been selected, AutoCAD's command line presents the grip options for the *Stretch* command. The user can toggle through the commands (*Stretch*, *Move*, *Rotate*, *Scale*, *Mirror*) by hitting the space bar.

The initial Grips prompt looks like this:

** STRETCH **
Specify stretch point or [Base point/Copy/Undo/eXit]:

By default, the user will stretch the object using the opposite endpoint as an anchor. Selecting the **Base point** option will allow the user to change the point at which he "holds" the object while stretching it. The default base point is the selected, or "hot," grip. The **Copy** option provides one of the great benefits of using grips—the user can leave the original object as it is and create stretched copies. (He can create multiple copies by default!)

> A nifty trick to using grips involves the **Shift** key. Holding it down while locating the "to point" of the stretched line will automatically start the **Copy** option. Continuing to hold it down after the first "to point" is selected will cause the Grips command to function in *offset* mode. That is, AutoCAD will determine an offset distance from the first "to point" selection; then it will create subsequent copies at the same interval.
>
> These techniques work as well for each of the five Grips commands!

Let's see what we can do with the *Stretch* Grips command.

WWW 17.2.1 Do This: Editing with Grips—Stretch

I. Be sure you are still in the *grips.dwg* file in the C:\Steps\Lesson17 folder. If not, open it now.

II. Follow these steps.

TOOLS	COMMAND SEQUENCE	STEPS
No Buttons are Available	**Command:**	**1.** Select the upper part of the star with a crossing window, as shown in Figure 17.2.1.1.

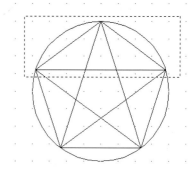

FIG. 17.2.1.1: Crossing Window

	Command:	**2.** Notice that grips appear on each of the individual objects and that each highlights (Figure 17.2.1.2).

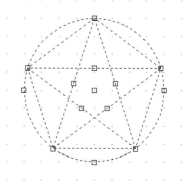

FIG. 17.2.1.2: Grips

	Command:	**3.** Select the topmost grip. Notice that it fills and turns red and that the command line presents the initial Grips options.

continued

TOOLS	COMMAND SEQUENCE	STEPS
	** STRETCH ** **<Stretch to point>/Base point/Copy/** **Undo/eXit:**	**4.** Pick a point three grid marks due north of the top of the star. (Turn on the **Snap** to help you.) Notice which objects stretch. Your drawing looks like Figure 17.2.1.4 (*zoom* or *pan* as necessary for a better view).

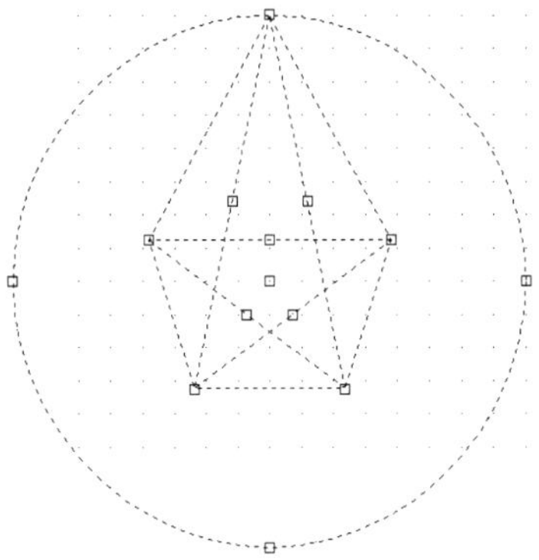

FIG. 17.2.1.4: Stretched with Grips

You have discovered the Grips method of resizing a circle. You have also seen that, although it was in the selection set, the horizontal line in the star did not change. This is because none of its grips was selected.

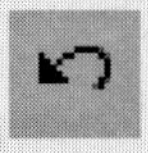	**Command:** *u*	**5.** *Undo* the last modification.
	Command:	**6.** Repeat Step 1.

continued

TOOLS	COMMAND SEQUENCE	STEPS
	Command:	7. Let's remove the circle from the selection set. Hold down the shift key and select it. Notice that its grips are still presented, but it is no longer highlighted (Figure 17.2.1.7).

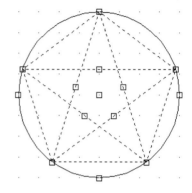

FIG. 17.2.1.7: Removed Circle

	Command: *u*	8. **Undo** the last selection.
	Command:	9. This time, select just the two lines that form the upper point of the star (Figure 17.2.1.9).

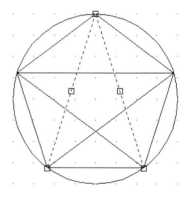

FIG. 17.2.1.9: Select These Lines

	** STRETCH ** **<Stretch to point>/Base point/Copy/Undo/eXit:** *c*	10. Select the topmost grip, then type *c* at the prompt to activate the **Copy** option.

continued

TOOLS	COMMAND SEQUENCE	STEPS
	** STRETCH (multiple) ** <Stretch to point>/Base point/Copy/Undo/eXit:	**11.** Notice how the prompt changes.
	** STRETCH (multiple) ** <Stretch to point>/Base point/Copy/Undo/eXit:	**12.** Going north, skip a grid mark then pick the next. Do this twice (see the drawing in Figure 17.2.1.20).
	** STRETCH (multiple) ** <Stretch to point>/Base point/Copy/Undo/eXit: *[enter]*	**13.** Hit *enter* to complete the procedure. (Or you can right-click and pick **Exit** from the cursor menu to complete a grips procedure.)
Esc	Command:	**14.** Hit the **Escape** key twice to clear the grips. Your drawing looks like Figure 17.2.1.20.

FIG. 17.2.1.20: Stretch and Copy

	Command: *u*	**15.** *Undo* the last modification.

We need to add some text to our star for clarity in this next exercise. Use the **Txt** layer and the default text size and style to add the text *My Star*, as shown in Figure 17.2b.

FIG. 17.2b: Add This Text

We can use the command line to access the various methods used by Grips. Simply use the spacebar to scroll through the options. But there is another way.

Any time the Grips prompts are on the command line—that is, any time you have selected a grip—you can right-click in the drawing area to access a cursor menu that provides the same options as the command line. The menu is shown in Figure 17.2c.

FIG. 17.2c
Grips' Cursor Menu

- The topmost option is the equivalent of its keyboard counterpart.
- The next section includes the five Grips commands (the same ones you will see when using the spacebar to toggle through the Grips commands on the command line).
- The third section shows the options available for the current Grips command.
- **Properties . . .** in the fourth section will cause the **Object Property Manager** to appear.
- **Go to URL . . .** is a network pick. In some AutoCAD drawings, you may find a *link* (a connection to a network or Internet site). This is one way of accessing that site. We discuss AutoCAD and the internet in *AutoCAD 2000: One Step at a Time—Advanced Edition*.
- **Exit**, of course, returns the user to the command prompt.

We will use this menu, as well as the spacebar approach to toggling Grips commands, in our next exercise.

Let's examine the ***Rotate*** and ***Mirror*** Grips procedures.

17.2.2 Do This: More Grips Editing

I. Be sure you are still in the *grips.dwg* file in the C:\Steps\Lesson17 folder. If not, open it now.

II. Follow these steps.

TOOLS	COMMAND SEQUENCE	STEPS
No Buttons are Available	**Command:**	1. Select all of the objects. Use a window or crossing window.
	Command:	2. Select the grip at the tip of the upper right star point. AutoCAD presents the grip command options.
	** STRETCH ** **Specify stretch point or [Base point/Copy/Undo/eXit]:** *[spacebar]*	3. Hit the spacebar twice to access the Grips *Rotate* command.
	** MOVE ** **Specify move point or [Base point/Copy/Undo/eXit]:** *[spacebar]*	
	** ROTATE ** **Specify rotation angle or [Base point/Copy/Undo/Reference/eXit]:** *c*	4. Type *c* to access the **Copy** option.
	** ROTATE (multiple) ** **Specify rotation angle or [Base point/Copy/Undo/Reference/eXit]:** *144*	5. Tell AutoCAD you wish to create a rotated copy at an angle of 144°.
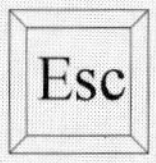 Esc	**** ROTATE (multiple) **** **Specify rotation angle or [Base point/Copy/Undo/Reference/eXit]:** *[enter]*	6. Hit *enter* to exit the command, then clear the grips (hit *escape* twice). Your drawing looks like Figure 17.2.2.6.

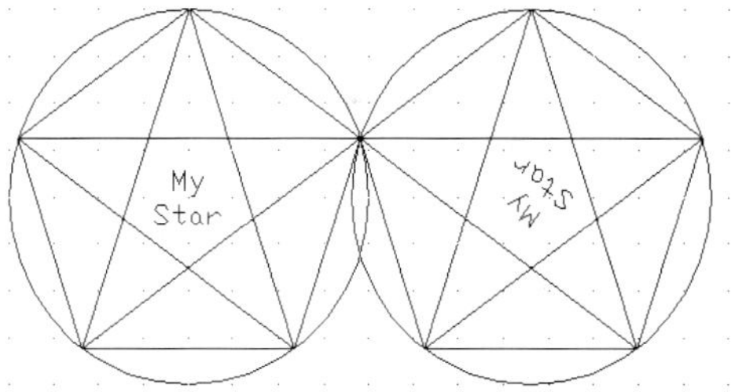

FIG. 17.2.2.6: Rotate and Copy

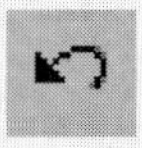	**Command:** *u*	7. *Undo* the last modification.
	Command:	8. Let's try to rotate the objects using the **Reference** option. Repeat Steps 1 and 2.

continued

TOOLS	COMMAND SEQUENCE	STEPS
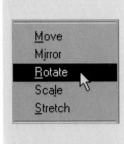	** STRETCH ** Specify stretch point or [Base point/Copy/Undo/eXit]: _rotate	9. Right-click in the drawing area and select the *Rotate* command.
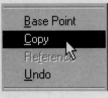	** ROTATE ** Specify rotation angle or [Base point/Copy/Undo/Reference/eXit]: _copy	10. Right-click again and select the **Copy** option.
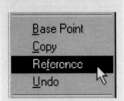	** ROTATE (multiple) ** Specify rotation angle or [Base point/Copy/Undo/Reference/eXit]: _reference	11. Right-click again and select the **Reference** option.
	Specify reference angle <0>: *[enter]*	12. Accept the default reference angle of *0*.
	** ROTATE (multiple) ** Specify new angle or [Base point/Copy/Undo/Reference/eXit]: *180*	13. Tell AutoCAD you want a rotated copy at 180°.

continued

TOOLS	COMMAND SEQUENCE	STEPS
	** ROTATE (multiple) ** Specify new angle or [Base point/ Copy/Undo/Reference/eXit]: *[enter]*	14. Hit *enter* to complete the command; clear the grips. Your drawing looks like Figure 17.2.2.14.

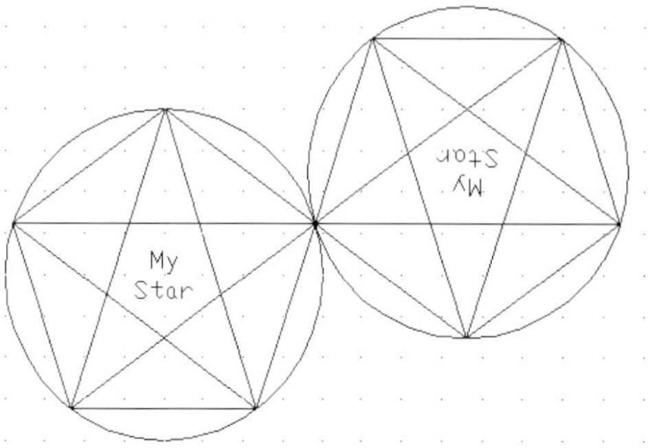

FIG. 17.2.2.14: Rotated with Grips

Remember the *Coprot.lsp* file we picked up in Lesson 7? This is AutoCAD's method for accomplishing what that Lisp routine did.

TOOLS	COMMAND SEQUENCE	STEPS
 ↶	Command: *u*	15. *Undo* the last modification.
	Command:	16. Now let's look at the *Mirror* Grips command. Repeat Step 1.
	Command:	17. Select the uppermost grip (at the upper tip of the star).
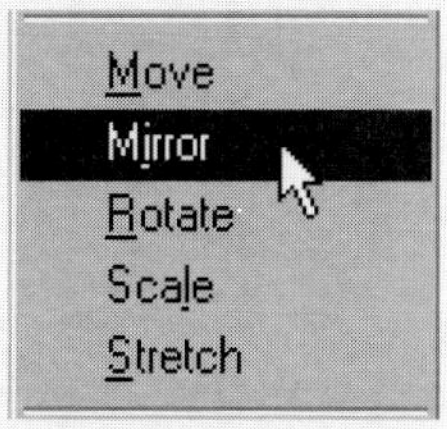 Move / **Mirror** / Rotate / Scale / Stretch	** STRETCH ** Specify stretch point or [Base point/ Copy/Undo/eXit]: _mirror	18. Right-click to display the Grips menu; pick the *Mirror* command.
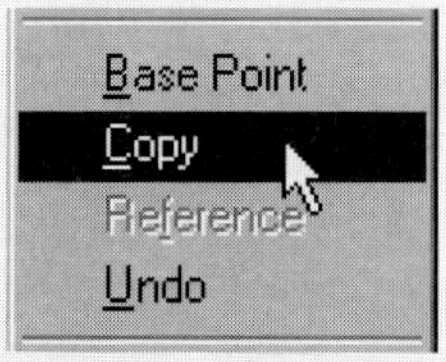 Base Point / **Copy** / Reference / Undo	** MIRROR ** Specify second point or [Base point/ Copy/Undo/eXit]: _copy	19. Right-click again and pick the **Copy** option.

continued

TOOLS	COMMAND SEQUENCE	STEPS
	** MIRROR (multiple) ** Specify second point or [Base point/Copy/Undo/eXit]:	20. Pick a point due east of the hot grip as shown. (Be sure to use the ortho.)

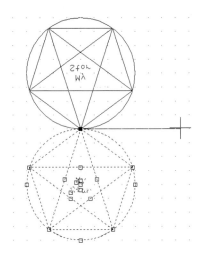

FIG. 17.2.2.20: Pick a Point Due East

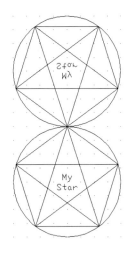	** MIRROR (multiple) ** Specify second point or [Base point/Copy/Undo/eXit]: _exit	21. Pick the **Exit** option on the Grips menu, then clear the grips. Your drawing looks like Figure 17.2.2.21.

FIG. 17.2.2.21: Mirrored Objects

(*Note:* These objects were mirrored with the **Mirrtext** sysvar set to 1. If yours is set to 0, the text will not mirror.)

	Command: *u*	22. *Undo* the last modification.

This last section will first walk you through the simple *Move* Grips command. Then we will use the *Scale* Grips command to make sized copies of objects. Let's proceed.

17.2.3 Do This: Move and Scale with Grips

I. Be sure you are still in the *grips.dwg* file in the C:\Steps\Lesson17 folder. If not, open it now.

II. Follow these steps.

TOOLS	COMMAND SEQUENCE	STEPS
No Buttons are Available	Command:	1. Select all of the objects. Use a window or crossing window.
	Command:	2. Select any one of the grips. (With the *Move* Grips command, all the objects with active grips will move.) I am using the grip at the topmost point of the star.
Move / Mirror / Rotate / Scale / Stretch	** STRETCH ** Specify stretch point or [Base point/ Copy/Undo/eXit]: _move	3. Right-click to display the Grips menu. Select the *Move* command.
Base Point / Copy / Reference / Undo	** MOVE ** Specify move point or [Base point/ Copy/Undo/eXit]: _base	4. Right-click again and select the **Base point** option.
	Specify base point: _cen of	5. Using OSNAPS, pick the center of the circle as the new base point.
	** MOVE ** Specify move point or [Base point/ Copy/Undo/eXit]: 6,4.5	6. Tell AutoCAD to move the objects to the absolute coordinate **6,4.5**. Notice that all of the objects with grips showing move even though only one grip has been selected.
💾	Command: *qsave*	7. Save the drawing, but do not exit.

continued

TOOLS	COMMAND SEQUENCE	STEPS
	Command:	**8.** Now let's scale the objects. If the grips are displayed, clear them.
	Command:	**9.** Select only the circle and the polygon (Figure 17.2.3.9).

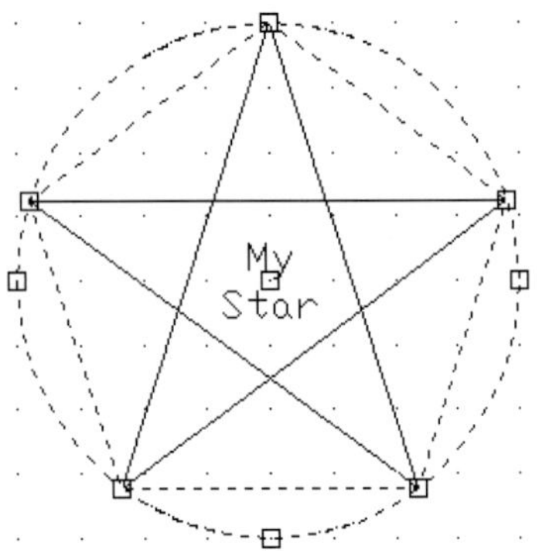

FIG. 17.2.3.9: Select the Circle and Polygon

TOOLS	COMMAND SEQUENCE	STEPS
	Command:	**10.** Select one of the grips. (I will use the bottom quadrant grip.)
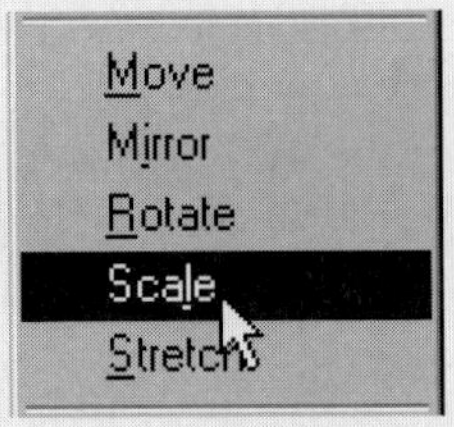 Move / Mirror / Rotate / **Scale** / Stretch	**** STRETCH **** **Specify stretch point or [Base point/ Copy/Undo/eXit]: _scale**	**11.** Call the Grips menu and pick the *Scale* command.
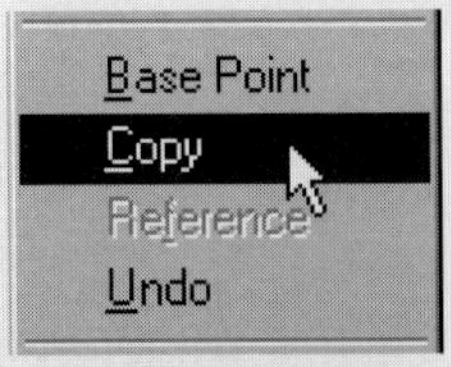 Base Point / **Copy** / Reference / Undo	**** SCALE **** **Specify scale factor or [Base point/ Copy/Undo/Reference/eXit]: _copy**	**12.** Call the Grips menu and pick the **Copy** option.
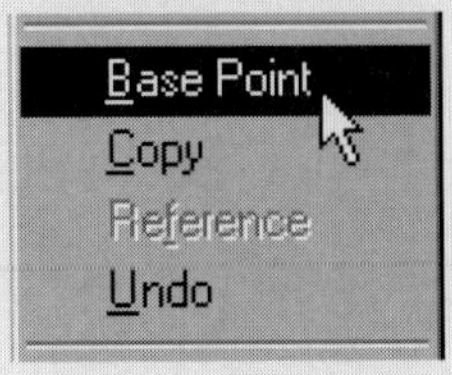 **Base Point** / Copy / Reference / Undo	**** SCALE (multiple) **** **Specify scale factor or [Base point/ Copy/Undo/Reference/eXit]: _base**	**13.** Call the Grips menu and pick the **Base Point** option.
	Specify base point: _cen of	**14.** Using OSNAPS, pick the center of the circle as the base point.

continued

TOOLS	COMMAND SEQUENCE	STEPS
	** SCALE (multiple) ** **Specify scale factor or [Base point/ Copy/Undo/Reference/eXit]:**	**15.** Make three scaled copies then clear the grips, as shown in Figure 17.2.3.15.
		FIG. 17.2.3.15: Scaled Copies
💾	**Command:** *qsave*	**16.** Save the drawing and exit.

So what do you think of Grips?

After learning the basic modifying commands in earlier lessons, Grips may be difficult at first. But practice, practice, practice! Grips will increase speed and productivity (excuse that nasty corporate word)!

17.3 Extra Steps

As your *Extra Step*, I suggest returning to any of the exercises in previous lessons (Lessons 8 and 10 would be especially useful) and experimenting with the settings and methods learned in this lesson.

While you are working, ask yourself:

- Is this any faster than the first time I did it?
- Is it any easier?
- Do I prefer this setting to AutoCAD's default?
- Do I prefer the basic or Grip method of modification?

Answering these questions will put you on your way to some real expertise with the software!

At this point in your training, you should begin to "customize" the software to your style of doing things. The user interface should begin to reflect your preferences for toolbars and scrollbars; the system variables should reflect your preferences for drawing styles and methods.

17.4 What Have We Learned?

Items covered in this lesson include:

- *AutoCAD Selection Tools:*
  - *Noun/Verb Selection*
  - *Shift to*
  - *Add*
  - *Press and Drag*
  - *Implied Windowing*

- *AutoCAD's Grip Tools:*
  - *Move*
  - *Stretch*
  - *Scale*
  - *Rotate*
  - *Mirror*

In this lesson, we learned some entirely different approaches to modifying AutoCAD drawings. Now you must decide which methods and settings you, as the CAD operator, will use.

You have completed the first three sections of this text and accomplished quite a bit. Tackle this lesson's exercise with confidence in your training and faith in your ability! Then approach *Part IV: Razzle Dazzle* knowing that you will soon have the knowledge necessary to succeed as a drafter or junior designer at any company in the United States!

17.5 EXERCISES

I have modified an earlier project for our first Lesson 17 exercise. This should give you some frame of reference when you compare different modifying commands and methods.

1. Open the *Cabin-adv.dwg* file in the C:\Steps\Lesson17 folder.
    1.1. Using the objects provided, the Grips commands, and your preferences in the Options dialog box, complete the drawing shown in Figure 17.5.1.

17.5 Exercises

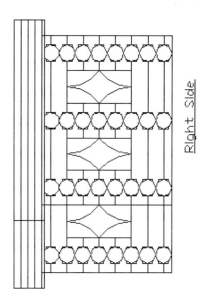
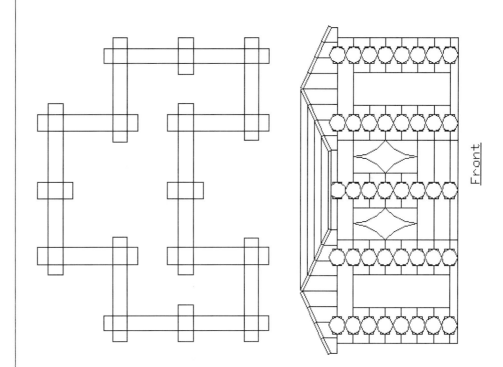

FIG. 17.5.1: *Cabin-adv.dwg*

  1.2. Save the drawing as *MyCabin-adv.dwg* in the C:\Steps\Lesson17 folder.
2. Start a new drawing from scratch.
  2.1. Use the following setup:
    2.1.1. Grid: $1/2$
    2.1.2. Snap and Layers: as needed
    2.1.3. Font: Times New Roman
    2.1.4. Dimension styles: as needed
  2.2. Save the drawing as *MyGasket* in the C:\Steps\Lesson17 folder.
  2.3. Create the drawing shown in Figure 17.5.2.

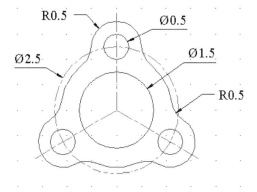

FIG. 17.5.2: *MyGasket.dwg*

3. Start a new drawing from scratch.
  3.1. Use the setup described in 2.1.
  3.2. Save the drawing as *MySlottedGuide* in the C:\Steps\Lesson17 folder.
  3.3. Create the drawing shown in Figure 17.5.3.

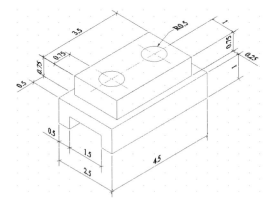

FIG. 17.5.3: *MySlottedGuide.dwg*

4. Start a new drawing from scratch.
    4.1. Repeat the setup you did in Exercises 2 and 3.
    4.2. Save the drawing as *MySpring* in the C:\Steps\Lesson17 folder.
    4.3. Create the drawing shown in Figure 17.5.4.

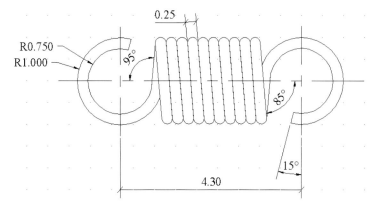

FIG. 17.5.4: MySpring.dwg

5. Create and dimension the drawing in Figure 17.5.5. The grid is set to $1/4$". Use appropriate layers.

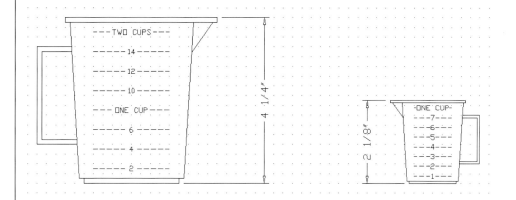

FIG. 17.5.5: Measuring Cups

6. Using the bottle detail shown in Figure 17.5.6a, create the drawing of the spice rack in Figure 17.5.6b. The grid is set to $1/8$". Fully dimension the drawing (not the bottles) and use appropriate layers.

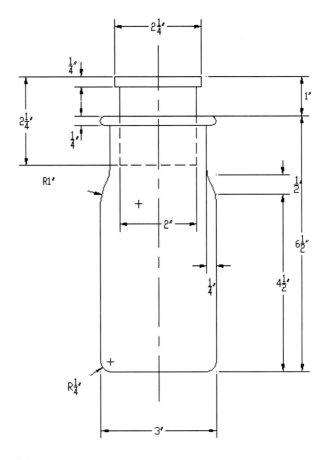

FIG. 17.5.6a: Bottle Detail

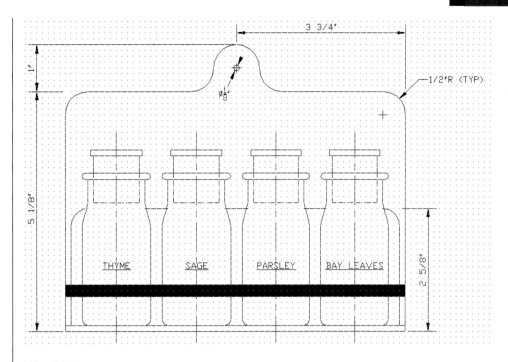

FIG. 17.5.6b: Spice Rack

17.6 REVIEW QUESTIONS

Write your answers on a separate sheet of paper.

1. Normal command procedures—giving a command then selecting the object of the command—can be called (Noun/Verb selection, Verb/Noun selection).

2. Access the Options dialog box by typing _____ at the command prompt.

3. Turn grips off by giving them a value of _____.

4. The system variable for Noun/Verb selection is _____.

5. Its default value is _____.

6. If the Use Shift to Add box in the Options dialog box has a check in it, the user must hold down the _____ key to select more than one object.

7. The sysvar for Use Shift to Add is _____.

8. Its default value is _____.

9. The sysvar for Press and Drag is _____.

10. Its default setting is _____.

11. _____ enables us to pick an open area of the drawing to automatically begin a selection window.

12. The sysvar for the previous command is _____.

13. Its default value is _____.

14. Picking the _____ button in the Options dialog box will reset all the defaults.

15. To change the size of the pickbox from the command line, type _____.

16. _____ are control points located on all objects, blocks and groups.

17. **to 21.** List the five modification commands available using grips.

22. The _____ command is also available as an option for each of the primary grip commands.

23. To disable (clear) grips, hit the _____ key twice.

24. To remove an object from a grips selection set, hold down the _____ key while selecting it.

25. The user toggles through the available grips commands by hitting the _____.

26. (T or F) There is no way to offset objects using grips.

27. Any time you have selected a grip, you can _____ in the drawing area to access a screen menu that provides the same options as the command line.

28. (T or F) Using the Move Grips command, all the objects with active grips will move.

PART IV

Razzle Dazzle

This part of our text contains these lessons:
18. Hatching and Section Lines
19. Many as One: Groups and Blocks
20. Advanced Blocks
21. Afterword: Getting an Edge

LESSON 18

Hatching and Section Lines

Following this lesson, you will:

➥ Know how to add Hatching and Section Lines to your AutoCAD drawing through:
  • The **Hatch** command
  • The **BHatch** command and dialog box
➥ Know how to edit hatch patterns using the **Hatchedit** command
➥ Know how to use the **Sketch** command

Remember all those templates I mentioned earlier—the ones I needed to draw ellipses? Remember how much simpler (and more accurate) it was to draw ellipses with AutoCAD?

Here is another drawing tool that puts those pieces of plastic to shame. No longer will the drafter need to spend hours drawing each line or symbol to show section lines, a brick façade, concrete, and so forth. With AutoCAD's hatch commands, the user can fill a large area with lines or symbols in a matter of seconds!

This fourth and final section of our text will cover some *razzle-dazzle* tools. These are the tools that pushed CAD systems ahead of the drawing board as the preferred tool in design work. We will begin with hatching in this lesson; then, in subsequent lessons, we will learn to create blocks of objects and of information.

Let's begin our study of Hatching.

18.1 The *Hatch* Command

There are actually a couple of ways to draw hatch patterns. (Section lines are created as a style of hatch pattern. Therefore, I will refer to both as hatch patterns.) There is a dialog box approach called Boundary Hatching; but we will begin with the command line approach as it presents some opportunities you will not find in the dialog box. The command sequence is

Command: *hatch* (or *–h*)
Enter a pattern name or [?/Solid/User defined] <ANSI31>: *[enter the pattern you wish to use]*
Specify a scale for the pattern <1.0000>: *[enter the scale for your hatch pattern]*
Specify an angle for the pattern <0>: *[enter the desired angle of the pattern]*
Select objects to define hatch boundary or <direct hatch>,
Select objects: *[select closed objects or objects that will form the closed boundaries of the hatching]*
Select objects: *[enter to complete the command]*
Command:

- There are four options presented at the initial hatch prompt.
  - The default option (the one within the angled brackets) is **ANSI31**. This is the standard hatch pattern most people use to show section lines.
  - A question mark entered here will call the following prompt:

    Enter pattern(s) to list <*>:

    Hit *enter* to view the available hatch patterns.
  - The **Solid** hatch pattern will produce the same results—but faster—as the *Solid* and *Donut* commands you learned in Lesson 11.

- The **User Defined** option allows the user to create a hatch pattern by defining the angle and spacing of hatch lines. If selected, AutoCAD presents these prompts:

    Specify angle for crosshatch lines <0>: *[enter the angle for the lines]*

    Specify spacing between the lines <1.0000>: *[enter the spacing between the lines]*

    Double hatch area? [Yes/No] <N>: *[double hatching means that the lines will be drawn according to the angle defined at the first prompt here and a second set of lines will be drawn perpendicular to the first]*

▌ The next prompt asks the user to enter a **Scale**. This may already be set if the user created the drawing using a wizard. But if it defaults to *1* and you wish to plot the final drawing to scale, enter the drawing scale factor here. (Refer to the Drawing Scales chart on the tear out reference card.)

▌ The **Angle** of the hatch pattern defaults to zero. This does *not* mean that the lines of the hatch pattern will be horizontal. It means that AutoCAD will draw the lines of the selected hatch pattern as they appear in the pattern's definition. Changing the angle here will cause AutoCAD to rotate the pattern by the angle specified.

▌ Next, AutoCAD requires the user to identify the boundaries of the hatch pattern. The user accomplishes this by simply selecting the objects to use as boundaries.

Or the user can hit *enter* to use the **direct hatch option**. This option allows the user to draw the boundaries (polylines) on the fly. The command sequence for the **direct hatch option** looks like this:

Retain polyline boundary? [Yes/No] <N>: *[tell AutoCAD if you wish to keep the boundary you are about to define]*

Specify start point: *[pick the start point]*

Specify next point or [Arc/Close/Length/Undo]: *[continue picking points on the boundary—this option repeats]*

Specify next point or [Arc/Close/Length/Undo]: *c [to close the boundary]*

Specify start point for new boundary or <apply hatch>: *[the user can hit enter to complete the command or hatch more than one area at a time]*

Command:

- The first option here — **Retain polyline?** <N> — allows the user to keep the polyline he will create to define the hatched area. The default is to delete it.
- The rest of the options here are standard *Pline* options.

Okay. Let's take a look at the *Hatch* command in action.

WWW 18.1.1 Do This: | Command Line Hatching

I. Open the *demo-hatch.dwg* file in the C:\Steps\Lesson18 folder. The drawing looks like Figure 18.1.1a.

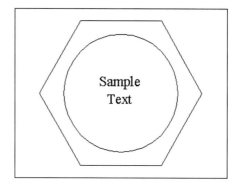

FIG. 18.1.1a: *Demo-hatch.dwg*

II. Follow these steps.

TOOLS	COMMAND SEQUENCE	STEPS
No Button Available (*Note*: The button on the Draw toolbar calls the **BHatch** command.)	Command: *–h*	1. Enter the **Hatch** command by typing *hatch* or *–h* at the command prompt.
	Enter a pattern name or [?/Solid/User defined] <ANSI31>: *[enter]* Specify a scale for the pattern <1.0000>: *[enter]* Specify an angle for the pattern <0>: *[enter]*	2. Accept the defaults for pattern, scale, and angle.

continued

TOOLS	COMMAND SEQUENCE	STEPS
	Select objects to define hatch boundary or \<direct hatch\>, **Select objects:** **Select objects:** *[enter]*	3. Select the outer rectangle, then hit *enter* to complete the command. Your drawing looks like Figure 18.1.1.3.

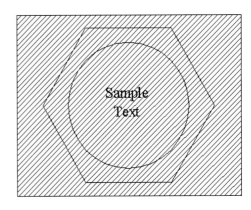

FIG. 18.1.1.3: *Default Hatching of Outer Object Only*

 Command: *u* — 4. Undo the last modification.

Command: *–h* — 5. Repeat Steps 1 and 2.

Select objects to define hatch boundary or \<direct hatch\>,
Select objects:
Select objects: *[enter]*

6. This time, select all of the objects before completing the command. Your drawing looks like Figure 18.1.1.6.

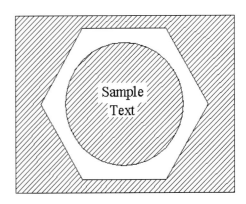

FIG. 18.1.1.6: *Default Hatching of All Objects*

Notice the results of the above hatching. AutoCAD has recognized multiple objects in the selection set and hatched so that the user can tell them apart. It has even recognized the text in the middle. This is AutoCAD's **Normal** style of hatching.

continued

TOOLS	COMMAND SEQUENCE	STEPS

There are three *styles* altogether. By default, AutoCAD uses the **Normal** style shown above. Now let's look at the other two—**Outer** and **Ignore**.

Command: *u*

7. Undo the last modification.

Command: *–h*

8. Repeat the *Hatch* command.

Enter a pattern name or [?/Solid/ User defined] <ANSI31>: *ansi31,o*

9. When prompted for the pattern, type *ANSI31,o*. (The *o* tells AutoCAD to hatch only between the two *Outermost* boundaries.)

Specify a scale for the pattern <1.0000>: *[enter]*

10. Accept the defaults for scale and angle.

Specify an angle for the pattern <0>: *[enter]*

Select objects to define hatch boundary or <direct hatch>, Select objects: Select objects: *[enter]*

11. Select all of the objects; complete the command. Your drawing looks like Figure 18.1.1.11. (Only the area inside the outer boundary has been hatched.)

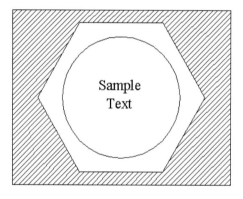

FIG. 18.1.1.11: Outer Hatching

Command: *u*

12. Undo the last modification.

Command: *–h*

13. Repeat the *Hatch* command.

Enter a pattern name or [?/Solid/ User defined] <ANSI31>: *ansi31,i*

14. When prompted for the pattern, type *ANSI31,i*. (The *i* tells AutoCAD to *ignore* any internal boundaries.)

continued

TOOLS	COMMAND SEQUENCE	STEPS
	Specify a scale for the pattern <1.0000>: *[enter]*	**15.** Accept the defaults for scale and angle.
	Specify an angle for the pattern <0>: *[enter]*	
	Select objects to define hatch boundary or <direct hatch>, Select objects: Select objects: *[enter]*	**16.** Select all of the objects; complete the command. Your drawing looks like Figure 18.1.1.16.

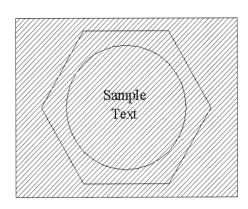

FIG. 18.1.1.16: *Ignore Hatching*

	Command: *u*	**17.** Undo the last modification.
	Command: *–h*	**18.** Let's define our own pattern. Repeat the *Hatch* command.
	Enter a pattern name or [?/Solid/User defined] <ANSI31>: *u*	**19.** Tell AutoCAD to use the **User defined** option.
	Specify angle for crosshatch lines <0>: *60*	**20.** Draw the lines for your **User defined** pattern at 60° . . .
	Specify spacing between the lines <1.0000>: *.25*	**21.** . . . and space them $1/4$ unit apart.
	Double hatch area? [Yes/No] <N>: *[enter]*	**22.** Do not double hatch the area yet.

continued

TOOLS	COMMAND SEQUENCE	STEPS
	Select objects to define hatch boundary or <direct hatch>, Select objects: Select objects: *[enter]*	23. Select the circle and complete the command. Your drawing looks like Figure 18.1.1.23.

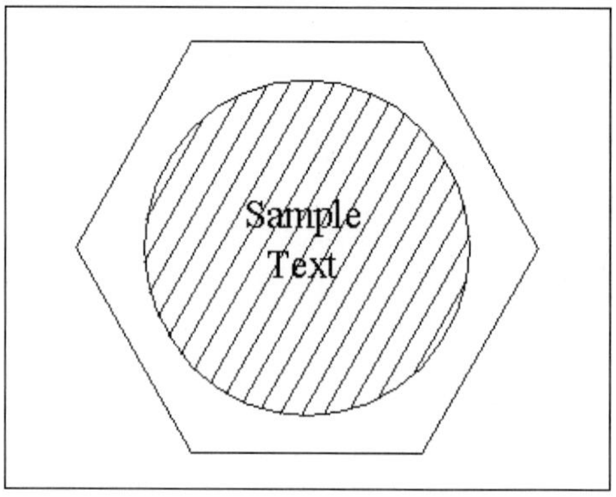

FIG. 18.1.1.23: User-Defined Hatching

	Command: *–h*	24. Repeat the *Hatch* command. (Do *not* hit *enter* as that will repeat the **Select objects:** prompt only. AutoCAD will use the **Style**, **Angle**, and **Spacing** you last defined.)
	Enter a pattern name or [?/Solid/User defined] <U>: *[enter]*	25. Define another style.
	Specify angle for crosshatch lines <60>: *[enter]*	26. Accept the default angle and spacing.
	Specify spacing between the lines <0.2500>: *[enter]*	
	Double hatch area? [Yes/No] <N>: *y*	27. This time, let's **Double hatch** an area.

continued

TOOLS	COMMAND SEQUENCE	STEPS
	Select objects to define hatch boundary or <direct hatch>, Select objects: Select objects: *[enter]*	**28.** Select the polygon *and* the circle. Then complete the command. Your drawing looks like Figure 18.1.1.28.

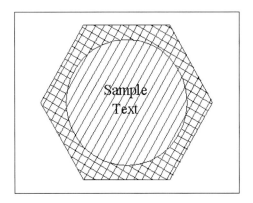

FIG. 18.1.1.28: Double Hatching

Notice that when double hatching is indicated, AutoCAD draws the second set of lines at 90° to the first.

	Command: *–h*	**29.** Now let's look at the **Solid** option. Repeat the **Hatch** command. (Do *not* hit *enter*.)
	Enter a pattern name or [?/Solid/ User defined] <U>: *s*	**30.** Enter *s* for the **Solid** option.
	Select objects to define hatch boundary or <direct hatch>, Select objects: Select objects: *[enter]*	**31.** Select the rectangle and the polygon. Then complete the command. Your drawing looks like Figure 18.1.1.31.

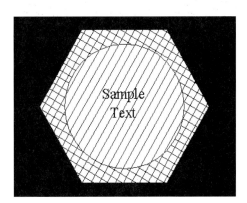

FIG. 18.1.1.31: Solid Hatching

continued

TOOLS	COMMAND SEQUENCE	STEPS
✏️	**Command:** *e*	**32.** Erase all the hatch patterns. Notice that a single pick selects all the lines in a pattern. (You may have to pick several places to find a pickable point in the solid pattern.)
	Command: *–h*	**33.** Let's change the **Angle** of the pattern and see what happens. Repeat the *Hatch* command.
	Enter a pattern name or [?/Solid/User defined] <SOLID>: *ansi31,o*	**34.** Use the **Outermost** hatch style and the **ANSI31** pattern.
	Specify a scale for the pattern <1.0000>: *[enter]*	**35.** Accept the **Scale**, but tell AutoCAD to use a 90° **Angle** for the hatching.
	Specify an angle for the pattern <0>: *90*	
	Select objects to define hatch boundary or <direct hatch>, **Select objects:** **Select objects:** *[enter]*	**36.** Select the rectangle and the polygon. Complete the command. Your drawing looks like Figure 18.1.1.36.

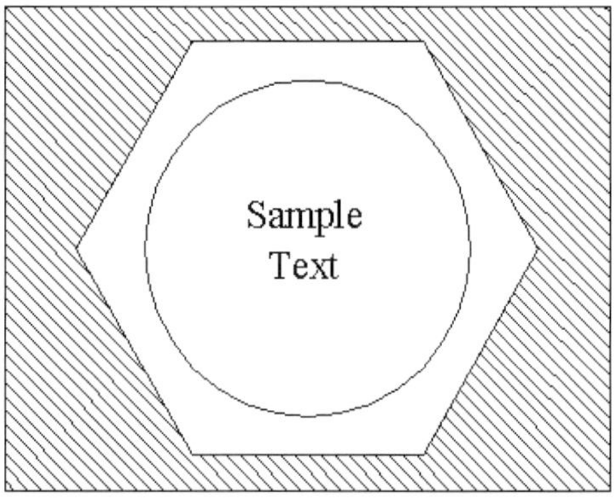

FIG. 18.1.1.36: Angled Hatching

	Command: *–h*	**37.** Now let's look at the **Scale** option. Repeat the *Hatch* command. (Do *not* hit *enter*.)
	Enter a pattern name or [?/Solid/User defined] <ANSI31,O>: *[enter]*	**38.** Accept the default pattern.

continued

TOOLS	COMMAND SEQUENCE	STEPS
	Specify a scale for the pattern <1.0000>: *2*	39. Change the **Scale** to *2*, and return the **Angle** to *0*.
	Specify an angle for the pattern <90>: *0*	
	Select objects to define hatch boundary or <direct hatch>, Select objects: Select objects: *[enter]*	40. Select the polygon and the circle, then complete the command. Your drawing looks like Figure 18.1.1.40.

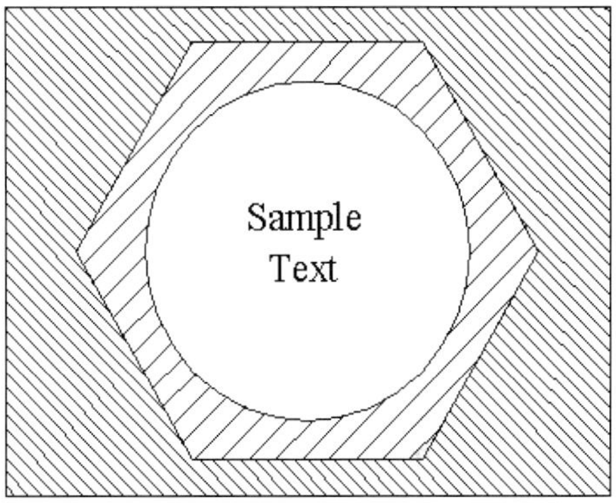

FIG. 18.1.1.40: Scaled Hatching

	Command: *–h*	41. We have one more thing to explore—the **direct hatch option**. Here we will create a hatch pattern within a temporary polyline. Repeat the *Hatch* command.
	Enter a pattern name or [?/Solid/User defined] <ANSI31,O>: *ansi31*	42. Use the **Normal ANSI31** pattern.
	Specify a scale for the pattern <2.0000>: *1*	43. Set the **Scale** to *1* and the **Angle** to *135*.
	Specify an angle for the pattern <0>: *135*	
	Select objects to define hatch boundary or <direct hatch>, Select objects: *[enter]*	44. This time, hit *enter* to access AutoCAD's **direct hatch option**.
	Retain polyline boundary? [Yes/No] <N>: *[enter]*	45. We do not want to retain the polyline.

continued

TOOLS	COMMAND SEQUENCE	STEPS
	Specify start point: <Ortho on> **Specify next point or [Arc/Close/Length/Undo]:** **Specify next point or [Arc/Close/Length/Undo]:** **Specify next point or [Arc/Close/Length/Undo]:** **Specify next point or [Arc/Close/Length/Undo]:** *c*	**46.** With the **Ortho *On***, draw a four-sided polyline around the text in the center of the circle. Be sure to *Close* the polyline.
	Specify start point for new boundary or <apply hatch>: *[enter]*	**47.** Hit *enter* to complete the command. Your drawing looks like Figure 18.1.1.47.

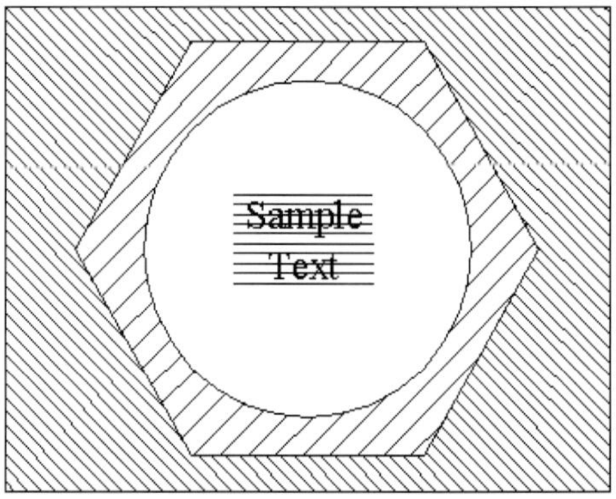

FIG. 18.1.1.47: Direct hatch

TOOLS	COMMAND SEQUENCE	STEPS
	**Command:** *e*	**48.** Erase all the hatch patterns.
	**Command:** *qsave*	**49.** Save the drawing, but do not exit.

> An important thing to remember about our exercises thus far is that each of the objects being hatched (circle, polygon, and rectangle) is *closed*. You can hatch lines and arcs as well, but they must also be closed.
>
> It is not at all unusual to get some fairly bizarre hatching results—particularly patterns that bleed outside of the selected boundaries. There are a number of quirks that can cause this, but the most common is an open boundary. If you are sure the boundary is closed and you still get some freaky result or if you simply do not want a boundary to show, use the **direct hatch option**.

The greatest benefit to creating hatch patterns using the *Hatch* command (that is not available using the *BHatch* dialog box) is the **direct hatch option**. This is an ideal tool for adding partial grating to platforms and walkways, or adding partial façade features (such as brickface) to a building elevation. Use this when you just need a sample of the pattern or when fully hatching an area might hinder the clarity of the drawing.

18.2 Boundary Hatching

The Boundary Hatch dialog box (Figure 18.2a), accessed by the *BHatch* command, provides user-friendly access to many of the *Hatch* command options. It also offers some major benefits that the *Hatch* command does not. Let's take a look at the dialog box.

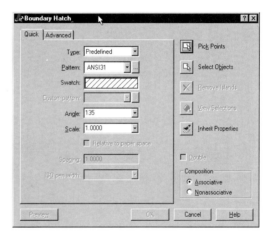

FIG. 18.2a: Boundary Hatch Dialog Box

The first tab you see (**Quick**) allows the user to set up the type, scale, and angle of hatching desired. The second tab (**Advanced**) enables the user to determine how the hatching will be applied.

Additionally, you will find a series of buttons down the right side of the dialog box. The first four buttons allow the user to identify where the hatch pattern will go. The fifth is sort of a cheater to set the properties on the **Quick** tab a little faster.

Let's take a more detailed look at the Boundary Hatch dialog box.

- We will begin with the **Quick** tab.
  - The **Type** control box allows the user to select a **Predefined** pattern, a **User-defined** pattern, or a **Custom** pattern.
    - The **Predefined** patterns accompany AutoCAD. You can see these in the **Swatch** box or by picking the button next to the **Pattern** control box.

- The **User-defined** option allows the user to set the **Angle** and **Spacing** of lines (using the control boxes on this tab), and to decide whether or not to **Double** the pattern as you did in the last exercise (using the **Double** check box to the right of the tabs). The **Spacing** text box and the **Double** check box are only available when the **User-defined** option has been selected. These correspond to their *Hatch* command counterparts.
- Selecting **Custom** in the **Type** control box allows the user to select a custom pattern. A *Custom* pattern is also predefined, but the definition is stored in a file other than the *Acad.pat* (the file where AutoCAD keeps hatch pattern definitions by default). The CAD guru will provide custom patterns if needed. If the **Custom** option is selected in the **Pattern** control box, the user must enter the name of the desired pattern in the **Custom Pattern** control box.
- The **Pattern** control box allows the user to scroll through the patterns one at a time; or you can pick the button to the right of the control box to display the Hatch Pattern Palette dialog box, which displays several patterns at a time.
- **Scale** and **Angle** control boxes are provided for the user to set these options as he did at the *Hatch* command prompt.
- The **Iso Pen Width** option is only available when an ISO pattern has been selected. This control box allows the user to select the width of the ISO lines.

▌ The **Advanced** tab (Figure 18.2b) offers the user some choices in how to apply the hatching.

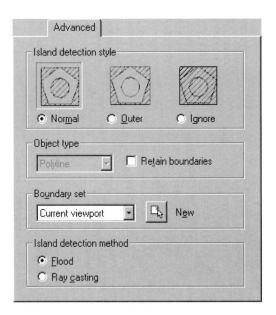

FIG. 18.2b: The Advanced Tab

- In the **Island detection style** frame, you find the same three styles you used on the command line (**Normal**, **Outer**, and **Ignore**). But thanks to the dialog box format, you can see samples of each.
- The **Object type** and **Boundary set** frames work together. Let me explain.

You will notice that, in larger drawings, it may take AutoCAD a while to locate boundaries when using the **Pick Points** method. This is because AutoCAD must search the drawing database for information on all visible objects. Using the **New** button in the **Boundary set** frame can really hasten this process. When picked, AutoCAD will prompt you to **Select objects** on the screen, then return to the **Advanced** tab of the Boundary Hatch dialog box. AutoCAD will then restrict its search to the objects selected. The user can make the new boundary set from the **Current viewport** or from an **Existing Set** by selecting your choice from the control box in the **Boundary set** frame.

> The tools available in the **Object type** and **Boundary set** frames can make a tremendous difference in the amount of time required for AutoCAD to hatch objects in larger drawings. However, you must decide if it is necessary to use these tools by considering how long you are willing to wait for the hatching and how long AutoCAD is taking to do the job.

When defining a new boundary set, the user should use a window or crossing window (other methods will work, but these are faster in a large drawing). In the **Object type** frame, the user can tell AutoCAD to retain the defined area as a polyline or a region (an object similar to a 2-dimensional solid). Simply place a check in the **Retain boundaries** box and then make your selection in the control box in the **Object type** frame.

- The last frame—**Island detection method**—is a bit deceiving. What it boils down to is simply: Do you want AutoCAD to detect islands (boundaries within boundaries) or not? If you do, leave the bullet next to **Flood**; if you do not, move the bullet to the **Ray casting** option.

▍ Once you have made your selections on the tabs, you must tell AutoCAD what to hatch using one of the buttons on the right side of the dialog box.

- You can find one of the greatest enhancements to hatching (and one of my personal favorites) in the **Pick Points** button. With this nifty tool, the user simply picks a point where he wants hatching. AutoCAD fans outward from that point to find the nearest boundaries. You just cannot get any simpler than this. (Again, you must be sure that you select inside a *closed* area.)
- The **Select Objects** button allows the user to select objects just as he did at the command line in the last exercise.
- Remember how the **Normal** style worked? AutoCAD hatched between the outer two boundaries, skipped hatching between the next two boundaries, hatched between the next two, and so forth. AutoCAD recognizes boundaries within the outermost boundary as *islands* and adjusts the hatching for clarity. The same holds true when the user selects the boundary using the **Pick Points** button. He can use the **Remove Islands** button to tell AutoCAD to ignore the islands and hatch over them. But the **Remove Islands** button gives him more control over which islands are ignored.
- The **View Selections** button highlights the currently selected boundaries to help you verify the selection.

- Use the **Inherit Properties** button when you do not know the style of pattern you want to match. That is, when you have used several styles in a drawing and wish to duplicate the settings of one of them, pick this button. AutoCAD will ask you to select the style you want to duplicate. It will then reset the Boundary Hatch dialog box options to match that style. (How is that for handy?)
- The **Preview** button (lower left corner of the dialog box) allows the user to see the hatched object *before* he completes the command. This way, the user can make any necessary adjustments before applying the hatching.
- The **Composition** frame in the lower right quadrant of the dialog box gives the user the chance to use *Associative* hatching. This means that hatch patterns will update automatically when the boundaries change. This option is *not* available at the command line.

The user can also tell AutoCAD to draw hatch lines as individual objects (**Nonassociative-** or preexploded). But be aware that exploded hatch patterns dramatically increase the size of a drawing and the amount of work involved in modifying the pattern. If the user does not create the pattern already exploded, he can always explode the pattern (using the *Explode* command) just as he could explode a polyline. (This option is available at the command line by placing an asterisk (*) in front of the name of the hatch pattern [as in *ANSI31].)

Let's take a look at hatching using the Boundary Hatch dialog box.

> In addition to the command line and toolbars, selecting **Hatch** under the Draw pull-down menu will access the *Bhatch* command.

18.2.1 Do This: Hatching with a Dialog Box

I. Be sure you are still in the *demo-hatch.dwg* file in the C:\Steps\Lesson18 folder. If not, open it now.

II. Follow these steps.

TOOLS	COMMAND SEQUENCE	STEPS
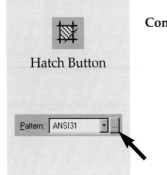 Hatch Button	Command: *h*	1. Open the Boundary Hatch dialog box by entering the **BHatch** command—type **bhatch**, **bh**, or **h** at the command prompt. Or you can pick the **Hatch** button on the Draw toolbar.
		2. Pick the **Pattern** button. The Hatch Pattern Palette dialog box appears (Figure 18.2.1.2).

continued

TOOLS	COMMAND SEQUENCE	STEPS

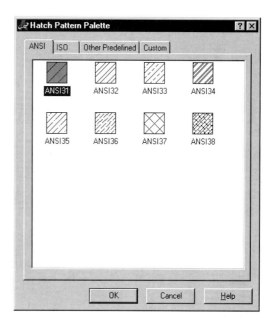

FIG. 18.2.1.2: Hatch Pattern Palette

Here you can select a pattern in one of two ways:

- double-click on it;
- click on the desired pattern, then pick the **OK** button.

Pick the tabs and use the scroll bar to see other patterns.

3. Double-click on the **ANSI33** pattern (first tab, first row, third column). AutoCAD returns to the Boundary Hatch dialog box. The ANSI33 pattern is now shown as current in the **Pattern** and **Swatch** control boxes.

4. Set the **Scale** to *1.5* and the **Angle** to *0*.

5. Pick the **Advanced** tab to see its options.

6. Set the **Style** to *Outer* in the **Island detection style** frame.

continued

TOOLS	COMMAND SEQUENCE	STEPS
 Pick Points Button		**7.** Pick the **Pick Points** button. AutoCAD returns to the graphics screen and prompts the user to **Select internal point:**.
	Select internal point:	**8.** Pick any point between the rectangle and the polygon.
	Selecting everything . . . Selecting everything visible . . . Analyzing the selected data . . . Analyzing internal islands . . . Select internal point: *[enter]*	**9.** AutoCAD tells you what it is doing, then repeats the prompt. Hit *enter* to complete the selection. AutoCAD returns to the Boundary Hatch dialog box.
Preview		**10.** Pick the **Preview** button to see the hatching before applying it. Your drawing looks like Figure 18.2.1.10.

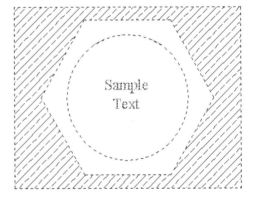

FIG. 18.2.1.10: Preview of Hatching

> *Note:* The Preview Hatch display is *not* always accurate. Most of the ANSI patterns preview fairly well; but many of the other, more complex, patterns tend to lose definition or not display at all. If the pattern does not display, but you are reasonably sure of your settings, go ahead and apply the pattern. You can always edit or erase the pattern and start again.

continued

TOOLS	COMMAND SEQUENCE	STEPS
	<Hit enter or right-click to return to the dialog> *[enter]*	11. Hit *enter* to return to the Boundary Hatch dialog box.
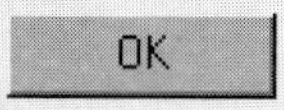 OK		12. Pick the **OK** button to complete the command. AutoCAD returns to the graphics screen and the command prompt. Your drawing looks like Figure 18.2.1.12.

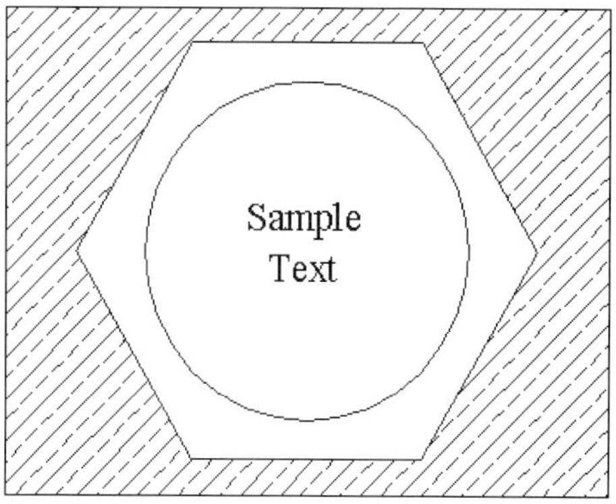

FIG. 18.2.1.12: Hatched Objects

| 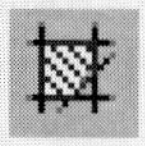 | Command: *h* | 13. Repeat Steps 1 through 15, but hatch between the polygon and the circle with the ANSI37 pattern. Use a **Scale** of **2.0** and an **Angle** of **45°**. Your drawing will look like Figure 18.2.1.13. |

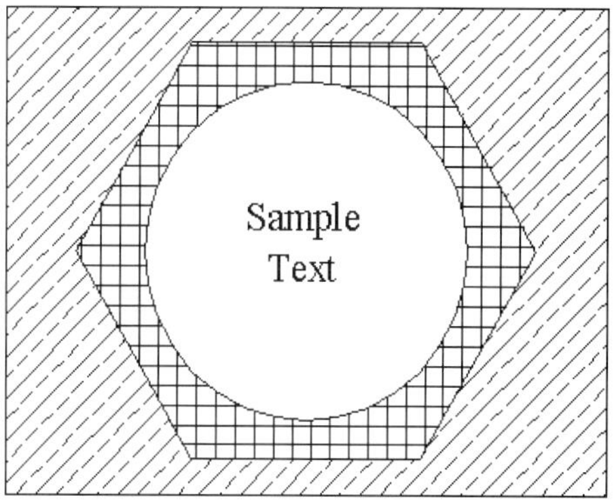

FIG. 18.2.1.13: More Hatched Objects

continued

TOOLS	COMMAND SEQUENCE	STEPS
	Command: *[enter]*	**14.** Now we will use the **Inherit Properties** button to hatch the circle using the ANSI33 pattern we used between the rectangle and the polygon. Repeat the *BHatch* command. Notice that the settings default to the last settings used.
 Inherit Properties Button		**15.** Pick the **Inherit Properties** button. AutoCAD returns you to the graphics screen and prompts **Select associative hatch object:**.
	Select associative hatch object:	**16.** Pick a line on the hatch pattern between the rectangle and the polygon.
	Inherited Properties: Name <ANSI33, O>, Scale <1.5000>, Angle <0>	**17.** Pick a point inside the circle.
	Select internal point:	
	Selecting everything . . .	**18.** Hit *enter*. AutoCAD returns to the Boundary Hatch dialog box. Notice that the settings have changed to match those used when you created the first hatching.
	Selecting everything visible . . .	
	Analyzing the selected data . . .	
	Analyzing internal islands . . .	
	Select internal point: *[enter]*	
		**19.** Change the **Angle** setting to *90°*.
 View Selections Button		**20.** Pick the **View Selections** button. AutoCAD returns to the graphics screen and highlights the boundaries you selected in the previous step.
	<Hit enter or right-click to return to the dialog> *[enter]*	**21.** Hit *enter* to return to the Boundary Hatch dialog box.

continued

TOOLS	COMMAND SEQUENCE	STEPS
[OK]		22. **Preview** the hatching if you wish, then **OK** it. Your drawing looks like Figure 18.2.1.22.
		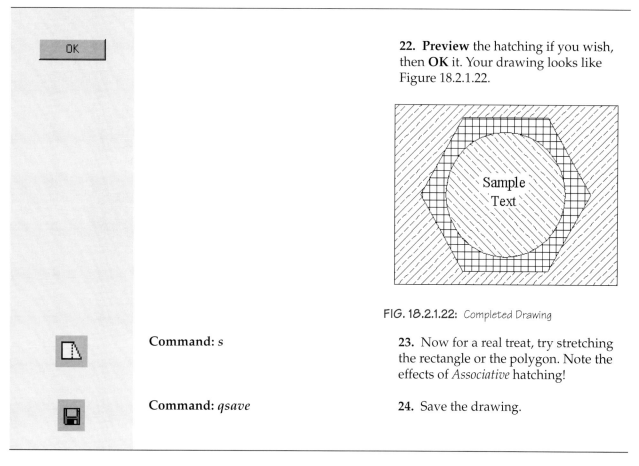 **FIG. 18.2.1.22:** Completed Drawing
[icon]	**Command:** *s*	23. Now for a real treat, try stretching the rectangle or the polygon. Note the effects of *Associative* hatching!
[icon]	**Command:** *qsave*	24. Save the drawing.

So, which do you prefer—the command line approach or the Boundary Hatch dialog box?

You probably think the dialog box is more user-friendly, and you are right. The **Pick Points** button makes a lot of difference when you are in a hurry or having difficulty selecting boundary objects. (But you can still use the **Select Objects** button to emulate the command line approach.) And you should not underestimate the value of seeing a pattern before selecting it. You may still run into the same problems with boundaries that are not quite closed or patterns that seem to bleed outside the boundaries. But if you do, use the command line and the **direct hatch** option.

> The next section requires a sample drawing included with the AutoCAD software. If you did not install AutoCAD's sample files, do so now.

Now let's take a look at the **Define Boundary Set** option.

18.2.2 Do This: Boundary Sets

I. Open the *zk147_2.dwg* file found in the C:\Steps\Lesson18 folder. The file is shown in Figure 18.2.2a.

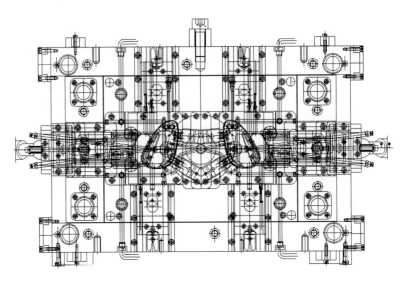

FIG. 18.2.2a: zk147_2.dwg

(This material has been reprinted with the permission of and under the copyright of Autodesk, Inc.

This reproduction is provided with RESTRICTED RIGHTS. Use, duplication or other disclosure by the US government is subject to restrictions set forth in FAR 52.227-19 (Commercial Computer Software-Restricted Rights) and DFAR 252.227-7013(c)(1)(ii) [rights in Technical Data and Computer Software], as applicable. ©1997 Autodesk, Inc.)

II. Follow these steps.

TOOLS	COMMAND SEQUENCE	STEPS
Model / Layout1		1. Notice that the drawing opens on the **Layout 1** tab (below the graphics area). We will work in Model Space, so pick on the **Model** tab to start. (This is a large drawing and the switch may take a few seconds to complete.)
🔍	**Command:** *z* **Specify corner of window, enter a scale factor (nX or nXP), or** **[All/Center/Dynamic/Extents/Previous/Scale/Window] \<real time>:** *90,355* **Specify opposite corner:** *370,90*	2. *Zoom* into the lower left portion of the drawing as indicated.

continued

TOOLS	COMMAND SEQUENCE	STEPS
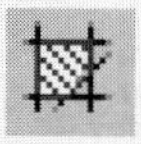	Command: *h*	**3.** Open the Boundary Hatch dialog box.
 Scale: 24		**4.** Use the default **Pattern** and **Angle**, but change the **Scale** to *24*.
	Select internal point: Selecting everything . . .	**5.** Use the **Pick Points** button to hatch the areas indicated in Figure 18.2.2.5. Pay close attention to how long it takes AutoCAD to find the boundaries.
	Selecting everything visible . . .	
	Analyzing the selected data . . .	
	Analyzing internal islands . . .	
	Select internal point:	
	Analyzing internal islands . . .	
	Select internal point: *[enter]*	

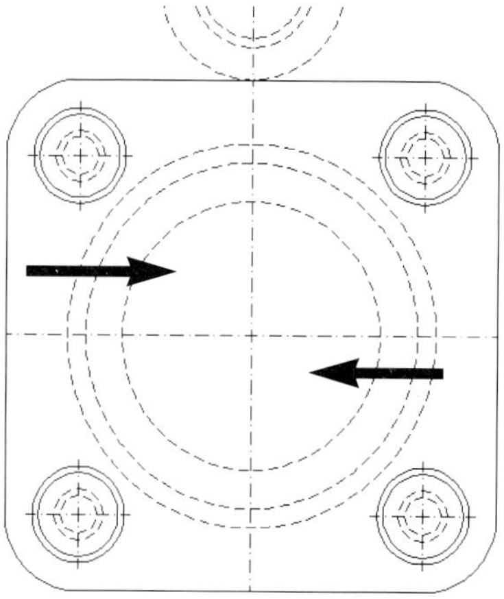

FIG. 18.2.2.5: Hatch These Areas

6. Complete the command.

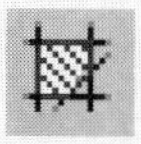

7. Repeat the *BHatch* command.

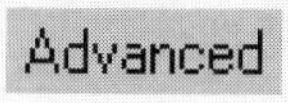

Command: *[enter]*

8. Pick the **Advanced** tab to display the Advanced options.

9. Pick the **New** button in the **Boundary Set** frame. AutoCAD returns to the graphics screen and prompts you to **Select objects:**.

continued

TOOLS	COMMAND SEQUENCE	STEPS
	Select objects: *190,320* Specify opposite corner: *290,85* Select objects: *[enter]* Analyzing the selected data . . . Select internal point: Analyzing internal islands . . . Select internal point: Analyzing internal islands . . . Select internal point: *[enter]*	**10.** Place a window as indicated. Then hit *enter* to complete the selection. AutoCAD returns to the Boundary Hatch dialog box.
[Pick Points icon]		**11.** Use the **Pick Points** method to hatch the areas indicated in Figure 18.2.2.11. Pay close attention to how long it takes AutoCAD to find the boundaries this time.

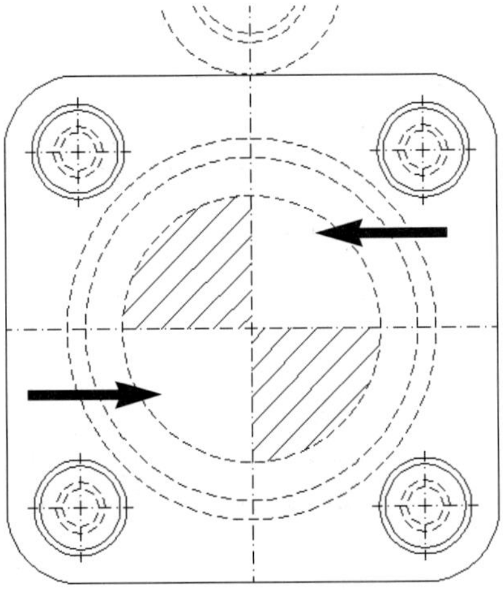

FIG. 18.2.2.11: Pick These Points

Did you notice the difference? Which was faster?

12. Complete the command.

| | Command: *quit* | **13.** Exit the drawing without saving. |

Novice operators frequently overlook this last technique—especially if they work on smaller drawings where boundary sets are not needed. But if you are going to be working on drawings larger than a half megabyte, it is a good idea to become comfortable with this approach to hatching.

18.3 Editing Hatched Areas

Remember how the *DDEdit* command, when used on MText, returned you to the Multiline Text Editor? Once there, editing the text was as easy—and followed the same rules and procedures—as creating the text. The *Hatchedit* command works the same way on Hatching. Simply enter the command and select the hatch pattern to be edited. AutoCAD returns you to the Boundary Hatch dialog box (but calls it the Hatchedit dialog box). Then you can proceed as though you were creating the hatching for the first time.

> The *Hatchedit* command is also available in the Modify pull-down menu. Or you can select the hatching to be edited and then select **Hatch Edit . . .** from the cursor menu.

Let's try it.

WWW 18.3.1 Do This: Hatch Editing

I. Reopen the *demo-hatch.dwg* file in the C:\Steps\Lesson18 folder.

II. Follow these steps.

TOOLS	COMMAND SEQUENCE	STEPS
Edit Hatch Button	Command: *he*	1. Enter the *Hatchedit* command by typing *hatchedit* or *he*. Or you can pick the **Hatchedit** button on the Modify II toolbar.
	Select associative hatch object:	2. Select the hatching inside the circle. AutoCAD presents the Hatch Edit dialog box (Figure 18.3.1.2).

continued

TOOLS	COMMAND SEQUENCE	STEPS

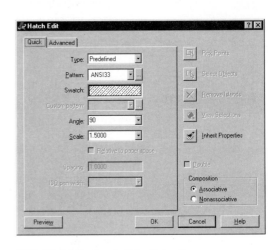

FIG. 18.3.1.2: Hatch Edit Dialog Box

3. Change the **Pattern** to **ANSI32** using the **Pattern** control box.

4. Change the **Angle** to zero.

5. **OK** the changes. Your drawing looks like Figure 18.3.1.5.

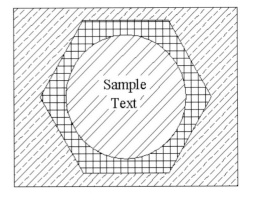

FIG. 18.3.1.5: Edited Hatch

Command: *qsave*

6. Save the drawing.

18.4 More Section Help: The *Sketch* Command

There is an old command that occasionally does not make it into textbooks. You might (justifiably) question its value (there are other commands that work better), but it can be useful in section drawings.

The command is *Sketch*. It provides the user with a computer method of sketching freehand. But consider this: Why would you want to do freehand sketching on a computer? Why use the most precise tool ever created to do rough outlines?

Frankly, there are no good answers to these questions. But let me show you how the *Sketch* command works, and maybe you will find an answer (and email it to me).

The command sequence is

Command: *sketch*

Record increment <0.1000>: *[enter or change the length of the line segments you will use to create your sketch]*

Sketch. Pen eXit Quit Record Erase Connect . *[begin drawing—left click to start the sketch; right click to stop; X to exit the command and complete the sketch]*

14 lines recorded. *[AutoCAD tells you how many lines it took to create your sketch]*

Command:

Let's look at the options.

- The *Sketch* command works by creating a series of line segments. The length, or **Record Increment**, of the line determines how soft the arcs of the sketch will appear. Shorter lines appear rounder. Unfortunately, the shorter the line segment, the more segments are required to complete the sketch. This results in a dramatic increase in the size of the drawing. Additionally, it becomes very difficult to modify a sketch with hundreds of line segments.

> The trick to drawing sketch lines is to set the system variable—**SKPoly**—to **1**. This way, the sketch lines are created as lwpolyline segments instead of individual lines. The user can then modify the entire object using the *Pedit* command.

- The **Pen** option acts as the default (even though it is not identified using normal AutoCAD conventions). A click of the left button activates the **Pen down** mode. The user then draws by moving the mouse. A second click of the same button activates the **Pen up** mode. The user can then move the mouse without drawing. Subsequent mouse clicks act as toggles between the two pen modes.

- The **eXit** option is the keyboard default when the user hits the *enter* key (right-clicking will not work here). It will cause the sketch to **Record**—become a part of the drawing—then leave the command.
- The **Quit** option allows the user to end the command without recording the current sketch.
- **Record** will make the sketch (the part completed thus far) a part of the drawing but will not leave the command.
- The **Erase** option allows the user to remove the current sketch back to the last **Record** entry.
- After the **Pen up** mode has been activated while in a *Sketch* command, the user can return to the endpoint of the last line segment drawn using the **Connect** option.

Let's see what we can do with the *Sketch* command.

18.4.1 Do This: Sketching

I. Start a new drawing from scratch.

II. Set the grid to *0.5*.

III. Follow these steps.

TOOLS	COMMAND SEQUENCE	STEPS
	Command: *rec* **Specify first corner point or [Chamfer/Elevation/Fillet/Thickness/Width]:** *2,2* **Specify other corner point:** *6,4*	1. Draw a rectangle as indicated.
	Command: *x*	2. *Explode* the rectangle.
	Command: *skpoly* **Enter new value for SKPOLY <0>:** *1*	3. Set the **SKPoly** system variable so that the *Sketch* command will draw lwpolylines.
No Button Available	**Command:** *sketch*	4. Enter the *Sketch* command. There is no hotkey or button to help you. (Be sure the **Ortho** is *Off* to avoid some really odd sketches.)

continued

TOOLS	COMMAND SEQUENCE	STEPS
	Record increment <0.1000>: *[enter]*	5. Accept the **Record Increment** default.
	Sketch. Pen eXit Quit Record Erase Connect . <Pen down>	6. Pick with the left mouse button and create a break line, as shown in Figure 18.4.1.6.

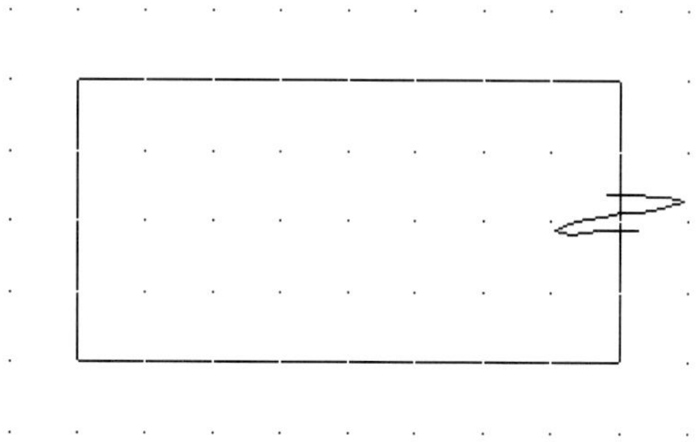

FIG. 18.4.1.6: Draw a Break Line

	<Pen up> **1 polyline with 15 edges recorded.**	7. Pick again with the left mouse button to stop drawing, then hit *enter* to complete the command. (*Note*: Your sketch may have a different number of edges than mine.)
	**Command:** *tr*	8. Trim the rectangle away from the break line. Your drawing looks similar to Figure 18.4.1.8.

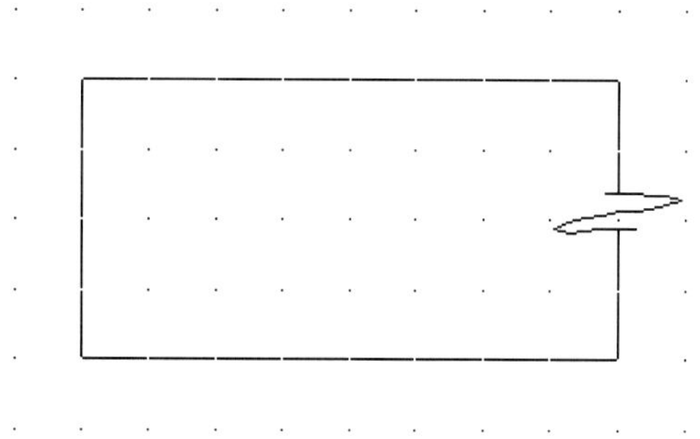

FIG. 18.4.1.8: Trim the Rectangle

continued

TOOLS	COMMAND SEQUENCE	STEPS
![icon]	Command: *h*	9. Hatch the rectangle using the ANSI31 pattern. Your drawing looks like Figure 18.4.1.9.
		FIG. 18.4.1.9: Hatched with a Broken Line
	Command: *quit*	10. Exit the drawing without saving.

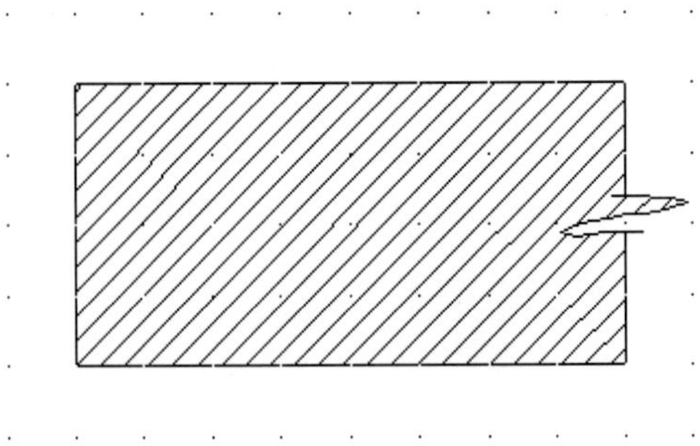

Again, there are better ways of creating break lines—splines, polylines, or even lines. But you never know when a need will arise that only the *Sketch* command can fill!

18.5 Extra Steps

Take a few minutes to experiment with the *Sketch* command. Try sketching your mouse or some other nearby object. Then try sketching it with pencil and paper. Which looks better? Repeat the experiment with several objects.

18.6 What Have We Learned?

Items covered in this lesson include:

- *Hatching Tools:*
  - *Hatch*
  - *BHatch*
  - *HatchEdit*
- *The Sketch Command*

By now you should be growing more comfortable with AutoCAD procedures and routines. You have passed beyond basic drawing and can create more complex objects in a few keystrokes or mouse picks. The one thing you still need—and need plenty of—is practice!

Work through the exercises at the end of this lesson. They may take a bit longer than earlier exercises, but that is because the drawings are becoming more complex. Additionally, I am expecting more from you while providing less.

Our next two lessons will teach you to use the first tools that make AutoCAD something more than a very expensive drafting tool. We will cover blocks in Lesson 19; we will attach *information* to blocks in Lesson 20.

Classroom acquired skills are to the mind what paint is to pottery—nothing but surface fluff that fades and disappears with the passing of time. For the skills to become actual knowledge, they must be tempered into the mind's clay through the repeated pounding of experience.

PRACTICE PRACTICE PRACTICE

Anonymous

18.7 EXERCISES

1. Use the following information to create the drawing in Figure 18.7.1d. Save the drawing as *MyTee* in the C:\Steps\Lesson18 folder.
    1.1. Layer Information:

LAYER NAME	STATE	COLOR	LINETYPE
0	On	7 (white)	CONTINUOUS
BORDER	On	5 (blue)	CONTINUOUS
CL	On	3 (green)	CENTER
DIM	On	5 (blue)	CONTINUOUS
HATCH	On	32	CONTINUOUS
MARKER	On	6 (magenta)	CONTINUOUS
OB	On	1 (red)	CONTINUOUS
TEXT	On	6 (magenta	CONTINUOUS

    1.2. Limits: 0,0 to 10.5,8
    1.3. Grid spacing = $1/4$"
    1.4. Text sizes = $3/8$", $1/4$", $3/16$", $1/8$"
    1.5. Font: Times New Roman
    1.6. Hatching information
        1.6.1. Pattern = ANSI32
        1.6.2. Scale = 1"
        1.6.3. Angle = 0

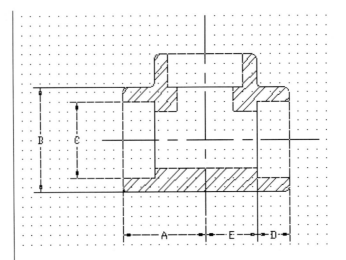

FIG. 18.7.1a

Dimensions (Inches)										
Nominal Pipe Size	1/8	1/4	1/2	3/4	1	1-1/4	1-1/2	2	2-1/2	3
A	7/8	7/8	1 1/8	1 5/16	1 1/2	1 3/4	2	2 3/8	3	3 3/8
B	29/32	29/32	1 5/16	1 9/16	1 3/4	2 7/32	2 1/2	3 1/32	3 5/8	4 5/16
C	.420	.555	.850	1.08	1.340	1.680	2	2.406	2.906	3.535
D	7/16	7/16	1/2	9/16	5/8	11/16	3/4	7/8	1 3/8	1 1/8
E	7/16	7/16	5/8	3/4	7/8	1 1/16	1 1/4	1 1/2	1 5/8	2 1/4

FIG. 18.7.1b

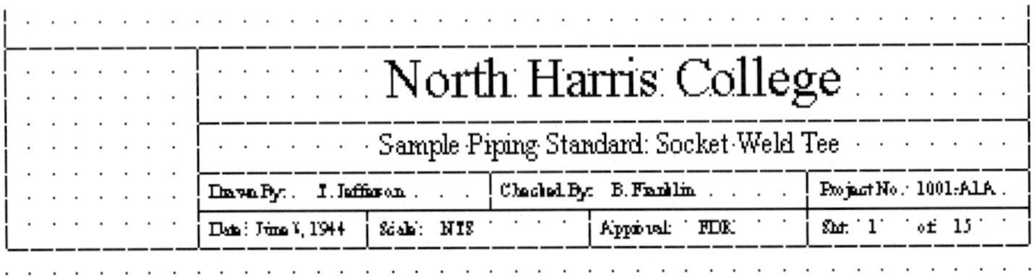

	North Harris College		
	Sample Piping Standard: Socket Weld Tee		
	Drawn By: T. Jefferson	Checked By: B. Franklin	Project No.: 1001-A1A
	Date: June 8, 1944 Scale: NTS	Approval: FDR	Sht: 1 of 15

FIG. 18.7.1c

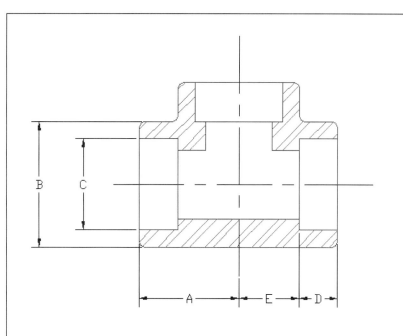

Dimensions (Inches)										
Nominal Pipe Size	1/8	1/4	1/2	3/4	1	1 1/4	1 1/2	2	2 1/2	3
A	7/8	7/8	1 1/8	1 5/16	1 1/2	1 3/4	2	2 3/8	3	3 3/8
B	29/32	29/32	1 5/16	1 9/16	1 3/4	2 7/32	2 1/2	3 1/32	3 5/8	4 5/16
C	.420	.555	.850	1.08	1.340	1.680	2	2.406	2.906	3.535
D	7/16	7/16	1/2	9/16	5/8	11/16	3/4	7/8	1 3/8	1 1/8
E	7/16	7/16	5/8	3/4	7/8	1 1/16	1 1/4	1 1/2	1 5/8	2 1/4

North Harris College

Sample Piping Standard: Socket Weld Tee

Drawn By: T. Jefferson	Checked By: B. Franklin	Project No.: 1001-A1A
Date: June 6, 1944	Scale: NTS	Approval: FDR	Sht: 1 of: 15

FIG. 18.7.1d: Socket Weld Tee

2. Start a new drawing from scratch.
   2.1. Create the following setup:
      2.1.1. Use a $1\frac{1}{2}"=1'-0"$ scale on an A-size sheet of paper ($8\frac{1}{2} \times 11$)
      2.1.2. Grid: 1" (snap as needed)
      2.1.3. These layers:

LAYER NAME	COLOR	LINETYPE
0	7 (white)	CONTINUOUS
BORDER	5 (blue)	CONTINUOUS
CL	6 (magenta)	CENTER2
CONST	1 (red)	CONTINUOUS
DIM	2 (yellow)	CONTINUOUS
HIDDEN	42	HIDDEN
OBJ1	4 (cyan)	CONTINUOUS
OBJ2	3 (green)	CONTINUOUS
OBJ3	1 (red)	CONTINUOUS
OBJ4	6 (magenta)	CONTINUOUS
TEXT	2 (yellow)	CONTINUOUS

   2.2. Use this information for the Grade hatching:
      Pattern: EARTH
      Scale: 8.0000
      Angle: 45
   2.3. Use this information for the sand hatching:
      Pattern: AR-SAND
      Scale: 0.5000
      Angle: 0
   2.4. Use the Times New Roman font. Large text should plot at $1/4"$; small text should plot at $1/8"$.
   2.5. Create the Exterior Slab drawing shown in Figure 18.7.2. Save it as *MySlab* in the C:\Steps\Lesson18 folder.

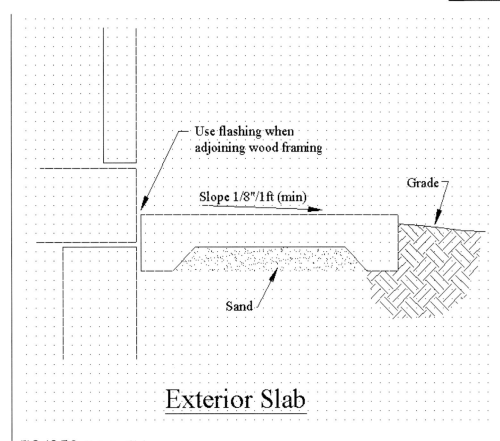

FIG. 18.7.2: Exterior Slab

3. Start a new drawing from scratch. Using the same information you used in the *Exterior Slab* drawing in Exercise 2, create the *Concrete Pier Footing* drawing in Figure 18.7.3.
   3.1. Use this information to hatch the concrete:
      Pattern: AR-CONC
      Scale: 0.5000
      Angle: 0
   3.2. Save the drawing as *MyFooting* in the C:\Steps\Lesson18 folder.

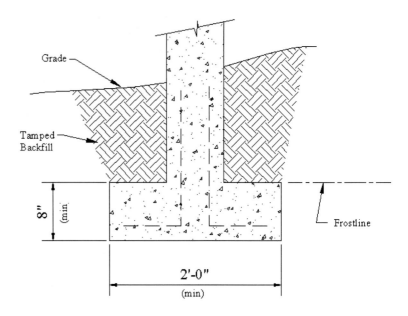

FIG. 18.7.3: Concrete Pier Footing

4. Open the *Cabin-hatch.dwg* file in the C:\Steps\Lesson18 folder. Use the following information to complete the drawing. The completed elevations are shown in Figs.18.7.4a and 18.7.4b. (See if you can add a chimney on your own.)

   4.1. Create these additional layers:

LAYER NAME	STATE	COLOR	LINETYPE
HATCH1	On	84	CONTINUOUS
HATCH2	On	22	CONTINUOUS
HATCH3	On	44	CONTINUOUS
HATCH4	On	214	CONTINUOUS

   4.2. Hatching information:
      4.2.1. Doors:
         4.2.1.1. Pattern: AR-RROOF
         4.2.1.2. Scale = $1/4$"
         4.2.1.3. Angle = 90°
         4.2.1.4. Layer = Hatch3
      4.2.2. Roof:
         4.2.2.1. Pattern: AR-RSHKE
         4.2.2.2. Scale = $1/16$"
         4.2.2.3. Angle = 0°
         4.2.2.4. Layer = Hatch3
      4.2.3. Facade:
         4.2.3.1. Pattern: AR-BRELM

- **4.2.3.2.** Scale = $1/8$"
- **4.2.3.3.** Angle = 0°
- **4.2.3.4.** Layer = Hatch2
- **4.2.4.** Curtains:
  - **4.2.4.1.** Pattern: STARS
  - **4.2.4.2.** Scale = 1"
  - **4.2.4.3.** Angle = 90°
  - **4.2.4.4.** Layer = Hatch4
- **4.2.5.** Gables:
  - **4.2.5.1.** Pattern: ANSI31
  - **4.2.5.2.** Scale = 1"
  - **4.2.5.3.** Angle = 45°
  - **4.2.5.4.** Layer = Hatch2
- **4.2.6.** Doorknobs are on the hatch1 layer.

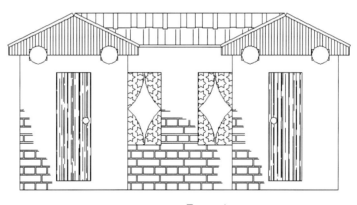

FIG. 18.7.4a: Front Elevation

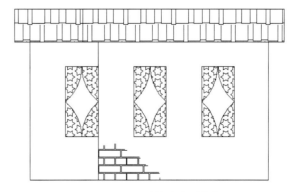

FIG. 18.7.4b: Side Elevation

5. Using what you have learned, create the drawings in Figures 18.7.5a to 18.7.5h. The grid, where shown, is 2".

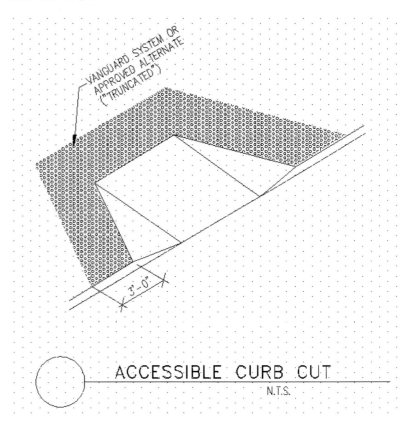

FIG. 18.7.5a: Curb Cut (Thanks to Tilco Vanguard of Snohomish, Washington, and Jon Julnes for allowing us to use this drawing.)

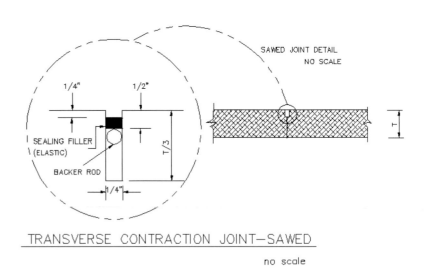

FIG. 18.7.5b: Transverse Contraction Joint

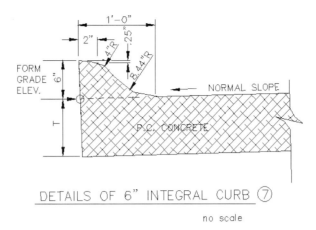

FIG. 18.7.5c: Integral Curb

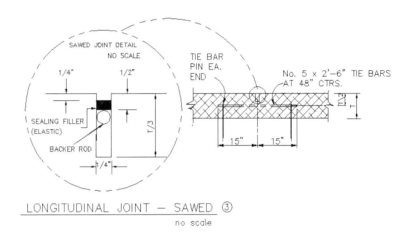

FIG. 18.7.5d: Longitudinal Joint

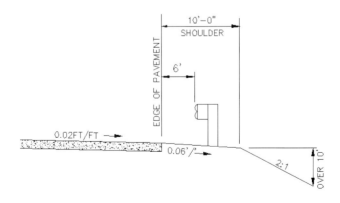

FIG. 18.7.5e: Guardrail Detail

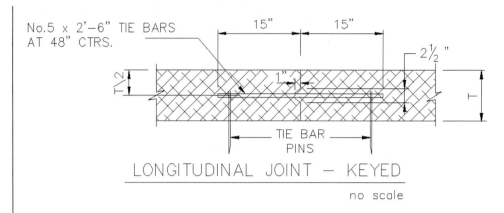

FIG. 18.7.5f: Longitudinal Joint

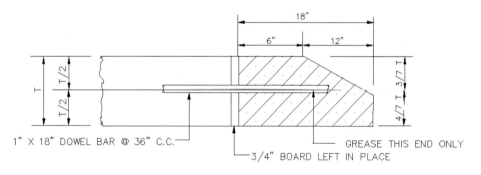

FIG. 18.7.5g: Concrete Header

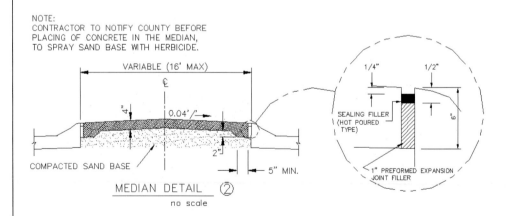

FIG. 18.7.5h: Median Detail (Special thanks to Randy Behounek at the Sarpy County Surveyors Office in Papillion, Nebraska, for permission to use Figures 18.7.5b to 18.7h drawings.)

18.8 REVIEW QUESTIONS

Write your answers on a separate sheet of paper.

1. Section lines are created as a style of _____.

2. The _____ hatch pattern will present the same results as the solid command.

3. and 4. The User Defined option of the hatch command allows the user to create a hatch pattern by defining the _____ and _____ of hatch lines.

5. (T or F) A hatch angle of 0° will draw the hatch lines horizontal.

6. The _____ option allows the user to create hatch boundaries on the fly.

7. to 9. The three AutoCAD hatch styles include: _____, which recognizes multiple objects and hatches every other one; _____ which hatches only between the outer two boundaries; and _____, which ignores all but the outermost boundary.

10. An important thing to remember about hatching is that the boundary around a hatch pattern must be _____.

11. and 12. Use the _____ option of the _____ command to draw partial grating on a platform or walkway.

13. Pick on the _____ button in the Boundary Hatch dialog box to see the Hatch Pattern Palette.

14. The Iso Pen Width option in only available when the user selects to use an _____ pattern.

15. _____ hatching means that the hatching will automatically update when the boundary changes.

16. Placing an asterisk (*) in front of the name of a hatch pattern at the command line will insert an _____ pattern.

17. Use the _____ button to match an existing hatch style.

18. Using the _____ button on the Boundary Hatch dialog box enables the user to simply pick a point within a boundary where hatching is required.

19. A boundary within a boundary is called an _____.

20. Use the _____ command to modify existing hatching.

21. The _____ command is a seldom used command for creating freehand sketches in AutoCAD.

22. and 23. The trick to drawing sketch lines is to create polylines by setting the _____ sysvar to _____.

24. What can the user do to turn classroom acquired skills into actual knowledge?

LESSON 19

Many as One: Groups and Blocks

Following this lesson, you will:

➥ Know how to create and manipulate Groups
➥ Know how to use blocks in a drawing
   • The **Block** command
   • The **WBlock** command
   • The **Insert** command
➥ Know how to create libraries of blocked objects
➥ Know how to share blocks between drawings using the AutoCAD Design Center

One of the great benefits of using a computer to create design plans lies in the ability of the operator to create something once, then duplicate or manipulate it as desired. We have used several modification commands, techniques, and procedures throughout the course of this text and have become fairly comfortable with the basic tools. But there are still two modification routines that can increase speed and make your work easier. These are the *Group* and *Block* commands.

The *Group* and *Block* commands enable the user to combine objects like lines, circles, and arcs into objects like chairs, tables, valves, and so forth. The user can then manipulate *a* chair (or some other object) rather than 8 or 10 lines. Using blocks, the user can even create a library of objects to use within one drawing or shared between any number of drawings.

In Lesson 19, we will look at the *Group* and *Block* commands. Although they often behave the same, each has its niche in the drawing world and it will be important to know when to use them. Then, in Lesson 20, we will learn why blocks are one of the most valuable design tools ever created.

19.1 Paper Dolls: The *Group* Command

Have you ever done a furniture layout for an office or an equipment layout for a shop? You may know then that the easiest way to accomplish the layout (in such a fashion that the equipment can be moved around as the plan develops) is to use the "paper dolls" approach. That is, you create a scaled drawing of the room or work area, make a blue line of it and then move scaled representations (paper dolls) of the equipment around the blue line until you are satisfied with the locations. Once you are satisfied, you tape the dolls to the blue line for reference and draw the objects in your layout.

It is easy work *if* you keep the windows closed and nobody slams a door. A slight breeze, however, sends people scrambling for the paper dolls and mumbling things that might make a sailor blush. Of course, this is only after the hours spent reminiscing about kindergarten while cutting paper dolls out of construction paper.

Once again, AutoCAD has created tools that provide the benefits of construction paper and tape (combined with a few new benefits) while removing the hassles. Enter *Groups*.

What is a group? Simply put, a group is a collection of objects treated as a single unit.

Let's take a look.

19.1.1 Do This: Working with Groups

I. Open the *groups.dwg* file in the C:\Steps\Lesson19 folder. The drawing looks like Figure 19.1.1a.

FIG. 19.1.1a: *groups.dwg*

II. Follow these steps.

TOOLS	COMMAND SEQUENCE	STEPS
✥	**Command:** *m*	1. We will switch the tree (in the upper left corner of the screen) with the easel pad (in the upper right corner). Enter the *Move* command.
	Select objects: 80 found, 1 group **Select objects:** *[enter]*	2. Select anywhere in the tree as shown in Figure 19.1.1.2, then complete the selection by hitting *enter*. 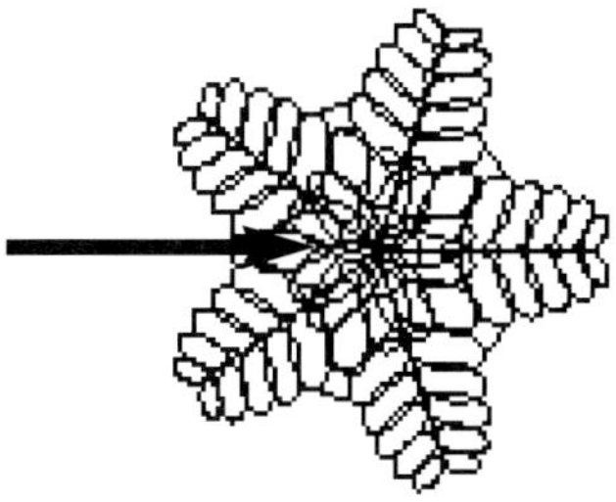

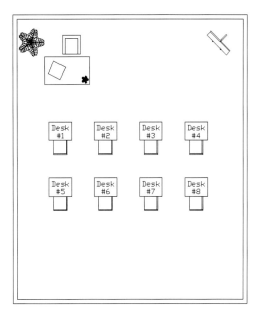

FIG. 19.1.1.2: Select the Tree

continued

TOOLS	COMMAND SEQUENCE	STEPS
		(Notice how easy it is to select these 80 objects.)
	Specify base point or displacement: *21',0*	**3.** Use the **displacement** method to move the tree 21 feet to the right.
	Specify second point of displacement or <use first point as displacement>: *[enter]*	
	Command: *m*	**4.** Repeat Steps 1 through 3 to move the easel pad 21 feet to the left. Your drawing looks like Figure 19.1.1.4.

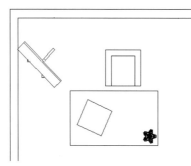

FIG. 19.1.1.4: Moved Groups

Notice that, in each instance, the objects were selected with a single pick and that AutoCAD indicated the number of objects *and the number of groups* found.

	Command: *ro*	**5.** Now rotate the easel pad to face the correct direction. This part of your drawing now looks like Figure 19.1.1.5.
	Current positive angle in UCS: ANGDIR= counterclockwise ANGBASE=0	
	Select objects: 14 found, 1 group	
	Select objects: *[enter]*	
	Specify base point:	
	Specify rotation angle or [Reference]: *90*	

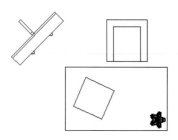

FIG. 19.1.1.5: Rotated Easel

continued

TOOLS	COMMAND SEQUENCE	STEPS
	Command: *qsave*	6. Save the drawing, but do not exit.

You can see how easy it is to manipulate grouped objects. And before you ask, yes, you can manipulate individual objects within a group (but *not* within a block).

Let's take a look at how you can create groups yourself, and at how you can work with individual parts of the group. When you enter the ***Group*** command, AutoCAD provides the Object Grouping dialog box to assist you (Figure 19.1a).

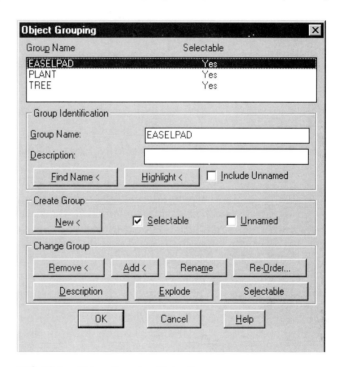

FIG. 19.1a: *Object Grouping Dialog Box*

- The upper list box shows the names of existing groups (under **Group Name**) and whether or not the group is currently **Selectable** (**Yes** or **No**). *Selectable* simply means that the group can be manipulated as a single object. You will want the group to be selectable to do what we did in the last exercise; but you will *not* want it selectable if you need to modify something within the group (like pruning the branches of the tree we moved). You cannot change the **Selectable** setting in the list box. We will see how to do that shortly.

- There are several useful items in the **Group Identification** frame.
  - Appearing in the appropriate text box, you will find the name and description (if any) of the group highlighted in the list box at the top of the dialog box. You will begin to create a new group by typing a new **Group Name** and

Description (if desired) into these boxes. (We will create some new groups in our next exercise.)
- To find the name of an existing group, begin by picking the **Find Name <** button. AutoCAD returns to the graphics screen and prompts the user to **Pick a member of a group**. It then displays the Group Member List dialog box (Figure 19.1b) showing a list of all the groups to which that object belongs.

FIG. 19.1b: Group Member List Dialog Box

- To find which objects belong to a certain group, begin by selecting the name of the group in the upper list box, then pick the **Highlight <** button. AutoCAD returns to the graphics screen, highlights all the objects belonging to that group, and presents a **Continue** button to return to the Object Grouping dialog box.
- Sitting inconspicuously in the lower right corner of the frame, you will find a check box next to the words **Include Unnamed**. When a group is copied, the copy is also a group. The user has not named the copy, but AutoCAD has assigned a temporary name to it. Placing a check in this box will tell AutoCAD to display the temporary names as well as the user-assigned names. The user can then rename the copies to something more appropriate.

▌ You will use items in the **Create Group** frame to create a new group. After typing a name for your new group in the **Group Name** text box (or placing a check in the **Unnamed** check box if you want AutoCAD to assign the name), pick the **New <** button here. AutoCAD will return to the graphics screen and prompt **Select objects for grouping:**. When you are finished selecting the objects, hit *enter* to return to the Object Grouping dialog box. Be sure there is a check in the **Selectable** check box if you want to manipulate the selected objects as a group.

▌ Use the seven buttons in the **Change Group** frame to modify a group. First, select the group to be modified in the list box. Then pick one of the buttons in the **Change Group** frame.

- With the **Remove <** and **Add <** buttons, the user can add or remove objects from the group.

- Use the **Rename** button to change the name of a group. For example, to change the temporary name assigned by AutoCAD to a more recognizable name, highlight the temporary name in the list box, type the new name in the **Group Name** text box, then pick the **Rename** button.
- Type some text into the **Description** text box, then pick the **Description** button to update the description of the highlighted group.
- The **Explode** button will remove the definition of the group from the drawing's database. All objects in the group will then behave as individual objects (not as part of a group).
- The **Selectable** button is a toggle for treating a group as a group or suspending the group definition so you can modify one or more of the objects within the group. When **Selectable** is toggled *Off*, the word **No** appears in the **Selectable** column of the list box, and all the objects within a group behave as individual objects. To manipulate the objects as part of the group, simply toggle **Selectable** back *On*.
- The **Re-order . . .** button presents the Order Group dialog box (Figure 19.1c). Here the user can change the order in which AutoCAD reads the objects in a group. Type the number of the object you wish to change in the **Remove from position** text box. Type the position to which you wish to place the object in the **Replace at position** text box. Type the number or range of objects to reorder in the **Number of objects** text box.

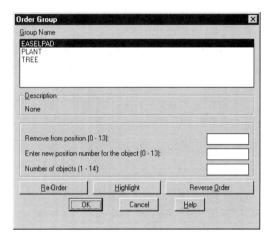

FIG. 19.1c: *Order Group Dialog Box*

Pick the **Reverse Order** button to simply reverse the order of the objects within the group.

If you are unsure of the position of the object you wish to reorder, pick the **Highlight** button. AutoCAD will present the Object Grouping counter shown in Figure 19.1d. Pick the **Next/Previous** buttons to scroll through the objects making up the group. The number of the object will be indicated at the bottom of the dialog box.

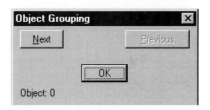

FIG. 19.1d: Object Grouping Dialog Box

> You will rarely (if ever) need the **Order** options of the *Group* command. In the few brief years since AutoCAD included this tool, I have not found a need or even a use for reordering objects in a group. However, this does not mean that you will not find a need or use. A corollary to AutoCAD's *Redundancy Rule* might be the *Overkill Rule*. Both rules simply and correctly state that *More is Better*. Or in other words, *If it might be useful, make it possible!*

Let's play with the *Group* command!

WWW 19.1.2 Do This: Creating and Manipulating Groups

I. Be sure you are still in the *groups.dwg* file in the C:\Steps\Lesson19 folder. If not, open it now.

II. Follow these steps.

TOOLS	COMMAND SEQUENCE	STEPS
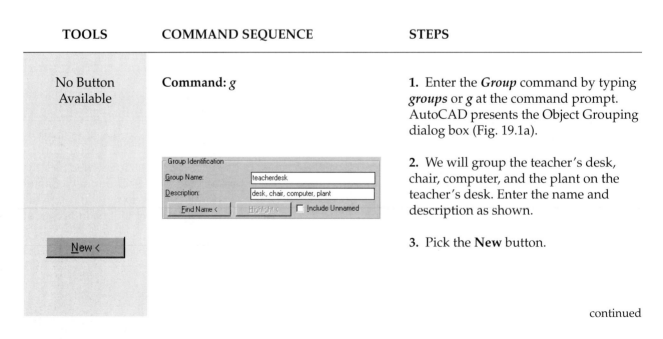	**Command:** *g*	1. Enter the *Group* command by typing *groups* or *g* at the command prompt. AutoCAD presents the Object Grouping dialog box (Fig. 19.1a).
		2. We will group the teacher's desk, chair, computer, and the plant on the teacher's desk. Enter the name and description as shown.
		3. Pick the **New** button.

continued

TOOLS	COMMAND SEQUENCE	STEPS
	Select objects for grouping: Select objects: *w* Specify first corner: Specify opposite corner: 86 found, 1 group Select objects: *[enter]*	4. Use a window to select the teacher's desk and the other objects (Figure 19.1.2.4). Hit *enter* to return to the dialog box. 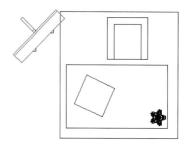 **FIG. 19.1.2.4:** Select These Objects 5. AutoCAD now lists the **Teacherdesk** in the list box as shown in Figure 19.1.2.5.

EASELPAD	Yes
PLANT	Yes
TEACHERDESK	Yes
TREE	Yes

FIG. 19.1.2.5: List Box Showing the TeacherDesk

TOOLS	COMMAND SEQUENCE	STEPS
OK		Pick the **OK** button to complete the command.
	Command: *m* Select objects: 86 found, 1 group Select objects: *[enter]* Specify base point or displacement: *13',0* Specify second point of displacement or <use first point as displacement>: *[enter]*	6. Move the **Teacherdesk** group next to the tree as indicated. Your drawing looks like Figure 19.1.2.6. 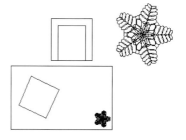 **FIG. 19.1.2.6:** Move the Teacherdesk Group
	Command: *g*	7. Return to the Object Grouping dialog box.

continued

TOOLS	COMMAND SEQUENCE	STEPS
		8. Now we will list all the groups to which an object belongs. Pick the **Find Name <** button.
	Pick a member of a group.	**9.** AutoCAD returns to the graphics screen and prompts the user to **Pick a member of a group**. Pick the plant on the teacher's desk.
		10. AutoCAD presents the Group Member List dialog box (Figure 19.1.2.10) showing that the object selected belongs to the **Plant** and **Teacherdesk** groups. You see now that groups can be *nested*—that is, one group can contain another group. Objects can also be shared by more than one group.

FIG. 19.1.2.10: Group Member List

Pick the **OK** button to return to the Object Grouping dialog box.

11. Let's see which objects are in the **Plant** group. Pick **Plant** in the list box above. Notice that the name appears in the **Group Name** text box (Figure 19.1.2.11).

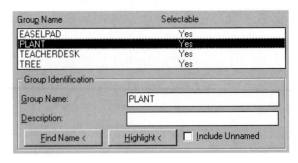

FIG. 19.1.2.11: Plant Group

continued

TOOLS	COMMAND SEQUENCE	STEPS
 [Highlight <]		**12.** Pick the **Highlight** button. AutoCAD returns to the graphics screen and highlights the plant on the teacher's desk.
 [Continue]		**13.** Pick the **Continue** button to return to the Object Grouping dialog box.
 [OK]		**14.** Pick the **OK** button to complete the command.
[copy icon]	**Command:** *co*	**15.** Let's give the teacher two new plants. Try to copy the **Plant** group. Notice that the entire desk highlights when the plant is selected. We must first make the desk unselectable.
 [Esc]		**16.** Hit the **Escape** key to cancel the command.
	Command: *g*	**17.** Return to the Object Grouping dialog box.
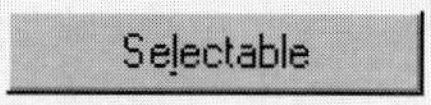 [Selectable]		**18.** Pick **Teacherdesk** in the list box and then pick the **Selectable** button in the **Change Group** frame. Notice that the word **No** appears in the **Selectable** column of the list box (Figure 19.1.2.18).

EASELPAD	Yes
PLANT	Yes
TEACHERDESK	**No**
TREE	Yes

FIG. 19.1.2.18: TeacherDesk Is Not Selectable

[OK]		**19.** Pick the **OK** button to complete the command.

continued

TOOLS	COMMAND SEQUENCE	STEPS
	Command: *co*	**20.** Now make two copies of the plant on the teacher's desk. Your drawing now looks something like Figure 19.1.2.20.

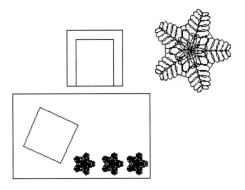

FIG. 19.1.2.20: Copy the Plants

	Command: *g*	**21.** Return to the Object Grouping dialog box.
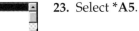		**22.** Place a check in the **Include Unnamed** check box in the **Group Identification** frame. Notice that two new groups appear in the list box—*A5 and *A6.
		**23.** Select *A5.
		**24.** Type the name *PlantB* into the **Group Name** text box.
Rename		**25.** Now pick the **Rename** button. AutoCAD renames the group.
		26. Repeat Steps 22 through 25 to rename *A6 to *PlantC*.
		27. Now let's add these new plants to the **Teacherdesk** group. Select the **Teacherdesk** group in the list box (you may need to scroll down a bit to find it).
Add <		**28.** Pick the **Add <** button in the **Change Group** frame.

continued

TOOLS	COMMAND SEQUENCE	STEPS
	Select objects to add to group . . . Select objects: 80 found, 1 group Select objects: 80 found, 1 group, 246 total Select objects: *[enter]*	**29.** AutoCAD returns to the graphics screen and prompts the user to **Select objects:**. Select the two new plants, then hit *enter* to return to the Object Grouping dialog box.
[Selectable button]		**30.** Pick the **Selectable** button to make the **Teacherdesk** group selectable again. Notice that the word **Yes** appears in the **Selectable** column of the list box.
[OK button]		**31.** Pick the **OK** button to return to the graphics screen.
[Save icon]	**Command:** *qsave*	**32.** Save the drawing, but do not exit.

We have seen how groups can benefit our work by simplifying object selection and manipulation. We have also seen how to manage our groups through some of the buttons in the **Change Group** frame of the Object Grouping dialog box. Now let's remove a group definition from a drawing.

WWW   19.1.3 Do This: Removing a Group Definition

I. Be sure you are still in the *groups.dwg* file in the C:\Steps\Lesson19 folder. If not, open it now.

II. Follow these steps.

TOOLS	COMMAND SEQUENCE	STEPS
No Button Available	**Command:** *g*	**1.** Enter the *Group* command by typing *groups* or *g* at the command prompt. AutoCAD presents the Object Grouping dialog box.
	[List box showing: EASELPAD Yes, PLANT Yes, PLANTB Yes, PLANTC Yes]	**2.** Select the **Easelpad** group in the list box.

continued

TOOLS	COMMAND SEQUENCE	STEPS
[Explode]		3. Pick the **Explode** button in the Change Group frame. Notice that **Easelpad** disappears from the list box.
[OK]		4. Pick the **OK** button to complete the command. AutoCAD returns to the graphics screen.
[move icon]	Command: *m*	5. Try to move the easel pad now. Notice that you cannot select the entire pad with a single pick. The pad is no longer a group.
[save icon]	Command: *qsave*	6. Save the drawing and exit.

I have frequently used groups in my capacity as a piping designer to manipulate (particularly moving and copying) items such as control stations and pipe configurations. Architects and interior designers can also use groups as we have done in these exercises. Other disciplines can make use of groups when dealing with layouts, circuitry, bolt holes, windows, and many other items.

The only limitations to groups are simple—their use is limited to a single drawing (you cannot share groups between drawings), the group definition is permanently lost once the group is exploded or erased from a drawing, and the user cannot attach information to the group. To overcome these limitations, use blocks instead of groups.

19.2 Groups with Backbone: The *Block* Commands

No single aspect of CAD has served as a stronger selling point than blocks. In fact, all quality CAD systems have blocks (or cells, or something that behaves like blocks). No other single command or routine can speed the drawing process as well, and none has a greater potential for cost cutting or streamlining design management.

What is a block? A block is a single object made up of several other objects. The user will manipulate blocks, like groups, as a single unit. But the user can share blocks *between* drawings. More importantly, *blocks can contain user-defined data*. We will discuss creation and manipulation of blocks here. In Lesson 20, we will discuss attaching data to blocks and exporting that data to other computer software (like spreadsheets or databases).

Unfortunately, you cannot make a block *selectable* or *not selectable* as you did with groups. So it is a good idea to be sure your block contains all the objects you want it to contain before you block it. If you must modify a block, follow this procedure:

1. Explode the block. (To remove the block definition—that is, to return blocked objects to individual objects—use the *Explode* command as you did with polylines.)
2. Make your modifications.
3. Redefine the block.* (Simply put, this means reblocking the objects. If you use the same block name, as you should, AutoCAD will inform you that a block of that name already exists and ask if it should be replaced. Respond *Yes*.)

*A useful quirk: When you redefine a block, all the blocks in the drawing going by that name are automatically updated to reflect the changes!

There are three block commands. *–Block* and *BMake* create blocks within a drawing. (*–Block* is used for command line interaction; *BMake* is used for dialog box interaction.) *WBlock* saves a block as a separate drawing file.

Groups of blocks are often assembled in useful packages called libraries (Figure 19.2a). A *library* is a group of blocks, or predefined drawings, used as inserts in a drawing. Use of blocked objects saves time the user might otherwise need to create the objects.

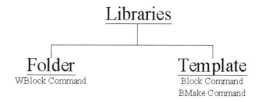

FIG. 19.2a: Libraries

Most libraries consist of drawings associated with a single design discipline. For example, an architectural library may contain drawings of doors, windows, toilets, tubs, and so forth, whereas a piping library will contain drawings of valves, elbows, tees, and so forth. Library creation has become quite a business alongside CAD industries.

Autodesk has created symbol libraries for several disciplines. For more on the Autodesk libraries, visit their web site:

www.autodesk.com/products/symbols/index.htm

There are two types of libraries (Figure 19.2a): *Template files*—associated with the *Block* and *BMake* commands—and *Folders*—associated with the *WBlock* command. Both libraries are useful (each has its pluses and minuses), but each project should use only one type.

> With AutoCAD 2000, we also have the ability to take a block directly from another drawing using the AutoCAD Design Center. We will cover this in Section 19.2.4.

- The project guru (CAD coordinator) should create the Folder Library—a collection of standardized drawings to be used on the contract—and store it on a network. The path to the folder should be defined in the *Support File Search Path* section of the **Files** tab of the Options dialog box. This will make insertion much easier. (*Note*: The Options dialog box is not a place for AutoCAD novices to experiment!)

    The Folder Library is easy for a single person to maintain. Adding, deleting or changing a project standard involves modifying only one file for each change.

- Blocks can be included as part of the templates used on a project (creating Template Libraries). Accessing blocks that are part of the original template is a bit faster than the Folder method, but the size of the template can become quite large (heavy with block definitions) before any actual drafting begins. Additionally, any changes in project standards require that all the templates be updated. Even if the templates are centrally stored on a network, this may involve several files.

The decision as to which type of library to use should be universal for the project and is best made by the guru and the design supervisor. This does not mean that the operator cannot create and use blocks on the fly. But the operator should take care not to duplicate (or overwrite) project standards.

Another thing that must concern the CAD operator, both in the creation and insertion of blocks, is how blocks relate to layers. This can sound complicated, so let's use a chart (after all, a chart is worth a thousand words).

BLOCKS AND LAYERS

IF THE OBJECTS USED TO MAKE THE BLOCK ARE CREATED ON:	WHEN INSERTED, THE BLOCK HAS THE CHARACTERISTICS OF:	WHEN INSERTED, THE BLOCK EXISTS ON:
Layer 0	The current layer	The current layer
Any other layer	The layer on which it was created	The current layer

Let me try to explain this by using some examples (and fewer than a thousand words).

Generally speaking, the user should always create objects for blocks on layer **0**. Then when inserting the block on another layer (let's say, layer **obj**), all the objects that went in to creating the block will appear on layer **obj**. When layer **obj** is frozen/turned off/locked, the block will be affected accordingly.

Text might be an exception. When part or all of the objects used to create a block exist on a layer other than **0** (say, **text**), those objects in the inserted block will appear with the characteristics of the layer on which those objects were created. This will be true regardless of the layer that is current (say, **obj**) when the block is inserted. Now, when the **obj** layer is frozen/turned off/locked, the block will be affected accordingly because the block exists on (was inserted on) layer **obj**. *Additionally*, when the **text** layer is frozen/turned off/locked, the block will also be affected because that is the layer on which it was created.

Seem complicated? Use the chart!

19.2.1 Template Library Creation

Let's begin our study with the *–Block* command. Remember; we use the *–Block* and *BMake* commands to create Template Libraries or to create blocks on the fly.

The command sequence is

> **Command:** *–block* **(or** *–b***)**
>
> **Enter block name or [?]:** *[enter the name you want to assign to the block]*
>
> **Specify insertion base point:** *[this can be deceiving—pick the point at which you want to hold the block when you insert it (much like the Move or Copy commands)]*
>
> **Select objects:** *[select the objects that will make up the block]*
>
> **Select objects:** *[enter to confirm your selection]*
>
> **Command:** *oops ["Now that is odd," you are thinking. The* **Oops** *command often follows the Block command because the Block command removes the selected objects from the drawing database. The* **Oops** *command returns them to your screen (it does not remove the block definition from the database).]*

When completed, the newly defined block has become part of the current drawing's database and can be inserted into this drawing at any time. Save this drawing as a template, and the block will be available to any drawing using this template.

Let's create some blocks.

WWW | **19.2.1.1 Do This:** | Creating Blocks on the Command Line

I. Open the *blocks.dwg* file in the C:\Steps\Lesson19 folder. The drawing has been set up (grid and limits) to assist in your block creation. (Toggle the snap as needed.)

II. Follow these steps.

TOOLS	COMMAND SEQUENCE	STEPS
	Command: *rec*	**1.** On layer **0**, draw the *sink* shown in Figure 19.2.1.1.1. (Do *not* draw the dimensions in this exercise. We will not need them in our blocks.)

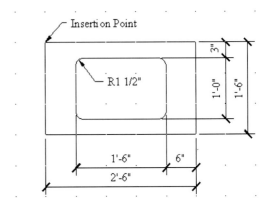

FIG. 19.2.1.1.1: Sink

No Button Available	**Command:** *–b*	**2.** Enter the **Block** command by typing *–block* or *–b* at the command line. Do not use the button on the Draw toolbar (it is for the **BMake** command).
	Enter block name or [?]: *sink*	**3.** Call the block *sink*.
	Specify insertion base point:	**4.** Select the intersection indicated in Figure 19.2.1.1.1.
	Select objects: **Select objects:** *[enter]*	**5.** Select both rectangles, then hit enter to complete the command. (*Note:* We will not use the **Oops** command since we will not need the objects again.)
		6. Repeat Steps 1 through 5 to create the block in Figure 19.2.1.1.6. Call it **WC**.

continued

TOOLS	COMMAND SEQUENCE	STEPS
	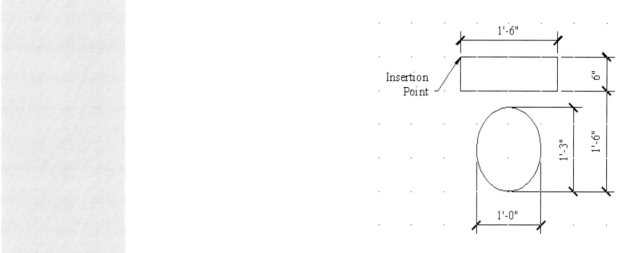 FIG. 19.2.1.1.6: WC	
💾	**Command:** *qsave*	7. Save the drawing, but do not exit.

Another approach to creating a block uses a dialog box. It is the **BMake** command. (The user may also enter the **Block** command without the dash to access the Block Definition dialog box.) **BMake** presents the Block Definition dialog box (Figure 19.2.1a) in which the user can enter the same information requested by the –**Block** command.

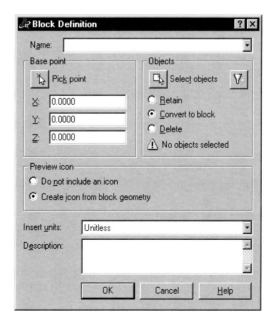

FIG. 19.2.1a: Block Definition Dialog

- Place the name of the block in the **Name:** control box.
- Although the user can enter an insertion point's X, Y, and Z coordinates in the **Base Point** frame, it is easier to use the **Pick point** button. AutoCAD prompts:

    Specify insertion base point:

    And the user picks a base point on the screen.
- Use the **Select Objects** button (in the **Objects** frame) to return to the graphics screen and select the objects to include in the block. Or the user may use **Quick Select** filters by picking the button next to the **Select objects** button.

    Once the objects have been selected, the user may specify what he wants AutoCAD to do with the objects after the block has been made. Simply place the bullet next to the appropriate option:

    - **Retain** is not available at the command line. A bullet here will eliminate the need to use the *Oops* command after creating your block. AutoCAD will retain the selected objects in their current state.
    - **Convert to block** will convert the objects from individual objects to the first insertion of the newly defined block.
    - **Delete** will behave like the command line—deleting the objects used to create the block.
- In the **Preview icon** frame, the user may opt to have AutoCAD create a preview image of the block. The icon will appear to the right of the frame.
- In the **Insert units** control box at the bottom of the dialog box, the user should define what units to use when inserting the block (preferably the units used by the drawing in which the block will be inserted).
- Lastly, at the bottom of the dialog box, the user may enter a description of the block in the **Description** text box.

Let's use the *BMake* command to create some more blocks.

19.2.1.2 Do This: Creating Blocks with a Dialog Box

I. Be sure you are still in the *blocks.dwg* file in the C:\Steps\Lesson19 folder. If not, please open it now.

II. Follow these steps.

TOOLS	COMMAND SEQUENCE	STEPS
		1. Create the drawing in Figure 19.2.1.2.1 (without the dimensions).

continued

TOOLS	COMMAND SEQUENCE	STEPS

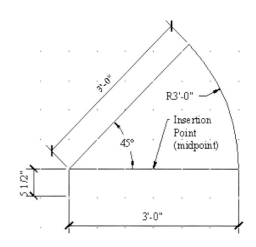

FIG. 19.2.1.2.1: 36" Door

| | Command: *b* | 2. Let's try the dialog box. Enter the *BMake* command by typing **bmake** or **b** at the command prompt. Or you can pick the **Make Block** button on the Draw toolbar. AutoCAD presents the Block Definition dialog box (Figure 19.2.1a). |

Make Block Button

3. Enter the name *door36* in the **Block name:** text box.

4. Pick the **Pick point** button. AutoCAD returns to the graphics screen.

Pick Point Button

| | Specify insertion base point: | 5. Pick the insertion point indicated in Figure 19.2.1.2.1. AutoCAD returns to the dialog box. |

6. Pick the **Select objects** button. AutoCAD returns to the graphics screen.

Select Objects Button

	Select objects:	7. Select the arc, the vertical lines and the angled line. Do *not* select the horizontal line. Confirm your selections. AutoCAD returns to the dialog box.
	Select objects:	
	Select objects: *[enter]*	

8. Place the bullet next to the **Delete** option in the **Objects** frame.

continued

TOOLS	COMMAND SEQUENCE	STEPS
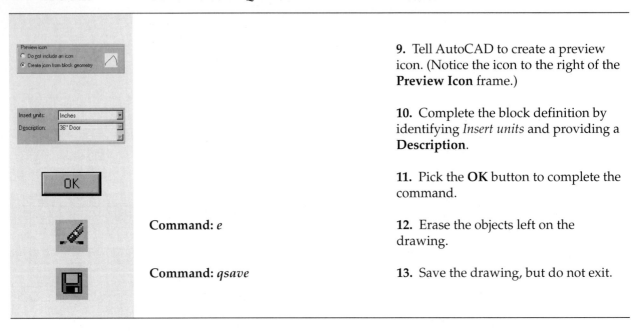		9. Tell AutoCAD to create a preview icon. (Notice the icon to the right of the **Preview Icon** frame.)
		10. Complete the block definition by identifying *Insert units* and providing a **Description**.
		11. Pick the **OK** button to complete the command.
	Command: *e*	12. Erase the objects left on the drawing.
	Command: *qsave*	13. Save the drawing, but do not exit.

In these last two exercises, we have begun a Template Library. We could continue to add dozens of drawings as needed. Then, when we have finished, we would save this drawing with the *.dwt* extension and use it as a template for any number of future drawings. But let's look at another approach to block libraries.

■ 19.2.2 Folder Library Creation

You will use the *WBlock* (for **Write**Block) command to create Folder Libraries. The procedure is very similar to creating blocks, but it also allows the user to convert Template Library files to Folder Library files.

The sequence for the command line is

> **Command:** *–wblock* (or *–w*)
>
> *[AutoCAD presents a standard Create File dialog box in which you will place the name of the new block or the name of an existing block that you wish to convert to a separate drawing file]*
>
> **Enter name of existing block or**
>
> **[= (block=output file)/* (whole drawing)] <define new drawing>:** *[hit enter to define a new block or "=" if you are converting an existing block to a separate drawing file]*
>
> **Specify insertion base point:** *[as with the Block command, pick the point at which you want to hold the block when you insert it]*
>
> **Select objects:** *[select the objects to include in the block]*
>
> **Select objects:** *[enter to complete the command]*

This looks as easy as the *Block* command, does it not? It really is. But let's take a look at that long, somewhat confusing first prompt (after the Create File dialog box).

T**he Enter name of existing block ... prompt** allows the user to enter the name of a block existing in the current drawing's database. AutoCAD will then create a separate file to be used in a Folder Library. If you have entered the block's name in the Create File dialog box, simply respond to this prompt with an "=" sign. The block will be converted to a separate drawing file.

AutoCAD 2000 also provides a dialog box interface to ease creation of WBlocks. Access it by entering the *WBlock* command (or the *W* hotkey). The Write Block dialog box appears in Figure 19.2.2a. Does it look familiar? It should. Two of the frames are identical to the Block Definition dialog box. Let's examine the other frames.

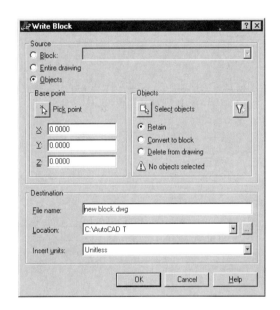

FIG. 19.2.2a

- In the **Source** frame, the user has three options:
  - A bullet in the **Block:** option opens the control box next to it. Here the user may select from a list of blocks existing in the drawing to be saved as separate drawing files.
  - Of course, the **Entire drawing** option allows the user to create a separate drawing file using all of the objects in the drawing.
  - The **Objects** option behaves like the *Block* command and allows the user to select objects just he did using the Block Definition dialog box.
- The **Base point** and **Objects** frames are identical to the same frames found on the Block Definition dialog box.
- In the **Destination** frame, the user:
  - names the new block (**File name:** text box)
  - identifies the destination of the new block (use the **Location:** control box or the **Browse** button to the right of the control box)

- identifies the **Insert units:** of the new block (just as he did in the Block Definition dialog box)

In our next exercise, we will create some WBlocks to begin a Folder Library. Then we will convert the blocks created in our Template Library to files we can use in our Folder Library.

19.2.2.1 Do This: Creating WBlocks

I. Be sure you are still in the *blocks.dwg* file in the C:\Steps\Lesson19 folder. If not, open it now.

II. Follow these steps.

TOOLS	COMMAND SEQUENCE	STEPS
	**Command:** *l*	1. On layer **0**, draw the *interior door* shown in Figure 19.2.2.1.1. (Again, do *not* draw the dimensions in this exercise. We will not need them in our blocks.)
		FIG. 19.2.2.1.1: Interior Door
	Command: *–w*	2. Enter the *–WBlock* command by typing *–wblock* or *–w* at the command prompt.

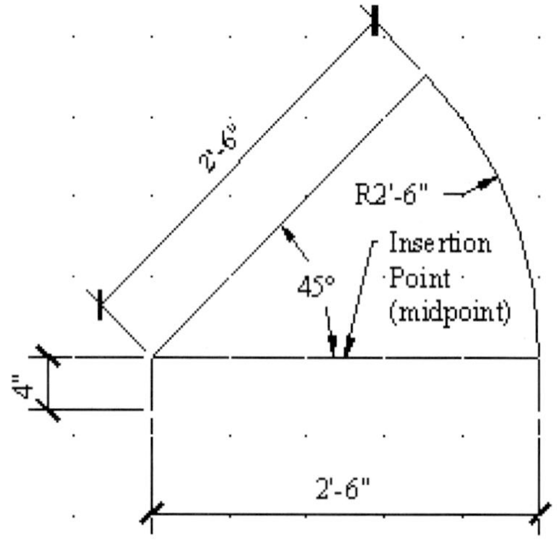

continued

TOOLS	COMMAND SEQUENCE	STEPS
		3. AutoCAD presents the Create Drawing File dialog box (which defaults to the Lesson 19 folder). Enter the name *door30,* as shown in Figure 19.2.2.1.3. Then pick the **Save** button.

FIG. 19.2.2.1.3: Create Drawing File Dialog Box

	Enter name of existing block or	4. Hit *enter* at the prompt.
	[= (block=output file)/* (whole drawing)] <define new drawing>: *[enter]*	
	Specify insertion base point:	5. Pick the insertion point identified in Figure 19.2.2.1.1.
	Select objects:	6. Select the arc, the vertical lines, and the angled line. Do *not* select the horizontal line. Confirm your selections.
	Select objects:	
	Select objects: *[enter]*	
	Command: *e*	7. Erase the horizontal line.
	Command: *qsave*	8. Save the drawing, but do not exit.

continued

TOOLS	COMMAND SEQUENCE	STEPS

9. Create the window drawing shown in Figure 19.2.2.1.9.

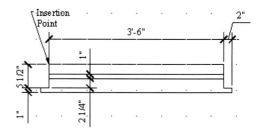

FIG. 19.2.2.1.9: Window

Command: *w*

10. We will use the dialog box to wblock the window. Enter the **WBlock** command by typing *wblock* or *w* at the command line.

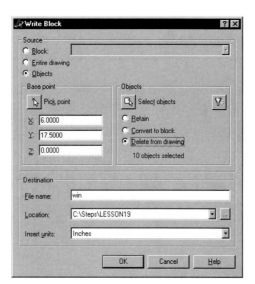

FIG. 19.2.2.1.11: Create the Win Write Block

11. AutoCAD presents the Write Block dialog box. Fill it in as shown in Figure 19.2.2.1.11 (your **Base point** may differ).

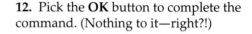

12. Pick the **OK** button to complete the command. (Nothing to it—right?!)

Command: *[enter]*

13. Now we will use the **WBlock** command to export our earlier blocks to our Folder Library. We will use the dialog box. Repeat the **WBlock** command.

continued

TOOLS	COMMAND SEQUENCE	STEPS

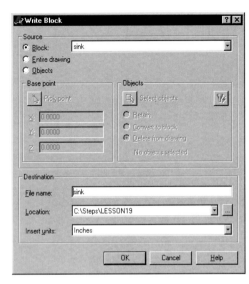

FIG. 19.2.2.1.14: Exporting a Block

14. (Refer to Figure 19.2.2.1.14.)

(a) Place the **Source** bullet in the **Block:** option and then select **Sink** from the control box. Notice that the **Base point** and **Objects** frames are no longer available (this information was provided when you defined the original block).

(b) Be sure the **Destination** information in your dialog box agrees with Figure 19.2.2.1.14.

(c) Pick the **OK** button to complete the command.

How Simple!

15. Repeat Steps 13 and 14 to send the remaining blocks (WC and DOOR36) to the Folder Library.

Command: *qsave*

16. Save the drawing.

■ 19.2.3 Using Blocks in a Drawing

Before we continue, we must know how to insert a block into a drawing once it has been created. There are two ways to do this—the *–Insert* command (for a command line interface) and the *Insert* command (for a dialog box interface). The command sequence for the *–Insert* command is

Command: *–insert* (or *–i*)

Enter block name or [?]: *[tell AutoCAD which block to insert—include a path if you wish to insert a file from a Folder Library]*

Specify insertion point or [Scale/X/Y/Z/Rotate/PScale/PX/PY/PZ/PRotate]: *[using coordinates or picking a point with the mouse, tell AutoCAD where to put the block]*

Enter X scale factor, specify opposite corner, or [Corner/XYZ] <1>: *[tell AutoCAD how to scale the block along the X–plane]*

Enter Y scale factor <use X scale factor>: *[tell AutoCAD how to scale the block along the Y–plane]*

Specify rotation angle <0>: *[tell AutoCAD how to orient the block]*

We are presented with a few options when using the *–Insert* command.

- Entering the question mark (?) at the first prompt will present this prompt:

  Enter block(s) to list <*>:

  Hit *enter* to display a list of all the blocks currently associated with the drawing.

- AutoCAD next asks the user for an insertion point. This sounds simple enough but the list of options that follow the prompt can be quite confusing. Let's look at these:
  - The first series of options (those without the "P" prefix) allows the user to scale the block as it is inserted. **Scale** and **Rotate** do just what they say. **X/Y/Z** options allow the user to scale along the selected axis.
  - The second series of options (those with the "P" prefix) does just what its counterparts without the "P" did, except that the scale or rotation is only for preview purposes. In other words, what you see is *not* what you get.

- The next prompt asks for an **X scale factor, specify opposite corner, or [Corner/XYZ]**
  - By default, AutoCAD will insert the block the same size it was when it was created. That is, the **X scale factor** and the **Y scale factor** will be *1* unless you set a scale at the last prompt. The user can resize the block after picking an insertion point by typing a value at the **X scale factor** prompt.
  - By typing *C* at this prompt (or by picking the opposite corner of the implied window), the user can size the block manually by locating opposite corners of the block.
  - The **XYZ** option enables the user to resize the block in three dimensions.

- At the **Y scale factor** prompt, the user can resize the block along the Y-plane independently of the X scale. Or the user can accept AutoCAD's default, which is the scale indicated for the **X scale factor**.

- The **Rotation** option, of course, allows the user to change the rotation angle of the insertion.

Some third-party libraries provide blocks using a single unit dimension (like creating a door that is one unit wide by one unit high). When inserting the block, the user indicates the size using the **X** and **Y scale factors**. By using this method, fewer blocks must be created but more work is involved in inserting them.

This is not always a good approach to blocks—especially if text is used in the block. Experience will show the user that time spent creating accurately sized blocks will almost always return "sevenfold" in time saved inserting them.

Let's insert some blocks from our new Folder Library into a drawing.

WWW | 19.2.3.1 Do This: | Inserting Blocks

I. Open the *cab-pln.dwg* file in the C:\Steps\Lesson19 folder. It looks like Figure 19.2.3.1a.

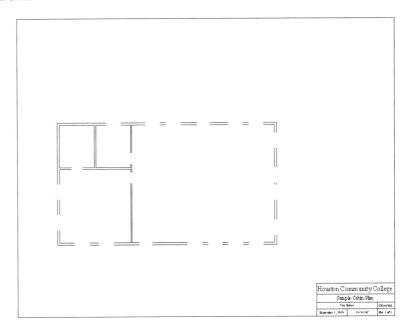

FIG. 19.2.3.1a: *cab-pln.dwg*

II. Follow these steps.

TOOLS	COMMAND SEQUENCE	STEPS
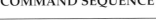	**Specify corner of window, enter a scale factor (nX or nXP), or [All/Center/Dynamic/Extents/Previous/Scale/Window] \<real time>:** *22',24'* **Specify opposite corner:** *36',13'*	**1.** Zoom in around the front door opening as indicated. Your drawing looks like Figure 19.2.3.1.1.  **FIG. 19.2.3.1.1:** *Zoom In*

continued

TOOLS	COMMAND SEQUENCE	STEPS
		**2.** Thaw the **Marker** layer. Notice that a node appears in the door opening (Figure 19.2.3.1.2). 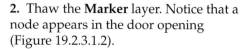 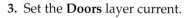 **FIG. 19.2.3.1.2:** A Node Appears in the Door
		**3.** Set the **Doors** layer current.
No Button Available	**Command:** *–i*	**4.** Enter the *–Insert* command by typing *–insert* or *–i* at the command prompt. There is not a button for the *–Insert* command. (The button on the Draw toolbar will call the Insert dialog box.)
	Enter block name or [?]: *c:\steps\lesson19\door36*	**5.** Tell AutoCAD to insert the *door36* file you created in the last exercise. Be sure to include the path.
	**Specify insertion point or [Scale/X/Y/Z/Rotate/PScale/PX/PY/PZ/PRotate]: _nod of**	**6.** Use the node as your insertion point.

continued

TOOLS	COMMAND SEQUENCE	STEPS
	Enter X scale factor, specify opposite corner, or [Corner/XYZ] <1>: *[enter]* Enter Y scale factor <use X scale factor>: *[enter]* Specify rotation angle <0>: *[enter]*	7. Accept the default **X** and **Y scale factors**, and the default **Rotation angle**. Your drawing looks like Figure 19.2.3.1.7. 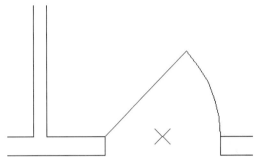 **FIG. 19.2.3.1.7:** Inserted Door Notice that, although we created the block on layer **0**, it exhibits the properties of the layer on which we inserted it (layer **Doors**).
	Command: *qsave*	8. Save the drawing, but do not exit.

The dialog box method of inserting blocks makes it a bit easier to identify the block to insert and to specify the insertion parameters. The *Insert* command calls the dialog box shown in Figure 19.2.3a.

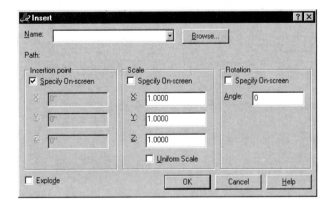

FIG. 19.2.3a: Insert Dialog Box

- The **Name:** control box will list all of the blocks currently associated with the drawing (by definition, previous insertion or template). To insert a wblock not currently associated with the drawing, use the **Browse . . .** button to locate the file. AutoCAD will show the path next to the **Path** statement.
- In the **Insertion point** frame, the user can identify the X/Y/Z coordinates for the insertion or (preferably) leave the check in the **Specify On-screen** box. Then the user can identify the insertion point on the screen.
- The **Scale** frame also has a **Specify On-screen** box. But it may be preferable (unless you must see the scaled block to be sure of its size) to enter the X/Y/Z scale in the text boxes provided. A check in the **Uniform Scale** box will ensure the X/Y/Z scales remain proportional to the original block definition.
- The **Rotation** frame also has a **Specify On-screen** box. Like the **Scale** data, it is up to the user to decide what is easiest to use.
- The last item in the Insert dialog box—the **Explode** check box—enables the user to place all the objects of a block without the definition of the block. In other words, you can insert the block preexploded.

> The *Insert* command can also be accessed from the Insert pull down menu.

Let's use the dialog box to insert some blocks into our drawing.

WWW 19.2.3.2 Do This: Using a Dialog Box to Insert Blocks

I. Be sure you are still in the *cab-pln.dwg* file in the C:\Steps\Lesson19 folder. If not, open it now.

II. Follow these steps.

TOOLS	COMMAND SEQUENCE	STEPS
	Command: *z*	1. Adjust the display on your screen to see the back door opening. It is the only opening on the back wall with a node in it.
Insert Block Button	Command: *i*	2. Enter the *Insert* command by typing *insert* or *i*. Or you can pick the **Insert Block** button on the Draw toolbar. AutoCAD presents the Insert dialog box (Fig. 19.2.3a).
Name: door36		3. Since we last used the *door36* block in this drawing, it is now part of the database and will be listed in the **Name** control box. If it is not, select it now.

continued

TOOLS	COMMAND SEQUENCE	STEPS
		4. Be sure the **Insertion point**, **Scale**, and **Rotation** are set as indicated in Figure 19.2.3.2.4.

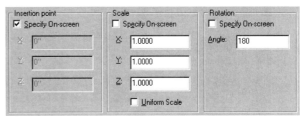

FIG. 19.2.3.2.4: Insertion Settings

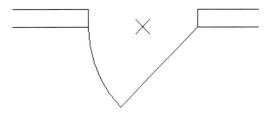		**5.** Pick the **OK** button. AutoCAD will continue the prompts on the command line.
	Specify insertion point: _nod of	**6.** Use the node as your **Insertion point**. AutoCAD will not prompt for **Scale** or **Rotation** because you entered that data in the dialog box. The door looks like Figure 19.2.3.2.6.

FIG. 19.2.3.2.6: Back Door

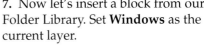		**7.** Now let's insert a block from our Folder Library. Set **Windows** as the current layer.
	**Command:** *i*	**8.** Repeat the *Insert* command.
		9. Pick the **Browse . . .** button. AutoCAD presents the Select Drawing File dialog box (Fig. 19.2.3.2.10).

continued

TOOLS	COMMAND SEQUENCE	STEPS

10. Go to the C:\Steps\Lesson19 folder and then select the *Win.dwg* file (Figure 19.2.3.2.10). Pick the **Open** button.

FIG. 19.2.3.2.10: Pick the Win file

[OK]

11. The *Win* file now appears in the **Name:** control box. Accept the insertion defaults and pick the **OK** button.

Specify insertion point: _endp of

12. Place the window in the opening to the left of the back door (Fig. 19.2.3.2.12). (Use the endpoint OSNAP when selecting the insertion point indicated.)

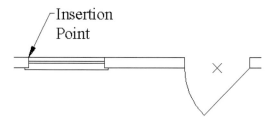

FIG. 19.2.3.2.12: Place the Window Here

Command: *qsave*

13. Save the drawing, but do not exit.

continued

TOOLS	COMMAND SEQUENCE	STEPS
		14. Place the rest of the windows and the interior doors and then freeze the **Markers** layer. When completed, your drawing will look like Figure 19.2.3.2.14.

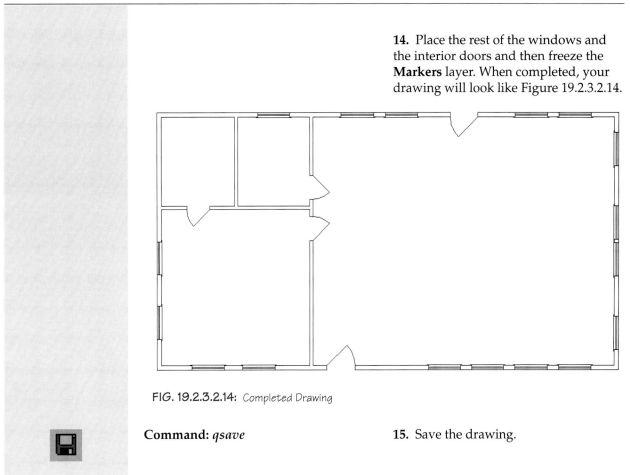

FIG. 19.2.3.2.14: Completed Drawing

	Command: *qsave*	15. Save the drawing.

■ 19.2.4 Sharing Blocks between Drawings Using the ADC

We first met the ADC (AutoCAD Design Center) back in Lesson 7 when we saw that we could share layer definitions between drawings. Well, sharing blocks between drawings is just as easy!

Let's give it a try.

19.2.4.1 Do This: Sharing Blocks

I. Open the *other cab-pln.dwg* file in the C:\Steps\Lesson19 folder. It looks just like the *cab-pln.dwg* file but has no blocks yet.

II. Follow these steps.

TOOLS	COMMAND SEQUENCE	STEPS
Tree Toggle View Button	**Command:** *adc* 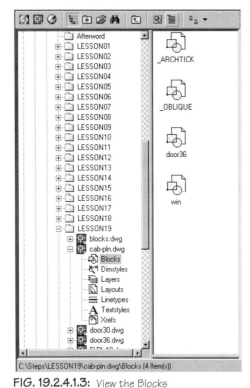 **FIG. 19.2.4.1.3:** View the Blocks	1. Open the AutoCAD Design Center. 2. Be sure you are using the Tree view (the **Tree Toggle View** button will be depressed). 3. (Refer to Figure 19.2.4.1.3.) **(a)** Follow the path to the C:\Steps\Lesson19\cab-pln.dwg file. **(b)** Select **Blocks**. Notice that the blocks associated with the file are listed in the right window. **(c)** Right-click on the *door36* block and select **Insert block . . .** from the cursor menu. **(d)** AutoCAD presents the Insert dialog box. Follow the procedures you learned in Exercise 19.2.3.2 to insert the block at the front door location.

This procedure saves quite a bit of time when the user creates blocks on the fly and finds later that he can use the same block in other drawings.

19.3 Extra Steps

Familiarize yourself with the following chart to help you know when to use Groups and when to use Blocks.

ABILITY	BLOCKS	GROUPS
Can be shared between drawings	Yes	No
The user can work on objects within the block/group	No	Yes
Can be treated as a single object	Yes	Yes
Can carry information (data)	Yes	No
Retrievable after erasure (without the *Undo* command)	Yes	No
May contain (other) blocks	Yes	Yes
May contain (other) groups	No	Yes

19.4 What Have We Learned?

Items covered in this lesson include:

- *Working with groups*
- *Folder & Temple libraries*
- *Block creation and insertion*
- *Using the ADC to share blocks between drawings*
- *AutoCAD Inquiry Commands:*
  - List
  - Block
  - BMake
  - Explode
  - WBlock
  - Insert

After 18 lessons of learning *How*, we have begun to answer the question *Why*. *Why* is CAD a better design tool than paper and pencil? *Why* should I spend thousands of dollars to do on a computer what I can do for a lot less on a board? *Why* must I educate myself (or in many cases, reeducate myself) to work with a computer?

Hopefully, the answers began to dawn on many of you in this last lesson.

We had already seen throughout the book that CAD drawings are neater and easier to read than many pen or pencil drawings. But up to the beginning of Lesson 18, it all seemed like a lot of work—often more than a comparable board drawing might require.

Then in Lesson 19, we found ways to manipulate large numbers of objects at once. We also found the CAD equivalent of drawing templates—*Blocks*. But with blocks, once created (or purchased), *no further drawing is necessary!* Additionally, a project (often a company and occasionally an entire industry) can expect standard symbols!

But believe it or not, we have only begun to scratch the surface of what we can accomplish using blocks. Wait till you see what we do in Lesson 20!

19.5 EXERCISES

1. Using the Folder Library we began in this lesson, place the doors and windows in the *flr-pln19.dwg* file located in the C:\Steps\Lesson19 folder. You will need to create some additional blocks to complete the drawing. The details are shown in Figures 19.5.1a, 19.5.1b, and 19.5.1c. The completed floor plan is shown in Figure 19.5.1d (the dimension layer has been frozen for clarity).

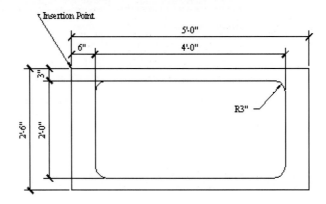

FIG. 19.5.1a: Tub

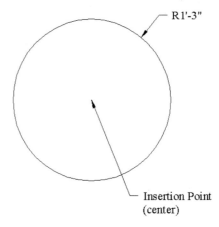

FIG. 19.5.1b: Water Heater

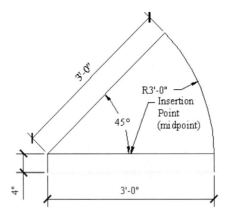

FIG. 19.5.1c: Door36I (Interior 36" door)

FIG. 19.5.1d: Completed Drawing

2. Start a new drawing from scratch.
   2.1. Set the limits for a full scale drawing on an 8 1/2" × 11" sheet of paper.
   2.2. Grid: 1/4"
   2.3. Create the following layers (all layers use a continuous linetype):

LAYER NAME	COLOR
BATTERY	56
BORDER	5 (blue)
COIL	4 (Cyan)
COPPERWIRE	3 (green)
GALVANOMETER	7 (white)
PIN	7 (white)
RESISTOR	6 (magenta)
SWITCH	1 (red)
TEXT	2 (yellow)

   2.4. Use the Times New Roman font. Text sizes are 1/4", 0.2", and 1/8".

2.5. Create the blocks shown in the *Legend* (Figure 19.5.2). (Remember to create all blocks on the layer 0.)

2.6. Create the drawing shown in Figure 19.5.2 (be sure to insert blocks on the appropriate layer). Save it as *MyBridge* in the C:\Steps\Lesson19 folder.

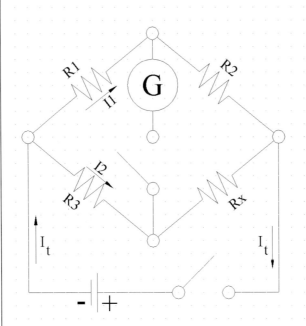
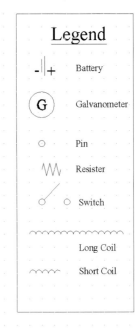

FIG. 19.5.2: MyBridge

3. Start a new drawing from scratch.
    3.1. Repeat the setup you used for Exercise 2 with the following exceptions:
        3.1.1. Add a layer called *Loop* using color 32.
        3.1.2. Grid: $1/2$
        3.1.3. Use the standard text style with $1/4$" and 0.2" text heights.
        3.1.4. Use the battery block you created in Exercise 2 and create a block for the wire loop.
    3.2. Create the *ElectroMagnet Circuitry* drawing shown in Figure 19.5.3. Save it as *MyMagnet* in the C:\Steps\Lesson19 folder.

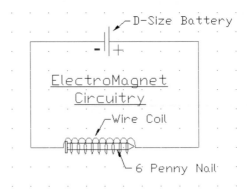

FIG. 19.5.3: Electromagnet Circuitry

4. Using what you have learned, create the two drawings (Figures 19.5.4c and 19.5.4d). Use blocks whenever possible. Be sure to use appropriate layers and complete dimensioning as shown. The details shown in Figures 19.5.4a and 19.5.4b will help. Both drawings should plot on a scale of $3/8" = 1' - 0"$ on an $11" \times 8 \frac{1}{2}"$ sheet of paper.

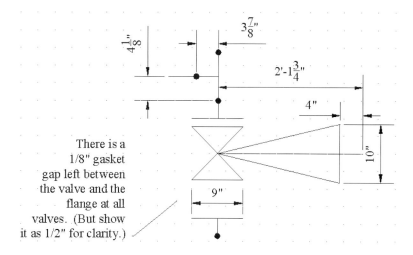

There is a 1/8" gasket gap left between the valve and the flange at all valves. (But show it as 1/2" for clarity.)

FIG. 19.5.4a

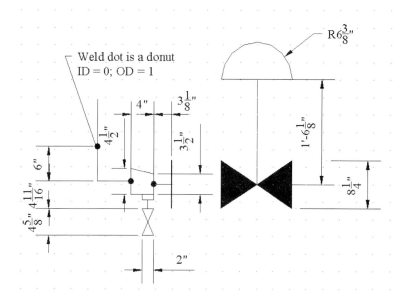

FIG. 19.5.4b

Lesson 19 Many as One: Groups and Blocks

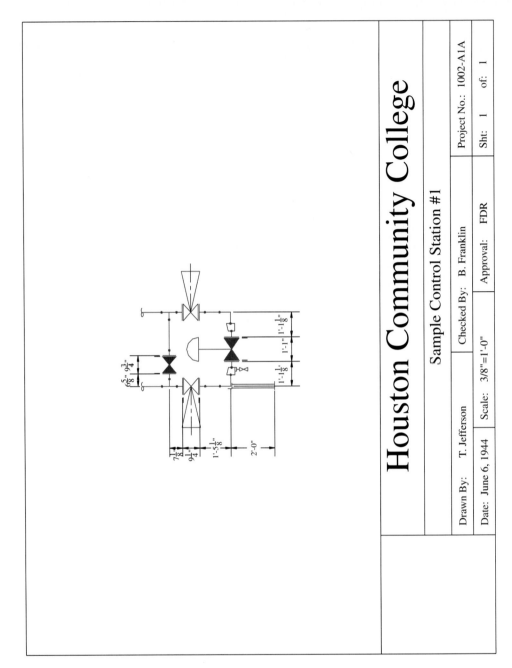

FIG. 19.5.4c: Control Station #1

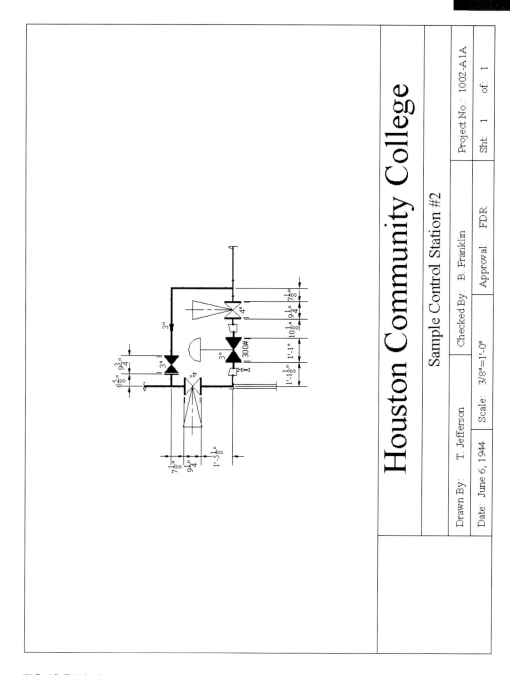

FIG. 19.5.4d: Control Station #2

5. Using blocks wherever possible, create the schematic drawing shown in Figure 19.5.5. Again, be sure to use appropriate layers. I used $1/4"$ grid marks.

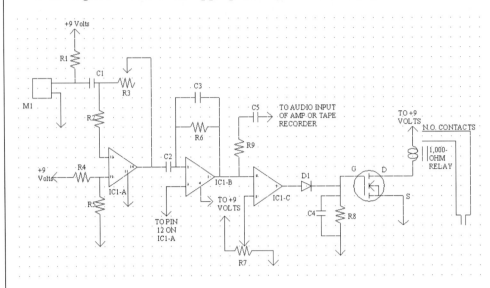

FIG. 19.5.5: Schematic

19.6 REVIEW QUESTIONS

Write your answers on a separate sheet of paper.

1. and 2. Two things that are made up of several objects but treated as one are _____ and _____.

3. Access the Object Grouping dialog box by typing _____ at the command prompt.

4. _____ simply means that a group can be manipulated as a single object.

5. To change one of the objects within a group, change the selectable setting to _____.

6. (T or F) Once a group is defined, the user cannot add or remove objects from it.

7. Pick the _____ button to see which objects are in a group.

8. Use the _____ button to change a temporary group name to something more appropriate.

9. The _____ button will remove the definition of the group from the drawing's database.

10. The _____ button is a toggle for treating a group as a group or suspending the group definition while you modify one or more of the objects within the group.

11. (T or F) Groups can be nested but objects can only belong to one group.

12. (T or F) Group use is limited to a single drawing.

13. (T or F) The group definition is permanently lost once the group is exploded or erased from a drawing.

14. (T or F) Block use is limited to a single drawing.

15. to 17. List the three block commands.

18. Groups of blocks are often assembled in useful packages called _____.

19. and 20. What are the two types of block library?

21. and 22. Accessing blocks that are part of a _____ library is faster than accessing blocks that are part of a _____ library.

23. If the objects used to make a block are created on layer *text*, they will have the characteristics of (layer text, the current layer) when inserted.

24. Generally speaking, the user should always create objects for blocks on layer _____.

25. The _____ command will return the original objects to the drawing after they have been blocked.

26. Use the _____ command to create folder libraries.

27. and 28. What are the two commands used to place a block in a drawing?

LESSON 20

Advanced Blocks

Following this lesson, you will:

➡ Know how to create and manipulate Block Attributes
  • The **–Attdef** and **Attdef** commands
➡ Know how to insert blocks with attributes
➡ Know how to control the display of attributes
  • The **Attdisp** and **Attreq** system variables
➡ Know how to edit attribute information
  • The **Attedit** and **DDAtte** commands
➡ Know how to redefine a block with attributes
  • The **Attredef** command
➡ Know how to extract attribute data to another program or for a bill or materials

One of the most important jobs a CAD operator may have involves the politics of convincing his supervisor (and often his supervisor's supervisor) of the importance of using AutoCAD as it was designed to be used. This will inevitably mean that the initial job setup will take more time than a non-CAD-oriented person might consider necessary. But the delay will be repaid "sevenfold" at the end of the project.

The operator might explain that AutoCAD is not, as is commonly believed, simply a very expensive drafting tool. Rather, AutoCAD should be considered the *backbone* of the overall project. Indeed, a design properly created in AutoCAD serves not only as an outline for construction, but reduces material talk-off and purchasing chores from weeks to minutes.

To cut large pieces of time from the end of the project, smaller pieces of time must be spent at the setup phase. Part of this time is required to create your libraries or to adjust purchased libraries to project standards. This adjustment should involve the addition of project specific *attributes* to your blocks. You will use these attributes to generate bills-of-materials and to share material information with Material Take-Off (MTO) and Purchasing programs.

This lesson will cover how to create and edit attributes, and how to share attribute data with other programs.

Let's begin.

20.1 Creating Attributes

What exactly is an attribute? Think of an attribute as a vessel that carries user-defined information. We have learned that all objects in a drawing contain information kept in the drawing's database. This information identifies the object using such things as: type, style, color, linetype, layer, position, etc. An attribute allows the user to attach his own information to a block and to retrieve that information later.

Like hatching and block creation, AutoCAD provides two methods for creating attributes—a command line method and a dialog box method.

■ 20.1.1 Defining Attributes at the Command Line: The –Attdef Command

The command sequence is

> **Command:** *–attdef*
> **Current attribute modes: Invisible=N Constant=N Verify=N Preset=N**
> **Enter an option to change [Invisible/Constant/Verify/Preset] <done>:** *[enter the appropriate capital letter to tell AutoCAD which mode, if any, you wish for this attribute]*

Enter attribute tag name: *[place an identifying tag or name of your attribute here]*

Enter attribute prompt: *[this is how AutoCAD will prompt the user when it places the block]*

Enter default attribute value: *[if you wish a default value, place it here]*

Specify start point of text or [Justify/Style]: *[These are standard text prompts]*

Specify height <0' − 4">:

Specify rotation angle of text <0>:

- The first line you see after entering the *–Attdef* command lists the four available attribute *modes*. By default, all modes are toggled *Off*. The next line prompts the user to **change** mode settings. The user can toggle one, two, three or all of the modes *On* or *Off* by typing the first letter of the mode. The prompts that follow may differ slightly depending on the mode settings.

  - An **Invisible** attribute holds data that will not show on the screen or drawing. This setting is ideal for vendor and pricing information. (Use visible data for information you wish to see on the drawing, such as valve sizes, window sizes, or title block data.)
  - A **Constant** attribute contains information that does not change. AutoCAD will not prompt for this information, and the value of a constant attribute cannot be edited after the block is inserted. The prompt sequence changes slightly, as shown here.

    Command: *–attdef*
    Current attribute modes: Invisible=N Constant=N Verify=N Preset=N
    Enter an option to change [Invisible/Constant/Verify/Preset] <done>: *c*
    Current attribute modes: Invisible=N Constant=Y Verify=N Preset=N
    Enter an option to change [Invisible/Constant/Verify/Preset] <done>: *[enter]*
    Enter attribute tag name: *[place an identifying tag or name of your attribute here]*
    Enter attribute value: *[enter the unchanging value]*
    Specify start point of text or [Justify/Style]:

  - When an attribute is created using the **Verify** mode, AutoCAD will first prompt the user to enter data for the attribute and then prompt the user to verify what was entered at the first prompt. **Verify** mode does not change the basic prompt sequence.
  - The user will not be prompted for any attribute values (except those requiring verification) when the block is created in the **Preset** mode. AutoCAD assumes default values for all the attributes when the block is inserted.

Let's create a block with attributes.

20.1.1.1 Do This: Create Attributes at the Command Line

I. Open the *blocks-pipe.dwg* file in the C:\Steps\Lesson20 folder (Figure 20.1.1.1a).

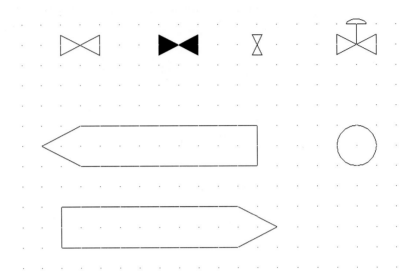

FIG. 20.1.1.1.1a: *blocks-pipe.dwg*

II. Follow these steps.

TOOLS	COMMAND SEQUENCE	STEPS
	**Command:** *z*	1. Zoom in around the upper left valve. It looks like Figure 20.1.1.1.1. **FIG. 20.1.1.1.1:** *Valve*
		2. Change the current layer to **Text**.
No Button Available	**Command:** *–att*	3. We are going to create five attributes to attach to this valve. Enter the *–Attdef* command by typing *–attdef* or *–att* at the command prompt. There is no toolbar button for this command.

continued

TOOLS	COMMAND SEQUENCE	STEPS
	Current attribute modes: Invisible=N Constant=N Verify=N Preset=N Enter an option to change [Invisible/Constant/Verify/Preset] <done>: *[enter]*	4. AutoCAD lists the current attribute mode settings and allows the user a chance to **change** one or all. We will go with the defaults for this first one. Hit *enter*.
	Enter attribute tag name: *size*	5. This attribute will carry size information for the valve, so use *size* as the **tag**.
	Enter attribute prompt: *What size is it?*	6. When we insert the block, how do we want AutoCAD to ask for the size data? Type this: *What size is it?*
	Enter default attribute value: *4"*	7. If we give our attribute a **default value**, the user will only have to hit *enter* to accept the value. (Of course, the user can enter a different value if desired.) We will use a default size of *4"*.
	Current text style: "TIMES" Text height: 0' – 4" Specify start point of text or [Justify/Style]: *c* Specify center point of text: Specify height <0' – 4">: *1/8* Specify rotation angle of text <0>: *[enter]*	8. Accept the current text style (Times) and center the attribute text at the snap above the center of the valve. Use a text height of $1/8$". Complete the command and AutoCAD places the tag of your new attribute, as shown in Figure 20.1.1.1.8

FIG. 20.1.1.1.8: Size Attribute

	Command: *qsave*	9. Save the drawing, but do not exit.

continued

TOOLS	COMMAND SEQUENCE	STEPS
No Button Available	**Command:** *–att*	10. Repeat Steps 3 through 8 to create a **Rating** attribute. All the modes for the attribute should be toggled off (*N*), the **tag** should be *Rating*, the prompt should read *What is the rating?* and the **default attribute value** should be *150#*. Center the tag below the valve as shown in Figure 20.1.1.1.10.

FIG. 20.1.1.1.10: Size and Rating Attributes

No Button Available	**Command:** *[enter]*	11. Now let's create a **Constant**, **Invisible** attribute. Repeat the *Attdef* command.
	Current attribute modes: Invisible=N Constant=N Verify=N Preset=N	12. Begin by setting the **Invisible** mode to **Y** by typing *I* for *invisible* at the prompt.
	Enter an option to change [Invisible/ Constant/Verify/Preset] <done>: *i*	
	Current attribute modes: Invisible=Y Constant=N Verify=N Preset=N	13. AutoCAD repeats the prompt this time listing the **Invisible** mode as **Y**. Set the **Constant** mode to *Y* as well by typing *C* for *constant*.
	Enter an option to change [Invisible/ Constant/Verify/Preset] <done>: *c*	
	Current attribute modes: Invisible=Y Constant=Y Verify=N Preset=N	14. AutoCAD repeats the prompt again—this time showing both the modes set to **Y**. Hit *enter* to proceed.
	Enter an option to change [Invisible/ Constant/Verify/Preset] <done>: *[enter]*	
	Enter attribute tag name: *vendor* **Enter attribute value:** *Vogt*	15. Identify the tag as *vendor* and set the **value** of the attribute to the vendor's name—*Vogt*.

continued

TOOLS	COMMAND SEQUENCE	STEPS
	Current text style: "TIMES" Text height: 0' − 0 1/8"	16. Hit *enter* at the text prompt to place the tag below the rating tag.
	Specify start point of text or [Justify/Style]: *[enter]*	
[save icon]	Command: *qsave*	17. Save the drawing, but do not exit.
	Command: *−att*	18. Repeat Steps 11 through 16 to create a *Type* attribute. Leave the modes set as they are now (creating an **Invisible**, **Constant** attribute). The **tag** should be *Type* and the **value** should be *Gate*. Place the tag beneath the vendor tag.
No Button Available	Command: *[enter]*	19. Now we will create an attribute that must be verified. Repeat the *−attdef* command.
	Current attribute modes: Invisible=Y Constant=Y Verify=N Preset=N	20. Toggle the **Constant** mode *Off* and the **Verify** mode *On*.
	Enter an option to change [Invisible/Constant/Verify/Preset] <done>: *c*	
	Current attribute modes: Invisible=Y Constant=N Verify=N Preset=N	
	Enter an option to change [Invisible/Constant/Verify/Preset] <done>: *v*	
	Current attribute modes: Invisible=Y Constant=N Verify=Y Preset=N	21. Hit *enter* to proceed.
	Enter an option to change [Invisible/Constant/Verify/Preset] <done>: *[enter]*	
	Enter attribute tag name: *price*	22. Set the **tag** to *price* and the **prompt** to *How much does it cost?* Do not provide a **default value** for this attribute.
	Enter attribute prompt: *How much does it cost?*	
	Enter default attribute value: *[enter]*	

continued

TOOLS	COMMAND SEQUENCE	STEPS
	Current text style: "TIMES" Text height: 0' – 0 1/8 " **Specify start point of text or [Justify/Style]:** *[enter]*	23. Hit *enter* at the text prompt to place the tag below the vendor tag. The drawing now looks like Figure 20.1.1.1.23.  FIG. 20.1.1.1.23: All Attributes
	Command: *qsave*	24. Save the drawing, but do not exit.
	Command: *wblock*	25. WBlock the valve and the attributes to the C:\Steps\Lesson20 folder. Call it *Gate* and use the midpoint shown in Figure 20.1.1.1.25 as the insertion point. (Also: retain the objects and use **Unitless** as the **Insert units** setting.) FIG. 20.1.1.1.25: Insertion Point
	Command: *qsave*	26. Save the drawing.

When an attributed block is inserted into a drawing, the attribute prompts will follow the standard insertion prompts. This can occur in one of two ways—command line prompts or the Enter Attributes dialog box. The method of prompting is tied to the method of block insertion, as we shall see.

Let's look at the command line method first; then we will look at the dialog box in Section 20.1.2.

> Whether or not AutoCAD prompts for attribute values at the command line or with a dialog box is controlled by the *Attdia* sysvar. A setting of *0* means you will receive prompts on the command line whereas as setting of *1* calls a dialog box.
>
> The command sequence looks like this:
>
> **Command:** *attdia*
>
> **Enter new value for ATTDIA <0>:** *1*

20.1.1.2 Do This: Inserting Attributed Blocks Using the Command Line

I. Open the *pid-20.dwg* file in the C:\Steps\Lesson20 folder. The drawing looks like Figure 20.1.1.2a.

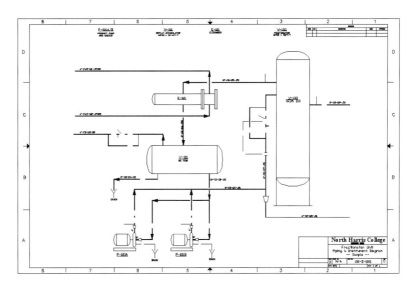

FIG. 20.1.1.2a: *pid-20.dwg*

II. Set the **Attdia** sysvar to *0*.

III. Follow these steps.

TOOLS	COMMAND SEQUENCE	STEPS
	Command: *v*	**1.** Restore the **Cont_Station1** view. The view looks like Figure 20.1.1.2.1 (without the arrow):  FIG. 20.1.1.2.1: Cont_Station1 View
		**2.** Set the current layer to **Va**.
	Command: *−i* **Enter block name or [?]:** *c:\steps\lesson20\gate* **Specify insertion point or [Scale/X/Y/Z/Rotate/PScale/PX/PY/PZ/PRotate]:** _endp of **Enter X scale factor, specify opposite corner, or [Corner/XYZ] <1>:** *[enter]* **Enter Y scale factor <use X scale factor>:** *[enter]* **Specify rotation angle <0>:** *[enter]*	**3.** Using the command line approach, insert the *Gate* block you created in the last exercise. Find it in the C:\Steps\Lesson20 folder. Insert it at the endpoint of the line referenced by the red arrow in Figure 20.1.1.2.1. Accept the default size and orientation.

continued

TOOLS	COMMAND SEQUENCE	STEPS
	Enter attribute values	4. Once you have accepted the rotation angle, AutoCAD asks you to **Enter attribute values**. Enter a price of *$55* and accept the default **rating** and **size**. (*Note:* You assigned these defaults in our last exercise.)
	How much does it cost?: *$55*	
	What is the rating? <150#>: *[enter]*	
	What size is it? <4">: *[enter]*	
		Notice that the prompts appear in the order in which you selected the attributes—unless you used a window to select. In that case, the prompts appear in reversed order from the order in which you created them.
	Verify attribute values	5. Remember that you used the **Verify** mode when creating the **Price** attribute. AutoCAD now asks you to verify the price you entered. (*Note:* The price you entered is the default. This makes it easy to simply confirm the entry.)
	How much does it cost? <$55>: *[enter]*	
	**Command:** *qsave*	6. Save the drawing, but do not exit. Your drawing looks like Figure 20.1.1.2.6.

FIG. 20.1.1.2.6: Inserted Block with Attribute Values

Notice that the **Type**, **Vendor**, and **Price** attributes are not visible. Notice also that the valve assumed the characteristics defined by the **Va** layer and the attribute text assumed the characteristics defined by this drawing's **Text** layer.

continued

TOOLS	COMMAND SEQUENCE	STEPS
	Command: *–i*	7. Repeat Steps 3 through 6 to place the other block valve. The drawing looks like Figure 20.1.1.2.7.

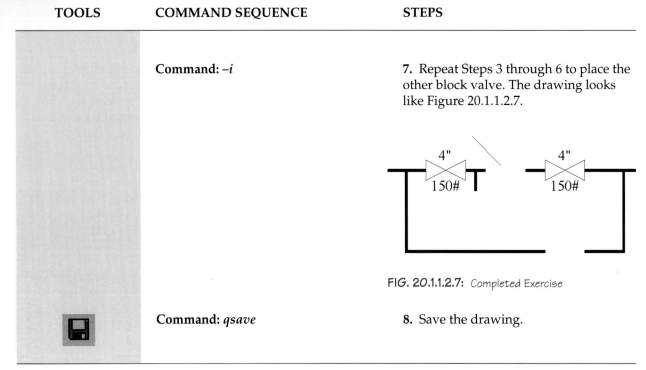

FIG. 20.1.1.2.7: Completed Exercise

	Command: *qsave*	8. Save the drawing.

20.1.2 Defining Attributes from a Dialog Box

After having used the command line approach to creating attributes, you will appreciate this simple, straightforward Attribute Definition dialog box (Figure 20.1.2a)!

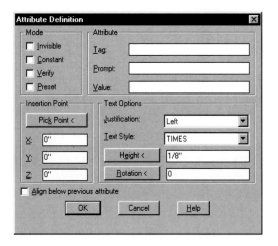

FIG. 20.1.2a: Attribute Definition Dialog Box

- As you can see, AutoCAD provides simple check boxes for the user to toggle the attribute's **Mode**.
- The **Attribute** information—**Tag**, **Prompt**, and **Value**—is easily inserted in the appropriate text box.

- The user may identify the **Insertion Point** by coordinate or use the **Pick Point <** button to select the insertion point on the screen.
- In the **Text Options** frame, the user uses control boxes to select the **Justification** and **Style** of the attribute's text. He can also place the **Height** and **Rotation** of the text into the appropriate text box or pick a button to select each from the screen.
- When creating several attributes, a check in the **Align below previous attribute** box will help keep the attributes organized.

Let's use the dialog box to create a block with attributes.

> The *DDAttdef* command can also be accessed from the Draw pull-down menu. Follow this path:
>
> *Draw—Block—Define Attributes . . .*

20.1.2.1 Do This: Defining Attributes with a Dialog Box

I. Reopen the *blocks-pipe.dwg* file in the C:\Steps\Lesson20 folder.

II. Follow these steps.

TOOLS	COMMAND SEQUENCE	STEPS
	Command: *z*	1. Zoom in around the *Globe* valve. It looks like Figure 20.1.2.1.1. FIG. 20.1.2.1.1: Globe Valve
No Button Available	Command: *att*	2. We will use the dialog box to create the same attributes for the globe valve that we created for the gate valve. Enter the *Attdef* command by typing *attdef* or *att* at the command prompt. AutoCAD presents the Attribute Definition dialog box (Fig. 20.1.2a).

continued

TOOLS	COMMAND SEQUENCE	STEPS

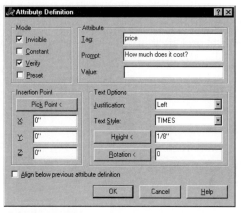

FIG. 20.1.2.1.3

3. Begin by creating the **Price** attribute. Fill in the **Mode**, **Attribute,** and **Text Options** frames of the dialog box, as shown in Figure 20.1.2.1.3.

4. Pick the **Pick Point <** button. AutoCAD returns to the graphics screen.

Start point:

5. Select the point two grid marks below the left end of the valve as the insertion point. AutoCAD returns to the dialog box.

6. Pick the **OK** button to complete the command. Your drawing looks like Figure 20.1.2.1.6.

PRICE

FIG. 20.1.2.1.6: Globe Valve Attributes

7. Repeat Steps 2 through 6 to create the additional attributes shown in the following chart.

		MODE				DEFAULT	
TAG	I	C	V	P	PROMPT	VALUE	TEXT JUSTIFY
Vendor	Y	Y	N	N		Jamesbury	
Type	Y	Y	N	N		Globe	
Rating	N	N	N	N	What is the rating?	150#	Centered
Size	N	N	N	N	What size is it?	3"	Centered

Your drawing now looks like Figure 20.1.2.1.7 (without the arrow):

continued

TOOLS	COMMAND SEQUENCE	STEPS
		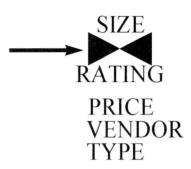 FIG. 20.1.2.1.7: Insertion Point
	Command: *w*	8. Wblock the valve and attributes under the name *Globe* to the C:\Steps\Lesson20 folder. Use the left endpoint of the polyline (indicated by the arrow) as the insertion point.
	Command: *qsave*	9. Save the drawing.

To edit or change an attribute definition *before* creating the block, use the standard text editor command (**DDEdit**). AutoCAD will present the dialog box shown in Figure 20.1.2b.

FIG. 20.1.2b: Edit Attribute Definition Dialog Box

Here you can change the **Tag**, **Prompt**, or **Default** value. You cannot change the *mode* of the attribute, however. You will have to create a new attribute for that.

Remember this handy method when you have several similar attributes to create. Simply copy your attributes to the blocks you wish to create, edit them as needed and then create the block!

Are dialog boxes not wonderful?
Let's look at another one. This one comes in handy when inserting the block.

20.1.2.2 Do This: Using a Dialog Box to Insert Attributed Blocks

I. Reopen the *pid-20.dwg* file in the C:\Steps\Lesson20 folder.

II. Set the *Attdia* sysvar to *1*.

III. Follow these steps.

TOOLS	COMMAND SEQUENCE	STEPS
	Command: *v*	1. If the **Cont_Station1** view is not current, restore it now.
	Command: *i*	2. Insert the *Globe* block from the C:\Steps\Lesson20 folder at the endpoint of the opening in the bypass line (indicated by the arrow). Accept the size and orientation defaults.

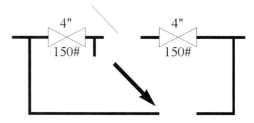

FIG. 20.1.2.2.2: Insert Here

3. AutoCAD presents the Enter Attributes dialog box. The box shows the command line prompts with their defaults. Notice that no prompts are given for the **Vendor** or **Type** attributes. Remember that these were created as **Constants** and cannot be changed.

Accept the defaults, but enter a price of **$75** in the appropriate text box (Figure 20.1.2.2.3).

FIG. 20.1.2.2.3: Enter these values

continued

TOOLS	COMMAND SEQUENCE	STEPS
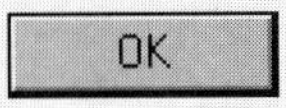 [OK]		4. Pick the **OK** button to complete the command. Your drawing looks like Figure 20.1.2.2.4.

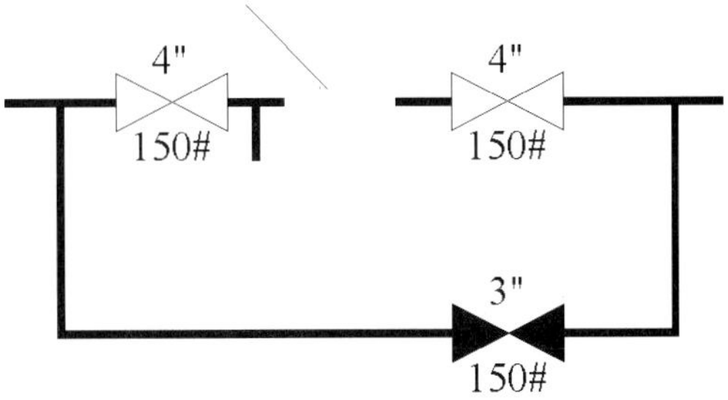

FIG. 20.1.2.2.4: Inserted Globe Valve

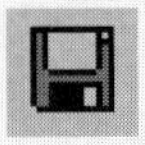	Command: *qsave*	5. Save the drawing, but do not exit.
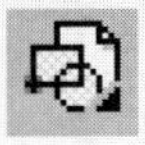	Command: *i*	6. Insert the control valve and the drain (both have already been created for you in the C:\Steps\Lesson20 folder) as shown. (The price of the control valve is $175; the price of the drain is $35.) The drawing looks like Figure 20.1.2.2.6.

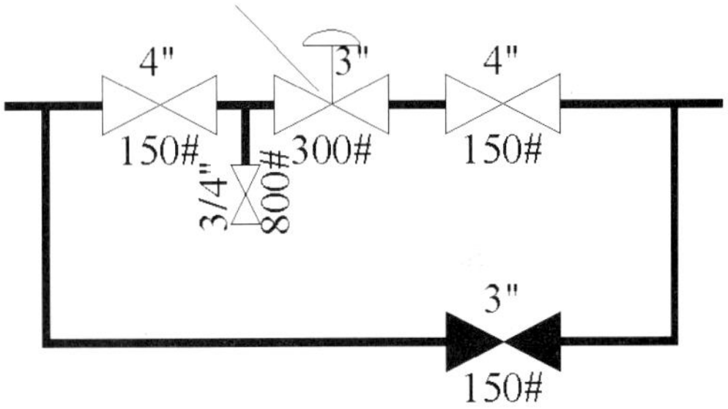

FIG. 20.1.2.2.6: Completed Control Station #1

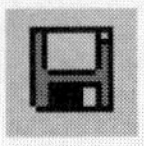	Command: *qsave*	7. Save the drawing, but do not exit.

continued

TOOLS	COMMAND SEQUENCE	STEPS
	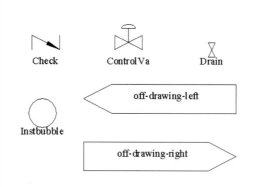	8. Use the blocks in Figure 20.1.2.2.8a to complete the drawing. (The blocks have already been created and can be found in the C:\Steps\Lesson20 folder *library*.) See the completed drawing in Figure 20.1.2.2.8b. Use the price list in the following chart to assign attribute values.

FIG. 20.1.2.2.8a: Blocks

SIZE/VALVE	GATE	GLOBE	CONTROL	CHECK
3/4"	$35			
4"	$85	$175	$275	
6"	$125			$195
8"	$385			
10"	$585			
8"	$385			
10"	$585			

 Command: *qsave* 9. Save the drawing, but do not exit.

AutoCAD includes a system variable—*Attreq*—that controls whether or not you will be prompted for attribute values. If the variable is set to *1*, AutoCAD prompts on the command line or with a dialog box (depending on the Attdia setting). If you set this sysvar to *0*, however, AutoCAD will not prompt at all for attribute values and you must use the attribute editor—*DDAtte* (Section 20.2)—to add the values.

20.2 Editing Attributes

Editing attribute values is as easy as recalling the Edit Attributes dialog box we used in Exercise 20.1.2.2. We do this using the *DDAtte* command. Let's try it.

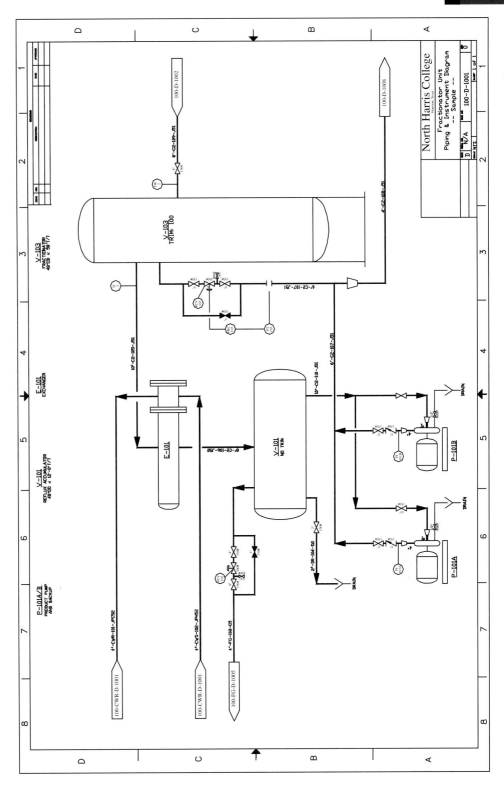

FIG. 20.1.2.2.8b: Completed PID

The *DDAtte* command can also be accessed from the Modify pull-down menu. Follow this path:

Modify—Attribute—Single . . .

20.2.1 Do This: Editing Attribute Values

I. Be sure you are still in the *pid-20.dwg* file in the C:\Steps\Lesson20 folder. If not, open it now.

II. Follow these steps.

TOOLS	COMMAND SEQUENCE	STEPS
	Command: *v*	**1.** If the **Cont_Station1** view is not current, restore it now.
Edit Attribute Button	**Command:** *ate*	**2.** When we inserted the gate valves, we assigned a price value of $55. According to the preceding chart, the price should be $85. Enter the *DDAtte* command by typing *ddatte* or *ate* at the command prompt. Or you can pick the **Edit Attribute** button on the Modify II toolbar.
	[Edit Attributes dialog box showing Block name: gate, How much does it cost? $85, What is the rating? 150#, What size is it? 4"] **FIG. 20.2.1.3:** Edit the PriceAttribute	**3.** Select anywhere on the left gate valve. AutoCAD presents the Edit Attributes dialog box. Change the *$55* entry to *$85*, as shown in Figure 20.2.1.3. Pick the **OK** button to complete the command.
	Command: *ate*	**4.** Repeat Steps 2 and 3 for the other gate valve.
	Command: *qsave*	**5.** Save the drawing, but do not exit.

An easy way to see the values of all the attributes attached to a block is to set the *Attdisp* sysvar to *On*. The command sequence is

Command: *attdisp*
Enter attribute visibility setting [Normal/ON/OFF] <Normal>: *on*

It can be useful to display all the attributes when searching for errors, but the display can quickly become quite crowded.

- The **Normal** option tells AutoCAD to display only those attributes created with the **Invisible** mode set to *N*.
- Turning *Off* the *Attdisp* tells AutoCAD to hide all the attributes regardless of the **Invisible** mode settings. This setting will speed regeneration. But be careful not to place objects where they will overlap the attributes when they are displayed again.

Editing the attributes themselves is quite a bit more difficult that editing just the attribute values. There is no dialog box to help, but AutoCAD does provide a means to change the value, position, height, angle, style, layer or color from the command line.

The command sequence is

Command: *attedit* (or *–ate*)
Edit attributes one at a time? [Yes/No] <Y>: *[enter]*
Enter block name specification <*>: *[enter]*
Enter attribute tag specification <*>: *[enter]*
Enter attribute value specification <*>: *[enter]*
Select Attributes: *[select the attributes to change]*
Select Attributes: *[enter to complete the selection]*
1 attributes selected. *[AutoCAD reports how many attributes were selected]*
Enter an option [Value/Position/Height/Angle/Style/Layer/Color/Next] <N>: *[tell AutoCAD what you want to do]*

- The first option—**Edit attributes one at a time?**—allows the user to change attributes globally (several at once) by typing *N* to override the *Y* default.
- The next three lines ask for some specifics about what you wish to edit. You can simply hit *enter* at each prompt if you will edit the attributes individually. The default is to accept all selected attributes for possible editing. But you can use these prompts to act as filters in a global selection set. For example, you can edit just the *Gate* valves in the previous exercise by responding *gate* to the **Enter block name specification** prompt.
- After the user selects the attributes to edit, AutoCAD presents a line of options to help specify what type of editing to perform.
  - The **Value** option allows the user to change the value of the attribute. Use this if you need to globally change a value. For example, say the price of 4" globe valves has just changed. You could edit your drawing globally to change the value of the **Price** attribute for all 4" globe valves. It is easier to use the *DDAtte* command to change individual values.

- The **Position** option comes in quite handy when the value of an attribute is physically too large to fit in the area you allotted for it (it overlaps something else). The *Move* command will move the entire block, but the **Position** option of the *Attedit* command allows the user to move just a single attribute.
- The **Angle** option is also quite handy for making attributes read properly despite the insertion rotation of the block.
- The other options—**Height**, **Style**, **Layer** and **Color**—allow the user to edit these aspects of attributes. If it is necessary to change any of these things, however, it may be better to redefine the block(s) so that it inserts properly in the first place.

Let's try some attribute editing.

20.2.2 Do This: Editing Attributes

I. Be sure you are still in the *pid–20.dwg* file in the C:\Steps\Lesson20 folder. If not, open it now.

II. Follow these steps.

TOOLS	COMMAND SEQUENCE	STEPS
No Button Available	**Command:** *v*	1. If the **Cont_Station1** view is not current, restore it now.
	Command: *–ate*	2. Our project engineer has determined that this control station requires a higher rating for the block and bypass valves. Enter the *–Attedit* command by typing *–attedit* or *–ate* at the command prompt.
	Edit attributes one at a time? [Yes/No] <Y>: *n*	3. We want to edit all the 150# valves. Tell AutoCAD that you do not want to **Edit attributes one at a time**.
	Performing global editing of attribute values. **Edit only attributes visible on screen? [Yes/No] <Y>:** *[enter]*	4. AutoCAD responds that this will be a global edit (more than one edit at once). Only the valves on the screen will be affected by this editing.
	Enter block name specification <*>: *[enter]*	5. We will be editing more than one type of block, so leave this default at the global setting.
	Enter attribute tag specification <*>: *rating*	6. We want to edit only the **Rating** attributes. If we tell AutoCAD this, it will not change any other type of attribute.

continued

TOOLS	COMMAND SEQUENCE	STEPS
	Enter attribute value specification <*>: *[enter]*	**7.** We do not need any further specifications for our edit.
	Select Attributes:	**8.** Select the three *150#* attributes showing on the screen, then confirm the selection.
	Select Attributes: *[enter]*	
	3 attributes selected.	
	Enter string to change: *150#*	**9.** AutoCAD wants to know what to change . . .
	Enter new string: *300#*	**10.** . . . and what to make it. Your drawing looks like Figure 20.2.2.10.

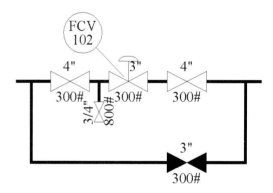

FIG. 20.2.2.10: Globally Edited Value

TOOLS	COMMAND SEQUENCE	STEPS
	Command: *qsave*	**11.** Save the drawing, but do not exit.
	Command: *–ate*	**12.** Notice that the **3"** attribute value slightly overlaps the control valve stem. Let's move it over a bit. Repeat the *–Attedit* command.
	Edit attributes one at a time? [Yes/No] <Y>: *[enter]*	**13.** This time we can speed through the first four options by simply accepting the defaults.
	Enter block name specification <*>: *[enter]*	
	Enter attribute tag specification <*>: *[enter]*	
	Enter attribute value specification <*> *[enter]*	

continued

TOOLS	COMMAND SEQUENCE	STEPS
	Select Attributes: **Select Attributes:** *[enter]*	**14.** Select the **3"** attribute value, then confirm the selection.
	1 attributes selected. **Enter an option [Value/Position/Height/Angle/Style/Layer/Color/Next] <N>:** *p*	**15.** AutoCAD reports the number of selected attributes then asks what you want to do. Tell it you wish to edit the **Position** of the attribute.
	Specify new text insertion point <no change>:	**16.** AutoCAD grabs the attribute at its insertion point and asks where you want to put it. Position it so that it does not touch the valve or the valve stem.
	Enter an option [Value/Position/Height/Angle/Style/Layer/Color/Next] <N>: *[enter]*	**17.** AutoCAD gives you the chance to change it again or move to the next selected attribute. Since you only selected the one attribute, an *enter* here will conclude the command. The valve looks something like Figure 20.2.2.17.

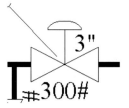

Fig. 20.2.2.17: Repositioned Attribute

| | **Command:** *–ate* | **18.** Reposition the remaining attributes to clarify this area. The drawing will look like Figure 20.2.2.18 when you are finished. |

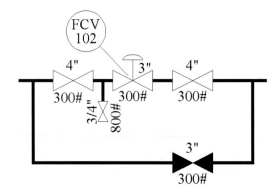

FIG. 20.2.2.18: Completed Drawing

continued

TOOLS	COMMAND SEQUENCE	STEPS
	Command: *qsave*	19. Save the drawing.

When it becomes necessary to redefine a block to change, add or remove attributes, use the *Attredef* command. Follow these steps:

1. Insert the block into an open area of the drawing.
2. Explode the block. This reveals all the attributes associated with it.
3. Make whatever changes are required to the attributes using the *DDEdit*, *Change* or *Properties* commands. Add or remove attributes as desired.
4. Enter the *Attredef* command and follow the prompts.

   Command: *attredef*

   Enter name of the Block you wish to redefine: *[enter the name of the block you are redefining]*

   Select objects for new Block ...

   Select objects: *[select the objects you want in the redefined block (the objects produced by exploding the old block and whatever objects you are adding)]*

   Select objects: *[confirm the selection]*

   Specify insertion base point of new Block: *[pick the same insertion point you used on the previous definition of the block—unless you wish to redefine that as well]*

5. AutoCAD will verify the attributes, redefine the block (for this drawing only), and update all existing instances of the block to reflect your changes.

This tool comes in handy when you discover a flaw in your block or attributes after having inserted several of them into your drawing.

20.3 The Coup de Grace: Using Attribute Information in Bills of Materials or Database Programs

This nifty trick is guaranteed to move you to the head of the class!

One of the most useful and timesaving devices available to AutoCAD users lies in AutoCAD's ability to save steps along the road to project completion. In this section of our lesson, we will discover how to use the information we have attached to our blocks. We will create a text file containing information acquired from our blocks. We will use this text file to create a bill of materials on our drawing. We will also create a database file using MS Access.

■ 20.3.1 Setting up the Attribute Extraction Template File

Before extracting information from your blocks, you must first tell AutoCAD exactly what information you wish to extract. To do this, you must create a *template file*. A template file is a text file that defines: the information you want, from where it will come, whether the information is in number or character format and how many spaces to allow for the information in the destination file.

It is not as complicated as it sounds. Information in a block comes in two forms—user-supplied data and AutoCAD-supplied data.

- User-supplied data include any information the user identified when creating the attribute or inserting the block. This includes all attribute values. The user identifies the attribute by its tag.

- AutoCAD-supplied data include all the other information stored in AutoCAD's database and related to the block. The user identifies the information with a tag assigned by AutoCAD. This includes the following 2-dimensional data:
  - **Name** of the block
  - The **Handle** of the block (the blocks identifier in AutoCAD's database)
  - **X** insertion coordinate
  - **Y** insertion coordinate
  - **Z** insertion coordinate
  - **X** scale factor of the block
  - **Y** scale factor of the block
  - **Z** scale factor of the block
  - **Layer** on which the block was inserted
  - The **Orientation** of the block

AutoCAD-supplied data also include a **Number** option that counts, or numbers, the block data in the extraction file.

(Refer to Figures 20.3.1a and 20.3.1b for the following explanation.)

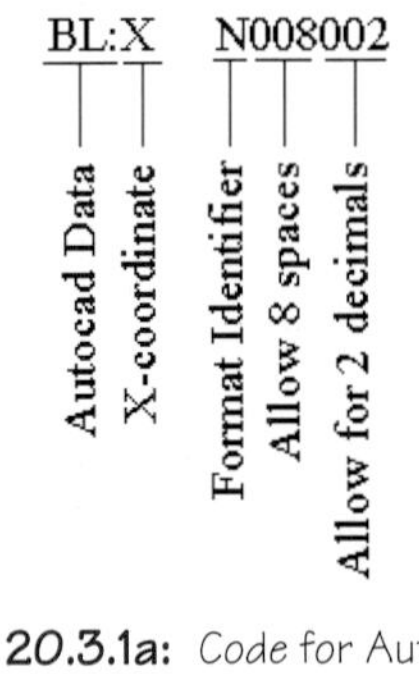

BL:X   N008002
- Autocad Data
- X-coordinate
- Format Identifier
- Allow 8 spaces
- Allow for 2 decimals

VENDOR   C010000
- User data tag
- Format Identifier
- Allow 10 spaces
- No decimal allowance

FIG. 20.3.1a: Code for AutoCAD Supplied Data in Numerical Format

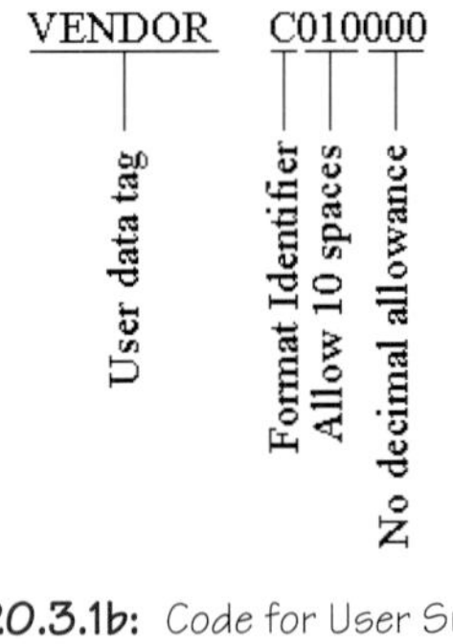

FIG. 20.3.1b: Code for User Supplied Data in a Character Format

AutoCAD-supplied data are preceded by a **BL:** in the template file. User-supplied data are not.

In the template file, you must identify the format of the information you are asking AutoCAD to find as either *character* (**C**) or *numerical* (**N**). Numerical information can contain only numbers—no $, %, #, or * symbols. All other formats are considered character.

Following the format identifier, tell AutoCAD how many spaces in the extraction file to allow for this information. This is given as a three-digit number.

Following the three-digit number identifying the number of spaces, tell AutoCAD how many decimal places to allow. This is a second three-digit number.

A sample file for the *PID* drawing might look like this:

```
BL:Name  C010000
BL:X     N008002
BL:Y     N008002
BLANK    C005000
TYPE     C010000
SIZE     C005000
RATING   C007000
VENDOR   C010000
PRICE    C010000
```

(The **Blank** tag will cause a space to appear between the AutoCAD supplied information and the user data.)

Note: It is crucial that you follow the last entry in the template file with a carriage return (**enter**) *but only a single carriage return*. AutoCAD is completely unforgiving about this. Without the return, the last tag will not be recognized; more than a single return will cause an **Invalid field specification** error.

Let's create a template file for our *PID* drawing.

20.3.1.1 Do This: Create an Attribute Extraction Template File

I. Open the Windows Notepad program. (Use the *Notepad* command at AutoCAD's command prompt.)

II. Type this code:

```
BL:NAME  C010000
BLANK    C005000
TYPE     C015000
SIZE     C005000
RATING   C007000
VENDOR   C010000
PRICE    C010000
```

III. Save the file as *MYPID* to the C:\Steps\Lesson20 folder (Notepad will automatically add the *.txt* extension).

■ 20.3.2 Extracting the Data

Extraction of block data is a simple command line process. The command is *Attext*; AutoCAD displays the Attribute Extraction dialog box shown in Figure 20.3.2a. As with most of AutoCAD's dialog boxes, we have a few frames of information to provide.

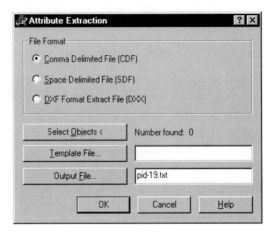

FIG. 20.3.2a: Attribute Extraction Dialog Box

▌ The options provided in the first frame allow the user to determine the format of the extraction file.

- The default—**Comma Delimited File (CDF)**—simply means that commas will separate the data in the destination file. Database programs—such as MS Access or DBase—readily recognize this format.
- A **Space Delimited File (SDF)**, of course, means that spaces will separate the data in the destination file. This is the easiest format for the novice to understand and the format in which you will create an extraction file for our bill-of-materials.
- A **DXF** file is designed for programmer use. We will ignore it here.

▌ The **Select Objects <** button allows the user to select the objects to include in the extraction. If selected, the user is prompted to **Select objects:**. If the user ignores the **Select Objects <** option, the extraction file will automatically include all the blocks in the drawing.

▌ The **Template File ...** button presents the standard Open File dialog box for the user to select the template file (the file you created in Exercise 20.3.1.1). Or the user can enter the file name (and path) into the text box next to the button.

▌ The **Output File ...** button also presents the standard Open File dialog box for the user to select the destination file for the extraction. Or the user can enter a file name into the text box for a new file.

There is an *Attext* command for the command line: *–Attext*. But the prompts are the same and the dialog box is easier to use. Still, feel free to experiment with it if you are more keyboard oriented.

Let's extract some data.

WWW 20.3.2.1 Do This: Extracting Attribute Data

I. Open the *pid-20.dwg* file in the C:\Steps\Lesson20 folder. If you have not completed this drawing, open the *pid-20-done.dwg* file.

II. Follow these steps.

TOOLS	COMMAND SEQUENCE	STEPS
	Command: *z*	1. Zoom out so you can see the entire drawing.
	Command: *attext*	2. Enter the *Attext* command.
	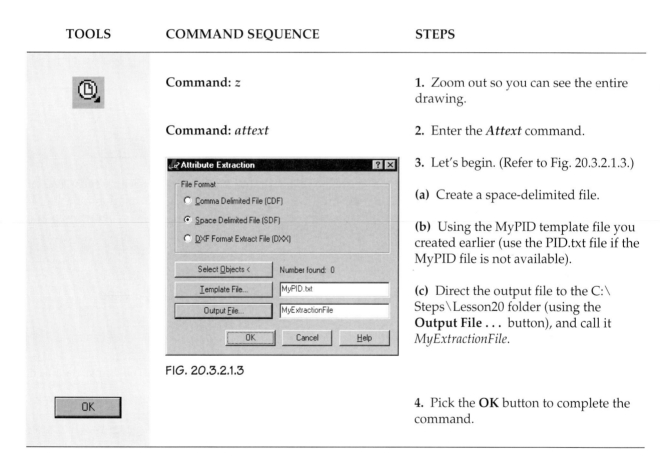 FIG. 20.3.2.1.3	3. Let's begin. (Refer to Fig. 20.3.2.1.3.) (a) Create a space-delimited file. (b) Using the MyPID template file you created earlier (use the PID.txt file if the MyPID file is not available). (c) Direct the output file to the C:\Steps\Lesson20 folder (using the **Output File . . .** button), and call it *MyExtractionFile*.
		4. Pick the **OK** button to complete the command.

III. Open the Windows Wordpad program. (Pick the **Start** button on your taskbar, then pick:
Programs . . . **Accessories** . . . **Wordpad**.)

IV. Open the *MyExtractionFile.txt* file you just created in the C:\Steps\Lesson20 folder. The file looks like Figure 20.3.2.1a.

```
GATE        Gate            4"      300#    Vogt        $85
GATE        Gate            4"      300#    Vogt        $85
GLOBE       Globe           3"      300#    Jamesbury   $75
CONTROLVA   Globe-Control   3"      300#    Jameson     $175
DRAIN       Gate            3/4"    800#    Vogt        $35
GATE        Gate            8"      150#    Vogt        $385
GATE        Gate            6"      150#    Vogt        $125
GATE        Gate            6"      150#    Vogt        $125
DRAIN       Gate            3/4"    800#    Vogt        $35
CONTROLVA   Globe-Control   4"      300#    Jameson     $275
GLOBE       Globe           4"      150#    Jamesbury   $175
GATE        Gate            2"      150#    Vogt        $55
GATE        Gate            6"      150#    Vogt        $125
CHECK       Check           6"      150#    Vogt        $195
GATE        Gate            10"     150#    Vogt        $575
DRAIN       Gate            3/4"    800#    Vogt        $35
DRAIN       Gate            3/4"    800#    Vogt        $35
GATE        Gate            10"     150#    Vogt        $575
CHECK       Check           6"      150#    Vogt        $195
GATE        Gate            6"      150#    Vogt        $125
```

FIG. 20.3.2.1a: Extracted Information

V. Repeat Steps 2 through 4, but this time create a *Comma-Delimited File* (enter **C** instead of **S** in Step 3) and save the file as *MyExtraction-Access.txt*.

You can now import the *SDF* file (using techniques we covered in Lesson 14) to create a valve list (bill of materials) for the drawing.

Wow! Is that a remarkable timesaver?!

But wait! Let's see how the Materials and Purchasing Departments can use the *CDF* file. (*Note:* In this section, I will import the attribute information into the MS Access program [Office 97 version]. If you do not have the Access program, you may skip this section.)

20.3.2.2 Do This: Importing Extracted Data into a DataBase Program

I. Using the MS Access program, create a new, blank database in the C:\Steps\Lesson20 folder. Name the database *MyPID.mdb*. The empty database opens (with no tables).

II. Follow these steps.

TOOLS	COMMAND SEQUENCE	STEPS

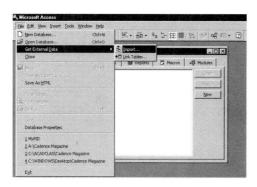

FIG. 20.3.2.2.1

1. Pick the following: **File** . . . **Get External Data** . . . **Import** . . . , as shown in Figure 20.3.2.2.1.

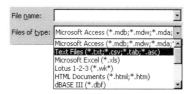

FIG. 20.3.2.2.2

2. Access presents the Import dialog box. Tell Access to find the *.txt* files, as shown in Figure 20.3.2.2.2.

FIG. 20.3.2.2.3

3. Select the *MyExtraction-Access.txt* file in the C:\Steps\Lesson20 folder. (See Figure 20.3.2.2.3.) (If this file is not available, use the *Extraction-Access.txt* file in the same folder.)
   Pick the **Import** button

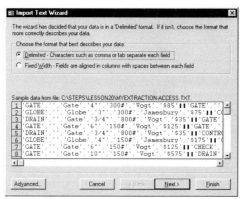

FIG. 20.3.2.2.4

4. Access begins the **Import Text Wizard**. (Figure 20.3.2.2.4). Be sure there is a bullet in the **Delimited** option button.
   Pick the **Next** button.

continued

TOOLS	COMMAND SEQUENCE	STEPS

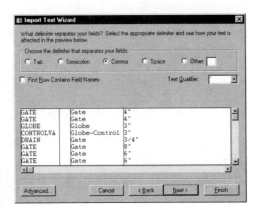

FIG. 20.3.2.2.5

5. The delimiter should default to **Comma** (Figure 20.3.2.2.5), but change the **Text Qualifier** to an apostrophe (use the control box).
  Pick the **Next** button.

FIG. 20.3.2.2.6

6. As we have no existing tables, we must store our data **In a New Table** (Figure 20.3.2.2.6).
  Pick the **Next** button.

FIG 20.3.2.2.7

7. Change the name of Field 3 to **Valve**.
  Continue to change the field names as follows: Field 4 to Size; Field 5 to Rating; Field 6 to Vendor; Field 7 to Price; and Field 1 to BlockName (Figure 20.3.2.2.7).
  Pick the **Next** button.

continued

TOOLS	COMMAND SEQUENCE	STEPS

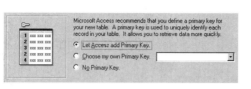

FIG. 20.3.2.2.8

8. Allow Access to create your primary field (Figure 20.3.2.2.8).
   Pick the **Next** button.

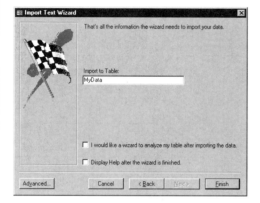

FIG. 20.3.2.2.9

9. Access now asks for the name of the new table. Call it *MyData* (Figure 20.3.2.2.9).
   Pick the **Finish** button.

10. Access tells you that the import was successfully completed and lists the MyData table in the database. Use Access' tools to remove the blank field (Field2).
    The database is now ready for the Materials people to count parts and the Purchasing people to order them.

20.4 Extra Steps

Refer to Lesson 14 as needed while importing the *SDF* file you created in exercise 20.3.2.1 into the *PID-20* drawing. There should be enough room in the lower left corner of the drawing for a Valve List.

20.5 What Have We Learned?

Items covered in this lesson include:

▪ *Creating, Inserting and Manipulating Block Attributes*
- *Attdef*
- *Attdia*
- *Insert*

▪ *Attribute display commands:*
- *Attdisp*
- *Attreq*

▪ *Editing Atrributes*
- *Attedit*
- *Ddatte*
- *Attredef*

▪ *Extracting Attribute Data*
- *Attext*

In Lesson 20, we learned what makes AutoCAD worth the price you (or your company) paid for it. Here we saw how we can shave weeks from a project by spending a day or two in additional setup time. This savings translates into increased profit for the company by cutting production time (thus increasing the amount of work possible in a given period of time). It means more money for the trained operator because it is his training that makes the savings possible.

What we have done in this lesson is to cover the methods and techniques available that can accomplish these savings. But it is often up to you, the CAD operator and designer, to explain, demonstrate, and even sell these possibilities to the people and companies for whom you work. Remember that CAD is still relatively new to many people. They are relying on your expertise to show them what can be done!

20.6 EXERCISES

We are going to recreate the project we did in Lesson 19, but this time our blocks will contain attributes that we will use to create a bill-of-materials for our drawing.

1. Begin by opening the *Blocks-arch.dwg* file in the C:\Steps\Lesson20 folder. Notice that the drawings have already been created for you.

    1.1. Create attributes for each (on the **Text** layer), as indicated in the chart that follows. Then WBlock each to the C:\Steps\Lesson20\Library folder. (Use the same insertion points you used in Lesson 19.)

BLOCK	TAG	MODE				DEFAULT PROMPT	TEXT VALUE	JUSTIFY
		I	C	V	P			
Door36-at	Type	Y	N	N	N	What type is it?	Flush	
Door36-at	Size	N	N	N	N	What size is it?		Center
Door36-at	Mfgr	Y	N	N	N	Who made it?	Ideal	
Door36-at	MfrID	Y	N	N	N	What is the ID?		
Door36i-at	Type	Y	N	N	N	What type is it?	Flush	
Door36i-at	Size	N	N	N	N	What size is it?		Center
Door36i-at	Mfgr	Y	N	N	N	Who made it?	Ideal	
Door36i-at	MfrID	Y	N	N	N	What is the ID?		
Door30-at	Type	Y	N	N	N	What type is it?	Flush	
Door30-at	Size	N	N	N	N	What size is it?		Center
Door30-at	Mfgr	Y	N	N	N	Who made it?	Ideal	
Door30-at	MfrID	Y	N	N	N	What is the ID?		
Win-at	Type	Y	N	N	N	What type is it?	DH	
Win-at	Size	N	N	N	N	What size is it?		Center
Win-at	Mfgr	Y	N	N	N	Who made it?	Anderson	
Win-at	MfrID	Y	N	N	N	What is the ID?		
Sink-at	Type	Y	N	N	N	What type is it?	Lav	

BLOCK	TAG	MODE I	MODE C	MODE V	MODE P	DEFAULT PROMPT	TEXT VALUE	JUSTIFY
Sink-at	Size	Y	N	N	N	What size is it?	18 × 30	
Sink-at	Mfgr	Y	N	N	N	Who made it?	American	
Sink-at	MfrID	Y	N	N	N	What is the ID?	W-POR-LA010	
Tub-at	Type	Y	N	N	N	What type is it?	Tub	
Tub-at	Size	Y	N	N	N	What size is it?	30 × 60	
Tub-at	Mfgr	Y	N	N	N	Who made it?	American	
Tub-at	MfrID	Y	N	N	N	What is the ID?	T/S030	
WC-at	Type	Y	N	N	N	What type is it?	WC	
WC-at	Mfgr	Y	N	N	N	Who made it?	American	
WC-at	MfrID	Y	N	N	N	What is the ID?	W-POR-WC001	
WH-at	Type	Y	N	N	N	What type is it?	WH	
WH-at	Size	Y	N	N	N	What size is it?	50gal	
WH-at	Mfgr	Y	N	N	N	Who made it?	American	
WH-at	MfrID	Y	N	N	N	What is the ID?	E52	

1.2. When inserting the blocks, include the following information.

BLOCK	TYPE	SIZE	MFR	MFRID
Door36-at	Flush	36"	Ideal	X-SOL−36
Door36i-at	Flush	36"	Ideal	I-HOL−36
Door30-at	Flush	30"	Ideal	I-HOL−30
Win-at	DH	42 × 24	Anderson	2-DH-S
Win-at	DH	42 × 35	Anderson	2-DH-M
Win-at	DH	42 × 74	Anderson	2-DH-L
Sink-at	Lav	18 × 30	American	W-POR-LA010
Tub-at	Tub	30 × 60	American	T/S030
WC-at	WC		American	W-POR−001
WH-at	WH	50 gal	American	E52

1.3. Open the *FlrPln 20.dwg* file in the C:\Steps\Lesson20 folder.
1.4. Create a layer called **Plumbing** for the fixtures. Make it **Red**.
1.5. Insert the blocks as shown in Figure 20.6.1.10.
1.6. Create a template file to be used to extract the information shown in the bill of materials. Save the file as *ArchTemp.txt* to the C:\Steps\Lesson19\Library folder.
1.7. Extract attribute data to a space delimited file called *MyData.txt* to the C:\Steps\Lesson20 folder.
1.8. Arrange the data as shown on the finished drawing (Figure 20.6.1.8).

```
                       Windows
Block        Type    Size      Manufacturer   ID
WIN-AT       DH      42x24     Anderson       2-DH-S
WIN-AT       DH      42x35     Anderson       2-DH-M
WIN-AT       DH      42x74     Anderson       2-DH-L
WIN-AT       DH      42x74     Anderson       2-DH-L
WIN-AT       DH      42x74     Anderson       2-DH-L
WIN-AT       DH      42x74     Anderson       2-DH-L
WIN-AT       DH      42x74     Anderson       2-DH-L
WIN-AT       DH      42x74     Anderson       2-DH-L
WIN-AT       DH      42x74     Anderson       2-DH-L
WIN-AT       DH      42x74     Anderson       2-DH-L
                        Doors
Block        Type    Size      Manufacturer   ID
DOOR30-AT    Flush   30"       Ideal          I-HOL-30
DOOR30-AT    Flush   30"       Ideal          I-HOL-30
DOOR30-AT    Flush   30"       Ideal          I-HOL-30
DOOR30-AT    Flush   30"       Ideal          I-HOL-30
DOOR36-AT    Flush   36"       Ideal          X-SOL-36
DOOR36-AT    Flush   36"       Ideal          X-SOL-36
DOOR36I-AT   Flush   36"       Ideal          I-SOL-36
DOOR36I-AT   Flush   36"       Ideal          I-SOL-36
DOOR36I-AT   Flush   36"       Ideal          I-HOL-36
DOOR36I-AT   Flush   36"       Ideal          I-HOL-36
DOOR36I-AT   Flush   36"       Ideal          I-HOL-36
DOOR36I-AT   Flush   36"       Ideal          I-HOL-36
                  Plumbing Fixtures
Block        Type    Size      Manufacturer   ID
TUB-AT       Tub     30x60     American       T/S030
TUB-AT       Tub     30x60     American       T/S030
SINK-AT      Lav     18x30     American       W-POR-LA010
SINK-AT      Lav     18x30     American       W-POR-LA010
SINK-AT      Lav     18x30     American       W-POR-LA010
WC-AT        WC                American       W-POR-WC001
WC-AT        WC                American       W-POR-WC001
WH-AT        WH      50gal     American       E-52
```

FIG. 20.6.1.8: Bill-of-Materials

1.9. Import the data into the lower right corner drawing as shown on the final drawing (Figure 20.6.1.10).
1.10. Save the drawing as *MyFlrPln20.dwg* in the C:\Steps\Lesson20 folder.

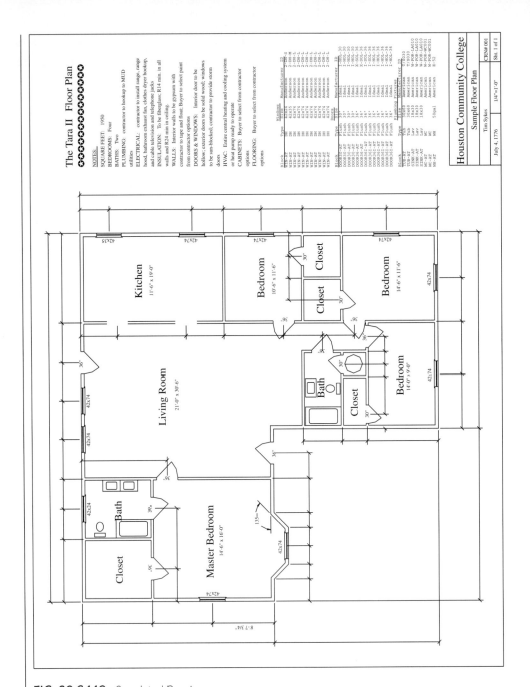

FIG. 20.6.1.10: Completed Drawing

2. Using what you have learned, create the next two drawings (Figures 20.6.2a and 20.6.2b) using attributes to create the lists (BOM and Valve List). If you have access to a database program (e.g., MS Access or DBase), create a database file of information for each drawing. Both are on 34" × 22" sheets of paper and are NTS.

20.6 Exercises

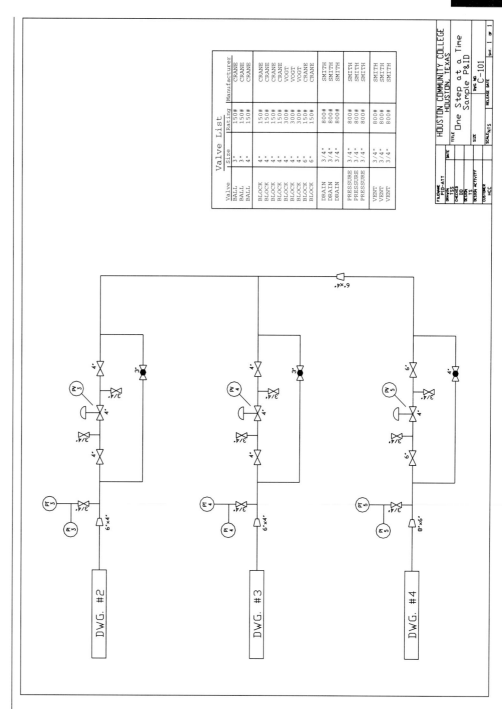

FIG. 20.6.2a: Sample PID

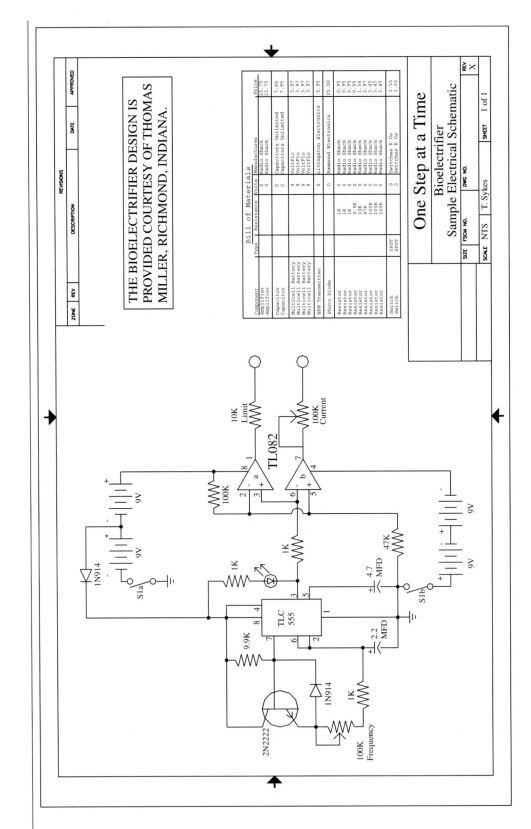

FIG. 20.6.2b: Bioelectrifier

20.7 REVIEW QUESTIONS

Write your answers on a separate sheet of paper.

1. An _____ allows a user to attach information to a block and retrieve that information later.

2. The command line command for defining attributes is _____.

3. An _____ attribute holds data that will not be shown on the screen or drawing.

4. A _____ attribute contains information that does not change.

5. When the attribute was created using the _____ mode, AutoCAD will first prompt the user to enter data and then prompt the user to verify the entered information.

6. AutoCAD assumes default values for all attributes when the block is inserted if the block was created using the _____ mode.

7. When an attributed block is inserted into a drawing, the attribute prompts will (precede, follow) the standard insertion prompts.

8. To edit or change the tag, prompt, or default value of an attribute definition before creating the block, use the _____ command.

9. and 10. Set the _____ sysvar to _____ to use the Enter Attributes dialog box.

11. and 12. If you find that AutoCAD is not prompting at all for attribute information, check to be sure that the _____ sysvar is set to _____.

13. To edit attribute values, enter the _____ command.

14. and 15. Setting the _____ sysvar to _____ will display all the attribute values assigned to blocks in the drawing.

16. and 17. Use the _____ option of the _____ command to change the position of an attribute.

18. Use the _____ command to change attribute values globally.

19. and 20. Use the _____ option of the _____ command to change the rotation of an attribute.

21. Use the _____ command to redefine a block to add or remove attributes.

22. Before extracting information from your blocks, you must first create a _____ file.

23. and 24. _____ and _____ supply information in a block.

25. It is crucial to follow the last line in your template file with a _____.

26. Extract attribute data from a drawing using the _____ command.

27. If you wish to extract data for a bill of materials, create a (CDF, SDF) file.

LESSON 21

Afterword: Getting an Edge

Following this lesson, you will:
➡ Know how to create your own hotkeys
➡ Be ready to tackle the Advanced Text!

You are about to enter a very competitive world. As an instructor, a mentor, and a guru (not to mention all around nice guy), I cannot let you go without giving you a bit of an edge over the next guy.

In this afterword, you will learn how to create your own hotkeys. What this means to you is simply that you will be able to enhance your drawing speed (by reducing keystrokes) with a tool most basic CAD operators do not have. I hope this will give you a "runging" start up the corporate ladder.

Additionally, I want to encourage you to pursue your training on to the next level. Although 2D CAD drafting is quick and fairly easy, it should not be considered more than a transition from the drafting *board* to the drafting *program*. I will tempt you in the

Z-direction. That is, I will introduce you to 3-dimensional drafting (in this case—3D Solid Modeling).

Let's begin with hotkeys.

21.1 Creating Your Own Hotkeys: The Acad.PGP File

AutoCAD provides two methods of creating hotkeys (or aliases). One—AutoLISP—requires some programming knowledge. The other—the *.pgp* (for ProGram Parameters) file—is a simple text file and quite easy for even the AutoCAD novice to manage.

With AutoLISP, the programmer can create some complex programs and access them with a few simple keystrokes. In fact, many of the features currently available in AutoCAD began as Lisp routines. Some commands still exist as Lisp routines that are automatically loaded when needed. Unfortunately, it would take a separate book to explain the workings of Lisp.

The user, on the other hand, can add to or modify the *.pgp* file with a rudimentary knowledge of the Notepad text editor that ships with MS Windows. The user is limited to single commands in the *.pgp* file. But as you will see, many commands would be tedious without the hotkeys defined in it.

> !!
> !!!!!!!!!!!!!!!! Always Make a Backup Copy of a File Before editing It !!!!!!!!!!!!!!!!
> !!

The *.pgp* file is divided into two main sections—External Commands (commands that work outside AutoCAD in the Windows environment) and Command Aliases (hotkeys). (*Note:* A semicolon in the *.pgp* file begins a *comments* line. AutoCAD will not read this line.)

- You will rarely need to use the External Commands section of the *.pgp* file. Most of what you will want to do can be accessed from the **Start** button on Windows' taskbar. However, AutoCAD still provides several legacy "shell" commands dating back to the MS-DOS releases. These include *Del* (for deleting files), *Dir* (for listing the contents of a directory or folder), and *Edit* (for accessing the MS-DOS text editor).

The format for creating a hotkey for an external command is

<Command Name>,<Program>,<Flag>,<Prompt>,

Example: WP,WORDPAD,   1,*File to Open: ,0

- The **Command Name** is the hotkey the user enters at the command prompt to execute the external command
- The **Program** is what the hotkey will execute. Precede most Windows programs with the word *Start*.

- The **Flag** is a code telling AutoCAD how to run the program. These are the possible entries:
  - *0* = starts the program and immediately returns to AutoCAD without waiting for the program to complete
  - *1* = starts the program but waits for it to finish before returning to AutoCAD
  - *2* = starts the program and runs it minimized
  - *3* = starts the program but does not wait for it to finish before returning to AutoCAD
  - *4* = starts the program and runs it hidden (you probably will not use this one)
  - *5* = starts the program and runs it hidden while returning to AutoCAD
  - *8* = allows the command to work with long file names
- The **Prompt** is what, if anything, you wish AutoCAD to ask the user before executing the program. For example, if you are creating a hotkey to open Notepad, you might want AutoCAD to ask the user which file to open. (*Note:* An asterisk before the prompt allows the user to use spaces in the response.)

A comma must follow the prompt whether you use an optional return code or not.

▌ Creating a hotkey for an AutoCAD command is really quite simple. The format looks like this:

<Command Name>,   *<AutoCAD Command>

Example: *AX,*   **attext*

- As with the external command, the **Command Name** is the hotkey the user enters at the command prompt.
- The **AutoCAD Command**, of course, is the command you wish to execute with the hotkey. You are limited to a single command. That is, *Zoom* will work but *Zoom Window* will not.

That is all there is to it!

Let's create some hotkeys of our own. The *Acad.pgp* file can be found in the *Support* subfolder (of the AutoCAD folder).

21.1.1 Do This: Create Some Hotkeys

I. *Make a Copy of the Acad.pgp file!*

> *Warning: You are about to change one of AutoCAD's program files. You must make a backup copy first! If you are not comfortable enough with the Windows operating system to make a backup copy of a file, please skip this section. You may not be able to fix a mistake. With a backup copy of the file, all you need do to fix a mistake is to delete the* Acad.pgp *file and rename the backup file to* Acad.pgp.

II. Follow these steps.

TOOLS	COMMAND SEQUENCE	EXPLANATION
No Button Available	**Command:** *notepad*	1. Access the **Notepad** program from within AutoCAD.
	File to edit: *c:\acad2000\support\acad.pgp*	2. AutoCAD asks which file you want to edit. Tell it to edit the *Acad.pgp* file. (*Note:* your path may vary depending on which folder contains the AutoCAD program on your computer.) **Notepad** opens the file in a window atop the AutoCAD window.
		3. A hotkey should never, when it can possibly be avoided, require more than two keystrokes. We will change the hotkey for **Notepad** to *NP*. Scroll down till you see the Notepad line. It looks like Figure 21.1.1.3.

```
NOTEPAD,   START NOTEPAD, 1,*File to edit: ,
```

FIG. 21.1.1.3: The Notepad Line

4. Change the line as shown in Figure 21.1.1.4.

```
NP,        START NOTEPAD, 1,*File to edit: ,
```

FIG. 21.1.1.4: The New Notepad Line

5. Save the file and exit **Notepad**.

No Button Available	**Command:** *reinit*	6. You must now reinitialize the file for AutoCAD to recognize the changes. Enter the *Reinit* command. AutoCAD presents the Re-initialization dialog box shown in Figure 21.1.1.6.

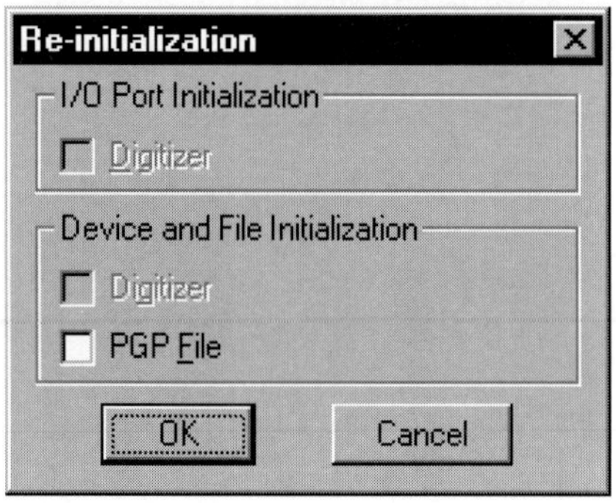

FIG. 21.1.1.6: Re-initialize Dialog Box

continued

TOOLS	COMMAND SEQUENCE	STEPS
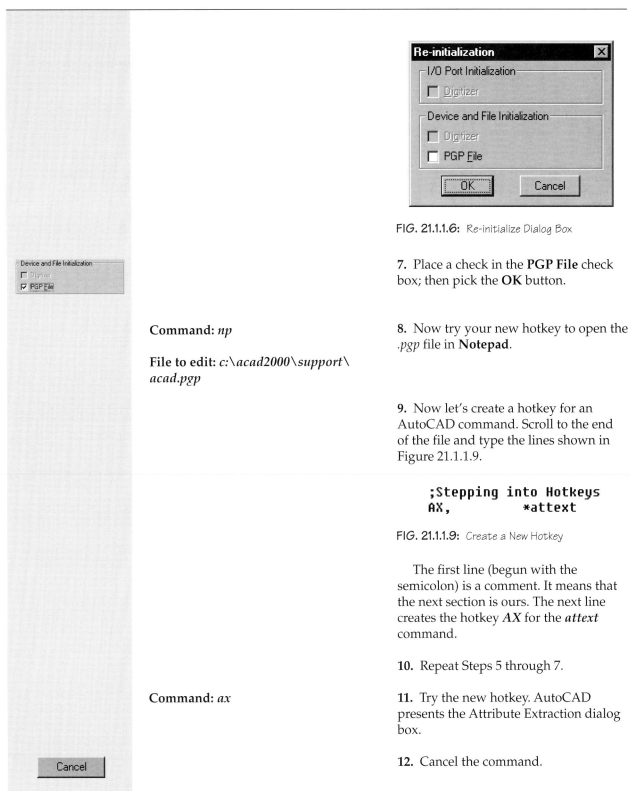		

FIG. 21.1.1.6: Re-initialize Dialog Box

7. Place a check in the **PGP File** check box; then pick the **OK** button.

Command: *np*

File to edit: *c:\acad2000\support\acad.pgp*

8. Now try your new hotkey to open the *.pgp* file in **Notepad**.

9. Now let's create a hotkey for an AutoCAD command. Scroll to the end of the file and type the lines shown in Figure 21.1.1.9.

```
;Stepping into Hotkeys
AX,          *attext
```

FIG. 21.1.1.9: Create a New Hotkey

The first line (begun with the semicolon) is a comment. It means that the next section is ours. The next line creates the hotkey *AX* for the *attext* command.

10. Repeat Steps 5 through 7.

Command: *ax*

11. Try the new hotkey. AutoCAD presents the Attribute Extraction dialog box.

12. Cancel the command.

continued

TOOLS	COMMAND SEQUENCE	STEPS
		13. Repeat Step 9 to create the hotkey *RI* for the *Reinit* command.

Some additional things to remember when editing the *Acad.pgp* file include:

1. Always try to limit hotkeys to one or two keystrokes. On rare occasions, three keystrokes may be acceptable but never more. Remember: the idea is to speed things up.
2. Whenever possible, use keystrokes that can be entered with one hand (preferably the left for right-handers or the right for left-handers). Try to leave the mouse hand free.
3. Whenever possible, use keystrokes that correspond to the command. For example, use *M* for *Move* or *O* for *Open*. Using *M* for *Break* would not make sense and would be difficult for the user to remember.
4. Each hotkey requires a tiny bit of memory that might otherwise be used to run AutoCAD. Be conservative, not frivolous.
5. Read the *Acad.pgp* file to learn what hotkeys are already available. Do not waste memory overwriting existing hotkeys.
6. Hotkeys for accessing Windows programs like **Notepad** or **Paint** may seem like a good idea. But ask yourself if they are necessary when the user can access them from the **Start** button. If your users are keyboard-oriented, they may be; if they are mouse-oriented, they may not.

21.2 The Z-Direction

"... oh my God!—it's full of stars!"
Dave Bowman on looking into the Obelisk
—Arthur C. Clarke's *2001: A Space Odyssey*

You may feel a bit like Dave once you have peered into the third dimension.

It would be impossible to give you anything but a taste of 3D drafting in these few pages—there is so much to learn. What I hope to do, then, is to tempt you into some exploration on your own (or through our next book: *AutoCAD: One Step at a Time—Advanced Edition*).

Creating objects with depth (a Z dimension) does not have to be difficult—especially when you are using 3D modeling tools. Rather than working with rectangles, polygons and circles (2-dimensional objects), you will work with boxes, pyramids, and cylinders (3-dimensional objects). Rather than working with a host of modifying commands, you will work with three basic editing tools—subtract, union, and intersect. (But you will still need the host of modifying tools, so do not forget them.) And rather than entering X,Y coordinates, you will enter X,Y,Z coordinates.

The first new tool you will need is one that allows you to see in 3-dimensional space on a 2-dimensional screen. AutoCAD calls this tool *VPoint*. It prompts you like this:

Current view direction: VIEWDIR=0.0000,0.0000,1.0000
Specify a view point or [Rotate] <display compass and tripod>:

The **ViewDir** sysvar tells you that you are currently looking at the drawing from a spot one unit above the 0,0 location of the drawing (+1 on the Z-axis). This is the standard plan view you have used throughout 2D drafting. Next, AutoCAD prompts you for a new viewpoint. Enter a coordinate from which you want to view the drawing (we will see this in our exercise). If you hit *enter* at the prompt, AutoCAD will provide a compass and tripod to help position you; but the coordinate approach is much more accurate.

We will need some other tools as well—such as boxes and cylinders. But we will look at these, as we need them.

> The solid modeling tools can also be found in the Draw pull-down menu. Follow this path:
>
> *Draw—Solids—[tool]*
>
> The solid editing tools can also be found in the Modify pull-down menu. Follow this path:
>
> *Modify—Solids Editing—[tool]*

Let's take a look at Solid Modeling. (We will draw a yo-yo.)

21.2.1 Do This: Exploring Solid Modeling

I. Start a new drawing from scratch.

II. Create two layers: Obj1 (cyan) and Obj2 (red). Make Obj1 current.

III. Follow these steps.

TOOLS	COMMAND SEQUENCE	EXPLANATION
No Button Available	**Command:** *–vp*	1. We will begin by adjusting our viewpoint (adjusting where we stand to view our drawing). Enter the *Vpoint* command by typing *vpoint* or *–vp* at the command prompt.

continued

TOOLS	COMMAND SEQUENCE	STEPS
	Specify a view point or [Rotate] <display compass and tripod>: *1,–1,1*	**2.** Tell AutoCAD that you wish to take a step (*1*) to the right (positive X), a step back (negative Y), and a step up (positive Z). You will view the model from this vantage as you draw it. Notice that the UCS icon changes (Figure 21.2.1.2) to show you where you stand in relation to coordinate 0,0,0 (the icon's position).

FIG. 21.2.1.2: The UCS Icon

Notice also that the crosshairs remain flat to the 2-Dimensional X,Y plane.

3. Open the **Solids** toolbar (Figure 21.2.1.3)—right-click on any existing toolbar and select **Solids** from the cursor menu

FIG. 21.2.1.3: Solids Toolbar

continued

TOOLS	COMMAND SEQUENCE	STEPS
No Button Available	**Command:** *isolines* **Enter new value for ISOLINES <4>:** *8*	4. Set the **Isolines** sysvar to *8*. This controls how many surface lines AutoCAD will use to display our model.
 Cylinder Button	**Command:** *cylinder* **Current wire frame density: ISOLINES=8** **Specify center point for base of cylinder or [Elliptical] <0,0,0>:** *3,3,0*	5. To paraphrase the sculptor, "we will start with a block and remove everything that is not a yo-yo." Insert a cylinder at coordinates *3,3,0*. (Type *cylinder* at the command prompt. Or you can use the **Cylinder** button on the Solids toolbar.)
	Specify radius for base of cylinder or [Diameter]: *d* **Specify diameter for base of cylinder:** *2.25*	6. Give the cylinder a diameter of *2.25*".
	Specify height of cylinder or [Center of other end]: *1.125*	7. And a height (Z-value) of *1.125*". Your drawing looks like Figure 21.2.1.7.

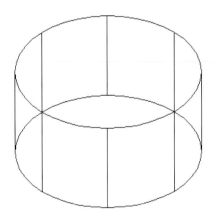

FIG. 21.2.1.7: Cylinder

Box Button	**Command:** *box*	8. It does not look much like a yo-yo yet, does it? Define the space in the middle with a box. (Place the box on the **Obj2** layer.) Begin by entering the *Box* command (type *box* at the command prompt or pick the **Box** button on the Solids toolbar).

continued

TOOLS	COMMAND SEQUENCE	STEPS
	Specify corner of box or [CEnter] <0,0,0>: *1.875,1.875,.5*	9. Locate the corner of the box as indicated.
	Specify corner or [Cube/Length]: *l*	10. Specify the length, width and height as indicated.
	Specify length: *2.25*	Your drawing looks like Figure 21.2.1.10.
	Specify width: *2.25*	
	Specify height: *.125*	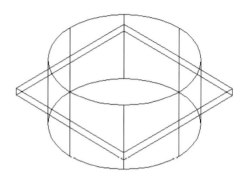
		FIG. 21.2.1.10: Box and Cylinder
	Command: *cylinder*	11. Now add the hub (on the **Obj1** layer) using the *Cylinder* command again. Your drawing looks like Figure 21.2.1.11
	Current wire frame density: ISOLINES=8	
	Specify center point for base of cylinder or [Elliptical] <0,0,0>: *3,3,0*	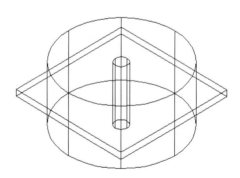
	Specify radius for base of cylinder or [Diameter]: *d*	
	Specify diameter for base of cylinder: *.25*	
	Specify height of cylinder or [Center of other end]: *1.125*	**FIG. 21.2.1.11:** All the Right Parts

continued

TOOLS	COMMAND SEQUENCE	STEPS
 Subtract Button		12. Now the sculptor must remove the parts that are "not a yo-yo." Begin by opening the **Solids Editing** toolbar (Figure 21.2.1.12)  FIG. 21.2.1.12: Solids Editing Toolbar
	Command: *subtract*	13. Now enter the *Subtract* command by typing *subtract* at the command line or picking the **Subtract** button on the Solids Editing toolbar.
	Select solids and regions to subtract from . . .	14. Tell AutoCAD you want to remove something *from* the large cylinder.
	Select objects: *[select the large cylinder]*	
	Select objects: *[enter]*	

continued

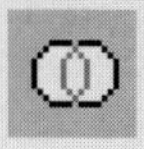

TOOLS	COMMAND SEQUENCE	STEPS
	Select solids and regions to subtract ...	**15.** Now tell AutoCAD to remove the box. Your drawing looks like Figure 21.2.1.15.
	Select objects: *[select the box]*	
	Select objects: *[enter]*	
		FIG. 21.2.1.15: Box Removed
Union Button	**Command:** *union*	**16.** Next, we will join the two sides of our yo-yo to the hub. Enter the **Union** command by typing *union* at the command line or picking the **Union** button on the Solids editing toolbar.
	Select objects:	**17.** Select the two cylinders. Your drawing looks like Figure 21.2.1.17.
	Select objects: *[enter]*	
		FIG. 21.2.1.17: A Solid Yo-Yo

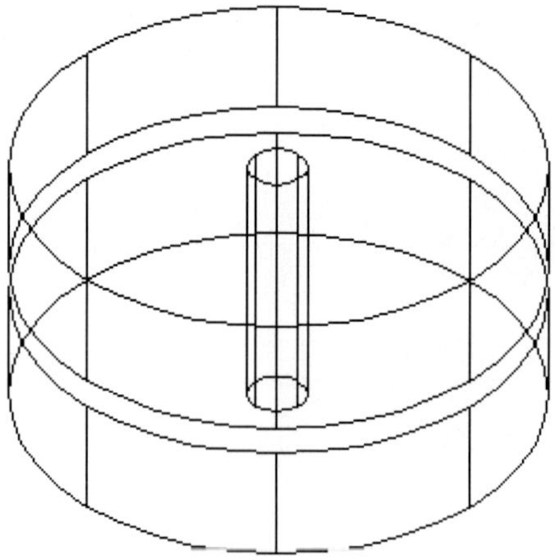

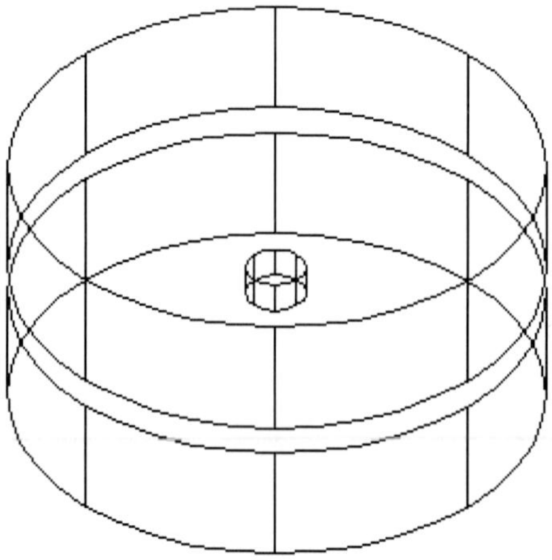

continued

TOOLS	COMMAND SEQUENCE	STEPS
	**Command:** *f*	18. Okay. So it still is not quite a yo-yo. Let's whittle it into shape. Use the *Fillet* command.
	Current settings: Mode = TRIM, Radius = 0.5000 **Select first object or [Polyline/Radius/Trim]:** *[select the top circle of the upper hemisphere]* **Enter fillet radius <0.5000>:** *[enter]* **Select an edge or [Chain/Radius]:** *[enter]* **1 edge(s) selected for fillet.**	19. Notice that the prompts are slightly different for filleting a 3-dimensional object. Accept the $1/2$" radius and select the top circle of the upper half of the yo-yo to fillet. AutoCAD prompts again for the fillet radius—hit *enter* to accept. We have already selected an edge and do not need another. Hit *enter* to confirm our selection.
	Command: *f*	20. Repeat Steps 18 and 19 for the bottom of the yo-yo. Your drawing looks like Figure 21.2.1.20.

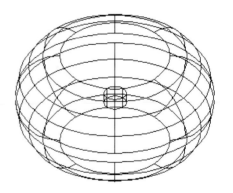

FIG. 21.2.1.20: Filleted Sides (Now It Is Starting to Look Like a Yo-Yo)

| | **Command:** *save* | 21. Save the drawing as *MyYoYo.dwg* in the C:\Steps\Lesson21 folder, but do not exit. |

I know—had you known it was that easy . . .

What do you mean, "it still doesn't look like a yo-yo"? Well, a yo-yo (like beauty) is in the eye of the beholder. What you see can be used by a CAD-CAM system to build our yo-yo. But AutoCAD can help the display look more like a yo-yo!

Let's see what we can do. First, we will stand it up using the 3-dimensional version of AutoCAD's *Rotate* command — *Rotate3d*. This is only one of several 3-dimensional specific modifying tools you will study in the advanced text.

21.2.2 Do This: Rotating Your Model

I. Be sure you are still in the *MyYoYo.dwg* file in the C:\Steps\Lesson21 folder. If not, open it now.

II. Follow these steps.

TOOLS	COMMAND SEQUENCE	EXPLANATION
No Button Available	**Command:** *rotate3d*	**1.** Enter the *Rotate3d* command by typing *rotate3d* at the command prompt.
	Current positive angle: ANGDIR= counterclockwise ANGBASE=0 **Select objects:** **Select objects:** *[enter]*	**2.** After presenting some information about your setup, AutoCAD asks what you wish to rotate. Select the yo-yo. Notice that the entire yo-yo is selected with a single pick. (The *Union* command you used in the last exercise joined the hub cylinder to the outer cylinder. The yo-yo is now a single object.)
	Specify first point on axis or define axis by **[Object/Last/View/Xaxis/Yaxis/Zaxis/ 2points]: _qua of** **Specify second point on axis: _qua of**	**3.** In a 2-dimensional drawing, you rotated about a point. In 3-dimensional space, you must identify the *axis* about which you wish to rotate. Select the southern quadrant (lower left of the object) of the bottom half of the yo-yo. Then select the northern (opposite) quadrant of the same half.

continued

TOOLS	COMMAND SEQUENCE	STEPS
	Specify rotation angle or [Reference]: *90*	**4.** Tell AutoCAD to rotate the object 90°. Does it look more like a yo-yo now (Figure 21.2.2.4)?

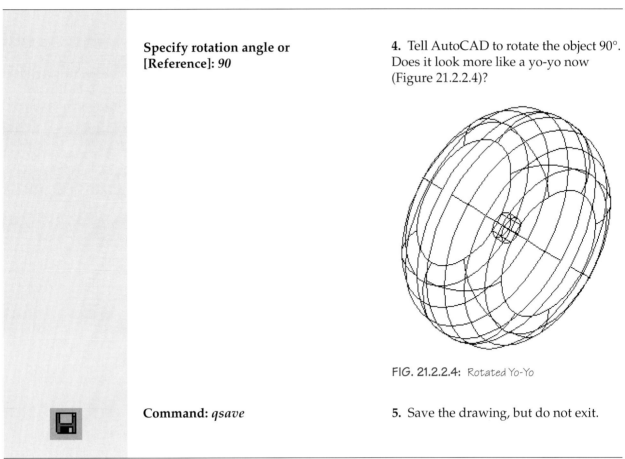

FIG. 21.2.2.4: Rotated Yo-Yo

	Command: *qsave*	**5.** Save the drawing, but do not exit.

Although it looks transparent, our yo-yo is a solid object. AutoCAD has been showing it in wireframe to speed our drawing procedures. But now we can fill it in and see what it really looks like.

AutoCAD has three approaches to making a solid object look solid (and even making it look like the final product): **Hide**, **Shade** and **Render**.

Let's look at each.

21.2.3 Do This: Displaying Your Model

I. Be sure you are still in the *MyYoYo.dwg* file in the C:\Steps\Lesson21 folder. If not, open it now.

II. Follow these steps.

TOOLS	COMMAND SEQUENCE	EXPLANATION
No Button Available	**Command:** *hide*	**1.** Enter the *Hide* command by typing *hide* at the command prompt. AutoCAD hides (or removes from the display) the lines that would normally be behind other objects. Your drawing looks like Figure 21.2.3.1.
		FIG. 21.2.3.1: Hidden Lines Removed
	Command: *re*	**2.** *Regen* the drawing to see the hidden lines again.

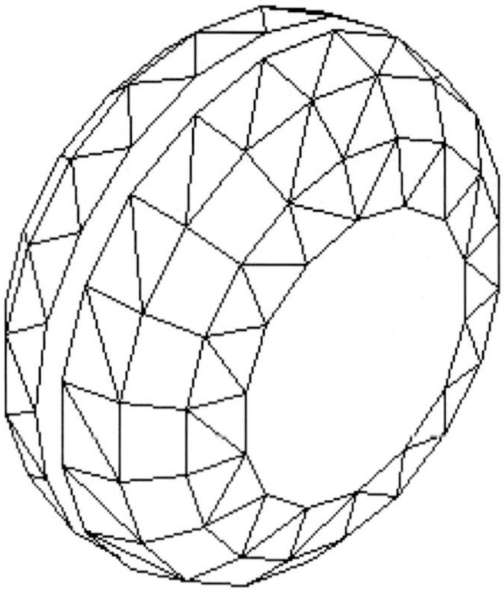

continued

TOOLS	COMMAND SEQUENCE	STEPS
No Button Available	**Command:** *shade*	**3.** The next approach—*Shade*—looks better and adds some color to the display. Enter the **Shade** command at the command prompt. Your drawing looks like Figure 21.2.3.3.

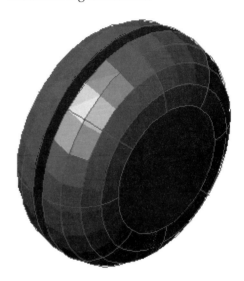

FIG. 21.2.3.3: Shaded Yo-Yo

Notice that the shade command replaces the UCS icon with a more colorful icon.

In earlier releases of AutoCAD, it was necessary to regen the drawing before continuing to work on the drawing. But with AutoCAD 2000, it is now possible to continue to modify our object(s) while they are shaded.

To remove the shading, enter the **Shademode** command and follow these steps:

> **Current mode: Flat+Edges**
> **Enter option [2D wireframe/3D wireframe/Hidden/Flat/**
> **Gouraud/fLat+edges/gOuraud+edges] <Flat+Edges>:** *2d*

	Command: *render*	**4.** The last approach to improving the display—the **Render** command—greatly surpasses the others in final product (although *not* in ease of use, I am afraid). Enter the **Render** command at the command prompt. Or you can open the Render toolbar and pick the Render button.
Render Button		

continued

TOOLS	COMMAND SEQUENCE	STEPS
Render	Rendering Type: Photo Raytrace	**5.** AutoCAD presents the Render dialog box, which we will examine in some detail in the advanced text. For now, set the Rendering Type: to Photo Ray Trace and pick the Render button to continue.

6. AutoCAD takes a few moments to render a drawing; but it is worth the wait. Your drawing now looks like Figure 21.2.3.6. |

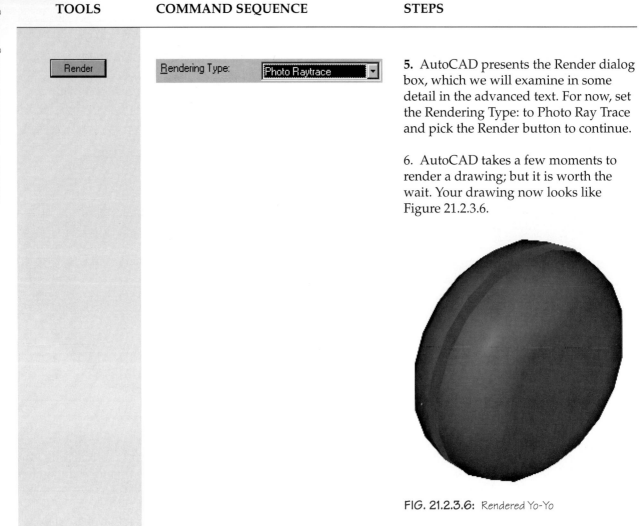

FIG. 21.2.3.6: Rendered Yo-Yo

Okay. *Now* it looks like a yo-yo. But wait! We have not quite finished. We want to see our yo-yo from different angles. Try this next exercise!

21.2.4 Do This: Razzle Dazzle

I. Be sure you are still in the *MyYoYo.dwg* file in the C:\Steps\Lesson21 folder. If not, open it now.

II. Shade the drawing as you did in the last exercise.

III. Follow these steps.

TOOLS	COMMAND SEQUENCE	EXPLANATION
		1. Open the 3D Orbit toolbar (Figure 21.2.4.1)  **FIG. 21.2.4.1:** 3d Orbit Toolbar
		2. Prepare yourself to be amazed.
 3D Continuous Orbit Button	**Command:** *3dcorbit*	3. Enter the *3DCOrbit* command by typing *3dcorbit* at the command prompt. Or you can pick the **3D Continuous Orbit** button on the 3D Orbit toolbar.
		4. Notice how the cursor has changed. Pick a point in the center of the yo-yo and drag (hold down the left mouse button) slightly to the right. Release the button and watch! (I am afraid that no image can do justice to the animation now showing on your screen. So I have not put a picture here. But as you watch, think of what Dave Bowman was feeling as he said, "My God, it's full of stars!")

continued

TOOLS	COMMAND SEQUENCE	STEPS
	Press ESC or ENTER to exit, or right-click to display shortcut-menu. *[enter]*	**5.** You can allow the yo-yo to rotate continuously. Hit *enter* or *esc* to stop it when it gets to the rotation shown in Figure 21.2.4.5.  **FIG. 21.2.4.5:** Rotated Yo-Yo

I once worked with a man who was new to AutoCAD. Every now and then, as the office became quiet and everyone was caught up in their own little computer worlds, this man would suddenly jump up and exclaim, "Wow! Look what I made it do!"

21.3 Looking Ahead

Obviously, there is a lot more to 3D drafting than what you have seen in our last section. We barely scratched the surface of possibilities. But has the curiosity bug bitten you?

If you choose to pursue your training, you will discover fascination after fascination. In our next book, we will cover 3-dimensional drafting in depth. And we will discover sharing drawings through X-refs (externally referenced drawings), the wonders of that other space—Paper Space, how to use other types of graphics (bitmaps, windows metafiles, and others) in our AutoCAD drawings, how to attach material information to our rendered drawing (imaging a yo-yo made of clear glass or granite), and much, much more! And we will use the same, simple, step-by-step approach to which you have now become accustomed.

> *It isn't the destination that matters so much as the road you take to get there.*
>
> Anonymous
>
> *Treat it as an adventure . . . enjoy it!*
>
> W.P. Sykes, MD

21.4 Extra Steps

- Print out a copy of the *Acad.pgp* file. Read through it to familiarize yourself with the various command aliases. Try the external commands identified in the *.pgp* file to see what each does. Make a list of hotkeys you would like to create or change. (I usually change the *C* hotkey to *Copy* rather than *Circle*. I use the *Copy* command more often than I use the *Circle* command. [I change *Circle* to *CC*].)
- Using your yo-yo drawing, explore the toolbars we opened in this lesson. See what else you can do!

21.5 What Have We Learned?

Think back now to what you knew when you first opened this book. Try to remember some of the many steps you have taken to get where you are now. Remember how difficult you considered the Cartesian Coordinate System? Remember wrestling with dimensions until you just did not care how far it was from one wall to the next? Remember how sick you became of trying to mirror that silly star? These things are easier now. With practice, they will become easier still.

Where do you go from here?

You have learned enough in this book to begin entry level CAD operation at most companies. You will still need to acquaint yourself with the specific organization's methods, symbols, and pull-down menus (in addition to lunchtime, quitting time, and permissible radio stations).

But there are still many things that AutoCAD can do of which you are still unaware. Chief among these, of course, is the entire spectrum of 3-dimensional commands and techniques. Additionally, there are such tools as paperspace, XRefs, AutoCAD's Internet interface, and much, much more.

I suggest to my students that they first review the material in this text until they are comfortable with it. Then pick up the next text, *AutoCAD: One Step at a Time — Advanced*, to continue training.

21.6 EXERCISES

1. Add/change hotkeys in the *Acad.pgp* file for the following AutoCAD commands (place your hotkeys in your own section at the end of the PGP file):

HOTKEY	COMMAND	HOTKEY	COMMAND	HOTKEY	COMMAND
AE	Attedit	MLS	Mlstype	DTE	Dimtedit
LM	Limits	MLE	Mledit	SK	Sketch
PPP	Plot	AD	Attdisp	ATD	Attdia
	ATT	Attext			

2. Add/change hotkeys in the *Acad.pgp* file for the follow external programs:

HOTKEY	PROGRAM
WX	Explorer
WP	WordPad

Appendix A Drawing Scales

Scale (= 1')	Scale Factor	Dimensions of Drawing when final plot size is:				
		8" × 11"	11" × 17"	17" × 22"	22" × 34"	24" × 36"
1/16"	192	136' × 176'	176' × 272'	272' × 352'	352' × 544'	384' × 576'
3/32"	128	90'8 × 117'4	117'4 × 181'4	181'4 × 234'8	234'8 × 362'8	256' × 384'
1/8"	96	68' × 88'	88' × 136'	136' × 176'	176' × 272'	192' × 288'
3/16"	64	45'4 × 58'8	58'8 × 90'8	90'8 × 117'4	117'4 × 181'4	128' × 192'
1/4"	48	34' × 44'	44' × 68'	68' × 88'	88' × 136'	96' × 144'
3/8"	32	22'8 × 29'4	29'4 × 45'4	45'4 × 58'8	58'8 × 90'8	64' × 96'
1/2"	24	17' × 22'	22' × 34'	34' × 44'	44' × 68'	48' × 72'
3/4"	16	11'4 × 14'8	14'8 × 22'8	22'8 × 29'4	29'4 × 45'4	32' × 48'
1"	12	8' × 6'11	11' × 17'	17' × 22'	22' × 34'	24' × 36'
1 1/2"	8	5'8 × 7'4	7'4 × 11'4	11'4 × 14'8	14'8 × 22'8	16' × 24'
3"	4	34" × 44"	3'8 × 5'8	8' × 6'11	7'4 × 11'4	8' × 12'
(1" =)						
10'	120	85' × 110'	110' × 170'	170' × 220'	220' × 340'	240' × 360'
20'	240	170' × 220'	220' × 340'	340' × 440'	440' × 680'	480' × 720'
25'	300	212'6 × 275'	275' × 425'	425' × 550'	550' × 850'	600' × 900'
30'	360	255' × 330'	330' × 510'	510' × 660'	660' × 1020'	720' × 1080'
40'	480	340' × 440'	440' × 680'	680' × 880'	880' × 1360'	960' × 1440'
50'	600	425' × 550'	550' × 850'	850' × 1100'	1100' × 1700'	1200' × 1800'
60'	720	510' × 660'	660' × 1020'	1020' × 1320'	1320' × 2040'	1440' × 2160'
80'	960	680' × 880'	880' × 1360'	1360' × 1760'	1760' × 2720'	1920' × 2880'
100'	1200	850' × 1100'	1100' × 1700'	1700' × 2200'	2200' × 3400'	2400' × 3600'
200'	2400	1700' × 2200'	2200' × 3400'	3400' × 4400'	4400' × 6800'	4800' × 7200'

Appendix B Function Keys and Their Uses

Key	Function	Key	Function	Key	Function
F1	AutoCAD's Help	F5	Isoplane toggle	F9	Snap toggle
F2	Toggle between graphics and text screen	F6	Toggle coordinate indicator on or off	F10	Polar tracking toggle
F3	OSNAP settings	F7	Grid toggle	F11	Object tracking toggle
F4	Tablet (Digitizer) toggle	F8	Ortho toggle	F12	Unassigned

AutoCAD Hotkeys

Command	Hotkey	Command	Hotkey	Command	Hotkey
ADCenter	ADC	Ellipse	EL	OSNAP	OS, –OS
Align	AL	Erase	E	Pan	P, –P
Area	AA	Explode	X	Pedit	Pedit
Array	AR	Extend	EX	Pline	PL
Attdisp	ATT, –ATT	Fillet	F	Point	PO
Attedit	ATE, –ATE	Filter	FI	Polygon	POL
Bhatch	BH, H	Grid	F7	Properties	PROPS
Block	B, –B	Grips	GR	Purge	PU
Break	BR	Group	G, –G	qsave	^S
Chamfer	CHA	Hatch	–H	Rectang	REC
Change	–CH	HatchEdit	HE	Redraw	R
Circle	C	Insert	I	Regen	Re
color	COL	Layer	LA, –LA	Rotate	RO
Copy	CO, CP	Leader	LE	save	^S
DDEdit	ED	Lengthen	LEN	Scale	SC
DDUnits	UN	Line	L	Snap	SN
Dimaligned	DAL	Linetype	LT, –LT	Solid	SO
Dimangular	DAN	List	LI	Spell	SP
Dimbaseline	DBA	LTScale	LTS	Spline	SPL
Dimcontinue	DCO	Lweight	LW	Splinedit	SPE
Dimdiameter	DDI	MatchProp	MA	Stretch	S
Dimedit	DED	Measure	ME	Style	ST
Dimlinear	DLI	Mirror	MI	Text	DT
Dimordinate	DOR	MLine	ML	Trim	TR
Dimradius	DRA	Move	M	View	V, –V
Dist	DI	MText	MT, T	WBlock	W, –W
Divide	DIV	Offset	O	Xline	XL
Donut	DO	open	^O	Zoom	Z
Dsettings	DS, SolidEdit	Ortho	F8		

806

Appendix C Answers to Review Questions

Lesson 1

1. Scale
2. Units
3. Sheet Size
4. 3"
5. 144' × 96'
6. Dialog Box
7. Wizard
8. Quick Setup
9. Advanced Setup
10. Use a Template
11. Open a Drawing
12. Right
13. Pull-Down Menus
14. Standard Toolbar
15. Object Properties
16. Taskbar/Min/Max/Exit Buttons
17. Cursor
18. Scroll bars
19. Status line
20. Graphics area
21. Command Prompt
22. Space Tabs
23. UCS Icon
24. Draw Toolbar
25. Modify Toolbar
26. Counterclockwise
27. East
28. Units
29. Area
30. Toolbar
31. Command Line
32. UCS
33. WCS
34. Save (or saveas)
35. Qsave
36. Quit
37. X-Box In Upper Right Corner
38. Save
39. Open
40. New
41. Ddunits
42. Limits
43. True
44. Enter
45. Paperspace
46. Multiple Document
47. Partialopen
48. Partiaload

Lesson 2

1. Geometric Shapes
2. Lines
3. Circles
4. Line
5. L
6. Erase
7. E
8. Esc
9. C
10. Polyline
11. Enter
12. Fillet
13. 1'6-$\frac{1}{2}$
14. Enter
15. Line
16. Rectangle
17. Undo
18. Redo
19. Erase
20. Crossing Window
21. 3
22. X, Y, Z
23. Y
24. X
25. 3
26. Absolute
27. Relative
28. Polar
29. Absolute
30. Relative
31. Polar
32. R
33. Regen
34. T
35. Selection Box
36. Window
37. Crossing Window

Lesson 3

1. Print
2. Plot
3. Plottermanager
4. Next
5. System Printer
6. F
7. Add a Plotter
8. What you see is what you get
9. Plot Style Table
10. Stylemanager
11. Use a PCP or PC2 file
12. General
13. Plot Style Table Editor
14. F
15. Convert to Grayscale
16. Plotter
17. Plotstyle
18. Page
19. Pagesetup
20. T
21. Plot Device
22. Plot
23. Partial Preview
24. Drawing Orientation
25. Limits
26. Extents
27. Plot scale
28. Partial Preview
29. Add a Plotter Wizard shortcut
30. Add a Plot Style Table Wizard shortcut

Lesson 4

1. Aspect
2. Snap
3. Half
4. Style
5. Snap
7. Ortho
8. Grid Toggle
9. Ortho Toggle
10. Snap Toggle
11. Polar Tracking Toggle
12. Object Tracking Toggle
13. Status
14. 30°
15. 90°
16. 330° Or −30°
17. CNTL+E
18. F5
19. DSettings
20. Eyeball
21. OSNAPS
22.–51. (see page 95)
47. Shift
48. Tab

807

Lesson 5	Lesson 6	Lesson 7	Lesson 8
1. Z	1. C	1. T	1. Trim
2. Either	2. Diameter	2. Cecolor	2. Cutting Edges
3. Zoom All	3. R Or Radius	3. Object Properties	3. F
4. Zoom Dynamics	4. Enter	4. Bylayer	4. Edgemode
5. Zoom Extents	5. Axis Endpoint 1	5. Other	5. Extend
6. Zoom Previous	6. Axis Endpoint 2	6. Ddcolor	6. Cutting
7. .5 ×	7. Other Axis	7. White	7. Boundary
8. 2 ×	8. Arc	8. Red	8. Change
9. Realtime	9. Isometric	9. Green	9. Break
10. Center	10. Pellipse	10. Blue	10. @
11. Pan	11. F	11. Yellow	11. Fillet
12. Right Mouse	12. Pull-Down Menus	12. Cyan	12. Trimmode
13. Aerial View	13. Arc	13. Magenta	13. Chamfer
14. View	14. Circle	14. Full Color Pallette	14. Distance
15. Ddview	15. Ellipse	15. Color	15. Angle
16. Ddview	16. Save	16. Dash	16. 3,0
17. Mtext	17. Polygon	17. Load	17. Hit Enter
18. F	18. End	18. Acad.Lin	18. F
19. T	19. Angle	19. Control	19. Align
20. 4"	20. Length	20. Ltscale	20. T
21. 90	21. Counterclockwise	21. Factor	21. Multiple
22. ASCII	22. D Or Direction	22. F	22. Copy
23. %%C	23. Hit Enter	23. On/Off Toggle	23. Move
24. Ddedit	24. Circle	24. Frozen/Thawed Toggle	24. Running Osnap
25. Ed	25. Edge	25. Locked/Unlocked Toggle	25. Fillet
26. Standard	26. 1024	26. Color pick	26. Chamfer
27. F	27. Inscribed	27. Plot Toggle	27. Named Views
28. 30	28. Circumscribed	28. Wildcard	28. Break
29. Qtext	29. Pol	29. All	29. Ortho
30. Autolisp	30. Savetime	30. Current	30. Extend
31. (Load "C:/Steps/Lesson05/Text")	31. Acad.Sv$	31. Layer Control	31. Trim
32. Named Views	32. Sysvars	32. Layers	32. Windows
33. Zoom Window		33. Set Layer Current	33. Cut & Paste
34. Zoom Previous		34. Match Properties	
35. Zoom Extents		35. Properties	
36. Zoom All		36. Ddchprop	
		37. Chprop	
		38. Properties	
		39. ADC	

Lesson 9	Lesson 10	Lesson 11	Lesson 12
1. Multi-segmented 2. Lwpolyline 3. Width 4. Half-Width 5. Specify next point 6. Arc 7. T 8. T 9. List 10. Dist 11. Area 12. Id 13. List 14. Area 15. List 16. Id 17. Dist 18. Redraw 19. F 20. Fit 21. Spline 22. T 23. T 24. Edit Vertiex 25. Pedit 26. Explode 27. List 28. Area 29. Polyline 30. Dist 31. Id 32. Explode	1. Copy 2. Offset 3. Array 4. Offset 5. Perpendicular 6. Mirror 7. F 8. T 9. Polar 10. Text 11. Mirrtext 12. Stretch 13. Lengthen 14. Crossing window or crossing-polygon 15. Percent 16. Lengthen 17. Total 18. Lengthen 19. Stretch 20. Rotate 21. Scale 22. Mirror 23. Mi 24. Offset 25. O 26. Lengthen 27. Len 28. Array 29. Ar 30. Scale 31. Sc 32. Stretch 33. S 34. Rotate 35. Ro	1. CAM (computer aided manufacturing) 2. Divide 3. Measure 4. Node 5. Point 6. 1 or blank 7. regen 8. PDMode 9. Solid 10. Donut 11. Fillmode 12. 0 13. wp 14. cp 15. Fence 16. Remove 17. QSelect	1. Spline 2. Object 3. Fit Tolerance 4. Surveyor 5. Fit points 6. Fit data 7. Refine 8. Weight 9. NURBS 10. XLine 11. Ray 12. XLine 13. F 14. XLine

Lesson 13	Lesson 14	Lesson 15	Lesson 16
1. Zero 2. No 3. Scale 4. Style 5. Multiline Style 6. MLStyle 7. Name 8. Add 9. Element Properties 10. Multiline Properties 11. Save 12. OK 13. Element Properties 14. .mln 15. Joints 16. T 17. MLEdit 18. Explode 19. First	1. Word Processor 2. WordPad 3. MText 4. Justify 5. Downward 6. Right 7. T 8. .txt 9. .rtf 10. Bold 11. Italicize 12. Underline 13. Color 14. Symbol 15. Properties 16. Replace 17. All 18. .cus	1. Arrow 2. Dimension 3. Extension Line Extension 4. Extension line 5. Extension line origin offset 6. Dimension line 7. Decimal 8. Associative 9. Normal 10. F 11. Dimaso 12. Explode 13. Dimlinear 14. Dimangular 15. Dimradius 16. Dimdiameter 17. Dimcontinue 18. Dimaligned 19. Dimbaseline 20. Dimordinate 21. Autocad 22. Crosshairs 23. Angle 24. Dimangular 25. F 26. Dimcontinue 27. Last 28. First 29. Leader 30. Dimordinate 31. Dimedit 32. Dimtedit 33. DDEdit 34. Oblique 35. Dimedit 36. Dimaligned 37. Dimaligned 38. Dimangular 39. Dimbaseline 40. Dimcontinue 41. Dimdiameter 42. Dimedit 43. Dimlinear 44. Dimtedit 45. Leader 46. Dimordinate	1. Dimvar 2. DDim 3. T 4. Parent 5. Linear 6. Radial 7. Diameter 8. Ordinal 9. Angular 10. Leader 11. Lines & Arrows 12. Scale Factor 13. Parallel 14. Horizontal Justification 15. F 16. T 17. Measurement Scale 18. Primary Units 19. F 20. Purge 21. Variables 22. –Dimstyle

Lesson 17	Lesson 18	Lesson 19	Lesson 20
1. Verb/Noun Selection	1. Hatch Pattern	1. Groups	1. Attribute
2. Options	2. Solid	2. Blocks	2. Attdef
3. 0	3. Angle	3. Group	3. Invisible
4. Pickfirst	4. Spacing	4. Selectable	4. Constant
5. 1	5. F	5. No	5. Verify
6. Shift	6. Direct Hatch	6. F	6. Preset
7. Pickadd	7. Normal	7. Highlight	7. Follow
8. 1	8. Outer	8. Rename	8. DDEdit
9. Pickdrag	9. Ignore	9. Explode	9. Attdia
10. 0	10. Closed	10. Selectable	10. 1
11. Implied Windowing	11. Direct Hatch	11. F	11. Attreq
12. Pickauto	12. Hatch	12. T	12. 1
13. 1	13. Pattern	13. T	13. DDAtte
14. Default	14. ISO	14. F	14. ATTDisp
15. Pickbox	15. Associative	15. Block	15. On
16. Grips	16. Exploded	16. BMake	16. Position
17. Stretch	17. Match Properties	17. WBlock	17. Attedit
18. Move	18. Pick Points	18. Libraries	18. Attedit
19. Rotate	19. Island	19. Template Files	19. Angle
20. Scale	20. Hatchedit	20. Folders	20. Attedit
21. Mirror	21. Sketch	21. Template	21. Attredef
22. Copy	22. SKPoly	22. Folder	22. Template
23. Esc	23. 1	23. Layer Text	23. Autocad
24. Shift	24. Practice	24. 0	24. The User
25. Spacebar		25. OOPS	25. Enter (Carriage Return)
26. F		26. WBlock	26. Attext
27. Right-click		27. Insert	27. SDF
28. T		28. DDInsert	

Appendix D MText Keystrokes

	These Keys:	Do This:
Position	← → ↑ ↓	Move the cursor through the text one space at a time
	Ctrl + ← Ctrl + →	Move the cursor through the text one word at a time
	Home	Move the cursor to the beginning of the line
	End	Move the cursor to the end of the line
	Page Up Page Down	Move the cursor through the document up to 28 lines at a time
	Ctrl + Home	Move the cursor to the beginning of the document
	Ctrl + End	Move the cursor to the end of the document
	Ctrl + Page Up Ctrl + Page Down	Move the cursor to the top or bottom of the screen
Selection	Ctrl + A	Selects all the text in the mtext object
	Shift + ← Shift + →	Selects / deselects text one character at a time
	Shift + ↑ Shift + ↓	Selects / deselects text one line at a time
	Ctrl + Shift + ← Ctrl + Shift + →	Selects / deselects text one word at a time
Action	Delete	Deletes the character to the right of the cursor
	Backspace	Deletes the character to the left of the cursor
	Ctrl + Backspace	Deletes the word to the left of the cursor
	Esc	Leaves the text editor without saving the changes
	Ctrl + C	Copy the selected text to the Windows clipboard
	Ctrl + V	Paste text from the Windows clipboard into the text editor
	Ctrl + X	Remove the selected text and place it on the Windows clipboard
	Ctrl + Z	Undo the last edit
	Enter	Starts a new paragraph

Appendix E Dimension Variables

Dimension Variable	Default Setting	Explanation
DIMADEC	−1	Angular decimal places (set from 0–8); a setting of −1 tells AutoCAD to use the DIMDEC setting
DIMALT	Off	Use of alternate units
DIMALTD	2	Decimal places in alternate units
DIMALTF	25.4	Scale factor in alternate units
DIMALTTD	2	Decimal places in tolerances of alternate units
DIMALTTZ	0	Suppression of zeros in alternate units (0 or 1)
DIMALTU	2	Units format for alternate units (except angular)
DIMALTZ	0	Suppression of zeros in tolerance values
DIMAPOST	""	Identifies a text suffix/prefix to be used with alternate dim
DIMASO	on	Associative dimensioning (on or off)
DIMASZ	0.18	Arrowhead size
DIMAUNIT	0	Format for angular dimensions
DIMBLK	""	Specify a block to place on the dimension line rather than an arrowhead
DIMBLK1 DIMBLK2	"" ""	Specifies blocks for both ends of the dimension line (if the DIMSAH variable is **on**)
DIMCEN	0.09	Controls what, if any, center marks are placed in circles and arcs (center-lines are drawn if the number is less than 0; center marks are drawn if the number is above 0; no center marks/lines are drawn if the value equals 0)
DIMCLRD	0	Color for dimension lines, arrowheads, and leaders
DIMCLRE	0	Color for extension lines
DIMCLRT	0	Color for dimension text
DIMDEC	4	Primary units decimal places
DIMDLE	0.0	Extension of dimension line beyond extension line
DIMDLI	0.38	Spacing of baseline dimension lines
DIMEXE	0.18	Extension of extension line beyond dimension line
DIMEXO	0.0625	Distance from origin to beginning of extension line

Dimension Variable	Default Setting	Explanation
DIMFIT	3	This controls whether a dimension line will be placed between the extension lines if there is enough space. Settings are: 0—arrows and text; 1—just text; 2—just arrows; 3—best fit (puts text and arrows between the extension lines as space is available); 4—leader (when no space is available for the text, it is placed aside and connected to the dimension line with a leader line); 5—no leader (same as 4 but will not use a leader line).
DIMGAP	0.09	Space around the text when it is placed inside the dimension line or between bottom of text and dimension line when text is placed above the line
DIMJUST	0	Horizontal position of dimension text. Possible settings include: 0—centered between extension lines; 1—next to the first extension line; 2—next to the second extension line; 3—next to and aligned with the first extension line; 4—next to and aligned with the second extension line
DIMLFAC	1.0	Global scale factor for dimensioning
DIMLIM	off	Places dimension limits as default dimension text
DIMPOST	""	Identifies a text prefix or suffix for the dimension text
DIMRND	0.00	Rounds the dimension distances to this value
DIMSAH	off	Determines whether blocks will be used instead or arrowheads
DIMSCALE	1.0	Sets an overall scale factor for all dimension variables requiring size settings
DIMSD1 DIMSD2	off off	Suppression of the first and second dimension line
DIMSE1 DIMSE2	off off	Suppression of the first and second extension line
DIMSHO	on	This actually controls whether the dimension will be redefined as the dimension is dragged
DIMSOXD	off	Controls whether dimension lines will be drawn outside the extension lines
DIMSTYLE	"Standard"	Current dimension style
DIMTAD	0	Vertical position of text. Possible settings are: 0—centered inside the dimension line; 1—place text above the dimension line; 2—on the side of the dimension line farthest from the defining point; 3—Japanese standard
DIMTDEC	4	Decimal places in tolerance value
DIMTFAC	1.0	Scale factor the text height of tolerance values
DIMTIH	on	Whether (**on**)or not (**off**) text inside the dimension lines will be horizontal
DIMTIX	off	Force text between the extension lines
DIMTM	0.0	Sets the minimum tolerance limits when either DIMLIM or DIMTOL is **on**
DIMTOFL	off	Force a dimension line between extension lines

Dimension Variable	Default Setting	Explanation
DIMTOH	on	Position of text outside dimension lines (**on** forces horizontal text)
DIMTOL	off	Add tolerances to dimension text
DIMTOLJ	1	Vertical justification of tolerance values (0—bottom; 1—middle; 2—top)
DIMTP	0.0	Upper tolerance limit
DIMTSZ	0.0	Size of ticks used instead of arrowheads
DIMTVP	0.0	Vertical position of text above/below dimension line
DIMTXSTY	"Standard"	Text style used for the dimension text
DIMTXT	0.18	Height of dimension text
DIMTZIN	0	Suppression of zeros in tolerance values. Possible values: 0—suppresses zero feet and zero inches; 1—includes zero feet and inches; 2—includes zero feet but suppresses zero inches; 3—includes zero inches but suppresses zero feet; 4—suppresses leading zeros in decimals; 8—suppresses trailing zeros in decimals; 12—suppresses both leading and trailing zeros in decimals.
DIMUNIT	2	Dimension units for everything but angular dimensions. Possible settings are: 1—Scientific; 2—Decimal; 3—Engineering; 4—Architectural (stacked); 5—Fractional (stacked); 6—Architectural; 7—Fractional; 8—windows settings
DIMUPT	off	If **on**, the cursor controls placement of both dimension and text; if **off**, the cursor controls placement only of the dimension
DIMZIN	0	Suppression of zeros in the primary units. . Possible values: 0—suppresses zero feet and zero inches; 1—includes zero feet and inches; 2—includes zero feet but suppresses zero inches; 3—includes zero inches but suppresses zero feet; 4—suppresses leading zeros in decimals; 8—suppresses trailing zeros in decimals; 12—suppresses both leading and trailing zeros in decimals.

INDEX

A

a, *See* Arc command
aa, *See* Area command
Absolute coordinates, 47–48
Absolute Units, setting point size to, 401
acad menu, 12
Acad.PGP file, 784–88
ADC, *See* AutoCAD Design Center (ADC)
ADCenter command, 243, 245–46
Add-a-Plotter Wizard, 58–63, 76
    Add Plotter-Finish dialog box, 63
    Add Plotter-Import Pcp or Pc2 dialog box, 62
    Add Plotter-Plotter Name dialog box, 62
    Add Plotter-System Printer dialog box, 61
    Begin dialog box, 60
    Calibrate Plotter General tab button, 65
    Calibrate Plotter Wizard, 65–68
    Device and Document Settings tab, Plotter Configuration Editor, 64
    Edit Plotter Configuration Add Plotter-Finish dialog box button, 63
    General tab, Plotter Configuration Editor, 64
    My Computer option, 60
    Network Plotter Server option, 60
    Plotter Configuration Editor, 63–64
    Ports tab, Plotter Configuration Editor, 64
    selecting printer/plotter type, 60
    System Printer option, 60
Add control point option, Splinedit command, 435
Add a Drawing button, 18
Adding a toolbar, 12
Add option:
    Area command, 328, 330
    Splinedit command, 435
Add Plot Style Table dialog box, 69
    Begin dialog box, 70
    File Name dialog box, 71
    Finish button, 72
    Pick Plot Style Table dialog box, 70
    Plot Style Table Editor, 71–76
Add Plotter-Finish dialog box, 63
Add Plotter-Import Pcp or Pc2 dialog box, 62
Add Plotter-Plotter Name dialog box, 62
Add Plotter-System Printer dialog box, 61
Add Style button, Plot Style Table Editor, 73
Add Vertex button, 491
Advanced blocks, 741–81
    -attdef command, 742–49
        command sequence, 742–43
        Constant mode, 743, 747
        Invisible mode, 743, 747
        Preset mode, 743
        size attribute, 745
        size and rating attributes, 746
        Verify mode, 743
    Attedit command, 761–65
        Angle option, 762
        command sequence, 761
        Edit attributes one at a time? option, 761–62
        Enter block name specification prompt, 761
        Position option, 762
        Value option, 761
    Attredef command, 765
    attribute data, extracting, 769
    attributed blocks:
        inserting using the command line, 749–52
        using a dialog box to insert, 756–58
    attribute extraction template file, setting up, 766–67
    attribute information, 752
        using in bills of materials/database programs, 765–73
    attributes:
        creating, 742–58
        creating at the command line, 744–48
        defining from a dialog box, 752–55
        editing, 758–65
    DDAtte command, 758–60
        Edit Attribute button, 760
        Edit Attributes dialog box, 760
    Edit Attribute Definition dialog box, 755
    extraction of block data, 768–73
        Attribute Extraction dialog box, 768
        Comma Delimited File (CDF), 768
        DXF file, 768
        importing extracted data into a database program, 770–73
        Output File . . . button, 768
        Select Objects < button, 768
        Space Delimited File (SDF), 768
        Template File . . . button, 768
    See also Blocks
Advanced lines, 467–506
    See also Multilines
Advanced modification techniques, 621–49
    Selection Settings, 622–29
Advanced Setup dialog box, 7
    Angle Direction options, 9
    Angle Measurement options, 9
    Angle option, 8
    Area options, 10
    Finish button, 10
    Unit Setup options, 8
    Use a Wizard button, 7
Advanced Setup option, Startup Wizard, 6
Advanced tab:
    Boundary Hatch dialog box, 666, 669, 675
        Boundary set frames, 666–67
        Island detection style frame, 666–67
        Object type frame, 666–67
Aerial View, 133–34
al, *See* Align command
Aliases, 34
Align command, 294–98
    command sequence, 294–95
    using, 295–98
Aligned Dimension button, 559
Aligning dimensions, 558–60
Align option, Text command, 141
All option, Zoom command, 128
Alternate Units tab, New Dimension Style: First Steps dialog box, 599, 608
ANGBASE system variable, 371–72
ANGDIR system variable, 371–72
Angle Direction options, Advanced Setup dialog box, 9
Angle Measurement options, Advanced Setup dialog box, 9
Angle method, Chamfer command, 287, 290
Angle option, 8
    Advanced Setup dialog box, 8
    Arc command, 182–83, 185
    DimTedit command, 573, 575
    Hatch command, 655
    Pline command, 316
Angles, measuring clockwise, 9
Ang option, XLine command, 444

Angular Dimension button, 551
Annotation Reuse frame, Leader command, 564
Annotation tab, Leader command, 563–64
Annotation Type frame, Leader command, 563
Answers to Review Questions, 807
Aperture box, 106
Apparent Intersection button, 95
Apply Constraints frame, Leader command, 565
Apply option, Dimstyle command, 613
Apply to: control box, Quick Select dialog box, 416
ar, *See* Array command
Arch/canopy assembly detail, 204
Architectural options, 8
Arc option, Ellipse command, 176
Arcs, 180–87
  Angle option, 182–83, 185
  Arc button, 181
  Arc command, 180
  CEnter option, 181–82, 185–86, 194–95
  chord Length option, 183, 186
  direction of, 184
  Direction option, 184
  ENd option, 184, 197
  Radius option, 184–85
Area, calculating, 328–31
Area command, 328–31
  command sequence, 328
Area options, Advanced Setup dialog box, 10
Array command, 357–61, 376, 378, 380, 381
  command sequence, 358
  Polar option, 358, 359–60
  Rectangular option, 358, 359–60
  using, 358–61
Arrowhead frame, Leader command, 564
Aspect option, Snap command, 90
Associative Dimensioning, 545–46
  converting Normal Dimensioning to, 546
-ate, *See* Attedit command
Attachment tab, Leader command, 564
-attdef command, 742–49
  command sequence, 742–43
  Constant mode, 743, 747
  Invisible mode, 743, 747
  Preset mode, 743
  size attribute, 745
  size and rating attributes, 746
  Verify mode, 743
Attdia command, 749
Attdia system variable, 749, 756

Attdisp system variable, 761
  Normal option, 761
  turning off, 761
Attedit command, 743, 761–65
  Angle option, 762
  command sequence, 761
  Edit attributes one at a time? option, 761–62
  Enter block name specification prompt, 761
  Position option, 762
  Value option, 761
Attext command, 769
-attext command, 765, 769
Attredef command, 765
  command sequence, 765
Attreq system variable, 758
Attribute data, extracting, 769
Attributed blocks:
  inserting using the command line, 749–52
  using command line to insert, 749–52
  using a dialog box to insert, 756–58
Attribute Extraction dialog box, 768, 787
Attribute extraction template file, setting up, 766–67
Attribute information, 752
  using in bills of materials/database programs, 765–73
Attributes:
  creating, 742–58
  at the command line, 744–48
  defining from a dialog box, 752–55
  editing, 758–65
AutoCAD Design Center (ADC), 242–46
  accessing, from the command line, 243
  ADCenter command, 245–46
  defined, 242
  Description button, 244
  Desktop button, 243
  Favorites button, 244
  Find button, 244
  History button, 243
  Load button, 244
  Open Drawings button, 243
  Preview button, 244
  sharing blocks between drawings using, 729–30
  Tree View button, 244
  Up button, 244
  using, 245–46
  Views button, 244
AutoCAD shortcut, 4
AutoCAD User Interface (AUI), 28
AutoCad warning box, 17
AutoCAD window, minimizing/

  maximizing, 11
AutoDESK, 11–13, 133
  libraries, 709
AutoLISP programming language, 156–58, 784
  frontslashes in, 157
Auto option, Undo command, 42
AV hotkey, 133

B
Back option, Undo command, 42
Backwards option, Text Style dialog box, 154
Baseline dimensions, 561–62
Base point frame, WBlock command, 717
Base point option, Move command, 293–94
Basics, AutoCAD, 1–208
Begin dialog box, 60
Begin option, Undo command, 42
BHatch command, 665–77
  Boundary Hatch dialog box, 665–68
  Hatch button, 668
  hatching with a dialog box, 668–73
  Hatch Pattern palette, 669
  Pick Points button, 670
  Preview button, 670
  Preview Hatch display, accuracy of, 670
Binoculars button, 523
Bisect option, XLine command, 444–45
Blank tag, 767
Block: option, WBlock command, 717
-Block command, 709, 711
  command sequence, 711
Block commands, 708–31
  -Block command, 709, 711
  Block Definition dialog box, 713
  BMake command, 709, 711–13
  -Insert command, 721–22
  Insert dialog box, 725–26
    Explode check box, 726
    Insertion point frame, 726
    Name: control box, 726
    Rotation frame, 726
    Scale frame, 726
  Make Block Button, 715
  Pick Point button, 715
  Select Objects button, 715
  WBlock command, 709, 716–21
Block data extraction, 768–73
  Attribute Extraction dialog box, 768
  Comma Delimited File (CDF), 768
  DXF file, 768
  importing extracted data into a database program, 770–73

Output File . . . button, 768
Select Objects < button, 768
Block option, Divide command, 395
Blocks:
　advanced, 741–81
　creating on the command line, 711–14
　creating with a dialog box, 714–16
　defined, 708
　inserting, 723–25
　and layers, 710
　libraries, 709
　making selectable/not selectable, 709
　redefining, 709, 765
　sharing, 729–30
　using in a drawing, 721–22
　See also Advanced blocks
BMake command, 709, 711–13
Bold button, MText command, 514–15, 517–18
Bottom Left option, Text command, 141
Bottom option, MLine command, 469
Boundary Hatch dialog box, 665–66
　Advanced tab, 666, 669, 675
　Boundary Set frames, 666–67, 675
　boundary sets, 673–77
　Composition frame, 668
　Define Boundary Set option, 673–77
　Inherit Properties button, 668
　Island detection style frame, 666, 669
　Object type frame, 666–67
　Pick Points button, 667, 673, 675–76
　Preview button, 668
　Quick tab, 665–66
　Remove Islands button, 667
　Select Objects button, 667
　View Selections button, 667
Boundary set frames, 666–67
Box button, 791
br, See Break command
Break command, 278–79, 545
　Break button, 281
　command sequence, 278–79
　trick with, 279
　using, 279–82
Broken polyline, 342
Browse button, 5
　WBlock command, 717
ByBlock option:
　Color command, 214
　Color Control box, 214
　Linetype command, 219
ByLayer option:
　Color command, 214
　Color Control box, 214
　Linetype command, 219

Lineweight Control Box, 232–33

C
c, See Circle command
Calibrate Plotter File Name dialog box button, 65
Calibrate Plotter General tab button, 65
Calibrate Plotter Wizard, 65–68
　Begin dialog box, 65
　File Name dialog box, 67
　Finish button, 68
　Measured Plot dialog box, 67
　Rectangle Size dialog box, 66
Cartesian coordinates, 47–49
　absolute coordinates, 47–48
　point, 47
　polar coordinates, 48
　relative coordinates, 48
　using, 49
　X-axis, 47
　Y-axis, 47
cecolor, See -color command
Center button, 95, 101, 102, 217
Centerlines, drawing, 224
CEnter option, Arc command, 181–82, 185–86, 194–95
Center option:
　DimTedit command, 573
　Ellipse command, 175–76
　Pline command, 316
　Zoom command, 128
Center OSNAP button, 217
-ch, See Change command
Chamfer command, 287–92
　Angle method, 287, 290
　Chamfer button, 288
　command sequence, 287
　Distance method, 287, 288
　first chamfer distance, 288
　Method option, 287
　Polyline method, 287
　second chamfer distance, 288
　Trim option, 287
　using, 288–92
Chamfered rectangles, drawing, 37, 39–40
Change command, 274–78
　Change Point option, 276
　command sequence, 275
Change Directories . . . button, Spell Checker, 528
Change group, 364–74
　Break command, 278–79
　Change command, 274–78
　Extend command, 270–74
　Fillet command, 282–87
　Lengthen command, 364, 366–71, 376
　Rotate command, 371–73
　Scale command, 373–74

Stretch command, 364–66
Trim command, 264–70
Change Group frame, Object Grouping dialog box, 700–01, 705–07
Change Properties dialog box, 250
Character Map dialog box, 511
Character tab, MText command, 511, 526
chord Length option, Arc command, 183, 186
CHProp command, 249–51
Circle, selecting center/quadrant of, 106–7
Circle button, 109, 171
Circle command, 108–13, 170–79, 198
Circles, 170–79, 357–61
　Circle button, 171
　Circle command, 108–13, 170–79, 198
　Ellipse command, 174–79
　ellipses, drawing, 175–77
　isometric circles, 174–79
　options, 170
　practice, 171–73
　radius, specifying, 173
　Specify center point option, 170
　squished circles, 174–79
　3P option, 170–71
　TTR (Tangent-Tangent-Radius) option, 170, 172, 173
　2P option, 170, 172
Close command, 14
Closed Cross button, 485
Closed Tee button, 488
Close option:
　Pedit command, 333, 335
　Pline command, 316
　Spline command, 428, 431
co, See Copy command
Color:
　ByBlock option, Color Control box, 214, 219
　ByLayer option, Color Control box, 214, 219
　-color command, 214–16, 218
　Color text box, 215
　drawing in, 216–18
　Full Color Palette, 215
　Select Color dialog box, 215, 217
Color assignments, plots, 74
Color button, 250
Color column, Layer Properties Manager, 235
-color command, 214–18
Color Control box, 216
　ByBlock option, 214
　ByLayer option, 214
　Object Properties toolbar, 214
Color text box, 215
Columns, 357–61

Comma Delimited File (CDF), 768
Command: prompts, 12–13
Command-mode Cursor menu, 14
Commands, 34, 50
Compare . . . button, Dimension Style Manager dialog box, 591
Composition frame, Boundary Hatch dialog box, 668
Concentrics, 354–57
Construction Line button, 446
Continuous option, Linetype command, 219
Control option, Undo command, 42
Coordinate Display box, 12
Coprot.lsp, 301
Copybase command, 304
Copy with Base Point, 304–05
Copy button, 12, 299
Copy command, 299–301, 398
   command sequence, 299
      Copy button, 12, 299
      Multiple option, 300
   using, 299–301
Copy/Paste commands, Windows, 301–304
Corner Joint button, 490
CR command, 301
Create File dialog box, 716–17
Create Group frame, Object Grouping dialog box, 700
Crosshairs, 11
CrossingPoly tool, 408
Crossing window selection, 45–46
Current button, Layer Properties Manager, 236
Current word frame, Spell Checker, 527
Cursor menu, 100
   calling, 6
Cursor menus, 13–14
   modes of, 14
Custom Properties Device and Document Settings tab button, Plotter Configuration Editor, 64
Cut All button, 492–93, 496
Cut button, 12
Cut (Cutclip) command, 305
Cut/Paste commands, Windows, 301–302
Cut Single button, 492
Cylinder button, Solids toolbar, 791
Cylinder command, 791–92

D

dal, *See* Dimaligned command
dan, *See* Dimangular command
datumPoint option, QDim command, 570
dba, *See* Dimbaseline command
dco, *See* Dimcontinue command
DDAtte command, 758–60
Edit Attribute button, 760
Edit Attributes dialog box, 760
DDEdit button, 148
DDEdit command, 148–50, 755
   command sequence, 148
   Edit Attribute button, 760
   and Modify pull-down menu, 760
DDim command, 570–606
   using, 592–93
DDPType, 401
   dialog box, 401
DDUnits command, 27
DDView command, 134–40
Decimal Degrees option, 8
Decurve option, Pedit command, 334, 337
ded, *See* Dimedit command
Default Cursor menu, 14, 43
   erasing an object using, 44
Default name, plots, 74
Define Boundary Set option, Boundary Hatch dialog box, 673–77
Define View Window button, 138
Degrees, determining, AutoCAD method of, 9
Delete option:
   Splinedit command, 435, 442
   View command, 136
Delete Vertex button, 491–92
DElta option, Lengthen command, 367, 368
Description button:
   AutoCAD Design Center (ADC), 244
   Object Grouping dialog box, 701
Deselect all option, 305
Desktop button, AutoCAD Design Center (ADC), 243
Destination frame, WBlock command, 717–18
Device and Document Settings tab, Plotter Configuration Editor, 64
di, *See* Dist command
Dialog box, defined, 5
Dialog-mode Cursor menu, 14
Dimaligned command, 558–60, 578–81
   Aligned Dimension button, 559
   command sequence, 558–59
   options, 559
   using, 559–60
dimang, *See* Dimangular command
Dimangular command, 550–53
   Angular Dimension button, 551
   command sequence, 550–51
dimbase, *See* Dimbaseline command
Dimbaseline command, 561–62
   Baseline Dimension button, 562
   command sequence, 561
   using, 561–62
dimcont, *See* Dimcontinue command
Dimcontinue command, 555–58
   command sequence, 555
   Continue Dimension button, 556
   using, 555–58
dimed, *See* Dimedit command
dimedit, *See* Dimedit command
Dimedit command, 573, 576–78
   command sequence, 576
   Dimension Edit button, 576
   Home option, 576
   New option, 576
   Oblique option, 576, 578
   Rotate option, 576
Dimension creation, 547–69
Dimensions, customizing, 589–619
   Child, setting up, 604–06
   DDim command, 570–606
      using, 592–93
   Dimension Style Manager dialog box, 590–91
   dimension styles, creating, 590–606
   Dimension Update button, 611–12
   floor plan, dimensioning, 606–11
   New Dimension Style dialog box, 591–93
   Purge command, 611
   simple repairs, 611–12
   Update command, 611
Dimension Edit button, 576
Dimensioning angles, 550–53
Dimension strings, 555
Dimension Style Manager dialog box, 590–91
   Compare . . . button, 591
   List: control box, 591
   Modify . . . button, 591
   New . . . button, 591
   Override . . . button, 591, 612
   Preview of: display box, 591
   SetCurrent button, 591
   Styles: listbox, 591
Dimension styles, creating, 590–606
Dimension Text Edit button, DimTedit command, 574
Dimension Update button, 611–12
Dimension variables, 815
Dimensioning, basic, 543–86
   aligning dimensions, 558–60
   Associative Dimensions, 545–46
   baseline dimensions, 561–62
   Dimaligned command, 558–60, 578–81
      Aligned Dimension button, 559
      command sequence, 558–59
      options, 559
      using, 559–60
   Dimangular command, 550–53
      Angular Dimension button, 551
      command sequence, 550–51
   Dimbaseline command, 561–62
      Baseline Dimension button, 562

command sequence, 561
  using, 561–62
Dimcontinue command, 555–58
  command sequence, 555
  Continue Dimension
    button, 556
  using, 555–58
Dimedit command, 573, 576–78
  command sequence, 576
  Dimension Edit button, 576
  Home option, 576
  New option, 576
  Rotate option, 576
dimension creation, 547–69
dimensioning angles, 550–53
dimension strings, 555
Dimlinear command, 547–50
  Angle option, 547
  command sequence, 547
  First extension line origin
    option or <select>
    object, 547
  Horizontal option, 547
  MText option, 547
  Rotated option, 548
  Second extension line
    origins, 547
  Text option, 547
  Vertical option, 547
Dimordinate command, 567–69
  Angle option, 567
  command sequence, 567
  MText option, 567
  Ordinate Dimension button, 568
  Text option, 567
  Xdatum option, 567–68
  Ydatum option, 567, 569
Dimradius command, 553–54
  command sequence, 553
  Radius Dimension button, 553
DimTedit command, 573–78
  Angle option, 573, 575
  Center option, 573
  command sequence, 573
  Dimension Text Edit button, 574
  Left option, 573–74
  Right option, 573–74
  using, 574–76
isometric dimensioning, 578–81
Leader command, 563–66
  Annotation Reuse frame, 564
  Annotation tab, 563–64
  Annotation Type frame, 563
  Apply Constraints frame, 565
  Arrowhead frame, 564
  Attachment tab, 564
  command sequence, 563
  Leader Line frame, 564
  Leader Lines & Arrows
    tab, 564–65
  Leader Settings dialog box,
    563–65
  Mtext options frame, 564
  Number of Points box, 564
  Quick Leader button, 565
    using, 565–66
  Linear Dimension button, 549
  linear dimensioning, 547–50
  Normal Dimensioning, 546
  ordinate dimensioning, 567–69
  placing leaders, 563–66
  QDim command, 569–73
    command sequence, 570
    datumPoint option, 570
    Edit option, 570
    Quick Dimension button, 570
    using, 570–72
  quick dimensioning, 569–73
  terminology, 544–46
dimlin, See Dimlinear command
Dimlinear command, 547–50
  Angle option, 547
  command sequence, 547
  First extension line origin option
    or <select> object, 547
  Horizontal option, 547
  MText option, 547
  Rotated option, 548
  Second extension line origins, 547
  Text option, 547
  Vertical option, 547
dimord, See Dimordinate command
Dimordinate command, 567–69
  Angle option, 567
  command sequence, 567
  MText option, 567
  Ordinate Dimension button, 568
  Text option, 567
  Xdatum option, 567–68
  Ydatum option, 567, 569
dimrad, See Dimradius command
Dimradius command, 553–54
  command sequence, 553
  Radius Dimension button, 553
Dimstyle command, 612–13
  Apply option, 613
  Save option, 612
  Status option, 612
  Variables option, 613
dimted, See DimTedit command
dimtedit, See DimTedit command
DimTedit command, 573–78
  Angle option, 573, 575
  Center option, 573
  command sequence, 573
  Dimension Text Edit button, 574
  Left option, 573–74
  Right option, 573–74
  using, 574–76
Dimvar, 590
Direct hatch option, hatch
  command, 655, 663–64
Direction option, Arc command, 184
Displacement method, Move
  command, 293–94
Display controls, 125–67
  DDView command, 134–40
  Pan command, 140
  View command, 134–40
    on the command line, 135–37
  Zoom command, 126–34
Distance, determining, 327–28
Distance method, Chamfer
  command, 287, 288
Dist command, 326–28
Dithering, plots, 74
div, See Divide command
Divide command, 394–99
  Block option, 395
  command sequence, 394–95
dli, See Dimlinear command
do, See Donut command
Docked toolbar, 12
Donut command, 405–08
  command sequence, 405
  using donuts, 406–08
dor, See Dimordinate command
Double bracket, preceding zoom
  prompt, 140
Double hatching, 660–61
dra, See Dimradius command
Drafting Settings dialog box, 92
  Object Snap tab, 93–94, 105
  Object Snap Tracking Settings
    frame, 93
  Polar Angle measurement frame, 93
  Polar Tracking tab, 93
  Snap and Grid tab, 92
Draw toolbars, 12
Drawing:
  printing, 80–83
  starting from scratch, 27–29
Drawing aids, 87–123
  Dsettings command, 91–94
  Grid command, 88–89, 90
  isometric drafting, 113–17
  Object Snap Tracking, 108, 110–13
  Ortho command, 88, 90
  OSNAPS, 94–108
  point filters, 108–10
  Polar Tracking, 89–90
  Snap command, 89–90
  See also Drafting Settings dialog
    box; OSNAPS
Drawing basics, 33–56
  Erase command, 41–44
  lines, 34–35
  rectangles, 34–35
  Redo command, 41–44
  Undo command, 41–44
Drawing lines, 35–38

Drawing orientation frame, Plot dialog box, 81
Drawing rectangles, 38–41
Drawing Scales Chart, 4, 10, 141, 807
Drawing Template File option, Save as type: box, 25
Drawing templates, creating, 24–25
Drawing tricks, 393–425
   Divide command, 394–99
   Donut command, 405–08
   Measure command, 394–99
   Point command, 399–402
   Solid command, 402–405
Drill Gizmo, 202
Driver, 58
Dsettings command, 91–94
DSViewer command, 133
dt, *See* Text command
.dwg files, 16
.dwt files, 16
.DXF files, 16, 768
DYnamic option, Lengthen command, 367, 370

E

e, *See* Erase command
ed, *See* DDEdit command
Edge option, Trim command, 265, 268
Edit Attribute button, 760
Edit Attribute Definition dialog box, 755
Edit Hatch button, 677
Editing drawings, 265–310
Editing polylines, 332–45
Editing text, 148–50
Edit menu, 11
Edit-mode Cursor menu, 14
Edit Multiline button, 485
Edit option, QDim command, 570
Edit Plotter Configuration Add Plotter-Finish dialog box button, 63
Edit Polyline button, 341
Edit Spline button, 436
Edit Text dialog box, 148–49
Edit Vertex option:
   Pedit command, 339–41
      Insert option, 340
      Move option, 340
      Next option, 340
      Previous option, 340
      Regen option, 340
      Tangent option, 340
      Width option, 340
el, *See* Ellipse command
Electrical schematic, 206
Electrical symbols, sample, 205
Element Properties dialog box, MLStyle command, 474–77
Elevate Order option, Splinedit command, 435, 439–40

Ellipse button, 175, 178
Ellipse command, 174–79, 199
   Arc option, 176
   Center option, 175–76
   Isocircle option, 177, 178
   Parameter option, 176–77
Ellipses, drawing, 175–77
ENd option, Arc command, 184, 197
End option, Undo command, 42
Endpoint button, 95, 97, 99, 102
End tangent option, Spline command, 428, 433
English settings, 5
Enter name of existing block . . . prompt, WBlock command, 717
Erase command, 35, 41–44, 381, 409, 413–15, 454, 623, 625, 628
   command sequence, 35, 41
   Erase button, 43
   erasing objects one at a time, 43–44
Erase option, Sketch command, 680
ESC (escape) key, 35
existing drawing, opening, 17–20
eXit option, Sketch command, 680
Explode button, 345
   Object Grouping dialog box, 701, 708
Explode check box, Insert dialog box, 726
Explode command, 345
Extend command, 270–74
   command sequence, 271
   Extend button, 271
   Select boundary edges option, 271–72
   Select object to extend option, 271, 273
Extension button, 95, 97, 98
Extension Lines frame, New Dimension Style: First Steps dialog box, 594–95
Extents option, Zoom command, 128
Extraction of block data, 768–73
   Attribute Extraction dialog box, 768
   Comma Delimited File (CDF), 768
   DXF file, 768
   importing extracted data into a database program, 770–73
   Output File . . . button, 768
   Select Objects < button, 768

F

F2 function key, 13, 325
F5 function key, 115, 116, 178
F8 function key, 431
f, *See* Fillet command
Favorites button, 18
   AutoCAD Design Center (ADC), 244
Fence tool, 408, 411, 412
File menu, 11, 15

Exit option, 16
Open command, 19
File Name dialog box, Calibrate Plotter Wizard, 67
File style, plots, 75
Fillet command, 282–87, 795
   command sequence, 282
   default radius for, 282
   Fillet button, 283
   Polyline option, 282
   Trim option, 282, 287
   using, 283–87
Filleted rectangles, drawing, 37–38
Fillmode system variable, 408
Find button, AutoCAD Design Center (ADC), 244
Find command, 150
Finding text, 150–51
Find Name < button, Object Grouping dialog box, 704
Find Next button, 150
Find/Replace button, 523
Find and Replace button, 12
Find and Replace dialog box, 532–34
   Find and Replace button, 533–34
   Find text string: control box, 532, 534
   Options . . . button, 532–33
   Options dialog box, 533
   Replace All button, 534
   Replace with: control box, 532
   Search in: control box, 532
   Search results: frame, 533
   Zoom to button, 534
Find/Replace tab, MText command, 512, 523
Find text string text box, 151
Fine Tuning frame, New Dimension Style: dialog box, 598
Fit Data option, Splinedit command, 435, 441–43
Fit option:
   Pedit command, 333, 336
   Text command, 141
Fit Options frame, New Dimension Style: dialog box, 597–98
Fit tab, New Dimension Style: dialog box, 597, 608
Fit Tolerance option, Spline command, 428, 432
Flyout toolbar buttons, defined, 95
Folder Libraries, 710
   creating, 716–18
Font, defined, 151
Font frame, 153
Font style options, 153
Form View tab, Plot Style Table Editor, 73
Freeze column, Layer Properties Manager, 234
From button, 95

Full Color Palette, 215
Full Preview, Plot offset frame
    button button, 82
Full scale, 4
Function keys, 90–91, 806

G

General Options frame of System
    tab, Options dialog box, 6
General tab:
    Plot Style Table Editor, 73
    Plotter Configuration Editor, 64
Geometric shapes, 169–208
    arcs, 180–87
    circles, 170–79
    polygons, 187–91
Graphics Area, 13
Graphics screen, toggling between
    text screen and, 13
Grayscale, plots, 74
Grid command, 88–89, 90
Grid Snap, 115
Grip command, 622–23
Grips command, 629–43
    assignment of grips, 630
    command sequence, 629
    editing with grips, 631–35
    Grips display, controlling, 630
    Mirror Grips procedure, 636,
        639–40
    Move Grips procedure, 641
    moving/scaling with grips,
        641–43
    and Noun/Verb Selection, 630
    Rotate Grips procedure, 636–38
    Scale Grips procedure, 642–43
    and Shift key, 631
    Stretch Grips procedure, 631–35,
        641–42
Grips menu, 14
Group command, 696–710
    creating/manipulating groups,
        702–709
    Object Grouping dialog box,
        699–704
        Add < button, 700
        Change Group frame, 700–01,
            705–07
        Create Group frame, 700
        Description button, 701
        Explode button, 701, 708
        Find Name < button, 704
        Group Identification frame,
            699–700, 706
        Group Member List dialog box,
            700, 704
        Group Name setting, 699
        Group Name text box, 704, 706
        Highlight button, 701, 705
        Included Unnamed check
            box, 706
        New < button, 702
        Order Group dialog box, 701
        Remove < button, 700
        Rename button, 701
        Re-order . . . button, 701
        Reverse Order button, 701
        Selectable button, 701, 707
    Order options, 702
    removing a group definition,
        707–08
    working with groups, 697–702
Group Identification frame, Object
    Grouping dialog box, 699–700,706
Group Member List dialog box,
    Object Grouping dialog box,
    700, 704
Group Name setting, Object
    Grouping dialog box, 699
Group Name text box, Object
    Grouping dialog box, 704, 706
Groups:
    defined, 696
    working with, 697–702
Guidelines, 444–55
    construction lines, 445–55
    Ray command, 445
    XLine command, 444, 446–49
        Ang option, 444
        Bisect option, 444–45, 454
        command sequence, 444
        Hor option, 444
        Offset option, 445, 447
        Ver option, 444

H

Hail Mary procedures, 43
Halfwidth option, Pline
    command, 316
Hatch button, 668
Hatch command, 653–65
    Angle option, 655
    ANSI31 option, 654
    command line hatching, 656–65
    command sequence, 654
    direct hatch option, 655, 663–64
    double hatching, 660–60
    Ignore hatching style, 658–59
    Normal hatching style, 657–58
    Outer hatching style, 658
    Outermost hatching style, 662
    Retain polyline? option, 655
    scaled hatching, 662–63
    solid hatching, 661
    Solid hatch pattern, 654
    User Defined option, 655, 659
    See also BHatch command
Hatched areas, editing, 677–78
Hatchedit command, 677–78
    Edit Hatch button, 677
Hatch Pattern palette, 669
he, *See* Hatchedit command

Height option, MText command, 509
Help button, MText command, 511
Hide command, 798
Highlight button, Object Grouping
    dialog box, 701, 705
Highlighted line, 35
History button, AutoCAD Design
    Center (ADC), 243
Home option, Dimedit command, 576
Hor option, XLine command, 444
Hotkeys, 34, 50, 806
    creating, 784–88
        for an AutoCAD command, 785
        Command Name, 784–90
        Flag, 784
        format for, 784
        Program, 784
        Prompt, 784
How to apply: frame, Quick Select
    dialog box, 416

I

ID command, 28, 331–32
    command sequence, 331
Ignore All button, Spell Checker, 527
Ignore button, Spell Checker, 527
Ignore hatching style, 658–59
Implied Windowing, 46
Implied Windowing selection, 629
    Pickauto system variable, 629
Import Text . . . button, MText
    command, 510, 516
Included Unnamed check box,
    Object Grouping dialog box, 706
Inherit Properties button, 668
    Boundary Hatch dialog box, 668
Inquiry commands, 324–32
    Area command, 328–31
    Dist command, 326–28
    ID command, 331–32
    List command, 324–26
Insert Block button, -Insert
    command, 726
Insert button, 95
-Insert command, 721–22, 756, 757
    command sequence, 721
    Insert Block button, 726
    Rotation option, 722
    Scale option, 722
    X scale factor, 722
    XYZ option, 722
    Y scale factor, 722
Insert dialog box, 725–26
    Explode check box, 726
    Insertion point frame, 726
    insertion settings, 727
    Name: control box, 726
    Rotation frame, 726
    Scale frame, 726
Insertion point frame, Insert dialog
    box, 726

Insert option, Edit Vertex, 340
Insert units: option, WBlock command, 718
Intersection button, 95, 104, 176
Invalid field specification error, 767
Island detection style frame, Boundary Hatch dialog box, 666, 669
Isocircle option, Ellipse command, 177, 178
Isometric block with isometric circles, 203
isometric circles, 174–79
   drawing, 178–79
Isometric dimensioning, 578–81
isometric drafting, 113–17
   isometric plane toggle, 114
   Snap command, 113–14
Isometric text, 156

J

Join option, Pedit command, 333
Justification options, MLine command, 468–69, 471
Justify option:
   MText command, 509
   Text command, 141

L

l, *See* Line command
la, *See* Layer command
Layer command, 233, 238–42, 397, 452
Layer Control dialog box, 251
Layer Properties Manager, 233–36, 238, 239
   Color column, 235
   Current button, 236
   details frame of, 238
   Freeze column, 234
   Linetype column, 235
   Lineweight column, 235
   Lock column, 235
   Name column, 234
   Named layer filters control box, 236
   New button, 235
   On column, 234
   Plot column, 235
   Plot Style column, 235
Layers, 213–63
   Change Properties dialog box, 250
   CHProp command, 249–51
   concept, 233
   control box, 247
   Layer command, 233, 238–42
   Layer Control dialog box, 251
   Layer Properties Manager, 233–36, 238, 239
   Layers button, 233, 238
   linetypes, 219–30
   lineweights, 230–32

Make Object's Layer Current button, 242
Matchprop command, 255
Named Layer Filters dialog box, 237–38
Object Property Manager, 249–54
   practice, 247–49
   simple command approach, 214–32
   using, 238–42
Layout name frame, Plot dialog box, 80
le, *See* Leader command
lead, *See* Leader command
Leader command, 563–66
   Annotation Reuse frame, 564
   Annotation tab, 563–64
   Annotation Type frame, 563
   Apply Constraints frame, 565
   Arrowhead frame, 564
   Attachment tab, 564
   command sequence, 563
   Leader Line frame, 564
   Leader Lines & Arrows tab, 564–65
   Leader Settings dialog box, 563–65
   Mtext options frame, 564
   Number of Points box, 564
   Quick Leader button, 565
   using, 565–66
Leader Line frame, Leader command, 564
Leader Settings dialog box, Leader command, 563–65
leaving a drawing session, 14–17
Left option, DimTedit command, 573–74
Lengthen command, 364, 366–71, 376–77
   command sequence, 366–67
   DElta option, 367, 368
   Dynamic option, 367, 370
   Percent option, 367, 369
   Total option, 367, 370
   using, 367–71
Libraries:
   defined, 709
   Folder Libraries, 710, 716–18
   Template Libraries, 710–11
   third-party, 722
Limits command, 27
   command sequence, 28
Linear Dimension button, 549
Linear Dimension frame:
   New Dimension Style: First Steps dialog box, 599–600
      Decimal separator: control box, 599
      Precision control box, 599
      Prefix box, 599

Primary Units tab, 600
Round off: number box, 599
Scale factor: number box, 599
Suffix box, 599
Unit format: option, 599
Zero Suppression number boxes, 600
Linear dimensioning, 547–50
Line and Arrows tab, New Dimension Style: First Steps dialog box, 593, 607
Line button, 35, 97, 217
Line command, 34–38, 97, 115–16, 140, 193–94
   and "C," 36
   command sequence, 34
   enter key, hitting a command prompt, 36
   Specify first point: prompt, hitting enter key at, 36
Line end styles, plots, 75
Line join styles, plots, 75
Line option, Pline (pl) command, 316
Lines, 34–35
   drawing, 35–38
   See also Advanced lines
Line Spacing option, MText command, 509
Line Spacing tab, MText command, 512
Linetype column, Layer Properties Manager, 235
-linetype command, 220
Linetype command, 219, 221
   ByBlock option, 219
   ByLayer option, 219
   Continuous option, 219
   -linetype command, 220
   Linetype Manager dialog box, 219, 221
Linetype Control dialog box, 251
Linetype filters box, 229, 236
Linetype Manager dialog box, 219, 221
   Current object scale, 229
   details section of, 227
   Global scale factor, 227–29
Linetype Manager List Box, 223
Linetype patterns, plots, 75
Linetypes, 219–30
   drawing with, 220–27
   Linetype command, 219, 221
      ByBlock option, 219
      ByLayer option, 219
      command line options, 220
      Continuous option, 219
      hotkey, 219
      -linetype command, 220
      Linetype Manager dialog box, 219, 221
   Linetype Control dialog box, 251

Linetype filters box, 229, 236
Linetype Manager dialog box, details section of, 227
Load or Reload Linetypes dialog box, 222
    File button, 222
    Linetype heading, 222
LTScale command, 225–29
plots, 75
Lineweight column, Layer Properties Manager, 235
Lineweight control box, New Dimension Style: dialog box, 593
Lineweights, 230–32
    changing, 231–32
    cursor menu, 232
    Lineweight Control Box, 232–33
        ByLayer option, 232–33
    Lineweight Settings dialog box, 231
    -lweight command, 230
    plots, 75
Lineweight Settings dialog box, 230–31
    Adjust Display Scale frame, 231
    Display Lineweight in Model Space check box, 231
    Lineweights frame, 231
    Units for Listing frame, 231
Lisp, 784
List: control box, Dimension Style Manager dialog box, 591
List command, 324–26
Load button, AutoCAD Design Center (ADC), 244
Load Multiline Styles dialog box, 477–79
    MLStyle command, 477–79
Load or Reload Linetypes dialog box, 222
    File button, 222
    Linetype heading, 222
Locate Point button, 332
Location: control box, WBlock command, 717
Location and Number, 292–304, 354–64
    Align command, 294–98
    Array command, 357–61
    Copy command, 299–301
    Mirror command, 361–64
    Move command, 292–94
    Offset command, 354–57
Lock column, Layer Properties Manager, 235
Look in: box, 19
Lookup button, Spell Checker, 527
ls, *See* List command
LTScale command, 225–29

Ltype gen option, Pedit command, 334, 339
-lweight command, 230
    command sequence, 290
Lwpolyline, 314

M

m, *See* Move command
ma, *See* Matchprop command
Macro, 11–12
    defined, 12
Make Object's Layer Current button, 242
Mark option, Undo command, 42
Matchprop command, 255, 305
    command sequence, 255
MDE (Multiple Document Environment), using to move/copy objects between drawings, 301–305
me, *See* Measure command
Measure command, 394–99
    command sequence, 395
Measured Plot dialog box, Calibrate Plotter Wizard, 67
Merged Cross button, 487
Merged Tee button, 489–90
Method option, Chamfer command, 287
Metric settings, 5
mi, *See* Mirror command
Midpoint button, 95, 104
Mirror command, 361–64
    command sequence, 361–62
    Mirror button, 362
    mirroring an object, 362–64
    mirror line, 362
    Mirrtext command, 364
Mirror Grips procedure, 636, 639–40
Mirrtext command, 364
MLEdit command, 483, 484–93
    Add Vertex button, 491
    Closed Cross button, 485
    Closed Tee button, 488
    Corner Joint button, 490
    Cut All button, 492–93
    Cut Single button, 492
    Delete Vertex button, 491–92
    Edit Multiline button, 485
    Merged Cross button, 487
    Merged Tee button, 489–90
    in Modify pull-down menu, 484
    Multiline Edit Tools dialog box, 484
    Open Cross button, 486
    Open Tee button, 489
    Weld All button, 493
MLine command, 468–72
    Bottom option, 469
    command sequence, 468

Justification options, 468–69, 471
    Scale option, 469, 472
    STyle option, 469
    Top option, 469
    Zero option, 469, 471
MLStyle command, 472–83
    Element Properties dialog box, 474–77
        Center Linetype option, 475
        Color button, 476
        Linetype button, 475
    in Format pull-down menu, 419
    Load Multiline Styles dialog box, 477–79
Model:
    displaying, 797–800
        Hide command, 798
        Render command, 799–800
        Shade command, 799
    rotating, 796–97
Model Space/Paper Space (Layout) tabs, 13
Modify . . . button, Dimension Style Manager dialog box, 591
Mouse, selecting/picking with, 6
Move command, 292–94, 398, 697
    Base point option, 293–94
    command sequence, 292–93
    displacement method, 293–94
    Move button, 293
    using, 293–94
Move Grips procedure, 641
Move option:
    Edit Vertex, 340
    Splinedit command, 435, 442
Move Vertex option, Splinedit command, 435, 437
Moving/scaling with grips, 641–43
mt, *See* MText command
MText command, 140, 507–87
    Bold button, 514–13, 517–18
    Cancel button, 510
    Character tab, 511, 526
    command sequence, 508–09
    cursor menu, editing with, 521–22
    in Draw pull-down menu, 512
    editing multiline text, 517–20
    Find/Replace tab, 512, 523
    Height option, 509
    Help button, 511
    Import Text . . . button, 510, 516
    Italics button, 514
    Justify option, 509
    keystrokes, 813
    Line Spacing option, 509
    Line Spacing tab, 512
    Multiline Text button, 513
    OK button, 510
    Properties tab, 511–12
    Replace button, 523

Rotation option, 509
search and replace text, 523–24
status line, 510
Style option, 509
Symbol button, 511, 525
symbol insertion, 525–26
Underline button, 517
using multiline text, 513–17
Width option, 509, 513–14
Mtext options frame, Leader command, 564
Multiline Edit Tools dialog box, 484
Multilines, 467–506
  creating, 469–72
  in Draw pull-down menu, 469
  MLine command, 468–72
    Bottom option, 469
    command sequence, 468
    Justification options, 468–69, 471
    Multiline Properties . . . button, 479, 481, 482
    Multiline Properties dialog box, 480–83
    Multiline Styles dialog box, 473, 474, 476
    Scale option, 469, 472
    STyle option, 469
    Top option, 469
    Zero option, 469, 471
  MLStyle command, 472–83
    Element Properties dialog box, 474–77
    in Format pull-down menu, 419
    Load Multiline Styles dialog box, 477–79
  Multiline button, 470
  multiline style creation, 473
Multiline Text button, 513
Multiline Text Editor, *See* MText command
Multiple Document Environment (MDE), using to move/copy objects between drawings, 301–305
Multiple-object selection, 45–46
Multisided figures, drawing, 187–91
MW command, 419
My Computer option, Add-a-Plotter Wizard, 60

N

Name: control box, Insert dialog box, 726
Name column, Layer Properties Manager, 234
Named layer filters control box, Layer Properties Manager, 236
Named Layer Filters dialog box, 237–38
Named Views button, 137, 142, 184
Nearest button, 95

Network Plotter Server option, Add-a-Plotter Wizard, 60
New < button, Object Grouping dialog box, 702
New button, 26, 302
New . . . button, Dimension Style Manager dialog box, 591
New button:
  Layer Properties Manager, 235
  Text Style dialog box, 152, 153
New command, 26, 302
New Dimension Style: dialog box, 593–95
  Alternate Units tab, 599, 608
  Arrowheads frame, 595
  baseline spacing, 594
  Center marks for circles frame, 595
  child, setting out, 604–05
  Color control box, 593
  Dimension Line frame, 593, 595
  Extension beyond ticks control box, 593–94
  Extension Lines frame, 594–95
    Color option, 594
    Extend beyond dim lines setting, 594–95
    Lineweight option, 594
    Offset from origin setting, 595
    Suppress options, 594
  Fine Tuning frame, 598
  Fit Options frame, 597–98
  Fit tab, 597, 608
  setting up, 598
  Linear Dimension frame, 599–600
    Decimal separator: control box, 599
    Precision control box, 599
    Prefix box, 599
    Primary Units tab, 600
    Round off: number box, 599
    Scale factor: number box, 599
    Suffix box, 599
    Unit format: option, 599
    Zero Suppression number boxes, 600
  Line and Arrows tab, 593, 607
  Lineweight control box, 593
  Preview frame, 593, 596
  Primary Units tab, 599, 608
  Scale for Dimension Features frame, 598
  Text Placement frame, 598
  Text tab, 596, 607
    Preview frame, 596
    Text Alignment frame, 596
    Text Appearance frame, 596
    Text Placement frame, 596
  Tolerance Format frame, 602
  tolerances, setting up, 603
  Tolerance tab, 601–602, 608

Alternate Unit Tolerance controls, 603
units, setting up, 601
New Dimension Style dialog box, 591–93
  New Style Name: text box, 591
  Start With: control box, 592
  Use for: control box, 592
New File button, 12
New Named Views button, View dialog box, 136–37
New option, Dimedit command, 576
New Page Setup Name text box, 79
New Text Style dialog box, 153
New View dialog box, 136–37
  Define View Window button, 138
  Define Window option, 137
  View name text box, 137
Next option, Edit Vertex, 340
Node, 394
Node button, 95, 101, 103, 178
None button, 95
Non-Uniform Rational B-Spline, 438
Normal Dimensioning, 546
  converting to Associative Dimensioning, 546
Normal hatching style, 657–58
Normal option, Attdisp system variable, 761
Notepad program, 786–88
Noun/Verb selection, 623–26
  Pickfirst system variable, 623
Np command, 787
Number of Points box, Leader command, 564

O

o, *See* Offset command
Object Grouping dialog box, 699–702
  Add < button, 700
  Change Group frame, 700–701, 705–07
  Create Group frame, 700
  Description button, 701
  Explode button, 701, 708
  Find Name < button, 704
  Group Identification frame, 699–700, 706
  Group Member List dialog box, 700, 704
  Group Name setting, 699
  Group Name text box, 704, 706
  Highlight button, 701, 705
  Included Unnamed check box, 706
  New < button, 702
  Order Group dialog box, 701
  Remove < button, 700
  Rename button, 701
  Re-order . . . button, 701
  Reverse Order button, 701
  Selectable button, 701, 707

Object Grouping option box, 629
Object option:
  Area command, 328, 330–31
  Spline command, 428, 430–31
object properties, listing, 324–26
Object Properties toolbar, Color Control box, 214
Object Property Manager, 249–54, 612
  changing the layer of objects, 253
  changing the linetype, 254
  Properties command, 252
  Quick Select button, 251–52, 254
  Quick Select dialog box, 253
  Select Objects button, 252
  using, 252–54
Object selection filters, 415–18
Object selection methods, 408–15
  Add, 409
  All, 409
  CrossingPoly, 408
  Fence, 408, 411, 412
  Last, 408
  Previous, 409, 413
  Remove, 409
  selection practice, 409–15
  WindowPoly, 408, 411
Object Selection Settings, *See* Selection Settings
Objects frames, WBlock command, 717
Object Snap settings, 105
Object Snap tab, Drafting Settings dialog box, 93–94, 105
Object Snap Tracking, 108, 110–13
Object Snap Tracking Settings frame, Drafting Settings dialog box, 93
Object type: control box, Quick Select dialog box, 416, 417
Object type frame, Boundary Hatch dialog box, 666–67
Oblique Angle text box, 154
Oblique option, Dimedit command, 576, 578
Offset command, 354–57, 380, 447
  command sequence, 354
  Offset button, 355
  Offset distance option, 354
  using, 354–57
Offset option, XLine command, 445
On column, Layer Properties Manager, 234
ON/OFF option, Snap command, 90
Open button, 18, 21
Open command, 20, 43
Open Cross button, 486
Open a Drawing button, 17–20
Open Drawing button, 5
Open a Drawing button: Browse button, 17

Look in: box, 19
Open button, 18
Open a File button, 18
Partiaload command, 22–24
Partialopen command, 20–22
Select a File: list box, 17–19
Select File dialog box, 17–19
Steps folder, 19
Open Drawings button, AutoCAD Design Center (ADC), 243
Open File box, 5
Open a File button, 18
Open File button, 12
Opening a drawing, 5
Open option:
  Pedit command, 333, 335
  Splinedit command, 435–36
Open Tee button, 489, 497
Operator control box, Quick Select dialog box, 416
Opposite copies, 361–64
Option buttons, 8
Options . . . button, Find and Replace dialog box, 532–33
Options command, 624
Options dialog box, 6
Order Group dialog box, 701
ordinate dimensioning, 567–69
Ortho command, 88, 90, 114, 277
Orthographic drawings, 113
Orthographic option, View command, 136
Ortho toggle, 115, 275
OSNAP button, 107
OSNAP command, 107–8
OSNAP Cursor menu, 100
OSNAPS, 94–108
  Apparent Intersection button, 95
  Center button, 95, 101, 102, 217
  cursor menu, 100
  defined, 94
  Endpoint button, 95, 97, 99, 102
  Extension button, 95, 97, 98
  From button, 95
  Insert button, 95
  Intersection button, 95, 104
  Midpoint button, 95, 104
  Nearest button, 95
  Node button, 95, 101, 103
  None button, 95
  OSNAP command, 107–8
  OSNAP Flyout toolbar, 94
  OSNAP Settings button, 95, 105
  OSNAP Settings dialog box, 105
  Parallel button, 95, 101, 102
  Perpendicular button, 95, 101
  practice, 96–104
  Quadrant button, 95
  running, 104–8, 171
  Snap From button, 101
  Tangent button, 95, 103

Temporary Track Point button, 95, 111
Tracking button, 94
OSNAP Settings button, 95, 105
OSNAP Settings dialog box, 105
  Options dialog box, 105–6
  Drafting tab, 106
Other menus Cursor menu, 14
Outer hatching style, 658
Outermost hatching style, 662
Output File . . . button, 768
Override . . . button, Dimension Style Manager dialog box, 591, 612

P

Pagesetup command, 76–79
Page Setup Name frame, 79
Page setup name frame, Plot dialog box, 80
Page Setup Wizard, 77–79
  Name control box, 78
  New Page Setup Name text box, 79
  Page Setup Name frame, 79
  Plot Device tab, 77
  Plot Style Table frame, 78
  Plotter configuration frame, 78
  Properties button, 78
  User Defined Page Setups dialog box, 79
Pan command, 134, 140
Paper Space tab, 13
Parallel button, 95, 101, 102
Parallels, 354–57
Parameter option, Ellipse command, 176–77
Partial Load dialog box, 23
Partiaload command, 22–24
Partialopen command, 20–22
Partial Open dialog box, 21–22
Partial Preview, Plot offset frame button button, 81–83
Paste as Block option, 304
Paste button, 12
Pasteclip command, 304
Paste to Original Coordinates option, 304
PDMode, 401, 419
pe, *See* Pedit command
Pedit command, 332–45
  command sequence, 333
PELLIPSE system variable, 177
Pen option, Sketch command, 679
Percent option, Lengthen command, 367, 369
Perpendicular button, 95, 101
.pgp file, 784
PGP File check box, 787
Pickadd system variable, 626
Pickauto system variable, 629
Pickbox system variable, 629

Pickdrag system variable, 627
Pickfirst system variable, 623
Pick Points button, 667, 670, 673, 675–76
   Boundary Hatch dialog box, 667
PID drawing, 127–28
   defined, 127
Piping symbols, sample, 205
pl, See Pline command
Placing leaders, 563–66
Pline command, 315–23
   Angle option, 316
   Arc option, 315–23
   CEnter option, 316
   CLose option, 316
   command sequence, 315
   Halfwidth option, 316
   Line option, 316
   Radius option, 317
   Second point option, 317
   Undo option, 317
   Width option, 316
Plot column, Layer Properties Manager, 235
Plot command, 82–83
Plot Device tab, Plot dialog box, 80
Plot dialog box, 80–83
   Drawing orientation frame, 81
   Layout name frame, 80
   Page setup name frame, 80
   Plot Device tab, 80
   Plot offset frame, 81–83
   Plot scale frame, 81–82
   Plot Settings tab, 81
Plot offset frame, 81–83
   Plot dialog box, 81–83
Plot Progress Indicator, 83
Plot scale frame, 81–82
   Plot dialog box, 81–82
Plot Settings tab, Plot dialog box, 81
Plot Style column, Layer Properties Manager, 235
Plot Style Manager Default name menu item, 69
Plot Style Table Editor, 71–76
   Adapative Adjustment row, 75
   Add Style button, 73
   Color row, 74
   Convert to grayscale row, 74
   Description row, 74
   Enable dithering row, 74
   Fill Style row, 75
   Form View tab, 73
   General tab, 73
   Line End Style row, 75
   Line Join Style row, 75
   Line Type row, 75
   Line Weight row, 75
   Name row, 74
   Screening row, 75
   Table View tab, 73
   Use assigned pen # row, 74
   Virtual pen # row, 75
Plot Style Table Wizard, 68, 69–76
   Add Plot Style Table dialog box, 69
   Begin dialog box, 70
   File Name dialog box, 71
   Finish button, 72
   Pick Plot Style Table dialog box, 70
   Plot Style Table Editor, 71–76
   Start from scratch option, 69
   Use an existing plot style table option, 69
   Use My R14 Plotter Configuration option, 69
   Use a PCP or PC2 file option, 69
Plotter:
   Add-a-Plotter Wizard, 58, 76
   Pagesetup command, 76–79
   Page Setup Wizard, 77–79
      Hints button, 78
      Name control box, 78
      New Page Setup Name text box, 79
      Page Setup Name frame, 79
      Plot Device tab, 77
      Plot Style Table frame, 78
      Plotter configuration frame, 78
      Properties button, 78
      User Defined Page Setups dialog box, 79
   plot style, setting up, 69–76
   Plot Style Manager Stylesmanager command menu item, 69
   Plot Style Table Wizard, 69, 76
   Plottermanager command, 58
   Plotters window, 58
   setting up, 58–68
   setting up the page, 76–79
   Stylesmanager command, 69
Plotter Configuration Editor, 63–64
   Device and Document Settings tab, 64
   General tab, 64
   Ports tab, 64
Plottermanager command, 58
Plotters window, 58
po, See Point command
Point, 47, 394
Point button, 400
Point command, 399–402
   command sequence, 399–400
   Point button, 400
   using points, 400–402
Point Filters, 108–10
Point Style, Format menu, 401
pol, See Polygon command
Polar Angle measurement frame, Drafting Settings dialog box, 93
Polar coordinates, 48
   syntax for, 48
Polar option, Array command, 358, 359–60
Polar Tracking, 89–90
Polar Tracking tab, Drafting Settings dialog box, 93
Polygon button, 217
Polygon command, 187–89, 198
   command sequence, 188
Polygons, 187–91
   Center of polygon option, 189
   Circumscribed about circle option, 189–90, 199
   drawing, 188–91
   Edge option, 190
   Inscribed in circle option, 189
   Polygon command, 187–88
Polyline button, 317
Polyline command, 317
Polyline method, Chamfer command, 287
Polyline option, Fillet command, 282
Polylines, 313–52
   area, calculating, 328–31
   broken, 342
   Current line-width setting, 318
   defined, 314
   distance, determining, 327–28
   drawing, 317–23
   editing, 332–45
   lightweight polylines (lwpolyline), 314
   moving the vertex, 344
   new vertex, adding, 343–44
   object properties, listing, 324–26
   Pedit command, 332–45
   Pline command, 315–23
      Arc option, 316–17
      Close option, 315
      Halfwidth option, 315
      Length option, 315
      selecting, 317
      Specify next point option, 316
      Undo option, 316
      Width option, 315
   simple editing tools, 334–41
   smart polylines, 314
Ports tab, Plotter Configuration Editor, 64
Precision option button, Unit Setup options, 8
Press and Drag selection, 627–29
   Pickdrag system variable, 627
Preview button, 668
   AutoCAD Design Center (ADC), 244
   BHatch command, 670
   Boundary Hatch dialog box, 668
Preview frame, 18

New Dimension Style: First Steps dialog box, 593, 596
Preview Hatch display, accuracy of, 670
Preview Icon frame, 716
Preview of: display box, Dimension Style Manager dialog box, 591
Previous option:
 Edit Vertex, 340
 Zoom command, 128
Previous tool, 409, 413
Primary Units tab, New Dimension Style: First Steps dialog box, 599, 608
Print button, 12
Print command, 82–83
Printer:
 Add-a-Plotter Wizard, 58–63
 Plottermanager command, 58
 setting up, 58–68
Print Preview button, 12
Project option, Trim command, 265
Prompting, by AutoCAD, 13
Properties box, Quick Select dialog box, 416, 417
Properties command, 252, 338, 339
Properties tab, MText command, 511–12
props, See Properties command
PT command, 418–19
Pull-Down Menu Bar, 11
Pumps view, 145
Purge command, 611
Purge option, Splinedit command, 435

Q

QDim command, 569–73
 command sequence, 570
 datumPoint option, 570
 Edit option, 570
 Quick Dimension button, 570
 using, 570–72
Qsave command, 14, 147, 173
QSelect command, 415, 417
Qtext command, 147–48
Quadrant button, 95
? option, View command, 136
Quick Dimension button, 570
Quick dimensioning, 569–73
Quick Leader button, Leader command, 565
Quick Select button, 251–52, 254
Quick Select dialog box, 253, 415–16
 Apply to: control box, 416
 How to apply: frame, 416
 Object type: control box, 416, 417
 Operator control box, 416
 Properties box, 416, 417
 Value control box, 416

Quick Setup option, Startup Wizard, 6, 10
Quick tab:
 Angle control boxes, 666
 Boundary Hatch dialog box, 665–66
  Custom patterns, 665
  Pattern control box, 666
  Predefined patterns, 665
  Scale control box, 666
  Type control box, 665
  User-defined patters, 665
 Iso Pen Width option, 666
Quit command, 14, 16, 24
Quit option, Sketch command, 680
Quitting without saving, 24

R

Radius Dimension button, 553
Radius option:
 Arc command, 184–85
 Pline command, 317
Ray command, 445, 452
 in Draw pull-down menu, 445
Realtime option, Zoom command, 126–28
Realtime Pan button, 132–33
Realtime Pan cursor, 133
Realtime Zoom button, 132–33
Realtime Zoom cursor, 132
Realtime zooming, 126–28
Record Increment option, Sketch command, 679, 681
Record option, Sketch command, 680
Rectangles, 34–35
 with chamfered corners, 37, 39–40
 drawing, 38–41
 filleted, 37–38
 Retang command, 37, 38–40, 240
Rectangle Size dialog box, Calibrate Plotter Wizard, 66
Rectangular option, Array command, 358, 359–60
Redefining blocks, 709, 765
Redo command, 41–44
 and Default Cursor menu, 43
 limitations of, 42
 Redo button, 42, 44
Redraw command, 50
Reference option:
 Rotate command, 371, 373
 Scale command, 373
Refine option, Splinedit command, 435, 438
Regen command, 50
Regen option, Edit Vertex, 340
Reinit command, 786
Re-initialize dialog box, 786–87
relative coordinates, 48
Relative to Screen, setting point size to, 401

Remove < button, Object Grouping dialog box, 700
Remove Islands button, 667
 Boundary Hatch dialog box, 667
Remove tool, 409
Rename button:
 Object Grouping dialog box, 701
 Text Style dialog box, 152
Render button, 799
Render command, 799–800
Re-order . . . button, Object Grouping dialog box, 701
Replace All button, 151, 534
Replace button, 523
Replace with text box, 151
Replacing text, 150–51
Reserve option, Splinedit command, 435
Restore option, View command, 135
Retain polyline? option, Hatch command, 655
Retang button, 38
Retang command, 37, 38–40
Reverse Order button, Object Grouping dialog box, 701
Rich Text Format (.rtf), 510
Right-click Cursor menus, 13–14
Right option, DimTedit command, 573–74
ro, See Rotate command
Rotate3d command, 796–97
Rotate command, 371–73
 command sequence, 371
 Reference option, 371
 Rotate button, 372
 Rotation angle option, 371
Rotate Grips procedure, 636–39
Rotate option:
 Dimedit command, 576
 Snap command, 90
Rotation angle option, Rotate command, 371
Rotation frame, Insert dialog box, 726
Rotation option:
 -Insert command, 722
 MText command, 509
Rows, 357–61
.rtf, 510
Running OSNAP button, 171

S

s, See Save command
Saveas command, 14, 15, 25
Save button, 16, 117
Save command, 14, 117
Save Drawing As dialog box, 15–16
Save File button, 12
Save in: box, 15
Save option, Dimstyle command, 612
Save as type: box, 16, 25
saving a drawing session, 14–17

sc, *See* Scale command
Scale, 4
Scale command, 373–74
   command sequence, 373
   Reference option, 373
   Scale button, 374
   Scale option, 374
Scaled hatching, 662–63
Scale for Dimension Features frame, New Dimension Style: First Steps dialog box, 598
Scale frame, Insert dialog box, 726
Scale Grips procedure, 642–43
Scale option:
   -Insert command, 722
   MLine command, 469, 472
   Scale command, 374
   Zoom command, 128
Scale to fit setting, Plot scale frame, 82
Screen, 11–13
   Coordinate Display box, 12
   Graphics Area, 13
   Status Line, 12
   User Coordinate System (UCS) icon, 12–13
   World Coordinate System (WCS), 13
Screening, plots, 75
Search in: control box, Find and Replace dialog box, 532
Search results: frame, Find and Replace dialog box, 533
Search Results list box, 150
Second point option, Pline command, 317
Selectable button, Object Grouping dialog box, 701, 707
Select Color dialog box, 215, 217, 250
Selected line, 35
Select a File: list box, 17–19
Select File dialog box, 5, 17–19
Selection Settings, 622–29
   Grip command, 622–23
   Implied Windowing, 629
   Noun/Verb selection, 623–26, 630
   Object Grouping option box, 629
   Pickbox Size frame, 629
   Press and Drag, 627–29
   Selection Modes frame, 629
   Use Shift to Add, 626–27
Select Objects < button, 768
Select Objects button, 252, 667, 715
   Boundary Hatch dialog box, 667
Select objects prompt, 46
Select Template dialog box, 26
SetCurrent button, Dimension Style Manager dialog box, 591
Settings option, Matchprop command, 255
Setup, changing, 27–29

Setup Wizard:
   using, 6–13
   See also Use a Wizard button
Shade command, 799
Shademode command, 799
Sharing blocks, 729–30
   Tree Toggle View button, 730
   using ADC, 729
Sheet size, changing, 28
Shift key, and Grips command, 631
Shortcuts, 38
"Show Startup dialog" check box, 6
Sketch command, 679–82
   command sequence, 679
   Erase option, 680
   eXit option, 680
   Pen option, 679
   Pen up mode, 680
   Quit option, 680
   Record Increment option, 679, 681
   Record option, 680
   sketching, 680–82
Slide guide, 204
Slotted holder, 203
Smart polylines, 314
sn, *See* Snap command
Snap command, 89–90, 113–14, 155, 178
   options, 90
Snap From button, 101
Snap and Grid tab, Drafting Settings dialog box, 92
Snap to Quadrant button, 190, 294
so, *See* Solid command
Solid command, 402–407
   command sequence, 402
   creating solids, 402–405
Solid hatching, 661
Solid hatch pattern, 654
Solid modeling, 789–96
Solids Editing toolbar, 793
   Subtract button, 793
Solids toolbar, 790
   Cylinder button, 791
Source frame, WBlock command, 717
sp, *See* Spell command
Space Delimited File (SDF), 768
spe, *See* Splinedit command
Specify center point option, circle command, 170
Specify first corner point option, Area command, 328
Specify next point or [Undo]: prompt, 117
Specify start point option, Text command, 141
Spell command, 526–31
   Change Directories . . . button, 528
   command sequence, 527
   Current word frame, 527
   Ignore All button, 527, 529

   Ignore button, 527, 529
   Lookup button, 527
   Suggestions text box, 527
   in Tools pull-down menu, 528
   using, 528–31
spl, *See* Spline command
Spline button, 430
Spline command, 428–34
   Close option, 428, 431
   command sequence, 428
   End tangent option, 428, 433
   Fit Tolerance option, 428, 432
   Object option, 428, 430–31
Splinedit button, 436
Splinedit command, 434–43
   Add control point option, 435
   Add option, 435
   command sequence, 434
   Delete option, 435, 442
   Elevate Order option, 435, 439–40
   Fit Data option, 435, 441–43
   in Modify pull-down menu, 435
   Move option, 435, 442
   Move Vertex option, 435, 437
   Open option, 435–36
   Purge option, 435
   Refine option, 435, 438
   Reserve option, 435
   Tangents option, 435, 441
   toLerances option, 435
   Weight option, 435, 438–39
Spline option, Pedit command, 333, 337, 428
Splines:
   changing, 434–43
   working with, 429–34
Splinetype system variable, 333
Squished circles, 174–79
st, *See* Style command
Standard default style, 151
Start from Scratch button, 5
Start tangent option, Spline command, 428, 433
Startup dialog box, 4–6, 10, 17, 24
   Open a Drawing button, 17–20
Status bar toggles, 91
Status Line, 12
Status option, Dimstyle command, 612
Steps folder, 15, 19
Stretch command, 364–66
   command sequence, 364
Stretch Grips procedure, 631–35, 641–42
Style, defined, 151
Style command, 151
Style Name frame, 153
Style option:
   MLine command, 469
   MText command, 509
   Snap command, 90
   Text command, 141

Styles: listbox, Dimension Style Manager dialog box, 591
Stylesmanager command, 69
Subtract button, Solids Editing toolbar, 793
Subtract command, 793
Subtract option, Area command, 328, 330
Suggestions text box, Spell Checker, 527
Sykes, W. P., 803
Symbol button, MText command, 511, 525
Symbol libraries, 709
System Printer option, Add-a-Plotter Wizard, 60
System tab, Options dialog box, 6

T

Table View tab, Plot Style Table Editor, 73
Tangent button, 95, 103
Tangent option, Edit Vertex, 340
Tangents option, Splinedit command, 435, 441
Template:
  creating, 24–25
  creating a new drawing using, 26–27
  Select Template dialog box, 26
  Use a Template button, 26
  using, 5
Template File . . . button, 768
Template Libraries, 710–11
Temporary Tracking button, 111
Temporary Track Point button, 95, 111
Text, 140–48
  DDEdit command, 148–50
  editing, 148–50
  finding/replacing, 150–51
  inserting, 142–48
  Qtext command, 147–48
  styles, creating, 151–56
  Text command, 140–48
  Text Tricks, 143–44
Text button, Draw toolbar, 140
Text command, 156
  Align option, 141
  Bottom Left option, 141
  command sequence, 140
  Fit option, 141
  hotkey, 140
  Justify option, 141
  Specify start point option, 141
  Style option, 141
  text options, 141
Text Finder dialog box, 150
Text Placement frame, New Dimension Style: dialog box, 598

Text Style dialog box, 151–54
  Backwards option, 154
  Effects section, 154
  Oblique Angle text box, 154
  Preview panel, 155
  Upside down option, 154
  Vertical option, 154
  Width Factor option, 154
Text tab:
  New Dimension Style: dialog box, 596, 607
    Preview frame, 596
    Text Alignment frame, 596
    Text Appearance framed, 596
    Text Placement frame, 596
Third-party software, 399
3D Continuous Orbit button, 801
3DCOrbit command, 801–802
3P option, circle command, 170–71
Title Bar, 11
Tolerance Format frame, New Dimension Style: dialog box, 602
toLerances option, Splinedit command, 435
Tolerance tab:
  New Dimension Style: dialog box, 601–602, 608
    Alternate Unit Tolerance controls, 603
Toolbars:
  adding, 12
  defined, 12
  docked, 12
  Object Properties toolbar, 214
  Solids Editing toolbar, 793
  Solids toolbar, 791
  undocked, 12
Top option, MLine command, 469
Total option, Lengthen command, 367, 370
tr, See Trim command
Tracking button, 94, 95
Tree Toggle View button, 730
Tree View button, AutoCAD Design Center (ADC), 244
Trim button, 266
Trim command, 264–70, 380, 411, 448, 451, 454, 545
  command sequence, 265
  cutting edges, 265, 267
  Edge option, 265, 268
  object to trim, 265
  Project option, 265
  Trim button, 266
  Undo option, 265
Trimmode system variable, 287
Trim option:
  Chamfer command, 287
  Fillet command, 282, 287
True coordinates, 28

TTR (Tangent-Tangent-Radius) option, circle command, 170, 172, 173
2P option, circle command, 170, 172
.txt, 510
Type option, Snap command, 90

U

UCS option, View command, 136
Underline button, MText command, 517
Undocked toolbar, 12
Undo command, 41–44
  command sequence, 41
  and Default Cursor menu, 43
  hotkey, 42
  options, 42
  Undo button, 42, 44
Undo option:
  Pedit command, 334
  Pline command, 317
  Trim command, 265
Union button, 794
Union command, 794, 796
Units, 4
  changing, 27
Unit Setup options:
  Advanced Setup dialog box, 8
  Use a Wizard button, 8
Up button, AutoCAD Design Center (ADC), 244
Update command, 611
Upside down option, Text Style dialog box, 154
User Coordinate System (UCS) icon, 12–13
User Defined option, Hatch command, 655, 659
User Defined Page Setups dialog box, 79
Use Shift to Add, 626–27
  Pickadd system variable, 626
Use a Template button, 5, 26
Use a Wizard button, 6, 7
  Advanced Setup dialog box, 7

V

v, See View command
Value control box, Quick Select dialog box, 416
Variables option, Dimstyle command, 613
Ver option, XLine command, 444
Vertex, defined, 340
Vertex locator, 340
Vertical option, Text Style dialog box, 154
View command, 134–40, 142
  on the command line, 135–37
  creating a view using a dialog box, 137–39

Define View Window button, 138
Delete option, 136
Named Views button, 137
New Named Views button, View dialog box, 136–37
New View dialog box, 136–37
  Define View Window button, 138
  Define Window option, 137
  View name text box, 137
Orthographic option, 136
? option, 136
Restore option, 135
setup time, 136
UCS option, 136
View dialog box, 136–37
Window option, 135
View dialog box, 136–37
ViewDir system variable, 789
View menu, 11, 12
Views button, AutoCAD Design Center (ADC), 244
View Selections button, 667
  Boundary Hatch dialog box, 667
Virtual pen, plots, 75
VPoint (-vp) command, 789–96

W

Warning box, AutoCAD, 17
WBlock command, 709, 716–21, 748, 755
  Base point frame, 717
  Block: option, Source frame, 717
  Browse button, 717
  command sequence, 716
  Destination frame, 717–18
  Enter name of existing block . . . prompt, 717
  Entire drawing option, Source frame, 717
  Insert units: option, 718
  Location: control box, 717
  Objects frames, 717
  Objects option, Source frame, 717
  Source frame, 717
  using, 718–21
Web Search button, 18
Weight option, Splinedit command, 435, 438–39
Weld All button, 493
Welding symbols, sample, 205
Width Factor option, Text Style dialog box, 154
Width option:
  Edit Vertex, 340
  MText command, 509, 513–14
  Pedit command, 333, 335
  Pline command, 316
Window option, View command, 135
Window placement, and Zoom command, 129–30
WindowPoly tool, 408, 411
Window selection, 45
Wizard:
  Add-a-Plotter Wizard, 58–63, 76
  Calibrate Plotter Wizard, 65–68
  defined, 6
  Page Setup Wizard, 77–79
  Plot Style Table Wizard, 69–76
  Setup Wizard, 6–13
  Use a Wizard button, 6, 7
  using, 6
Word processor, defined, 508
World Coordinate System (WCS), 13
WYSIWYG, 68

X

X-axis, 47
X button, AutoCAD window, 14, 16
XLine command, 444, 446–49, 453–54
  Ang option, 444
  Bisect option, 444–45
  command sequence, 444
  in Draw pull-down menu, 445
  Hor option, 444
  Offset option, 445
  Ver option, 444
X scale factor, -Insert command, 722
XYZ option, -Insert command, 722

Y

Y-axis, 47
Y scale factor, -Insert command, 722

Z

z, *See* Zoom command
Z-direction, 788–89
Zero option, MLine command, 469, 471
Zoom All button, 130, 193
Zoom Center button, 132
Zoom command, 23, 126–34, 142, 375, 379, 397
  All option, 128
  Center option, 128
  cursor menu, 128
  Extents option, 128
  hotkey, 126
  options, 126–28
  practice, 129–32
  Previous option, 128
  Realtime option, 126–28
  Realtime Pan button, 132–33
  Realtime Pan cursor, 133
  Realtime Zoom button, 132–33
  Realtime Zoom cursor, 132
  Scale option, 128
  Zoom All button, 130
  Zoom Center button, 132
  Zoom Dynamics button, 131
    exchanger, 131
    location box, 131
    sizing box, 131
  Zoom Extents button, 130–31
  Zoom In button, 130
  Zoom In option, 128
  Zoom Out option, 128
  Zoom Previous button, 130
  Zoom Window button, 129–30
Zoom cursor menu, 128
Zoom Dynamics button, 131
Zoom Extents button, 131
Zoom In button, 130, 151
Zoom In option, Zoom command, 128
Zoom Limits button, 286
Zoom Out option, Zoom command, 128
Zoom Previous button, 130, 377
Zoom Realtime button, 127
Zoom to button, Find and Replace dialog box, 534
Zoom Window button, 129

SITE LICENSE AGREEMENT AND LIMITED WARRANTY

READ THIS LICENSE CAREFULLY BEFORE OPENING THIS PACKAGE. BY OPENING THIS PACKAGE, YOU ARE AGREEING TO THE TERMS AND CONDITIONS OF THIS LICENSE. IF YOU DO NOT AGREE, DO NOT OPEN THE PACKAGE. PROMPTLY RETURN THE UNOPENED PACKAGE AND ALL ACCOMPANYING ITEMS TO THE PLACE YOU OBTAINED THEM. THESE TERMS APPLY TO ALL LICENSED SOFTWARE ON THE DISK EXCEPT THAT THE TERMS FOR USE OF ANY SHAREWARE OR FREEWARE ON THE DISKETTES ARE AS SET FORTH IN THE ELECTRONIC LICENSE LOCATED ON THE DISK:

1. GRANT OF LICENSE and OWNERSHIP: The enclosed computer programs and data ("Software") are licensed, not sold, to you by Prentice-Hall, Inc. ("We" or the "Company") and in consideration of your payment of the license fee, which is part of the price you paid and your agreement to these terms. We reserve any rights not granted to you. You own only the disk(s) but we and/or licensors own the Software itself. This license allows you to use and display the enclosed copy of the Software on a local area network at a single campus or branch or geographic location so long as you comply with the terms of this Agreement. You may make one copy for back up only.

2. RESTRICTIONS: You may not transfer or distribute the Software or documentation to anyone else. Except for backup, you may not copy the documentation or the Software. You may not reverse engineer, disassemble, decompile, modify, adapt, translate, or create derivative works based on the Software or the Documentation. You may be held legally responsible for any copying or copyright infringement which is caused by your failure to abide by the terms of these restrictions.

3. TERMINATION: This license is effective until terminated. This license will terminate automatically without notice from the Company if you fail to comply with any provisions or limitations of this license. Upon termination, you shall destroy the Documentation and all copies of the Software. All provisions of this Agreement as to limitation and disclaimer of warranties, limitation of liability, remedies or damages, and our ownership rights shall survive termination.

4. LIMITED WARRANTY AND DISCLAIMER OF WARRANTY: Company warrants that for a period of 60 days from the date you purchase this Software (or purchase or adopt the accompanying textbook), the Software, when properly installed and used in accordance with the Documentation, will operate in substantial conformity with the description of the Software set forth in the Documentation, and that for a period of 30 days the disk(s) on which the Software is delivered shall be free from defects in materials and workmanship under normal use. The Company does not warrant that the Software will meet your requirements or that the operation of the Software will be uninterrupted or error-free. Your only remedy and the Company's only obligation under these limited warranties is, at the Company's option, return of the disk for a refund of any amounts paid for it by you or replacement of the disk.

THIS LIMITED WARRANTY IS THE ONLY WARRANTY PROVIDED BY THE COMPANY AND ITS LICENSORS, AND THE COMPANY AND ITS LICENSORS DISCLAIM ALL OTHER WARRANTIES, EXPRESS OR IMPLIED, INCLUDING WITHOUT LIMITATION, THE IMPLIED WARRANTIES OF MERCHANTABILITY AND FITNESS FOR A PARTICULAR PURPOSE. THE COMPANY DOES NOT WARRANT, GUARANTEE OR MAKE ANY REPRESENTATION REGARDING THE ACCURACY, RELIABILITY, CURRENTNESS, USE, OR RESULTS OF USE, OF THE SOFTWARE.

5. LIMITATION OF REMEDIES AND DAMAGES: IN NO EVENT, SHALL THE COMPANY OR ITS EMPLOYEES, AGENTS, LICENSORS, OR CONTRACTORS BE LIABLE FOR ANY INCIDENTAL, INDIRECT, SPECIAL, OR CONSEQUENTIAL DAMAGES ARISING OUT OF OR IN CONNECTION WITH THIS LICENSE OR THE SOFTWARE, INCLUDING FOR LOSS OF USE, LOSS OF DATA, LOSS OF INCOME OR PROFIT, OR OTHER LOSSES, SUSTAINED AS A RESULT OF INJURY TO ANY PERSON, OR LOSS OF OR DAMAGE TO PROPERTY, OR CLAIMS OF THIRD PARTIES, EVEN IF THE COMPANY OR AN AUTHORIZED REPRESENTATIVE OF THE COMPANY HAS BEEN ADVISED OF THE POSSIBILITY OF SUCH DAMAGES. IN NO EVENT SHALL THE LIABILITY OF THE COMPANY FOR DAMAGES WITH RESPECT TO THE SOFTWARE EXCEED THE AMOUNTS ACTUALLY PAID BY YOU, IF ANY, FOR THE SOFTWARE OR THE ACCOMPANYING TEXTBOOK. SOME JURISDICTIONS DO NOT ALLOW THE LIMITATION OF LIABILITY IN CERTAIN CIRCUMSTANCES, THE ABOVE LIMITATIONS MAY NOT ALWAYS APPLY.

6. GENERAL: THIS AGREEMENT SHALL BE CONSTRUED IN ACCORDANCE WITH THE LAWS OF THE UNITED STATES OF AMERICA AND THE STATE OF NEW YORK, APPLICABLE TO CONTRACTS MADE IN NEW YORK, AND SHALL BENEFIT THE COMPANY, ITS AFFILIATES AND ASSIGNEES. This Agreement is the complete and exclusive statement of the agreement between you and the Company and supersedes all proposals, prior agreements, oral or written, and any other communications between you and the company or any of its representatives relating to the subject matter. If you are a U.S. Government user, this Software is licensed with "restricted rights" as set forth in subparagraphs (a)-(d) of the Commercial Computer-Restricted Rights clause at FAR 52.227-19 or in subparagraphs (c)(1)(ii) of the Rights in Technical Data and Computer Software clause at DFARS 252.227-7013, and similar clauses, as applicable.